D1631131

WITHDRAWN
FROM
UNIVERSITIES
AT
MEDWAY
LIBRARY

Advances in

ECOLOGICAL RESEARCH

VOLUME 40

Advances in Ecological Research

Series Editor: HAL CASWELL

Biology Department
Woods Hole Oceanographic Institution
Woods Hole, Massachusetts

Advances in

ECOLOGICAL RESEARCH

VOLUME 40

High-Arctic Ecosystem Dynamics in a Changing Climate

Ten years of monitoring and research at Zackenberg Research Station, Northeast Greenland

Edited by

HANS MELTOFTE, TORBEN R. CHRISTENSEN, BO ELBERLING, MADS C. FORCHHAMMER and MORTEN RASCH

AMSTERDAM • BOSTON • HEIDELBERG • LONDON
NEW YORK • OXFORD • PARIS • SAN DIEGO
SAN FRANCISCO • SINGAPORE • SYDNEY • TOKYO
Academic Press is an imprint of Elsevier

Academic Press is an imprint of Elsevier
84 Theobald's Road, London WC1X 8RR, UK
Radarweg 29, PO Box 211, 1000 AE Amsterdam, The Netherlands
Linacre House, Jordan Hill, Oxford OX2 8DP, UK
30 Corporate Drive, Suite 400, Burlington, MA 01803, USA
525 B Street, Suite 1900, San Diego, CA 92101-4495, USA

First edition 2008

Copyright © 2008 Elsevier Ltd. All rights reserved.

No part of this publication may be reproduced, stored in a retrieval system or transmitted in any form or by any means electronic, mechanical, photocopying, recording or otherwise without the prior written permission of the Publisher.

Permissions may be sought directly from Elsevier's Science & Technology Rights Department in Oxford, UK: phone (+44) (0) 1865 843830; fax (+44) (0) 1865 853333; email: permissions@elsevier.com. Alternatively you can submit your request online by visiting the Elsevier web site at http://elsevier.com/locate/permissions, and selecting *Obtaining permission to use Elsevier material.*

Notice
No responsibility is assumed by the publisher for any injury and/or damage to persons or property as a matter of products liability, negligence or otherwise, or from any use or operation of any methods, products, instructions or ideas contained in the material herein. Because of rapid advances in the medical sciences, in particular, independent verification of diagnoses and drug dosages should be made.

ISBN: 978-0-12-373665-9
ISSN: 0065-2504

For information on all Academic Press publications
visit our website at books.elsevier.com

Printed and bound in China

08 09 10 11 12 10 9 8 7 6 5 4 3 2 1

Working together to grow
libraries in developing countries

www.elsevier.com | www.bookaid.org | www.sabre.org

ELSEVIER BOOK AID International Sabre Foundation

Contents

The Study Area at Zackenberg

HANS MELTOFTE AND MORTEN RASCH

Present-Day Climate at Zackenberg

BIRGER ULF HANSEN, CHARLOTTE SIGSGAARD, LEIF RASMUSSEN, JOHN CAPPELEN, JØRGEN HINKLER, SEBASTIAN H. MERNILD, DORTHE PETERSEN, MIKKEL P. TAMSTORF, MORTEN RASCH AND BENT HASHOLT

Permafrost and Periglacial Geomorphology at Zackenberg

HANNE H. CHRISTIANSEN, CHARLOTTE SIGSGAARD, OLE HUMLUM, MORTEN RASCH AND BIRGER U. HANSEN

Snow and Snow-Cover in Central Northeast Greenland

JØRGEN HINKLER, BIRGER U. HANSEN, MIKKEL P. TAMSTORF, CHARLOTTE SIGSGAARD AND DORTHE PETERSEN

Hydrology and Transport of Sediment and Solutes at Zackenberg

BENT HASHOLT, SEBASTIAN H. MERNILD, CHARLOTTE SIGSGAARD, BO ELBERLING, DORTHE PETERSEN, BJARNE H. JAKOBSEN, BIRGER U. HANSEN, JØRGEN HINKLER AND HENRIK SØGAARD

Soil and Plant Community-Characteristics and Dynamics at Zackenberg

BO ELBERLING, MIKKEL P. TAMSTORF, ANDERS MICHELSEN, MARIE F. ARNDAL, CHARLOTTE SIGSGAARD, LOTTE ILLERIS, CHRISTIAN BAY, BIRGER U. HANSEN, TORBEN R. CHRISTENSEN, ERIC STEEN HANSEN, BJARNE H. JAKOBSEN AND LOUIS BEYENS

Inter-Annual Variability and Controls of Plant Phenology and Productivity at Zackenberg

SUSANNE M. ELLEBJERG, MIKKEL P. TAMSTORF, LOTTE ILLERIS, ANDERS MICHELSEN AND BIRGER U. HANSEN

High-Arctic Plant–Herbivore Interactions under Climate Influence

THOMAS B. BERG, NIELS M. SCHMIDT, TOKE T. HØYE, PETER J. AASTRUP, DITTE K. HENDRICHSEN, MADS C. FORCHHAMMER AND DAVID R. KLEIN

Phenology of High-Arctic Arthropods: Effects of Climate on Spatial, Seasonal, and Inter-Annual Variation

TOKE T. HØYE AND MADS C. FORCHHAMMER

Effects of Food Availability, Snow and Predation on Breeding Performance of Waders at Zackenberg

HANS MELTOFTE, TOKE T. HØYE AND NIELS M. SCHMIDT

Vertebrate Predator–Prey Interactions in a Seasonal Environment

NIELS M. SCHMIDT, THOMAS B. BERG,
MADS C. FORCHHAMMER, DITTE K. HENDRICHSEN,
LINE A. KYHN, HANS MELTOFTE AND TOKE T. HØYE

Lake Flora and Fauna in Relation to Ice-Melt, Water Temperature and Chemistry at Zackenberg

KIRSTEN S. CHRISTOFFERSEN, SUSANNE L. AMSINCK,
FRANK LANDKILDEHUS, TORBEN L. LAURIDSEN
AND ERIK JEPPESEN

Population Dynamical Responses to Climate Change

MADS C. FORCHHAMMER, NIELS M. SCHMIDT, TOKE T. HØYE, THOMAS B. BERG, DITTE K. HENDRICHSEN AND ERIC POST

Solar Ultraviolet-B Radiation at Zackenberg: The Impact on Higher Plants and Soil Microbial Communities

KRISTIAN R. ALBERT, RIIKKA RINNAN, HELGE RO-POULSEN, TEIS N. MIKKELSEN, KIRSTEN B. HÅKANSSON, MARIE F. ARNDAL AND ANDERS MICHELSEN

High-Arctic Soil CO_2 and CH_4 Production Controlled by Temperature, Water, Freezing and Snow

BO ELBERLING, CLAUS NORDSTRØM, LOUISE GRØNDAHL, HENRIK SØGAARD, THOMAS FRIBORG, TORBEN R. CHRISTENSEN, LENA STRÖM, FLEUR MARCHAND AND IVAN NIJS

Spatial and Inter-Annual Variability of Trace Gas Fluxes in a Heterogeneous High-Arctic Landscape

LOUISE GRØNDAHL, THOMAS FRIBORG, TORBEN R. CHRISTENSEN, ANNA EKBERG, BO ELBERLING, LOTTE ILLERIS, CLAUS NORDSTRØM, ÅSA RENNERMALM, CHARLOTTE SIGSGAARD AND HENRIK SØGAARD

Zackenberg in a Circumpolar Context

MADS C. FORCHHAMMER, TORBEN R. CHRISTENSEN, BIRGER U. HANSEN, MIKKEL P. TAMSTORF, NIELS M. SCHMIDT, TOKE T. HØYE, JACOB NABE-NIELSEN, MORTEN RASCH, HANS MELTOFTE, BO ELBERLING AND ERIC POST

Contributors to Volume 40

PETER J. AASTRUP, *National Environmental Research Institute, Department of Arctic Environment, University of Aarhus, Frederiksborgvej 399, P.O. Box 358, DK-4000 Roskilde, Denmark.*

KRISTIAN ALBERT, *Institute of Biology, Department of Terrestrial Ecology, University of Copenhagen, Øster Farimagsgade 2D, DK-1353 Copenhagen K, Denmark.*

SUSANNE L. AMSINCK, *National Environmental Research Institute, Department of Freshwater Ecology, University of Aarhus, Vejlsøvej 25, DK-8600 Silkeborg, Denmark.*

CLAUS ANDREASEN, *Greenland National Museum and Archives, Box 145, DK-3900 Nuuk, Greenland.*

MARIE F. ARNDAL, *Forest & Landscape, Hørsholm Kongevej 11, DK-2970 Hørsholm, Denmark.*

CHRISTIAN BAY, *CVU Øresund, Tuborgvej 235, DK-2400 Copenhagen NV, Denmark.*

OLE BENNIKE, *Geological Survey of Denmark and Greenland, Øster Voldgade 10, DK-1350 Copenhagen K, Denmark.*

THOMAS B. BERG, *Naturama, Dronningemaen 30, DK-5700 Svendborg, Denmark.*

LOUIS BEYENS, *Department of Biology, University of Antwerp, Universiteitsplein 1, BE-2610 Wilrijk, Belgium.*

HANS HENRIK BRUUN, *Department of Ecology, Lund University, S-22362 Lund, Sweden.*

JENS BÖCHER, *The Natural History Museum of Denmark, University of Copenhagen, Universitetsparken 15, DK-2100 Copenhagen Ø, Denmark.*

JOHN CAPPELEN, *Danish Meteorological Institute, Lyngbyvej 100, DK-2100 Copenhagen Ø, Denmark.*

HANNE H. CHRISTIANSEN, *Department of Geology, The University Centre in Svalbard, UNIS, P.O. Box 156, N-9171 Longyearbyen, Norway.*

JENS HESSELBJERG CHRISTENSEN, *Danish Meteorological Institute, Lyngbyvej 100, DK-2100 Copenhagen Ø, Denmark.*

TORBEN R. CHRISTENSEN, *GeoBiosphere Science Centre, Physical Geography and Ecosystems Analysis, Lund University, Sölvegatan 12, S-22362 Lund, Sweden.*

KIRSTEN S. CHRISTOFFERSEN, *Freshwater Biological Laboratory, University of Copenhagen, Helsingørgade 51, DK-3400 Hillerød, Denmark.*

ANNA EKBERG, *GeoBiosphere Science Centre, Department of Physical Geography and Ecosystems Analysis, Lund University, Sölvegatan 12, S-22362 Lund, Sweden.*

BO ELBERLING, *Department of Geography & Geology, University of Copenhagen, Øster Voldgade 10, DK-1350 Copenhagen K, Denmark, and Department of Arctic Technology, UNIS, P.O. Box 156, N-9171 Longyearbyen, Norway.*

SUSANNE M. ELLEBJERG, *National Environmental Research Institute, Department of Arctic Environment, University of Aarhus, Frederiksborgvej 399, P.O. Box 358, DK-4000 Roskilde, Denmark.*

MADS C. FORCHHAMMER, *National Environmental Research Institute, Department of Arctic Environment, Centre for Integrated Population Ecology, University of Aarhus, Frederiksborgvej 399, P.O. Box 358, DK-4000 Roskilde, Denmark.*

BENT FREDSKILD, *Botanical Museum, University of Copenhagen, Gothersgade 130, DK-1123 Copenhagen K, Denmark.*

THOMAS FRIBORG, *Department of Geography & Geology, University of Copenhagen, Øster Voldgade 10, DK-1350 Copenhagen K, Denmark.*

LOUISE GRØNDAHL, *National Environmental Research Institute, Department of Arctic Environment, University of Aarhus, Frederiksborgvej 399, P.O. Box 358, DK-4000 Roskilde, Denmark.* Present address: *National Survey and Cadastre, National Geodata Bank, Rentemestervej 8, DK-2400 Copenhagen, Denmark.*

BIRGER U. HANSEN, *Department of Geography & Geology, University of Copenhagen, Øster Voldgade 10, DK-1350 Copenhagen K, Denmark.*

ERIC STEEN HANSEN, *Botanical Museum, University of Copenhagen, Gothersgade 130, DK-1123 Copenhagen K, Denmark.*

BENT HASHOLT, *Department of Geography & Geology, University of Copenhagen, Øster Voldgade 10, DK-1350 Copenhagen K, Denmark.*

DITTE K. HENDRICHSEN, *Department of Biology, University of Copenhagen, Universitetsparken 15, DK-2100 København Ø, Denmark.*

JØRGEN HINKLER, *Department of Geography & Geology, University of Copenhagen, Øster Voldgade 10, DK-1350 Copenhagen, Denmark .*

OLE HUMLUM, *Institute of Geosciences, University of Oslo, Box 1047, Blindern, N-0316 Oslo, Norway.*

TOKE THOMAS HØYE, *National Environmental Research Institute, Department of Arctic Environment, University of Aarhus, Frederiksborgvej 399, P.O. Box 358, DK-4000 Roskilde, Denmark.* Present address: *Department of Wildlife Ecology and Biodiversity, Grenåvej 14, DK-8410 Rønde, Denmark.*

KIRSTEN B. HÅKANSSON, *Biological Institute, Department of Terrestrial Ecology, University of Copenhagen, Øster Farimagsgade 2D, DK-1353 Copenhagen K, Denmark.*

LOTTE ILLERIS, *Biological Institute, University of Copenhagen, Øster Farimagsgade 2D, DK-1353 Copenhagen K, Denmark.*

BJARNE H. JAKOBSEN, *Department of Geography & Geology, University of Copenhagen, Øster Voldgade 10, DK-1350 Copenhagen K, Denmark.*

ERIK JEPPESEN, *National Environmental Research Institute, Department of Freshwater Ecology, University of Aarhus, Vejlsøvej 25, DK-8600 Silkeborg, Denmark.*

DAVID R. KLEIN, *Institute of Arctic Biology, University of Alaska, Fairbanks, Alaska, 99775, USA.*

LINE A. KYHN, *National Environmental Research Institute, Department of Arctic Environment, University of Aarhus, Frederiksborgvej 399, P.O. Box 358, DK-4000 Roskilde, Denmark.*

FRANK LANDKILDEHUS, *National Environmental Research Institute, Department of Freshwater Ecology, University of Aarhus, Vejlsøvej 25, DK-8600 Silkeborg, Denmark.*

TORBEN L. LAURIDSEN, *National Environmental Research Institute, Department of Freshwater Ecology, University of Aarhus, Vejlsøvej 25, DK-8600 Silkeborg, Denmark.*

REBEKKA LUNDGREN, *Institute of Biology, Department of Population Biology, University of Copenhagen, Universitetsparken 15, DK-2100 Copenhagen Ø, Denmark.*

FLEUR MARCHAND, *University of Antwerp, Department of Biology, Campus Drie Eiken, Universiteitsplein 1, B-2610 Wilrijk, Belgium.*

HANS MELTOFTE, *National Environmental Research Institute, Department of Arctic Environment, University of Aarhus, Frederiksborgvej 399, P.O. Box 358, DK-4000 Roskilde, Denmark.*

SEBASTIAN H. MERNILD, *University of Alaska Fairbanks, International Arctic Research Center and Water & Environmental Research Center, Fairbanks, Alaska 99775-0292, USA.*

ANDERS MICHELSEN, *Institute of Biology, Department of Terrestrial Ecology, University of Copenhagen, Øster Farimagsgade 2D, DK-1353 Copenhagen K, Denmark.*

TEIS N. MIKKELSEN, *Risoe National Laboratory, Biosystems Department, Technical University of Denmark, Frederiksborgvej 339, P.O. Box 49, DK-4000 Roskilde, Denmark.*

JACOB NABE-NIELSEN, *National Environmental Research Institute, Department of Arctic Environment, Centre for Integrated Population Ecology, University of Aarhus, Frederiksborgvej 399, P.O. Box 358, DK-4000 Roskilde, Denmark.*

IVAN NIJS, *University of Antwerp, Department of Biology, Campus Drie Eiken, Universiteitsplein 1, BE-2610 Wilrijk, Belgium.*

CLAUS NORDSTRØM, *Danish Meteorological Institute, Lyngbyvej 100, DK-2100 Copenhagen Ø, Denmark.* Present address: *National Environmental*

Research Institute, Department of Atmospheric Environment, University of Aarhus, P.O. Box 358, DK-4000 Roskilde, Denmark.

DORTHE PETERSEN, *ASIAQ – Greenland Survey, Box 1003, DK-3900 Nuuk, Greenland.*

MARIANNE PHILIPP, *Department of Biology, Universitetsparken 15, DK-2100, Copenhagen Ø, Denmark.*

ERIC POST, *Penn State University, Biology Department, 208 Mueller Lab, University Park, PA 16802-5301, USA.*

MORTEN RASCH, *Danish Polar Center, Strandgade 102, DK-1401 Copenhagen K, Denmark.*

LEIF RASMUSSEN, *Danish Meteorological Institute, Lyngbyvej 100, DK-2100 Copenhagen Ø, Denmark.*

ÅSA RENNERMALM, *Department of Civil and Environmental Engineering, Princeton University, Princeton, New Jersey, 08544, USA.*

RIIKKA RINNAN, *Institute of Biology, Department of Terrestrial Ecology, University of Copenhagen, Øster Farimagsgade 2D, DK-1353 Copenhagen K, Denmark.*

HELGE RO-POULSEN, *Institute of Biology, Department of Terrestrial Ecology, University of Copenhagen, Øster Farimagsgade 2D, DK-1353 Copenhagen K, Denmark.*

NIELS M. SCHMIDT, *National Environmental Research Institute, Department of Arctic Environment, Centre for Integrated Population Ecology, University of Aarhus, Frederiksborgvej 399, P.O. Box 358, DK-4000 Roskilde, Denmark.*

CHARLOTTE SIGSGAARD, *Department of Geography & Geology, University of Copenhagen, Øster Voldgade 10, DK-1350 Copenhagen K, Denmark.*

MARTIN STENDEL, *Danish Meteorological Institute, Lyngbyvej 100, DK-2100 Copenhagen Ø, Denmark.*

LENA STRÖM, *GeoBiosphere Science Centre, Physical Geography and Ecosystems Analysis, Lund University, Sölvegatan 12, S-22362 Lund, Sweden.*

HENRIK SØGAARD, *Department of Geography & Geology, University of Copenhagen, Øster Voldgade 10, DK-1350 Copenhagen K, Denmark.*

LISE LOTTE SØRENSEN, *Risoe National Laboratory, Wind energy department, DK-4000 Roskilde, Denmark.*

MIKKEL SØRENSEN, *SILA, the Greenland Research Centre at the National Museum of Denmark, Nationalmuseet SILA, Frederiksholms kanal 12, DK-1220 Copenhagen K, Denmark.*

MIKKEL P. TAMSTORF, *National Environmental Research Institute, Department of Arctic Environment, University of Aarhus, Frederiksborgvej 399, P.O. Box 358, DK-4000 Roskilde, Denmark.*

Preface

Climate is one of the most fundamental and ubiquitous natural drivers affecting organisms and the environment in which they are embedded throughout the entire surface of our planet. Indeed, climate connects geophysical as well as biological processes in time and space as no other single driving parameter does, and what we observe in terms of ecosystem functioning today is closely linked to the current climate as well as how it has developed through the recent past. We know that a range of different plant and animal species as well as many physical parameters do respond to both long- and short-term changes in climate, but our current understanding of how an entire ecosystem with its complex mixture of physical and biological compartments responds is rather limited. This book presents, to our knowledge, the first synthesis of such comprehensive interdisciplinary knowledge collected within a single arctic ecosystem over 10 consecutive years.

From ice and sediment cores we know that the past climate has been highly variable, and that fluctuations, even dramatic decadal shifts in, for example, temperature, occurred long before the establishment of human civilization. However, notwithstanding the impacts of the natural climate dynamics, it is now a reasonably well-established fact that during the most recent decades, human activities have contributed to global warming through the combustion of fossil fuels and land-use changes. The climatic consequences of these changes have been most pronounced in the Arctic, and will continue to be so in the future. This pertains to different feedback mechanisms, of which changes in albedo due to melting of snow and ice are among the most important. Diminishing snow-cover and sea ice decrease the albedo and thereby further enhance the warming of land and sea.

As a contribution to the fast-growing area of research into the effects and feedbacks of climate change, Zackenberg Research Station was established in high-arctic Northeast Greenland in 1995 with the purpose of describing an entire ecosystem and monitor how structure and function respond to climate variability and change. Since then, concurrent changes in weather- and climate-related effects have been monitored in detail, supplemented annually by short-term, in-depth research projects.

This book presents the results of much of this work. It focuses on the physical and biological key elements and their interactions. Many results reported in this book have been published in the international scientific literature, whereas other aspects are presented here for the first time. In either case, results reported in the book are specifically merged into a larger perspective of arctic ecology and climate change.

The book may serve particularly well as a textbook on structure and functioning of a high-arctic ecosystem under climate change. However, we hope that any person enthused by the Arctic as well as climate change will find this book interesting. The High Arctic is one of the few places on Earth where pristine ecosystems can be studied. This book documents that such systems are highly sensitive to climate changes and that, even within a decadal time frame, species and systems display marked and, in some cases, rather dramatic responses.

Hans Meltofte, Torben R. Christensen,
Bo Elberling, Mads C. Forchhammer
and Morten Rasch

Acknowledgments

The work at Zackenberg would not have been possible without the collaboration of many people and institutions each providing know-how and hard work in making the monitoring and research at Zackenberg a reality. Most important was the dedicated effort by the Danish Polar Center in establishing and running Zackenberg Research Station from the very beginning. Likewise, the continuous economic support by the Environmental Protection Agency of the Danish Ministry of Environment together with contributions from The Commission of Scientific Research in Greenland and Aage V. Jensen Charity Foundation has provided the very foundation for establishing the research station and running of the long-term monitoring programmes, GeoBasis and BioBasis. Similarly, the Greenland Home Rule backed the monitoring up by financing the ClimateBasis programme run by ASIAQ (Greenland Survey). Finally, we are pleased to acknowledge the efforts by a range of universities and other research institutions in supervising and supporting the monitoring programmes as well as initiating a large number of supportive in-depth research projects at Zackenberg. In particular, the University of Copenhagen and the National Environmental Research Institute, University of Aarhus, Denmark, have played important roles.

Also, the production of this book has drawn upon the support and collaboration of many people and institutions, too numerous to be mentioned here in detail. Most of these, however, are reflected through the affiliations of the team of authors. A special effort on standardising and improving all the illustrations for the book was made by Juana Jacobsen, Kathe Møgelvang, and Tinna Christensen at the Graphics Workshop of the National Environmental Research Institute, University of Aarhus, Denmark. The work of the executive editor of the book, Hans Meltofte, was generously funded by Aage V. Jensen Charity Foundation.

A large number of dedicated scientists have provided important and highly valuable input as referees. These were John Cappelen, Danish Meteorological Institute; Bent Christensen, University of Copenhagen, Denmark; Hugh French, University of Ottawa, Canada; Olivier Gilg, University of Helsinki, Finland; Inger Hanssen-Bauer, Norwegian Meteorological Institute, Norway; Richard Harding, Natural Environment Research Council, UK; Christian Hjort, University of Lund, Sweden; Ian D. Hodkinson, Liverpool John Moores University, UK; Ole Humlum, University of Oslo, Norway; Jim

Hurrell, National Center for Atmospheric Research, USA; Ingibjörg S. Jónsdóttir, Agricultural University of Iceland; Sven Jonasson, University of Copenhagen, Denmark; Niels Tvis Knudsen, University of Aarhus, Denmark; Johannes Kollmann, University of Copenhagen, Denmark; Anders Michelsen, University of Copenhagen, Denmark; John O'Brien, The University of North Carolina at Greensboro, USA; Gareth K. Phoenix, University of Sheffield, UK; Eric Post, Penn State University, USA; Leif Rasmussen, Danish Meteorological Institute; Milla Rautio, Univesité Laval, Canada; Hans Schekkerman, Dutch Centre for Avian Migration & Demography, The Netherlands; Pavel Tomkovich, Moscow Lomonosov State University, Russia; Ingrid Tulp, Institute for Marine Resources and Ecosystem Studies, The Netherlands; Nicholas Tyler, University of Tromsø, Norway; Bernd Wagner, Baltic Sea Research Institute, Germany; Patrick J. Webber, Michigan State University, USA; Jeff Welker, University of Alaska, USA; and Jon Børre Ørbæk, The Research Council of Norway.

We are most grateful to them all.

Executive Summary

This book portrays the numerical and analytical complexity of climatic changes, ecosystem function, and, especially, how these interact. Within a single ecosystem, more than 1500 physical and biological parameters have been measured annually by the monitoring programme at Zackenberg over a 10-year period (Meltofte *et al.* a[1]). This vast amount of data forms the analytical core of the following 21 chapters. Although presented as independent studies, the chapters of this book form a well-defined synthesis of climate–ecosystem dynamics across *c.* 13 scientific themes. Indeed, each chapter often relates to several themes; for example, the chapter by Elberling *et al.* (a) integrates permafrost, soil, vegetation and gas flux, while the chapter by Ellebjerg *et al.* integrates the dynamics of snow, radiation and vegetation. Below, we present the main conclusions from all chapters in a thematic context embracing a decadal perspective of the structure, function and feedback of a high-arctic ecosystem in relation to climate.

The Zackenberg study area in central Northeast Greenland has a high-arctic climate, with July–August mean temperatures between 3 and 7 °C, and winter mean temperatures below −20 °C during the polar night. Cyclonic activity is most intensive and frequent during winter, and the annual precipitation averaged 261 mm during the study years 1996–2005, with 90% falling as snow and sleet. A warming of 2.25 °C in the annual mean has been recorded since 1991, and the five warmest years within the last century were all within the last 10 years. The annual precipitation has increased by 1.9 mm per year over the last 50 years (Hansen *et al.*).

The Zackenberg area was deglaciated prior to 10,000 years BP. After deglaciation, the region warmed to mean July temperatures exceeding the current mean by 2–3 °C at the peak of the Holocene thermal maximum. Subsequent cooling culminated during the Little Ice Age, which peaked about 100–200 years ago. Immigration of plants and animals started immediately after the deglaciation, with warmth-dependent species of plants and insects occurring during the Holocene Thermal Optimum. A few hardy plant and invertebrate species may have survived in ice-free refugia during the last glacial maximum, but most Holocene species immigrated

[1]All references in this Executive Summary are to chapters in the book. References marked "a" refer to the first chapter with the quoted author and so forth.

from the Canadian Arctic Archipelago and northwestern Europe following deglaciation (Bennike *et al.*).

In Northeast Greenland, pre-historic man arrived from Canada about 4500 years ago (Bennike *et al.*; Klein *et al.*), but the region was depopulated several times before the last Inuit finally disappeared from Northeast Greenland some time after 1823 AD, perhaps as a consequence of poor hunting conditions during the peak of the Little Ice Age (Bennike *et al.*).

Primarily because of an expected reduction in sea ice along the coast, climate models predict that the future climate at Zackenberg may potentially converge to the low-arctic climate presently found on the southeast coast of Greenland; that is, towards milder and windier winters together with increased precipitation during both winter and summer (Hinkler *et al.*; Stendel *et al.*). Specifically, within the next few decades (2021–2050), annual precipitation is predicted to increase 40% in East Greenland, whereas a 60% increase is expected for the following period, 2051–2080. The same models predict that temperature will increase by 3.2 °C in winter, 4.6 °C in spring and 1.1 °C in autumn over the next 10–40 years, while summer temperatures will decrease by 0.6 °C. As with precipitation, much larger increases in temperature are expected by 2051–2080. The latter implies that the number of days per year with positive average temperatures will increase from the observed *c.* 80 during 1961–1990 to an average of 248 at the end of the 2051–2080 period (Stendel *et al.*).

Given the complexity with which ecological processes may respond to climate changes embracing many uncharted multiplicative interactions (Forchhammer *et al.* b), it is difficult to evaluate the effects of such extreme climatic changes. Therefore, the statements given here on future changes pertain mainly to the next few decades. Although models predict more snow, the predicted increase in spring temperatures may counteract this and lead to no change or only a minor shift in the average timing of snowmelt over most of the area, while in snow accumulation areas, snowmelt may be delayed by 2–3 weeks as compared to present conditions (Hinkler *et al.*). However, the climate may also become much more variable than observed today, especially with respect to the length of the snow-free season as well as the occurrence of thaw events in winter (Stendel *et al.*). Presently, the timing of snowmelt (i.e., the date with 50% snow-cover) varies by about 1 month around a mean of June 21 (Hinkler *et al.*).

The thickness of permafrost at Zackenberg has been modelled to be around 200–400 m. During 1996–2005, the average thickness of the active layer in late August was between 45 and 80 cm, with the deepest thaw during the latest years. With the future scenarios for climate change in the region, the average thickness of the active layer on the valley floor and mountain slopes may increase by between 8–12 cm and 20–30 cm, respectively, depending on soil type, water content and exposure. However, the active-layer thickness may decrease in snow accumulation areas (Christiansen *et al.*). Deeper active layers may result in increased frequency of active-layer processes such as solifluction and active-layer detachment sliding in the future. Similarly, coastal erosion may increase in summers with longer ice-free periods, as already observed at Zackenberg (Christiansen *et al.*).

During the study years, 1996–2005, the mean annual runoff in the main river, Zackenbergelven, with its catchment area of *c.* 514 km^2, corresponded to 380 mm precipitation. With an annual evapotranspiration of more than 100 mm, this gives a deficit of more than 200 mm as compared to annual precipitation at the research station. This deficit is most likely due to a negative mass balance of glaciers and snow patches within the catchment area, which cover about 20% of the area but contribute about half the runoff in Zackenbergelven. The calculated mean annual suspended sediment transport was at least 38,900 t, corresponding to 76 t per km^2 catchment area. However, the inter-annual variation was considerable, and during single extreme events several times more sediment was delivered than during whole years with normal transport. In the future, runoff from the land is likely to increase following increased precipitation and spring temperatures, but this may be counterbalanced by a possible positive mass balance of glaciers within the catchment area (Hasholt *et al.*).

A thicker active layer will lead to increased release of plant nutrients (Elberling *et al.* a), which may facilitate plant growth and invasion of new species, such as shrubs, depending on water availability. Likewise, new insect pollinators and herbivores are likely to invade the region, while some of the present species will disappear. This will have pronounced effects on ecosystem structure and dynamics (Elberling *et al.* a; Forchhammer *et al.* a,b; Klein *et al.*). Since snow-cover is one of the main determinants of plant community distribution, these effects will be highly differentiated between habitats and plant communities. Some areas will see expanding plant cover, for example, due to more melt-water, while others may suffer from delayed snowmelt or, in contrast, from water stress in late summer due to earlier snowmelt (Elberling *et al.* a; Ellebjerg *et al.*).

Timing of snowmelt together with air temperature during early summer were the major controlling factors for the timing of plant growth, that is, the date of culmination of standing biomass in all six vegetation types characteristic of the Zackenberg area (Ellebjerg *et al.*). During the years 1999–2006, a negative trend in green plant biomass was probably related to drying of the upper soil layers due to earlier snowmelt and higher evapotranspiration during recent years (Ellebjerg *et al.*; Hasholt *et al.*; see further below).

Timing of snowmelt had the most significant effect on the onset of flowering in all plant species studied, followed by the effects of incoming photosynthetic active radiation (PAR) and temperature during the pre-flowering period (the time from snowmelt to flowering). Shrubs appeared to take more advantage of early snowmelt than herbs in their timing of flowering, while some species developed open seed capsules faster in warm summers (Ellebjerg *et al.*). Most species increased the number of flowers in years following a warm growth season, but some species also responded to temperatures during the pre-flowering period. In most species, there was an effect of temperature one year and growth the following year, in that higher temperatures resulted in reduced green biomass according to the far red normalized difference vegetation index (NDVI-FR) 1 year later. This was probably due to increased current-year flowering diminishing resources allocated to growth the following year.

Increased plant biomass and deeper snow-cover will benefit collared lemmings *Dicrostonyx groenlandicus*, if the length of the snow-covered period remains unchanged. A later build up of the snow pack will shorten the period of snow protection for the lemmings (Berg *et al.*; Schmidt *et al.*; see further below). Musk oxen *Ovibos moschatus* will also benefit from more plant biomass, and they will even benefit from a prolonged snow-free period in autumn. On the contrary, both lemmings and musk oxen may suffer from icing of the vegetation and ice crusts in the snow pack, destroying the vegetation and preventing access to forage. Here especially, thaw events in winter may be detrimental to both of these herbivores (Berg *et al.*).

Emergence of insects and other arthropods was closely related to timing of snowmelt, whereas the temperature in early summer had only a secondary effect (Høye and Forchhammer). Consequently, as demonstrated for flowers, the emergence of arthropods displayed considerable spatial variation across habitats, mediated by concurrent variation in snow-cover. During the second half of the study period, 1996–2005, earlier flowering and invertebrate emergence moved these phenological events closer to summer solstice, at which time incoming radiation is at its maximum (Ellebjerg *et al.*; Høye and Forchhammer). Increased variability in timing of snowmelt and summer temperatures in the future will mean more varying conditions for the arthropods. For example, many species may disappear from late-melting snow accumulation areas, while others may benefit from higher spring temperatures in exposed and early snow-free areas (Høye and Forchhammer).

The abundance of arthropod prey for the adult waders (shorebirds) prior to and during egg-laying turned out to be the main determinant of their egg-laying phenology (Meltofte *et al.* b). However, in years with extensive spring snow-cover, that is, >75% in early June, snow-cover was the most important factor determining the egg-laying phenology of waders. Because snow-free spring-feeding area, and thereby early season snow-cover, also is an important determinant of population densities of waders in high-arctic Greenland, increased variability in the distribution of spring snow-cover in the future may result in more variable breeding success in these species (Meltofte *et al.* b).

Since species vary in their phenological response to snowmelt and temperature, there is an increasing risk of inter-trophic mismatch with increasing inter-annual variability in snowmelt and early season temperatures. The risk is more pronounced in species with a narrow period of occurrence and a high degree of host specialisation than in generalist species occurring over more extended periods of the summer (Høye and Forchhammer).

A potentially important factor in shaping the population dynamics of birds and mammals in high-arctic Greenland is predator–prey interactions. In particular, the collared lemming and its predators are important, not only for this complex itself but also for other vertebrates and even plants (Forchhammer *et al.* a; Klein *et al.*; Schmidt *et al.*). At Zackenberg, lemmings show the same cyclic population fluctuations as in many other arctic and sub-arctic populations, and predictions from a lemming–predator population model suggest that the annual fluctuations in lemming numbers primarily are driven by a 1-year delay in stoat *Mustela erminea* predation and stabilising predation from the generalist predators; at Zackenberg, mainly the arctic fox *Alopex lagopus* (Forchhammer *et al.* a; Schmidt *et al.*). However, as also indicated by the last 10 years of observations at Zackenberg, the coupling between the stoat and the lemming population is relatively weak. Because the impact of predators is markedly different between summer and winter, model simulations suggest that the length of the snow-free season may be particularly important for the periodicity of lemming population dynamics at Zackenberg. Hence, a delay in onset of permanent snow-cover appears to prolong the lemming cycles and make them more unstable (Schmidt *et al.*).

The interaction between the lemming and its predators exerts a pervasive influence on the dynamics of the other vertebrate species related to additional predator–prey as well as their competitive interactions. Whereas most mammalian and avian species displayed direct density dependence, indicative of intra-specific competition, only two avian species exhibited dynamics influenced by delayed density dependence. The influence of snow-cover on inter-annual population dynamics was related primarily to the resident species. Specifically, the dynamics of collared lemming and stoat were positively affected, whereas the dynamics of arctic fox and musk ox were negatively affected (Forchhammer *et al.* a).

Some of the most sensitive ecosystems to summer season length in the High Arctic are found in lakes and ponds. Here, ice-melt in spring in particular governs onset of primary production and—together with inflow of organic and inorganic nutrients—the total productivity of the system (Christoffersen *et al.*). If a more snow-rich and maritime climate in the future results in delayed melting of lake and pond ice, this will lead to reduced productivity. On the contrary, more summer precipitation may lead to enhanced in-wash of nutrients and sediments to the fresh water systems. This will result in increased productivity and possibly increased frequency of die out of fish due to oxygen depletion under the winter ice, but it will also provide protection to invertebrates and fish from harmful UV radiation (Christoffersen *et al.*).

Present-day ambient ultraviolet-B (UV-B) radiation was shown to be a significant physiological stress factor to plants at Zackenberg, with possible consequences for primary productivity. This is complementary to studies at sub-arctic Abisko and high-arctic Svalbard, and in accordance with results from Antarctica (Albert *et al.*). To what extent the demonstrated UV-B effects on important high-arctic plant species will persist in the future and thereby increase the potential impact on herbivores mediated through consumer–resource interactions will depend on the extension of stratospheric ozone depletion and climate change.

Timing of snowmelt as well as temperature during the growth season along with the concomitant thaw depth in the soil and plant growth intensity are the primary factors controlling summer CO_2 uptake and CH_4 release in this high-arctic tundra (Elberling *et al.* b; Grøndahl *et al.*). On a landscape scale, CO_2 and CH_4 fluxes are highly dependent on surface hydrology, characteristics of the active layer and the resulting plant composition, but all vegetated surfaces act as sinks during summer (Elberling *et al.* b; Grøndahl *et al.*). Microbial activity continues down to at least -18 °C in winter, which results in an outburst of CO_2 at snowmelt in spring with a total contribution of up to 25% of the annual release of CO_2 from soil respiration (Elberling *et al.* b).

In the future, the large amounts of organic carbon in the soil stored in terrestrial ecosystems may be exposed to degradation due to thicker active layers. This will result in enhanced soil respiration, but different parts of the landscape are expected to respond differently to the same climate changes. Here, the onset of low air temperatures and snow-cover during autumn is critical (Elberling *et al.* b). The CH_4 emission rates from the fen areas at Zackenberg are comparable to what has been found at lower latitudes, and the emissions may increase in the future as a result of increased precipitation and snowmelt (higher water tables in fens and grasslands) (Grøndahl *et al.*).

Winter precipitation and spring snow-cover at Zackenberg are related non-linearly to the inter-annual fluctuations in the North Atlantic Oscillation (NAO), where the high and low phases of the NAO display differential influence on snow conditions (Forchhammer *et al.* b). This has consequences for the ecosystem, and the effects of NAO-mediated changes in snow are seen in a range of organisms at Zackenberg, embracing growth and reproduction of plants, breeding phenology of animals, population dynamics, inter-trophic interactions, community stability and ecosystem feedback dynamics. The influence of the NAO on the ecosystem at Zackenberg sets the observed climate effects here in a circumpolar perspective and comparisons indicate, for example, that over the last 30–40 years, the annual growth of *Salix arctica* at Zackenberg has become less synchronized with the growth of dwarf shrub species at other arctic and alpine localities. Similarly, the indirect bottom-up effect of climate on musk oxen through plants observed at Zackenberg varies across arctic musk ox and caribou populations related to differential influences of the NAO (Forchhammer *et al.* b).

Taking all these ecosystem elements together, the combined influence of the timing of snowmelt and the length of the growing season, which is related to the large-scale climate dynamics mediated by the NAO (Forchhammer *et al.*, b), appears to be of major importance for functioning and feedback dynamics of this high-arctic ecosystem. Furthermore, since timing of snowmelt is the result of both snow depth and early spring temperature, both these factors may be highly influential. Indeed, according to regional models, both snow precipitation and spring temperatures will increase in the future (Hinkler *et al.*; Stendel *et al.*). Predictions from such models indicate that the average timing of snowmelt will not change, whereas the variability in the timing of snowmelt is expected to increase considerably (Stendel *et al.*), which may be manifested in the future structure and functioning of the high-arctic ecosystem at Zackenberg as already indicated in several ecosystem components at Zackenberg (Forchhammer *et al.* b).

The question is, however, whether our modelling has already been overtaken by reality! We predict that the Zackenberg region in central Northeast Greenland will develop in the direction of present-day conditions further down the east coast of Greenland, but during recent years, particularly 2002–2005, snowmelt has been so early and summer temperatures so high that it exceeds these projections. This may be the result of the expected increase in variability, but still we have seen a number of phenological events in plants and animals occurring extremely early for Northeast Greenland. Only continued monitoring and research will answer this key question and reveal its consequences.

Hans Meltofte, Torben R. Christensen, Bo Elberling,
Mads C. Forchhammer and Morten Rasch

Introduction

HANS MELTOFTE, TORBEN R. CHRISTENSEN, BO ELBERLING, MADS C. FORCHHAMMER AND MORTEN RASCH

SUMMARY

Our continuously changing global environment requires continuous and detailed monitoring for us to understand how ecosystems are structured and function in response to climatic changes. Understanding the arctic ecosystems is of particular importance (Oechel *et al.*, 1997). Indeed, rather than in boreal and temperate regions, the forecasted climatic changes will be first and most pronounced in the Arctic. Hence, performing long-term monitoring of an arctic ecosystem provides us with the unique ability to not only give "early warnings" of climate change impacts but also, and perhaps even more important, predict how and where in the ecosystem these will be most pronounced and with what consequences for stability, structure and function.

Since 1995, Zackenberg Ecological Research Operations (ZERO) has monitored annually over 1500 variables concurrently across the physical and biological compartments of a single high-arctic terrestrial ecosystem in central Northeast Greenland. This makes ZERO the most integrated and comprehensive long-term monitoring and research programme presently operating in the Arctic.

This book explores the complex physical and ecological long-term dynamics of a high-arctic terrestrial ecosystem. Since the book is based on data from ZERO, this introductory chapter presents the structural and organisational foundation for ZERO. Following our introduction are four chapters providing the climatic and ecological background together with a presentation of the

ADVANCES IN ECOLOGICAL RESEARCH VOL. 40 0065-2504/08 $35.00
© 2008 Elsevier Ltd. All rights reserved
DOI: 10.1016/S0065-2504(07)00001-3

study area. The rest of the book is devoted entirely to the physical, ecological and ecosystem processes.

I. THE SCENE: AMPLIFIED CLIMATIC CHANGES IN THE ARCTIC

This book is about changes in climate and how these changes are perceived and perpetuated across multiple trophic levels within a single high-arctic terrestrial ecosystem at Zackenberg Research Station in central Northeast Greenland (74°30′N, 20°30′W). Indeed, the variation in climate may very well be one of the major stochastic drivers influencing ecosystem structure and functioning through short-term seasonal forcing as well as through long-term climatic regime shifts (Walther *et al.*, 2002; Forchhammer and Post, 2004).

The retrospective data of climatic changes embracing thousands of years derived from long-term glacial profiles provide us with a unique history of dramatic climatic shifts in the Northern Hemisphere (Andersen *et al.*, 2004). In fact, long before the human industrialization started, considerable changes in Northern Hemisphere temperature have been recorded within time spans from a few years (Johnsen *et al.*, 2001) to cyclic changes over thousands of years (Petit *et al.*, 1999). However, during recent decades human industrial activities have most likely contributed to a global warming by burning of fossil fuels and land use changes, which have resulted in net increase of CO_2 in the atmosphere (IPCC, 2007). While the link between observed changes in the atmosphere's chemical composition and global climate changes is debated, the modelling of future climate changes is not less so. One predictive condition seems to be supported unanimously, however, and that is that any climatic effects of changes following the greenhouse warming will be most pronounced in the Arctic (McBean *et al.*, 2005).

The Arctic is a massive heat sink, since much more heat is exported in the form of outgoing low-frequency heat radiation than is imported by high-frequency solar radiation and inflow of warm air and water masses from the south (Weller, 2000). The balance between in- and outgoing radiation is very sensitive to the albedo of the land and sea surfaces (McBean *et al.*, 2005). Specifically, snow- and ice-cover plays a dominating role, in that 70–90% of the incoming radiation is reflected into space from snow- and ice-covered surfaces as compared to about 15% from tundra (Callaghan *et al.*, 2004). Consequently, any change in snow and ice cover will have pronounced effects on the climate of the Arctic—and the Globe (McBean *et al.*, 2005). Indeed, northern latitudes have, on average, experienced an increase in the amount of snow during the twentieth century (Bamzai, 2003; McBean *et al.*, 2005), but spring temperatures have also increased during recent decades, resulting in earlier snowmelt and thereby a longer period with exposed ground absorbing solar radiation (Dye, 2002; Dye and Tucker, 2003). For example, during the last decades, the annual mean temperature in the Arctic has increased by 0.4 °C per decade, most of which has been in winter and spring, which is considerably more than the 0.25 °C per decade reported for lower latitudes (McBean *et al.*, 2005). Concurrently, the cover of winter sea ice in the Arctic has thinned and retreated by 2.9% per decade (McBean *et al.*, 2005). Hence, climatic changes affecting the high-arctic ecosystems may be viewed as an integrative measure consisting of several simultaneously acting and, indeed, interacting weather factors such as changes in temperature, snow and sea ice (Stendel *et al.*, 2008, this volume).

Integrative measures of climate are often portrayed in the dynamics of large-scale patterns of pressure and circulation anomalies spanning vast geographical areas connecting local weather patterns in different part of the Arctic (Hurrell, 1995; Hurrell *et al.*, 2003). For example, the winter of 1995/1996 was unusually cold and dry far into arctic Russia, whereas northern Canada and West Greenland experienced warm winters with an unusually large amount of snow. These regional weather differences were related to the same atmospheric anomaly: a strong negative phase of the North Atlantic Oscillation (NAO) (Kushnir, 1999). Using large-scale integrative measures of climate change combined with local or regional weather observations is central to our investigation of ecosystem responses to climatic changes for two reasons. First, whereas local weather measurements vary in response to global climate change, large-scale atmospheric systems, such as the NAO, *are* a major component of global change (Hurrell *et al.*, 2003). Hence, the combined use is particularly necessary when ecosystem responses have to be interpreted in a global context. Second, notwithstanding the importance of ecosystem's response to local weather changes (Forchhammer and Post, 2004), simple integrated large-scale measures of climate like the NAO have proven to be extremely useful in linking climate-mediated effects on different species and evolutionary distinct taxa (Forchhammer, 2001; Walther *et al.*, 2002; Hallett *et al.*, 2004), which obviously will improve our ability to describe and model whole ecosystem responses to climate as well as how such responses perpetuate across trophic levels and, eventually, the degree of collective ecosystem feedback to the atmosphere.

The multitude of ecosystem responses also carries important feedback mechanisms with them in a changing climate. The organic rich soils in the wet tundra are subject to possible extra releases of CO_2, if the climate gets warmer and drier, and possible increases in emissions of methane, a very powerful greenhouse gas, if the climate gets warmer and wetter. These have been shown as important feedback mechanisms in sub-arctic mires (Christensen *et al.*, 2004; Malmer *et al.*, 2005). Both these feedback effects, however, amplifying climate warming, should be balanced against the reverse effects of decreased methane emissions in a drier scenario and increased CO_2 uptake in the wet scenario. Furthermore these scenarios of feedback effects on climate are even more complicated by the necessary inclusion of the associated effects of changes in energy exchange as a result of vegetation changes. Such a whole ecosystem balance view of all potential feedback mechanisms in action in a changing climate has rarely been documented for any ecosystem of the world because of a lack of comprehensive data sets. But at Zackenberg there is a possibility to combine data sets and get closer to a full answer as to how a composite arctic landscape will respond and provide feedback mechanisms in a changing climate.

II. THE BACKGROUND: THE NEED FOR LONG-TERM MONITORING AND RESEARCH IN HIGH-ARCTIC GREENLAND AND THE CHOICE OF ZACKENBERG

Mounting evidence clearly suggests a discernable human influence on the climate (IPCC, 2007). Although the relative roles of anthropogenic and natural forcing still remain to be clarified, the recent decades of warming of the Earth and the Arctic in particular, urgently, calls for a far better understanding of the interactions between climate and ecosystems. Indeed, our present knowledge of how climate influences ecosystem structure and functioning, and, eventually, to what extent this results in changed feedback forcing from ecosystem to the atmosphere may be characterised as inferior. The complexity embodied in ecosystem responses to climate change requires multiple concurrent data collection on several trophic levels over long time span in order to differentiate cascading climatic effects from interactions inherent to the ecosystem (Forchhammer, 2001; Petersen *et al.*, 2001).

The Arctic is poorly populated. Hence, long-term data exist only from a very few places—particularly in the High Arctic—where weather stations and military facilities built during or after World War II often constitute the only permanent habitations. In high-arctic Greenland, no interdisciplinary ecosystem-based long-term monitoring and research took place until Zackenberg Research Station was established in 1995 (see Box 1). The initiative to establish a permanent research facility in high-arctic Greenland was taken in 1986 by a group of arctic researchers from the University of Copenhagen. The idea was presented to the board for the National Park in North and East Greenland and later to the committee, which made preparations for the establishment of a Danish Polar Center (Udvalget vedrørende Dansk Polarcenter, 1990). Obviously, when potential man-made climate change became a political issue during the late 1980s, the idea of a permanent research facility devoted to ecosystem monitoring and research gained much more impetus.

Box 1

History of the Area

The name Zackenberg was given to the mountain west of the research station by The Second German North Pole Expedition 1869–1870, the first expedition to winter in Northeast Greenland (Koldewey, 1874). It was named after a mountain in Tyrol, Austria, displaying similar jagged peaks (German: zacken) as the northern edge of "our" Zackenberg. A stylized section of this jagged edge is included in the logo for ZERO (Box Figure 1).

Box Figure 1 The logo of Zackenberg Ecological Research Operations, ZERO, symbolising the "zero-line" before modern "Global Change" followed by future climatic and ecological perturbations in the form of the jagged "graph," which is a stylised section of the jagged northern edge (German: zacken) of the mountain Zackenberg.

Before this first wintering expedition, Inuit people of the Thule Culture lived in the area probably from the fourteenth century to the nineteenth century (Bennike *et al.*, 2008, this volume). The last Inuit to be seen in Northeast Greenland was on the south coast of Clavering Ø, just south of Zackenberg, in 1823 (Clavering, 1830). Few Inuit remains are found in Zackenbergdalen although larger ruins are found in neighboring areas (Bennike *et al.*, 2008, this volume).

Besides a few summer visits by expeditions, Zackenbergdalen remained "untouched" by humans until Danish trappers established a small hut in 1930, after which trapping, mainly for arctic fox pelts, continued in the area until 1960—from 1945 in a proper trapping station by the name Zackenberg (Mikkelsen, 1994). Also Norwegian trappers utilised the area, but mainly for catching arctic char in river Zackenbergelven, which is one of the most fish-rich rivers in Northeast Greenland (Mikkelsen, 1994).

Meanwhile, a number of modern expeditions had utilised the fine conditions at Zackenberg as bases for efforts further north, such as the Danish Peary Land Expeditions 1947–1950 (Martens *et al.*, 2003) and The British North Greenland Expedition 1952–1954 (Simpson, 1957), both using the iceberg-free Young Sund for operations of seaplanes (Box Figure 2).

In 1944, a weather station was established at Daneborg, 23 km southeast of Zackenberg, and in 1951 the base of the military dog

(continued)

Box 1 *(continued)*

sledge patrol, Sirius, was moved to the same site. About 20 men lived at Daneborg until the weather station was closed down in 1975. Today, Sirius is the only populated station within a 300 km radius of Zackenberg Research Station. When Zackenberg was chosen for the research station, one of the benefits was the close cooperation which we could obtain with the experienced personnel at Sirius regarding ship and airplane transport, supervision of the research station during off-season, etc. This expectation has been fulfilled.

Box Figure 2 Danish Navy Catalinas together with the expedition ship Godthaab at Zackenberg during the Danish Peary Land Expeditions 1946–1950. Photo: The archive of Eigil Knuth in the Queen's Hand Library.

The Danish Polar Center was established in 1989, and the establishment of a permanent monitoring and research facility in high-arctic Greenland became the core target for the center from the very beginning (Dansk Polarcenter, 1991). Most of high-arctic Greenland is protected within the National Park of North and East Greenland (see map in Meltofte and Rasch, 2008, this volume), which is the largest national park in the world and classified as a Man and Biosphere Reserve with specific obligations for monitoring and research. In 1991, an expedition was organised to survey the area close to the military facility at Daneborg (74°17′N, 20°13′W) for suitable sites for a research station. The expedition included six representatives from all the relevant themes of natural science together with the chief logistician of the Danish Polar Center. The expedition members unanimously pointed at the valley Zackenbergdalen at 74°30′N, 20°30′W (Figure 1) as the most appropriate site for the station (Andersson *et al.*, 1991).

Zackenbergdalen offers research and monitoring opportunities for almost all branches of natural sciences, embracing almost all landscape features, habitats and species known for high-arctic Greenland, as well as offering access to nearby sites representing north–south and coastal–inland gradients (see Meltofte and Rasch, 2008, this volume). And, rather important, from a logistic point of view, Zackenbergdalen has a large gravel plateau centrally on the valley floor perfectly suited for an airstrip, and situated close to the main river offering water supply. Finally, Zackenbergdalen is situated about 450 km north of Ittoqqortoormiit, the northernmost human settlement in East Greenland, and, hence, exposed to no or very low levels of anthropogenic disturbance throughout the year.

Figure 1 Manipulated IKONOS satellite photo of Zackenbergdalen seen from the south showing the delineation of the prime study area (zone 1A). The research station and runway is at the S centrally in the valley.

The choice turned out to be satisfying and today Zackenbergdalen houses the Zackenberg Research Station base for one of the largest and most comprehensive interdisciplinary ecosystem monitoring programmes in the Arctic, Zackenberg Basic. The first 10 years of monitoring and research has fulfilled our high expectations. We have witnessed a considerable variation from year to year in almost all climatic, geophysical and ecological parameters monitored by Zackenberg Basic.

Similar research and monitoring is taking place in the adjacent marine ecosystem, recently published by Rysgaard and Glud (2007).

III. THE CONCEPT: THE DEVELOPMENT AND RUNNING OF ZERO

Zackenberg Ecological Research Operations' (ZERO's) primary function is to coordinate the monitoring programme Zackenberg Basic with research and logistics activities at Zackenberg Research Station and to secure ZERO's well-defined goal of producing high-quality interdisciplinary ecosystem-based monitoring and research. The cooperation in ZERO is formalised through the ZERO Working Group, which includes representatives from (1) the institutions involved in the monitoring programme Zackenberg Basic, (2) the science community embracing the different disciplines in Zackenberg Basic and (3) the Danish Polar Center, which is responsible for the logistic operations of ZERO.

The pivotal component of ZERO is the monitoring programme Zackenberg Basic, which monitors more than 1500 variables concurrently across the physical and biological compartments of a high-arctic ecosystem. It consists of four sub-programmes, ClimateBasis, GeoBasis, BioBasis and MarineBasis, which are maintained by the following Danish and Greenlandic institutions: Asiaq/Greenland Survey (ClimateBasis), the National Environmental Research Institute (GeoBasis, BioBasis and MarineBasis), the Greenland Institute of Natural Resources (MarineBasis) and the University of Copenhagen (GeoBasis).

Zackenberg Basic is considered by the Danish Environmental Protection Agency as a Danish contribution to the climate change effects monitoring component of the Arctic Monitoring and Assessment Programme (AMAP), and is the main funding agency of the monitoring programme. The running of Zackenberg Research Station is carried out through funding from the Danish Ministry of Science, Technology and Innovation. The total annual funding for the Zackenberg Research Station and Zackenberg Basic was DKK 5.6 million in 2005.

From the very initiation of Zackenberg Basic it has been a major objective to secure that the huge amount of data material collated by the monitoring programmes should be available to anyone being interested in using the data. Today, it is possible to download all data from Zackenberg Basic at the homepage www.zackenberg.dk.

Since 1995, Zackenberg Basic has become affiliated with or contributed to a large number of international organisations, programmes and networks including AMAP, Conservation of Arctic Flora and Fauna (CAFF), Arctic Climate Impact Assessment (ACIA), International Tundra Experiment (ITEX), Global Runoff Data Centre (GRDC), Global Terrestrial Observing System (GTOS), Circumpolar Active Layer Monitoring (CALM) programme, Arctic Coastal Dynamics (ACD), Arctic Birds Breeding Conditions Survey (ABBCS), Scandinavia/North European Network of Terrestrial Field Bases (SCANNET), European Network for Arctic–Alpine Research (ENVINET) and Circumarctic Environmental Observatories Network (CEON).

IV. THE BOOK: IDEA AND ORGANISATION

The central idea of this book is with an integrated approach to draw together findings from a wide range of traditional research disciplines within the bio-hydro-geosphere specifically related to the ecosystem-based monitoring at Zackenberg Research Station and to organise them in a coherent way. Many of these research results have previously been reported separately in reports and as papers in peer-reviewed journals, but a synthesis of such research-based knowledge has not been made previously. In this book, the value and relevance of the research findings has been optimised by a close engagement and dialogue between the researchers and research groups. This dialogue has not focussed on a site-specific description of an ecosystem, but rather on a general analytical approach of the variation in structural and functional patterns as well as processes observed through 10 years of field observations of a single high-arctic ecosystem. This work has led to several bio-geosphere-integrated process models for the Zackenberg region based on the observed physical, geographical and ecological patterns, which are reported here. Such integrated process models represent the first step in a system science approach with a focus on terrestrial ecosystem processes, although both marine and atmospheric processes and interactions are included. The main questions addressed in this book embrace the aspects of (1) ecosystem structure, functioning and dynamics; (2) degree of non-linearity and thresholds in ecosystem patterns and processes; (3) feedback dynamics between ecosystem and atmosphere; and (4) the linkage between a single high-arctic ecosystem and global dynamics (Table 1).

Table 1 Core questions addressed in this book

How is a high-arctic ecosystem structured and how does it function?
- How do intra- and inter-annual changes in weather affect the dynamics of central elements of a high-arctic ecosystem?
- How and to what extent are climatic effects on a single element at one trophic level perpetuated to the entire ecosystem?
- What are the structural and functional consequences of direct and indirect climatic effects?

How can critical thresholds and interactions within and across trophic levels in an arctic system be identified and quantified?
- What and where are the key thresholds for change?
- How will climate change interact with other drivers of environmental changes?
- What controls the magnitude and frequency of extreme geophysical events in the arctic system?
- What controls the distribution limits and population abundance of species and ecosystems?

Assessing vulnerability to change and the capacity to adapt:
- When and how often are critical thresholds likely to be exceeded in the future?
- Will species adapted to harsh conditions prove inflexible, when the environment becomes more benign?
- How vulnerable are biological systems?

How arctic baseline information from Zackenberg obtained during the last decade can improve the understanding of terrestrial feedback conditions to climatic conditions:
- What is the natural background level of climatic and sea ice variability in the region?
- What are the critical dynamics and limitations of arctic ecosystems?

Placing arctic climate change in the Earth System context:
- Is the Arctic a net sink or source of greenhouse gases?
- How do global and arctic changes interact?
- What are short-term effects and what are long-term effects?

Such an integrated assessment is not just an exercise in piecing together elements of knowledge from different disciplines, but to redesign research activities from the beginning. Furthermore, an integrated assessment must rely on insights from all natural sciences, allowing the resulting common methodology to deliver qualitatively different research findings. This mode of research is being increasingly recognised as essential and necessary in relation to questions of global change and impacts resulting from global change in the Arctic. This book represents a unique step in that direction. Thus, through various extrapolation methods most chapters will analyse and discuss observations from the last 10 years at Zackenberg in a long-time perspective (50+ years).

Finally, by writing this book, experience and information have been exchanged between research groups. This has identified needs for new data collection and improved methods used for obtaining data. Thus, the long-term observation strategies for Zackenberg have been evaluated, and this book is central for future recommendations.

REFERENCES

Andersen, K.K., Azuma, N., Barnola, J.-M., Bigler, M., Biscaye, P., Caillon, N., Chappellaz, J., Clausen, H.B., Dahl-Jensen, D., Fischer, H., Flückiger, J., Fritzsche, D., *et al.* (2004) *Nature* **431**, 147–151.

Andersson, T.I.H., Böcher, J., Fredskild, B., Jakobsen, B.H., Meltofte, H., Mogensen, G.S. and Muus, B. (1991) *Rapport om muligheden for placering af en naturvidenskabelig forskningsstation ved Zackenberg, Nationalparken i Nord- og Østgrønland.* Københavns Universitet and Dansk Polarcenter.

Bamzai, A.S. (2003) *Int. J. Clim.* **23**, 131–142.

Callaghan, T.V., Björn, L.O., Chernov, Y., Chapin, T., Christensen, T.R., Huntley, B., Ims, R.A., Johansson, M., Jolly, D., Jonasson, S., Matveyeva, N., Panikov, N., *et al.* (2004) *Ambio* **33**, 459–468.

Christensen, T.R., Johansson, T., Malmer, N., Åkerman, J., Friborg, T., Crill, P., Mastepanov, M. and Svensson, B. (2004) *Geophys. Res. Lett.* **31**, 1–4L04501, doi:10.1029/2003GL018680.

Clavering, D.C. (1830) *Edinburgh New Philosophical Journal,* New Series, **9**, 1–30.

Dansk Polarcenter (1991) *Zackenberg—en forskningsstation i Grønlands Nationalpark, Nordøstgrønland.* Danish Polar Center, Copenhagen.

Dye, D.G. (2002) *Hydrol. Proces.* **16**, 3065–3077.

Dye, D.G. and Tucker, C.J. (2003) *Geophys. Res. Lett.* **30**(7), 1405, doi:10.1029/2002GL016384.

Forchhammer, M.C. (2001) In: *Climate Change Research* (Ed. by A.M.K. Jørgensen, J. Fenger and K. Halsnæs), pp. 219–236. Gads Forlag, Copenhagen.

Forchhammer, M.C. and Post, E. (2004) *Popul. Ecol.* **46**, 1–12.

Hallett, T.B., Coulson, T., Pilkington, J.G., Clutton-Brock, T.H., Pemberton, J.M. and Grenfell, B.T. (2004) *Nature* **430**, 71–75.

Hurrell, J.W. (1995) *Science* **269**, 676–679.

Hurrell, J.W., Kushnir, Y., Ottersen, G. and Visbeck, M. (2003) *The North Atlantic Oscillation. Climatic Significance and Environmental Impact.* American Geophysical Union, Washington DC.

IPCC (2007) In: *Climate Change 2007: The Physical Science Basis* (Ed. by Solomon, S., Qin, D., Manning, M., Chen, Z., Marquis, M., Averyt, K.B., Tignor, M. and Miller, H.L.), 996 pp. Cambridge University Press, Cambridge, United Kingdom and New York, NY, USA.

Johnsen, S.J., Jensen, D.D., Gundestrup, N., Steffensen, J.P., Clausen, H.B., Miller, H., Masson-Delmotte, V., Sveinbjornsdottir, A.E. and White, J. (2001) *J. Quaternary Sci.* **16**, 299–307.

Koldewey, K. (1874) *Die zweite deutsche Nordpolarfahrt in dem Jahren 1869 und 1870 unter Führung des Kapitän Karl Koldewey.* F.A. Brockhaus, Leipzig.

Kushnir, Y. (1999) *Nature* **398**, 289–291.

Malmer, N., Johansson, T., Olsrud, M. and Christensen, T.R. (2005) *Glob. Change Biol.* **11**, 1895–1909.

Martens, G., Jensen, J.F., Meldgaard, M. and Meltofte, H. (2003) *Peary Land.* Forlaget Atuagkat, Nuuk.

McBean, G., Alekseev, G., Chen, D., Førland, E., Fyfe, J., Groisman, P.Y., King, R., Melling, H., Vose, R. and Whitfield, P.H. (2005) In: *Arctic Climate Impact Assessment* (Ed. by C. Symon, L. Arris and B. Heal), pp. 21–60. Cambridge University Press, Cambridge.

Mikkelsen, P.S. (1994) *Nordøstgrønland 1908–60. Fangstmandsperioden.* Dansk Polarcenter, Copenhagen.

Oechel, W.C., Callaghan, T., Gilmanov, T., Holten, J.I., Maxwell, B., Molau, U. and Sveinbjörnsson, B. (1997) *Global Change and the Arctic Terrestrial Ecosystem.* Springer, Berlin Heidelberg, New York.

Petersen, H., Meltofte, H., Rysgaard, S., Rasch, M., Jonasson, S., Christensen, T.R., Friborg, T., Søgaard, H. and Pedersen, S.A. (2001) In: *Climate Change Research* (Ed. by A.M.K. Jørgensen, J. Fenger and K. Halsnæs), pp. 303–330. Gads Forlag, Copenhagen.

Petit, J.R., Jouzel, J., Raynaud, D., Barkov, N.I., Barnola, J.-M., Basile, I., Bender, M., Chappellaz, J., Davis, M., Delaygue, G., Delmotte, M., Kotlyakov, V.M., *et al.* (1999) *Nature* **399**, 429–436.

Rysgaard, S. and Glud, R.N. (Eds.), (2007) In: *Meddr. Grønland,* Biosci. **58**, p. 214.

Simpson, C.J.W. (1957) *North ice: The British North Greenland Expedition.* Hodder and Stoughton, London.

Udvalget vedrørende Dansk Polarcenter (1990) *Betænkning om Dansk Polarcenter.* Betænkning nr. 1191 Afgivet af Udvalget vedrørende Dansk Polarcenter.

Walther, G.R., Post, E., Convey, P., Menzel, A., Parmesan, C., Beebee, T.J.C., Fromentin, J.M., Hoegh-Guldberg, O. and Bairlein, F. (2002) *Nature* **416**, 389–395.

Weller, G. (2000) In: *The Arctic, Environment, People, Policy* (Ed. by M. Nuttall and T.V. Callaghan), pp. 143–160. Harwood Academic Publishers, Newark, New Jersey.

Arctic Climate and Climate Change with a Focus on Greenland

MARTIN STENDEL, JENS HESSELBJERG CHRISTENSEN AND DORTHE PETERSEN

SUMMARY

Paleoclimatic evidence suggests that the Arctic presently is warmer than during the last 125,000 years, and it is very likely[1] that the increase in concentration of greenhouse gases in the atmosphere has an effect, which is larger in the Arctic than elsewhere on the globe (Christensen *et al.*, 2007a). In recent years, concerns about the stability of the Greenland Ice Sheet, the fate of arctic sea ice and a possible weakening of the thermohaline circulation (THC) under future warming conditions have led to increased research

[1]The term "likelihood" is used here as in the Fourth Assessment Report of the Intergovernmental Panel on Climate Change (IPCC AR4). According to the definition in this report, "very likely" corresponds to a likelihood of more than 90%.

ADVANCES IN ECOLOGICAL RESEARCH VOL. 40
0065-2504/08 $35.00
DOI: 10.1016/S0065-2504(07)00002-5
© 2008 Elsevier Ltd. All rights reserved

activities, including an assessment of arctic climate and climate change (ACIA, 2005), the fourth assessment report (AR4) of the International Panel on Climate Change (IPCC, 2007) and a large number of research project related to the International Polar Year (IPY).

Assessments of climate variability and change with a focus on the Arctic in general and Greenland in particular have to consider uncertainties related to the paucity of reliable observations and, for projections of future climate, the large natural variability that makes it difficult to detect an anthropogenic climate signal. Further uncertainties are due to the underlying emission scenarios as well as model uncertainties and deficiencies including insufficient horizontal resolution. Most of these uncertainties can be addressed by considering large ensembles of model simulations instead of a single realization with only one model. In this chapter, we summarize the findings of two such approaches that have focused on the Arctic, ACIA (2005), based on 5 global circulation models (GCMs) and Christensen *et al.* (2007a), based on 21 AR4 GCMs of the most recent generation. In opposition to the ACIA models, the AR4 models no longer need a flux correction to keep their climate stable. We can thus also demonstrate the advance from ACIA to AR4. As the typical resolution of the AR4 models of 150–200 km is hardly adequate to realistically simulate many arctic processes, we also present results from a new transient simulation using a regional climate model (RCM) for Greenland and surrounding seas with a horizontal resolution of 25 km. Thus, differences related to model resolution can be addressed.

The ensemble of global models indicate an increase of 3 °C in global mean temperature by the end of the twenty-first century, whereas for the Arctic, temperature increases of up to 6 °C in the annual mean and up to 10 °C in winter are projected for the same period. The regional model gives locally much larger temperature increases of up to 18 °C in winter, which is related to the retreat of sea ice, in particular along Greenland's east coast. Precipitation is projected to increase everywhere in the Arctic with respect to present-day conditions, ranging from 5%–10% in the south to 35% in the High Arctic for the global models and more than 60% for the regional model. A considerably larger percentage of this precipitation than under present-day conditions is expected to fall as rain, along with an increase in snow depth in the northern half and a decrease in the southern half of Greenland. A substantial decrease in sea ice is also projected.

With a northward retreat of the ice edge, continentality is projected to decrease in Northeast Greenland in general and in the Zackenberg area in particular. This goes along with an increase in both temperature and precipitation, especially in winter. Consequently, climate conditions will develop in the direction of conditions in the Low Arctic with higher snow-cover, lengthening of the thawing season and increased variability including possible thaw periods during winter.

I. INTRODUCTION

The Arctic is the northern polar component of the global climate system. Caused by the Earth's inclination, its characteristics are very little or no irradiation in winter and very long days in summer. The annual mean temperature is below 0 °C over most of the region, and large regions of the Arctic are arid with an average annual precipitation of 100 mm or less. The distribution of irradiation causes numerous cryospheric features such as sea ice, permafrost and glaciers. Snow and ice have high reflectivity and low thermal conductivity, and a large amount of heat is required to melt them to liquid water. In terms of physical geography, the Arctic consists of a deep ocean covered with ice and surrounded by the land masses of Eurasia and North America, except for gaps at the Bering Strait and in the North Atlantic (ACIA, 2005). Of the Arctic Ocean's roughly 14 million km^2, more than half is covered by shallow shelves.

Both weather and climate in the Arctic can vary considerably, and in models as well as observations, the interannual variability in monthly temperatures has maximum values at high latitudes (Räisänen, 2002). Weather variability is caused by the poleward transport of warm water and the outflow of very cold air from the interior of the Arctic as well as interactions with lower latitudes. An assessment of arctic climate variability is quite complex due to numerous nonlinear interactions across all timescales between atmosphere, cryosphere, ocean, land surface and ecosystems. Large low-frequency variability in atmosphere and ice parameters (Polyakov *et al.*, 2003a,b) makes it difficult to detect and attribute climate change in the Arctic. There is evidence that natural multidecadal variability has played a role in the large high-latitude warming around the 1930s (Bengtsson *et al.*, 2004; Johannessen *et al.*, 2004) and the subsequent cooling until the 1960s. Rapid climate change in the Arctic is possible since snow and ice regimes are sensitive to relatively small temperature changes and because of the cold oceans' sensitivity to small changes in salinity. Both processes can amplify initially small signals and contribute to the large climate variability in arctic regions.

The nonlinearity of arctic climate and the fact that many processes still are poorly understood challenge climate models (ACIA, 2005). A further impediment is the fact that only few direct observations of arctic climate are available, which makes model evaluation a difficult task. Precipitation measurements in particular are difficult in cold regions, especially under windy conditions (e.g., Bogdanova *et al.*, 2002). Despite promising attempts to derive the zonation of permafrost from global climate models (Stendel and Christensen, 2002), the spatial resolution of these global circulation models (GCMs) is generally not sufficient to reliably simulate arctic climate. On the contrary, the Arctic is a large region so that only few attempts have been made to use regional climate models (RCMs) over a pan-arctic domain. When these models are driven by observed sea surface temperatures (SSTs) and sea ice boundaries, they show smaller biases in temperature and precipitation than GCMs. This may be due to biases originating from lower latitudes, which are advected into the model domain (e.g., Dethloff *et al.*, 2001; Wei *et al.*, 2002; Lynch *et al.*, 2003; Semmler *et al.*, 2005) or to a more realistic representation of snow-cover in RCMs (Stendel *et al.*, 2007). Across-model scatter in RCM simulations (Tjernström *et al.*, 2005; Rinke *et al.*, 2006) is related to differences in simulated sea ice cover (a reduction in ice extent leads to warming due to increased absorption of solar radiation at the surface). A recent development is the construction of coupled RCMs for atmosphere, ocean and ice (Maslanik *et al.*, 2000; Debernard *et al.*, 2003; Rinke *et al.*, 2003; Mikolajewicz *et al.*, 2005).

Numerous positive and negative feedback processes occur in the Arctic, covering several timescales. The most important positive feedbacks are the snow–ice albedo feedback, changes in the duration of time that the atmosphere is insulated from the Arctic Ocean by sea ice and the feedback between permafrost and methane hydrate (CH_4) as well as carbon dioxide (CO_2): an increase in the active layer thickness effectively means a downward propagation of the actual permafrost table, enabling partly decomposed organic material that was buried in the permafrost layer to ultimately release CH_4 and CO_2 into the atmosphere (Oechel and Vourlitis, 1994). Furthermore, huge amounts of CH_4 are stored as methane hydrate on the ocean floor at present. The present approximate equilibrium might become unstable, when the ocean water warms, releasing methane into the atmosphere—a potentially powerful positive climate feedback, which might further enhance warming. On the contrary, negative feedbacks can result from an increase in freshwater input (which increases the stratification of the upper ocean layers), from a reduction in the intensity of the thermohaline circulation (THC) and from a connection between vegetation and CO_2, where increased vegetation results in a reduction of albedo.

II. THE ARCTIC AS PART OF THE GLOBAL CLIMATE SYSTEM

In a climatological sense, winter circulation in the Arctic is dominated by low pressure over the oceans (Icelandic and Aleutian Lows) and high pressure over the continents (in particular the Siberian High). In summer, pressure gradients are much smaller, and we find only small pressure differences poleward of the subtropical high-pressure belt. Areas of considerable synoptic activity (storm tracks) are found in the North Atlantic and North Pacific, in particular in winter. They transport heat, moisture and momentum into the Arctic. These features are caused by the huge difference in solar irradiation (small to non-existent in winter and large, though with very little absorption due to the high albedo of snow and ice, in summer). Winter mean temperatures in the High Arctic are generally below −20 °C. Much lower temperatures are observed inland under shallow inversions (Alaska and, in particular, Siberia, where temperatures can drop to as low as −60 °C) and on the Greenland Ice Sheet.

The inner Arctic Ocean is a deep (2000–4000 m) basin surrounded by shelf seas with a typical depth of 100–300 m. The only deep connection to lower latitudes is via the Fram Strait (2600 m) off the east coast of Greenland. About 11% of global river runoff goes into the Arctic Ocean, which represents only 1% of global ocean volume (Shiklomanov *et al.,* 2000). Thus the general ocean circulation transports cold, dense deep waters out of the Arctic (Fram and Bering Straits, East Greenland and Labrador Currents) and warm, saline near surface waters into it (Norwegian-Atlantic and West Greenland Currents and, to a certain extent, Bering Strait).

Even though the Arctic Ocean covers only a small part of the globe, global effects can be triggered by positive feedbacks that weaken the THC. The THC transports warm, saline surface water poleward with a deep return flow of cold, less saline water. The overturning, taking place in the Greenland, Irminger and Labrador Seas (Broecker *et al.*, 1990), is very sensitive to the water density, which depends on the outflow of fresh water. An increase of this outflow (e.g., by melting of the Greenland Ice Sheet) is likely to weaken the THC and therefore the heat flux into high latitudes. Paleoclimatic evidence exists that abrupt changes (e.g., Dansgaard *et al.*, 1989) have occurred in the past.

A. Dynamical Links Between the Arctic and Lower Latitudes

Arctic climate is very variable, both on an interannual timescale and on the order of thousands to millions of years, and the Arctic interacts with regions further south via atmosphere, oceans and cryosphere. On short timescales, variability is due to the fact that the Arctic atmosphere is strongly influenced by the general circulation via the North Atlantic Oscillation (NAO) and the Northern Annular Mode (NAM) as well as the Pacific Decadal Oscillation (PDO). The strong positive feedback of ice/snow-cover and temperature, important also on short timescales, becomes dominant for longer timescales.

The NAO is defined as the difference of the normalized sea level pressure (SLP) anomalies between Lisbon and Stykkisholmur (Hurrell, 1995) or Gibraltar and Reykjavik (Jones *et al.*, 1997). The former extends back in time to 1864, the latter to 1821. The choice of stations (other combinations have been suggested as well) has only minor influence, since all NAO indices are highly correlated on interannual and longer timescales (Jones *et al.*, 2003). A good overview of recent NAO research is given in Hurrell *et al.* (2003).

The NAM, also known as Arctic Oscillation (AO), is defined as the amplitude of the pattern defined by the first empirical orthogonal function (EOF) of monthly mean winter SLP anomalies north of 20°N (Thompson and Wallace, 1998, 2000). There is an ongoing debate whether the NAO or the NAM is of larger physical relevance (Deser, 2000; Ambaum *et al.*, 2001; Ambaum and Hoskins, 2002), but both are closely related to each other in winter (December to March), when they have the strongest signature and explain the largest amount of SLP variance (a third for the whole domain, a fourth in the Arctic). We note that because the NAM index is derived from linear considerations, nonlinear variability cannot be described, so that the NAM should be interpreted as the linear approximation of this variability (Monahan *et al.*, 2003). A more in-depth discussion can, for example, be found in ACIA (2005) and references therein.

The PDO is defined as the amplitude of the pattern defined by the first EOF of SST in the Pacific north of 20°N (Mantua *et al.*, 1997; Deser *et al.*, 2004). As for the NAM, there is an ongoing debate whether the PDO can be interpreted as a mode of variability. Several authors have described the PDO as El Niño-related patterns of variability (see, e.g., Evans *et al.*, 2001; Newman *et al.*, 2003; Deser *et al.*, 2004), that is, the PDO is forced from the tropics. Other results point to contributions from the extratropics via oceanic processes (e.g., Gu and Philander, 1997; Barnett *et al.*, 1999; Vimont *et al.*, 2001; Deser *et al.*, 2003). Other authors have questioned whether at all there exists a coupled ocean-atmosphere mode of variability (Biondi *et al.*, 2001; Gedalof *et al.*, 2002). Furthermore, Schneider and Cornuelle (2005) have suggested that the PDO is not at all a mode of variability but rather should be regarded as a blend of El Niño and Aleutian Low variability.

Both NAM/NAO and PDO exhibit considerable interannual and interdecadal variability. A time series of the "extended winter" (December through March) NAO (Figure 1), following the definition by Hurrell (1995, plus updates), reveals low values in the 1960s, a positive trend within the following 20 years, very high positive values in the mid-1990s and a decline to near-average conditions in recent years (as mentioned above, a time series of the NAM would look very similar). Positive anomalies lead to an extension of the Icelandic Low into the High Arctic, anomalously warm temperatures over most of Europe and Russia and cold anomalies over the Chukchi Sea and western as well as parts of southeastern Greenland (Hanssen-Bauer, 2007). This implies that the effect of the NAM/NAO is generally opposite for western and most of eastern Greenland. This behaviour is referred to as the temperature seesaw (van Loon and Rogers, 1978), but has actually been

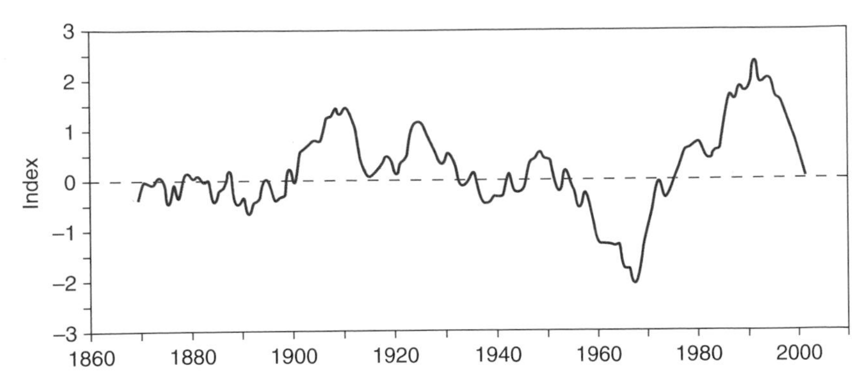

Figure 1 Ten-year running average (e.g., 1864–1873 for the first data point) of the extended winter (December through March) index of the North Atlantic Oscillation (NAO) for the period 1864–2005. The index is based on the difference of normalized sea level pressure (SLP) between Lisbon (Portugal) and Stykkisholmur/Reykjavik (Iceland). The SLP anomalies at each station were normalized by division of each seasonal mean pressure by the long-term mean (1864–1983) standard deviation. Normalization is used to avoid the series being dominated by the greater variability of the northern station. Positive values of the index indicate stronger-than-average westerlies over the middle latitudes (updated from Hurrell, 1995).

known much longer, at least since the second half of the eighteenth century.[2] The PDO showed a negative (cool) phase from the end of the 1940s to 1976, a sudden shift to positive (warm) conditions and prevailing positive values until the mid-1990s. In the positive phase, the Aleutian Low strengthens, and therefore warm anomalies are observed over western North America.

Model results indicate that the NAM/NAO responds to increased concentrations of greenhouse gases through tropospheric (Fyfe *et al.*, 1999; Shindell *et al.*, 1999; Gillett *et al.*, 2003; Miller *et al.*, 2006) and stratospheric processes (Shindell *et al.*, 2001; Sigmond *et al.*, 2004; Rind *et al.*, 2005a). Due to the close relationship between NAO variability and extratropical SST variations (Rodwell *et al.*, 1999), also changes in tropical SSTs (Hoerling *et al.*, 2001, 2004; Hurrell *et al.*, 2004) and the spatial structure of SST changes from the

[2]The Danish missionary Hans Egede Saabye kept a diary in Greenland for several years in the 1770s. The editor of this diary (Ostermann, 1942) cites Egede as follows: "In Greenland, all winters are severe, yet they are not alike. The Danes have noticed that when the winter in Denmark was severe, as we perceive it, the winter in Greenland in its manner was mild, and conversely." According to Ostermann, this relationship was common knowledge of the time. Further details of the history of research on the NAO, including a discussion of old Norse NAO observations around AD1230, can be found in Stephenson *et al.* (2003).

tropics to the extratropics (Rind *et al.*, 2005b) can contribute to an upward trend in the NAM/NAO. Multimodel analyses of the Fourth Assessment Report of the Intergovernmental Panel on Climate Change (IPCC AR4) models reveal that most of the models show a positive trend in the NAM (Rauthe *et al.*, 2004; Miller *et al.*, 2006) and/or NAO (Osborn, 2004; Kuzmina *et al.*, 2005; Stevenson *et al.*, 2006) for the twenty-first century.

Recently, Dethloff *et al.* (2006) showed that relatively minor improvements in the parameterisation of arctic sea ice and snow-cover in ECHAM4 with respect to observations exert a strong influence on the climate of most of the Northern Hemisphere. For high latitudes, these changes lead to an increase in sea ice and therefore lower temperatures under present-day conditions, whereas warmer mid-latitudes are simulated because of changes in the sub-polar westerlies and storm tracks, which resemble the negative phase of the NAO. For future conditions, however (Stendel and Christensen, in preparation), these parameterisation changes lead to a considerably faster temperature increase in the Arctic during the mid-twenty-first century than using the standard ECHAM parameterisation. It would be desirable to extend this investigation by including other models.

III. DESCRIPTION OF MODELS AND SCENARIOS

There are several kinds of uncertainties related to climate projections using simulations with coupled atmosphere-ocean GCMs. Apart from uncertainties in future greenhouse gas and aerosol emissions and their conversion to radiative forcings (which will not be discussed here), there are uncertainties in global and, in particular, regional climate responses to these forcings due, for example, to different parameterisations (discussed in detail by Stocker *et al.*, 2001). There is also large natural variability on the regional scale (consider, e.g., the NAO), so that it is difficult to determine which part of the response of a model is due to anthropogenic forcing and to natural variability (solar, volcanic, but also unforced), respectively. Additional uncertainties are caused by insufficient resolution of global models and different approaches to downscale their output to the regional scale.

This implies that there is no single "best" model to use in an assessment of Arctic (or Greenland) climate changes (Walsh *et al.*, in press). However, most of the uncertainties mentioned in the previous paragraph can be quantified by using ensembles of model simulations rather than one particular model. We will discuss two of these ensembles, the IPCC AR4 multimodel data set (MMD, Randall *et al.*, 2007)[3] and the ACIA data set (ACIA, 2005).[4]

One of the MMD models is ECHAM5-MPI/OM1 (Marsland *et al.*, 2003; Roeckner *et al.*, 2003; Jungclaus *et al.*, 2006). This model was used to force the RCM HIRHAM4 (Christensen *et al.*, 1996, 1998) for the Greenland domain. HIRHAM4 is based on the adiabatic part of the HIRLAM (High-Resolution Limited Area Model) short-range weather prediction model (Källén, 1996), which was jointly developed by a number of national weather services.[5] In HIRHAM4, the physical parameterisation of HIRLAM has been replaced by that of ECHAM5's predecessor ECHAM4 (Roeckner *et al.*, 1996), in order to construct a model that is suitable for long climate integrations, so that HIRHAM4 can be thought of as a high-resolution limited-area version of ECHAM4.

While high-resolution regional climate simulations to date mainly have been run as time-slice experiments, we here present results of a transient simulation at an (for Greenland) unprecedented high spatial resolution (25 km) covering the period 1950–2080. All the forcing data have been taken from the transient ECHAM5-MPI/OM1 run. Of course, it would be desirable to investigate an ensemble of RCM simulations (different RCMs forced by different GCMs, as done, e.g., in the PRUDENCE project, see Christensen *et al.*, 2007b). This was, however, impossible due to lack of computer capacity, and so we had to restrict ourselves to this particular configuration.

HIRHAM4 has been shown to be able to simulate present-day climate realistically (Christensen *et al.*, 1998). In general, predicted temperatures in RCMs follow the patterns of the driving GCMs. However, Dethloff *et al.* (2002) and Kiilsholm *et al.* (2003) show that HIRHAM4 shows closer agreement to observations than the forcing model (in both cases ECHAM4-OPYC3), which is mainly due to a more realistic treatment of

---

[3]The MMD consists (for the Arctic) of 21 state-of-the-art coupled atmosphere-ocean models, which participated in the fourth IPCC AR4. The simulations considered here are forced by the IPCC SRES A1B, A2, and B2 scenarios (Nakicenovic *et al.*, 2000) and cover the period 1990–2100.

[4]The ACIA data set consists of five coupled atmosphere-ocean models which participated in the third IPCC assessment report (TAR). These models are considered to give most realistic results over the pan-arctic domain. The procedure by which these models were chosen is described in detail in ACIA (2005). It was required that they realistically simulate present-day climate, which in particular means that models have to have an ice-free Barents Sea throughout the year and realistic snow-cover in summer (ACIA, 2005). The simulations considered here are forced by the IPCC SRES B2 scenario (Nakicenovic *et al.*, 2000) and cover the period 1990–2100.

[5]Denmark, Finland, the Netherlands, Iceland, Ireland, Norway, Sweden, and (in part) France and Spain.

precipitation and snow-cover (see also Christensen and Kuhry, 2000) in the regional model. Both the Dethloff *et al.* and Kiilsholm *et al.* papers also present results of coarse resolution (50 km) time-slice climate change projections for Greenland.

IV. ARCTIC CLIMATE VARIABILITY AND CHANGE

In Section IV.A, we will discuss present-day climate for the Arctic as a whole and projected change through the twenty-first century. To quantify model-related biases and problems, we will refer to the above-mentioned ensembles of models. Greenland and in particular the Zackenberg region are not discussed here, but rather in the subsequent sections.

A. Present-Day Climate

ACIA (2005) made the first assessment of temperature variations through the twentieth century for the pan-arctic domain. A time series of annual arctic land-surface temperature anomalies (not shown) reveals generally positive temperature trends for the first four decades of the twentieth century, a temperature decrease from about 1945 to 1965 and a subsequent temperature rise for the remainder of the twentieth and the beginning of the twenty-first centuries. This is elucidated in Figure 2, which shows temperature trends for several periods during the last roughly 100 years using the Global Historical Climatology Network (GHCN) data set (Peterson and Vose, 1997, updated), which is very similar to the data set from the Climatic Research Unit (CRU, Jones and Moberg, 2003). Details of the calculation of trends and the treatment of missing data (a common problem in the Arctic) are discussed by Peterson *et al.* (1999).

It becomes clear from Figure 2 that the only period characterised by widespread cooling was 1946–1965, but even during this period, large areas (notably southern Canada and southern Eurasia) had increasing temperatures. For the period from 1966 to 2003, trends are up to 2 °C/decade over large regions in northern Eurasia and northwestern North America, mostly in winter and spring (not shown). Given the scarcity of data prior to 1945, it is difficult to say whether the Arctic as a whole was warm in the 1930s and 1940s (as it was during the 1990s). The warming of the most recent 40 years goes along with a reduction in sea ice thickness (Rothrock *et al.*, 1999) and a decline in perennial sea ice cover (Comiso, 2002).

An assessment of arctic precipitation through the twentieth century is difficult due to the sparse data and measuring problems, in particular of snowfall under windy conditions (Goodison *et al.*, 1998). Trends calculated

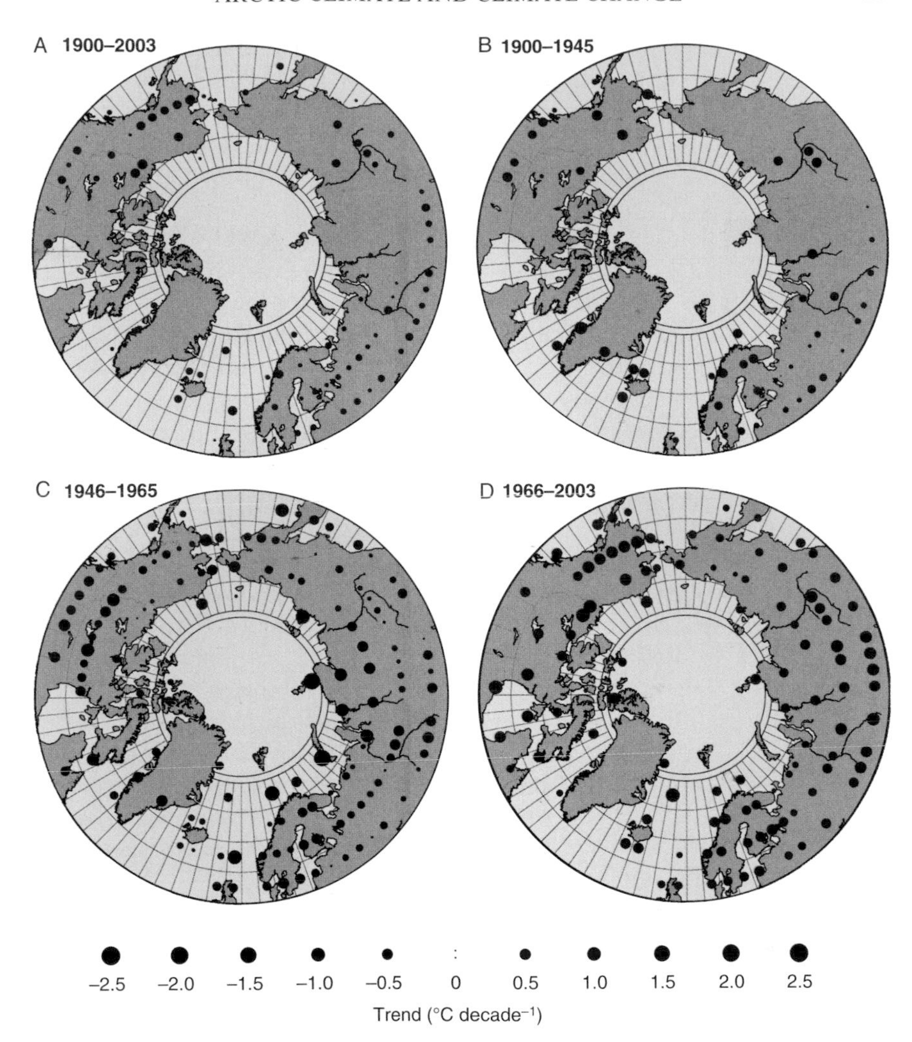

Figure 2 Annual land-surface air temperature trends for (A) 1900–2003, (B) 1900–1945, (C) 1946–1965 and (D) 1966–2003, calculated using the GHCN data set (updated from Peterson and Vose, 1997, and ACIA, 2005).

from the GHCN data set (Peterson and Vose, 1997) indicate that precipitation amounts have increased by 1–2% during the twentieth century, with maximum values in autumn and winter (Serreze *et al.*, 2000). Locally, much higher increases have been observed: 15% in northern Norway (Hanssen-Bauer *et al.*, 1997), 20% in Alaska (Karl *et al.*, 1993) and the permafrost-free part of Russia (Groisman and Rankova, 2001) and even 25% in Svalbard (Hanssen-Bauer and Førland, 1998).

A decrease in SLP in the central Arctic and a compensating pressure increase over the sub-polar oceans has been observed since 1979 (Walsh *et al.*, 1996). Observations from Russian coastal stations from the late nineteenth century suggest that this may be part of a longer oscillation (Polyakov *et al.*, 2003b).

B. Future Climate Evolution

Figure 3 shows a comparison of the temporal evolution of global and arctic annual mean temperatures for all ACIA models during the twenty-first century under the B2 emission scenario (Nakicenovic *et al.*, 2000). For all models, arctic temperature increase is at least a factor 2 larger than for the globe. For the period 2071–2090, a temperature increase of 2.8–4.6 °C relative to 1981–2000 is projected for the Arctic.

The projected warming is not uniformly distributed (Figure 4). As an example, we here show arctic temperature change in the MMD. Most of the models show maximum warming over the high-latitude continental areas and the adjacent shelf seas. An inspection of individual models (not shown) reveals values up to 10 °C over northeastern Siberia and the Arctic Ocean. For southern Greenland, most models show a temperature increase of 2–4 °C, whereas for northern Greenland a temperature increase of 5–7 °C is projected. Smallest values (even slight cooling in several models) are projected over the northern North Atlantic. This is consistent with a projected weakening of the THC (Meehl *et al.*, 2007). Since a large amount of the

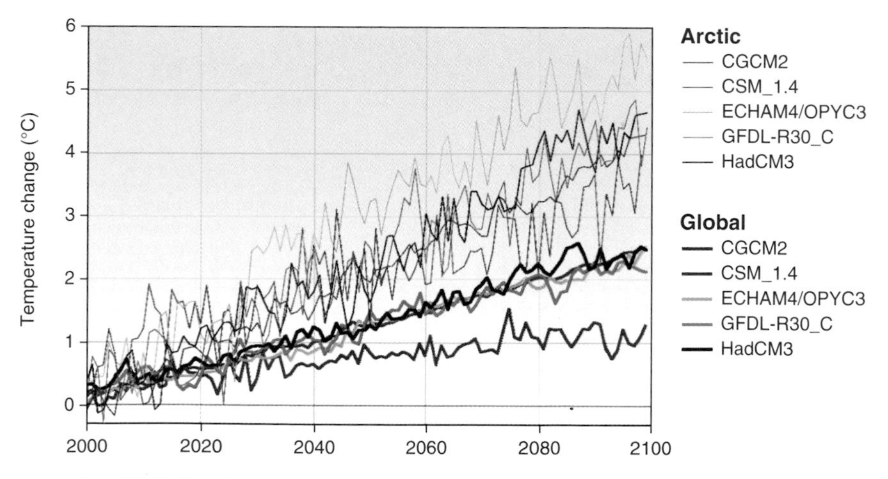

Figure 3 Global and arctic (60°–90°N) changes in annual mean surface air temperature (°C) with respect to the period 1981–2000 for five general circulation models, scenario B2 (from ACIA, 2005).

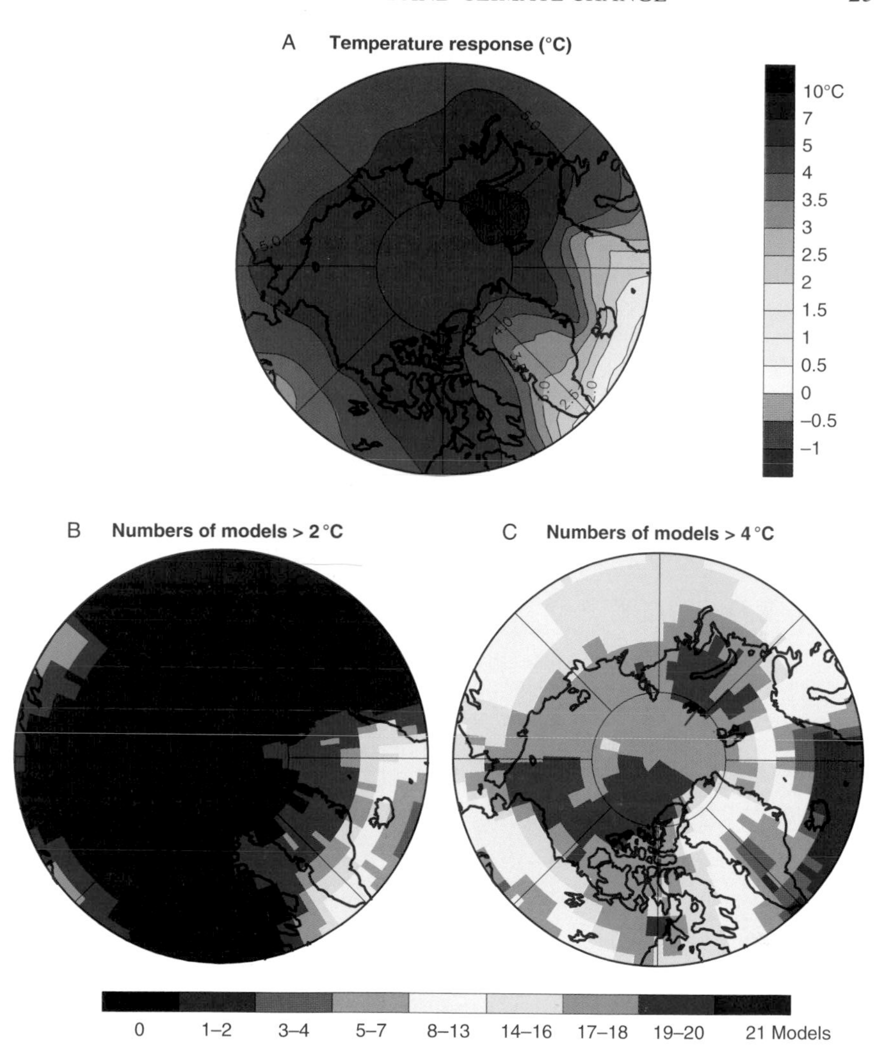

Figure 4 Annual surface air temperature change in the Arctic from 1980–1999 to 2080–2099 under the A1B scenario. Top: mean response, averaged over 21 MMD models; bottom left and right: number of MMD models that generate a warming greater than 2 °C and 4 °C, respectively (reproduced from Christensen *et al.*, 2007a, with kind permission from IPCC).

heating is mixed into the deeper ocean and because the THC weakens in most models under warming conditions, the simulated surface temperature changes are relatively small in this region.

Even though natural variability in the Arctic is large compared with other regions, signals in the MMD are large enough to emerge from the noise within the first third of the twenty-first century (Christensen *et al.*, 2007a), with the exception of Alaska (Chapman and Walsh, 2007). In contrast to these findings, ACIA (2005) concluded that in several regions of the Arctic it may remain difficult to detect the anthropogenic signal for the next few decades. This discrepancy has two reasons. First, the MMD ensemble is considerably larger than the ACIA ensemble, so that the noise is smaller in the MMD. Second, because of improvements from the models used in ACIA to those in the MMD (Christensen *et al.*, 2007a), the signal is larger in the latter. Consequently, the signal-to-noise ratio (i.e., the ratio between mean temperature change and standard deviation) is considerably higher in the MMD.

Looking more in detail, projected temperature changes in autumn and winter are, due to the retreat of sea ice, circulation changes and increasing wind speed (not shown), generally larger than for the other seasons (not shown). In summer, melting sea ice prevents temperatures over most of the Arctic Ocean from rising much above 0 °C. On the contrary, strong temperature increases are projected over land and over ocean regions, which are subject to disappearing or retreating sea ice.

Even though global mean precipitation increases only modestly in many models, a clear increase in precipitation is projected for the Arctic. For the period 2071–2090 under scenario B2, the ACIA 5-model average gives an enhancement of 12% compared with 1981–2000 (Figure 5). This (well-understood, see, e.g., Manabe and Wetherald, 1975) increase is caused by the increased temperature, which allows for more water vapor in the air and

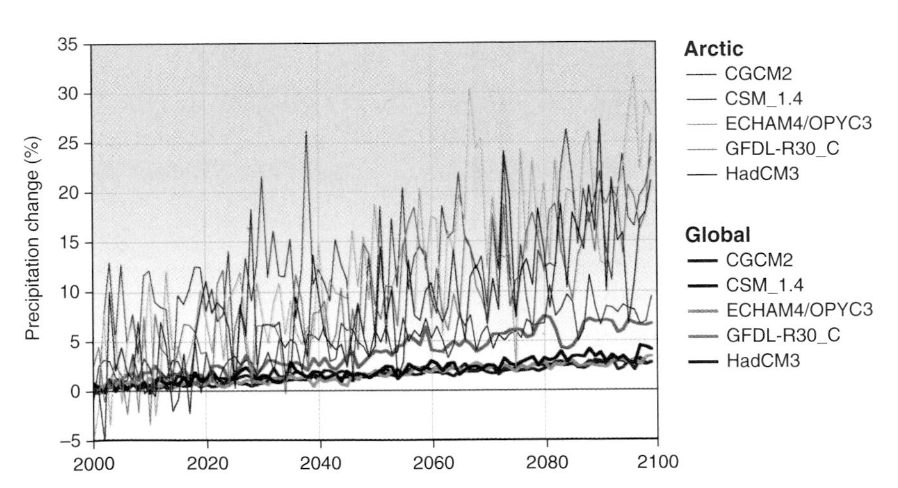

Figure 5 Global and arctic (60°–90°N) changes in annual mean precipitation (%) with respect to the period 1981–2000 for five general circulation models, scenario B2 (from ACIA, 2005).

therefore larger transport of moisture into high latitudes. Accordingly, the largest increase is predicted for the High Arctic in winter.

A feature that is common to most climate models is the decrease in arctic SLP in winter, that is, a shift towards the positive phase of the NAM/NAO. However, differences between individual models are large, and since a large fraction of the NAM/NAO variance is associated with internal atmospheric variability, differences between models do not necessarily reflect uncertainties in the simulated processes. As an example, a decrease of up to 10 hPa is simulated in winter for almost the entire region north of 60°N in ECHAM4 (not shown). At the same time, the intensity of the subtropical highs shows a slight increase, so that, consequently, we find an increase in wind speed (not shown) over Europe, most of the Russian Arctic and Alaska mainly in winter.

A number of studies with GCMs have been conducted, which indicate that the area of the Northern Hemisphere underlain by permafrost could be reduced substantially in a warmer climate (Anisimov *et al.*, 1997; Smith and Burgess, 1999; Nelson *et al.*, 2001; Stendel and Christensen, 2002). However, thawing of permafrost, in particular if it is ice rich, is subject to a time lag due to the large latent heat of fusion of ice. It is not possible to adequately model these processes with state-of-the-art models, since even the most advanced subsurface schemes rarely treat depths below 5 m explicitly. Moreover, in the transitional regions between areas with purely annually frozen ground and areas underlain by continuous permafrost, the heterogeneity basically prohibits a detailed simulation.

Therefore, a simple permafrost model based on a "surface frost index" was introduced by Anisimov and Nelson (1997). The concept was further developed by Stendel and Christensen (2002), who applied Stefan's solution (Stefan, 1891) for the heat transfer problem in a solid medium to simulated deep soil rather than air temperatures. By this approach, complications associated with the explicit parameterisation of snow-cover can be circumvented, and the inherent problem in many GCMs related to the lack of a specific description of freezing and thawing processes can be overcome. All these approaches need information about soil properties, vegetation and snow-cover, which are hardly realistic on a typical GCM grid. In order to close the gap between GCMs on the one hand and local permafrost dynamics on the other, the GCM output can be downscaled to drive an RCM, and then a sophisticated permafrost model is forced with the RCM output (Romanovsky *et al.*, 2004; Stendel *et al.*, 2007). This approach has the advantage that the permafrost model can operate on the same order of grid size as the driving RCM, so that further interpolations can be avoided.

In good agreement with observations (Pavlov, 1994; Oberman and Mazhitova, 2001; Romanovsky *et al.*, 2002; Osterkamp, 2003), the model gives typical increases in temperature at the permafrost level during the twentieth century in the order of 2–3 °C, whereas active layer depth has increased by 0.5 m (not shown). Romanovsky *et al.* (2004) project a further increase in temperature at the permafrost level by up to 7 °C and in active layer depth by up to 2 m (Figure 6) for the end of the twenty-first century for northern Alaska and large parts of northern Siberia. According to these results, large regions of the Arctic will suffer from permafrost degradation in the second half of the twenty-first century, with potentially severe consequences for ecosystems and infrastructure. Note that Greenland is not included in these simulations, since in the model configuration used by Romanovsky *et al.* (2004), the horizontal resolution (50 km) is not sufficient to resolve the narrow coastal region with permafrost. An assessment based on a new high-resolution (25 km) Greenland simulation (see below) is underway.

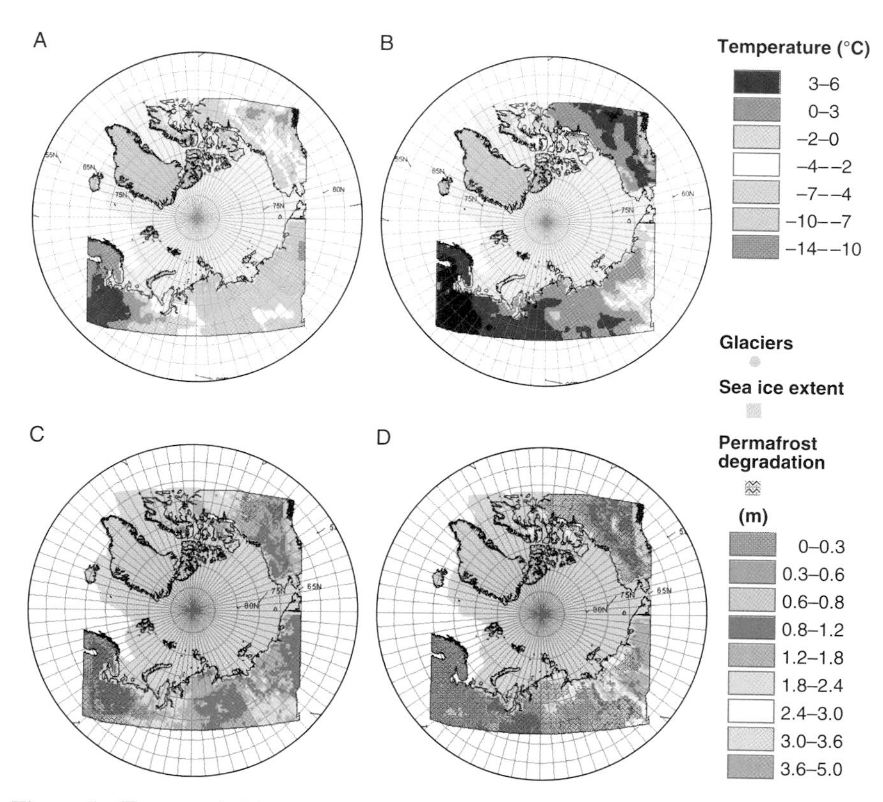

Figure 6 Top panel: Mean annual ground temperature, that is, temperature at the permafrost level for (A) present-day (average 1961–1990) and (B) future conditions (average 2095–2099) in the circumpolar domain using scenario B2. Bottom panel: Active layer thickness for (C) present-day (1961–1990) and (D) future conditions (average 2095–2099). Permafrost degradation (melting of permafrost from the surface) is denoted by cross-hatching (from Romanovsky *et al.*, 2004).

V. GREENLAND AS A KEY REGION FOR PRESENT AND FUTURE CLIMATE

As discussed, for example, in Meehl *et al.* (2007), there are quite large differences between global models in simulating arctic climate and climate change. While part of this difference is due to natural variability, poorly understood physical processes and uncertainties in feedbacks, a substantial part is related to the coarse resolution and, therefore, inadequate parameterisation of physical processes. It is therefore reasonable to apply a dynamical downscaling procedure to drive a limited-area RCM.

For practical reasons, such regional simulations have so far mostly been conducted as time slices, typically for a period towards the end of the twenty-first century and for present-day climate. In such a setup, the forcing fields are taken from a transient Atmosphere-Ocean General Circulation Model (AOGCM) simulation to drive the regional model, and the time-dependent concentrations of greenhouse gases as well as SSTs and sea ice concentrations are taken from the corresponding periods of the global model. Because of the large interannual variability in the Arctic, these time slices need to be long enough. For the entire Arctic, only two such data sets exist (Dorn *et al.*, 2003; Kiilsholm *et al.*, 2003), both run with HIRHAM4 forced with data from a transient GCM simulation with ECHAM4/OPYC3 (Stendel *et al.*, 2002).

To put local climate conditions in the Zackenberg region (discussed in detail in Hansen *et al.*, 2008, this volume) into a broader context, we first present observational evidence of present-day Greenland climate variability. For future climate evolution, results from a new transient HIRHAM simulation (Stendel and Christensen, 2007) covering Greenland and surrounding seas for the period 1950–2080 with a very high spatial resolution of 25 km using the A1B forcing (Nakicenovic *et al.*, 2000) are discussed. Finally, implications of climate change for the Zackenberg region are considered.

A. Present-Day Climate

Of particular importance for Arctic and Northern Hemisphere climate is the Greenland Ice Sheet, which forms by far the largest orographic feature in the Arctic. About 82% (1.7×10^6 km^2) of the area of Greenland is covered by a single sheet of ice with an average elevation of 1500 m and a maximum height of 3255 m (Bamber *et al.*, 2001). Because of the high topography, Greenland strongly enhances Northern Hemisphere meridional heat exchange (Kristjánsson and McInnes, 1999) and influences in particular the location of the Icelandic Low.

Temperature analyses by Putnins (1970) and, more recently, Cappelen *et al.* (2001) reveal a large degree of interannual and interdecadal variability, which generally increases from south to north, particularly in winter. Nevertheless, there have been overarching trends with all stations showing warming in the first three decades of the twentieth century, followed by cooling until the mid-1970s. Figure 7 shows the temperature evolution for several stations along the west and east coast of Greenland.

Following a compilation by Box (2002), winter temperatures for the period 1961– 1990 are between −23 °C in the north (Pituffik/Thule Air Base) and −5 °C in the south (Qaqortoq/Julianehåb). Spring and autumn temperatures range from −15 °C to 0 °C. Generally, summer temperatures are above freezing, with maximum mean monthly values near 10 °C in the south (Narsarsuaq, Kangerlussuaq/Søndre Strømfjord). The annual average temperature ranges from −11 °C in the north (Pituffik/Thule Air Base) and northeast (Danmarkshavn) to just above 0 °C in the south (Paamiut/Frederikshåb, Narsarsuaq). Except for Kangerlussuaq all the stations are situated in coastal areas away from the ice sheet. Much colder temperatures, however, are measured on the ice sheet. At Summit (elevation 3208 m), the average temperature in winter (1987–2001) is −41 °C, in spring −32 °C, in summer −15 °C and in autumn −32 °C, yielding an annual average temperature of −30 °C.

Greenland is generally rather dry (Cappelen *et al.*, 2001). In the north and on the ice sheet, precipitation is less than 100 mm/year. It is also comparably

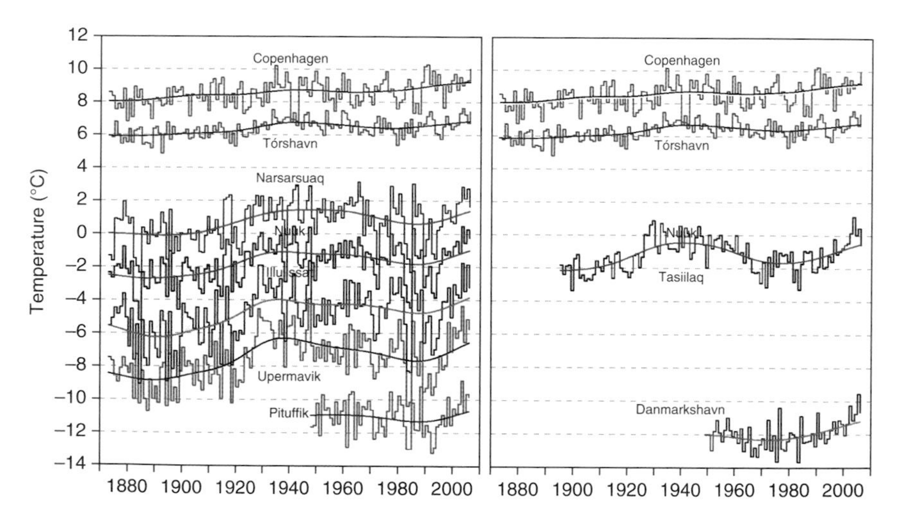

Figure 7 Annual mean temperatures 1873–2006 for selected stations in West Greenland (left panel) and East Greenland (right panel). Temperatures for Copenhagen and Torshavn (Faroe Islands) are given for comparison. A Gauss filter with filter width (standard deviation) 9 years (comparable to a 30-year running mean) has been applied to create the smooth curves (from Cappelen, 2007).

dry on the west coast (Aasiat/Egedesminde: 304 mm/year). Going further south, precipitation increases to 858 mm/year in Qaqortoq/Julianehåb (southwest coast) and even nearly 1000 mm/year on the southeast coast (Tasiilaq/Ammassalik). In the rugged terrain of the southeast coast, where strong winds prevail quite often, it is very difficult to estimate the amount of precipitation. Judged from the precipitation observations at Prins Christian Sund (2474 mm/year), there may be as much as 3000 mm of annual precipitation locally (J. Cappelen, personal communication). In the southwest, there is a summer maximum of precipitation, whereas along the southeast coast most precipitation falls in January, due to cyclogenesis and orographic enhancement.

B. Future Climate Evolution

The driving model for the 25 km HIRHAM simulation is a coupled atmosphere-ocean model, that is, the atmosphere is not forced by observed SSTs and sea ice concentrations, but rather by modeled ones. This implies that in order to obtain meaningful statements on variability and change, the considered periods need to be long enough, in particular in the Arctic, owing to its large internal variability. This is often not possible using time slices of future climate evolution. Here we compare two 30-year averages for periods in the twenty-first century (2021–2050 and 2051–2080) with modeled present-day conditions (1961–1990).

Figure 8 shows seasonal changes in 2 m temperature for the two future periods in Greenland. For 2021–2050, a general temperature increase of 3 °C in winter, 4 °C in spring and 2 °C in summer and autumn is found. In winter, locally larger values up to 6 °C along the west coast (between Sisimiut/Holsteinsborg and Upernavik) and 4 °C along the east coast (between Tasiilaq/Ammassalik and Ittoqqortoormiit/Scoresbysund) and further northeast to Svalbard are found. These are adjacent to regions covered with sea ice under present-day conditions, but not in future climate. On the ice sheet, the model projects an increase by 2 °C throughout the year. For the later period (2051–2080), winter temperature increases accelerate considerably, reaching 7–8 °C throughout the Arctic and 12 °C along the east coast, culminating in a more than 18 °C increase at Svalbard's northeast coast, which is a region prone also to large recent temperature anomalies. On the ice sheet, a temperature increase of 5–6 °C in winter and spring and 3–4 °C in summer and autumn is projected. The comparatively large temperature increase over the northern section of the ice sheet in spring is related to an increase in cloudiness and wind velocity (not shown) so that less low-level inversions can form. Summer temperature changes are small and more uniform as long as temperature remains near 0 °C due to the presence

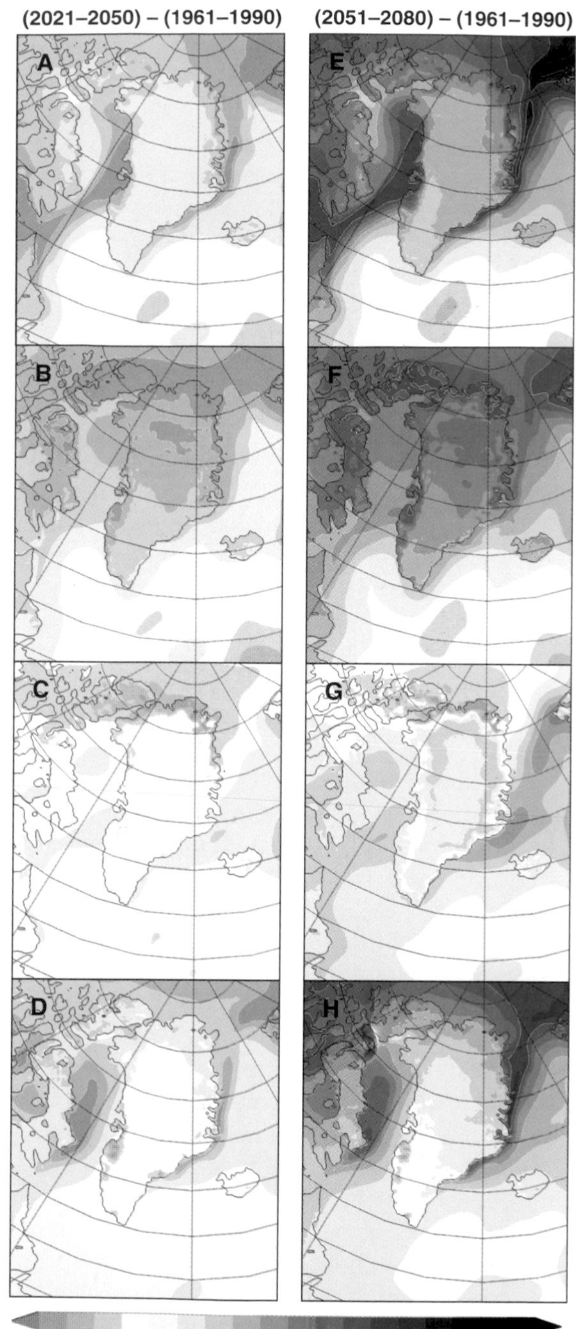
(2021–2050) – (1961–1990)
(2051–2080) – (1961–1990)
A
B
C
D
E
F
G
H
–2 –1 –0.5 0 0.5 1 2 3 4 5 6 7 8 10 12 14 16 18 20
Temperature change (°C)

of melting ice. The small temperature decrease along Greenland's north coast is related to changes in lower tropospheric wind (not shown). Generally, a large decrease in the number of extremely cold days, which here are defined as having an average temperature below −20 °C, is projected.

The climate simulation projects a retreat of sea ice by about 100 km for 2021–2050 compared with present-day conditions and by a further 300–400 km by 2051–2080 (Figure 9A–D). Particularly large changes are projected along the east coast. This is further elucidated by investigating the annual number of thawing days (Figure 9E and F), here defined as days with positive average temperature. Linear trends of more than 20 days/decade are visible along most of the east coast. Other stations in the southwest and northwest also show increases in the number of thawing days, though considerably smaller. In accordance to the huge winter temperature increase near Svalbard discussed above, several winters with no sea ice at all are projected for this region towards the end of the period (not shown).

Holland *et al.* (2006) discuss the possibility of abrupt transitions in September sea ice cover. In our simulation, there is a clear downward trend both for late winter (March) and late summer (September) ice cover. There are no indications of abrupt changes along the southern and western parts of Greenland's coast. For the northeastern part of Greenland, we find indications for transitions (Figure 9G), but more gradual than described in Holland *et al.* (2006).

In agreement to findings in ACIA (2005) and Christensen *et al.* (2007a), most of the Arctic region north of about 60°N is projected to gain an increase in precipitation, especially Siberia, Scandinavia (not shown) and the northeastern part of Greenland (Figure 10A and B). For 2021–2050, a precipitation increase by 15% over western and 40% over interior and eastern Greenland is projected. For 2051–2080, there is further increase by 30–40% over western and interior and about 60% over eastern Greenland. Looking more in detail, we find that there is a distinct decrease in both snowfall amount and frequency along the southern parts of Greenland's coast, whereas more snow is projected on the lower parts of the ice sheet, in particular along the east coast (not shown). An investigation of snowfall exceeding certain thresholds (Figure 10C–F) reveals that there will be a modest increase in extreme snowfall events by 2021–2050. For 2051–2080, this will only be the case in regions with an elevation of more than 1500 m due to the enhanced temperatures. In particular, there will be more extreme snowfall events on the ice sheet according to the model simulation. The hydrological cycle is expected to

Figure 8 Simulated mean near surface (2 m) temperature change in HIRHAM4 for the difference between the two 30-year periods 2021–2050 and 1961–1990 for (A) winter (December–February), (B) spring (March–May), (C) summer (June–August) and (D) autumn (September–November). (E)–(H) Same as (A)–(D), for 2051–2080 minus 1961–1990.

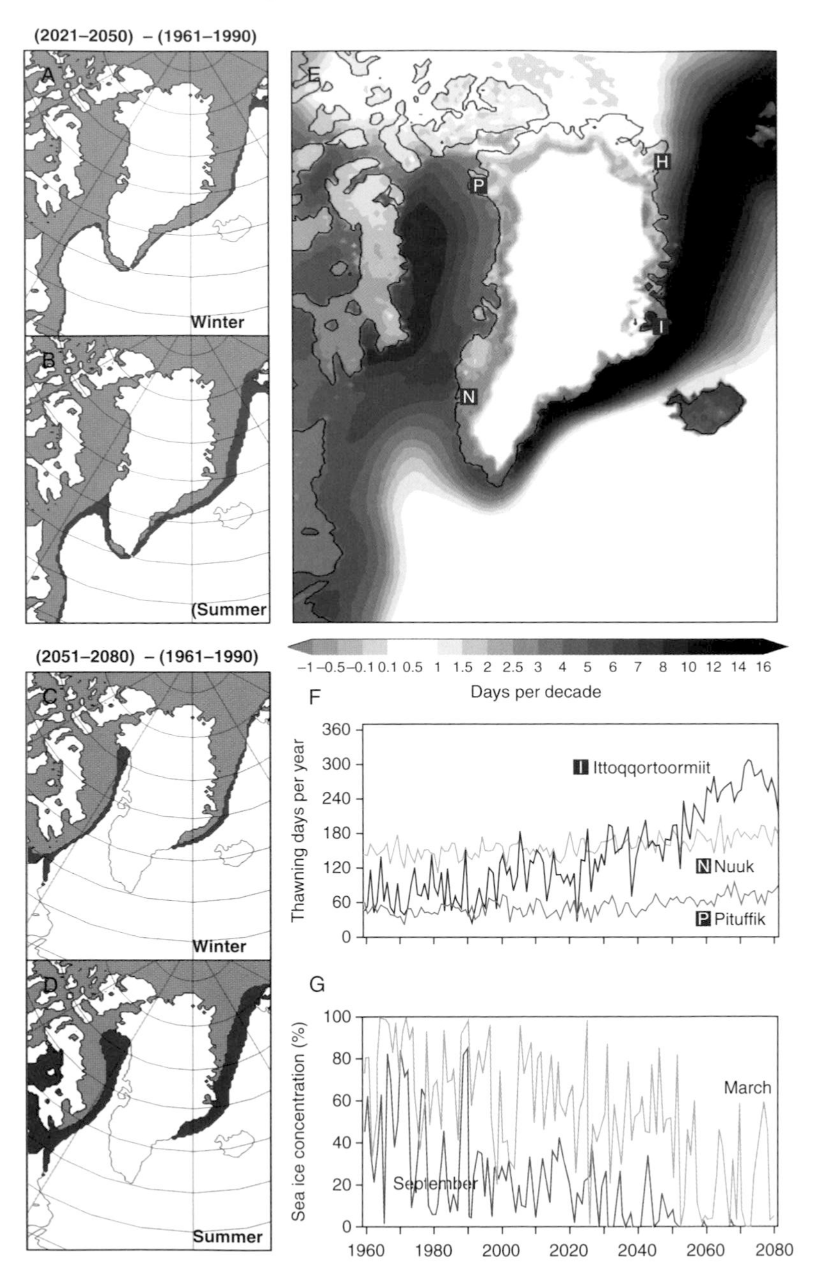

Figure 9 Changes in winter (A) and summer (B) sea ice cover for the periods 2021–2050 minus 1961–1990. (C) and (D) same as (A) and (B), for 2051–2080 minus 1961–1990. Grey shading denotes areas with sea ice both under present-day and future conditions, white means no sea ice. Red shading denotes areas which have sea ice

intensify, as an increase in temperatures allows for a higher absolute humidity in the atmosphere. This increased humidity leads to an enhanced occurrence of orographic precipitation (not shown). This as well as an increased runoff contribution (roughly a third from land in the few areas with bare land points) will contribute to an enhanced freshwater supply in general.

In the Arctic, the length of the growing season is one of the important factors impacting ecosystem processes. The timing of snowmelt in spring is determined by temperatures and snow depth. Even the most recent generation of GCMs allows only limited insight into changes in Greenland snow-cover, since in most models the ice sheet is treated as passive. Taking ECHAM5 as an example, this means that snow is allowed to accumulate on the ice sheet until a predefined maximum snow-cover of 10 m is reached. Since no ablation takes place, this means in practice that the ice sheet is covered by 10 m of snow throughout the year. Similarly, no thawing from the ice cap is considered. Other models behave similarly. The coastal region where snow is allowed to accumulate in the models is rather narrow over most of Greenland, which means that present-day GCMs generally are of limited use to assess changes in snow-cover over Greenland due to the coarse resolution. On the contrary, the high-resolution RCM simulation allows limited insight into the evolution of snow-cover outside the ice sheet (not shown). In all seasons, there is an increase in snow depth compared to the period 1961–1990 over the northern half of Greenland, including the Zackenberg region, and a decrease along the southwestern, southern and south-eastern coasts. The construction of coupled atmosphere-ice-ocean RCMs including three-dimensional thermomechanically coupled ice sheet models is a recent development, which will allow to quantify the response of the ice sheet to climate change and feedback mechanisms.

C. Local Climate Change in the Zackenberg Area

Zackenberg is situated in the southern part of the High Arctic. This region is characterised by continental climate with very cold winters and generally dry conditions. This is true even though the region is located near the coast, since

under present-day, but not under future conditions. Sea ice here means that the ice concentration in a grid cell is greater than zero. No attempt is made to assess ice thickness. (E) Linear trend (days/decade) in the number of thawing days for the period 1961–2080. A thawing day is defined as a day with an average temperature above 0 °C. (F) Time series 1961–2080 of the number of thawing days for model grid points near three selected locations: Nuuk/Godthåb (green line, denoted with "N"), Pituffik/Thule Air Base (blue line, denoted with "P") and Ittoqqortormiit/Scoresbysund (red line, denoted with "I"). (G) Sea ice concentration (in %) in March (blue curve) and September (red curve) for a grid point near Holm Land (81°N, 15°W; H on the map).

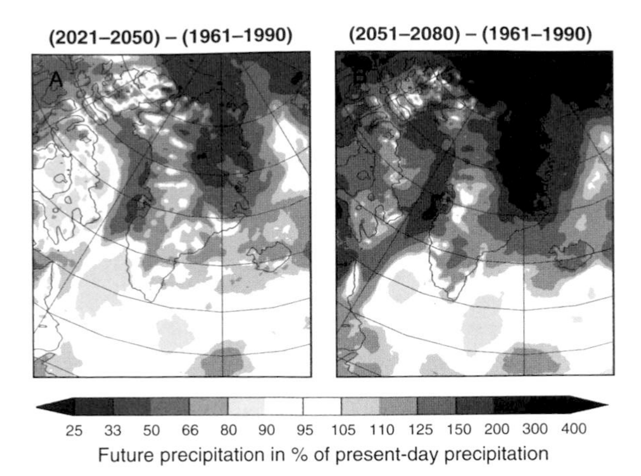

Figure 10 Change in annual precipitation, expressed as a percentage of present-day (1961–1990) precipitation for (A) 2021–2050 and (B) 2051–2080. Linear trend in the number of days (expressed as a percentage change per 100 years compared to present-day) with snowfall exceeding thresholds of (C) 0.1, (D) 1, (E) 10 and (F) 20 mm/day, respectively. White areas over the ice sheet denote regions where no such precipitation events are simulated under present-day conditions.

under present-day conditions the sea is frozen most of the time in a belt several hundred km wide. Changes in sea ice cover will therefore be decisive for local climate changes in the Zackenberg area.

Figure 1 shows considerable multidecadal variability in the strength of the AO in recent times, and many models suggest an increase in the positive phase of the NAM/NAO for the twenty-first century (Christensen *et al.*, 2007a). These changes will also affect Greenland. Positive indices of the NAM/NAO result in warmer, moister and windier conditions than normal over most of northern Europe and Northeast Greenland, whereas it is colder and drier than normal over the western part of Greenland. Even though the local effect of the NAM/NAO on temperatures at Zackenberg is rather small (see, e.g., Jones *et al.*, 2003; Hanssen-Bauer, 2007; and Hansen *et al.*, 2008, this volume), an increase in the strength of the NAM/NAO will favor a retreat of sea ice in the Zackenberg region. With a northward retreat of the ice edge, continentality will decrease in Northeast Greenland in general and in the Zackenberg area in particular. This goes along with an increase in both temperature and precipitation, especially in winter. Consequently, climate conditions will develop in the direction of conditions in the Low Arctic with deeper snow-cover, lengthening of the thawing season and increased variability including extended thaw periods during winter.

Table 1 shows the monthly mean temperatures for Zackenberg for the three periods 1961–1990, 2021–2050 and 2051–2080. The values shown are an average of the nearest two grid points, one of which is located over sea, whereas the other has a height of 560 m a.s.l. To compare with local measurements at Zackenberg, the lower tropospheric lapse rate (approximated by the quotient of temperature differences and geopotential differences between the pressure levels of 1000 hPa and 850 hPa) needs to be accounted for. In summer, temperature typically decreases by 0.5 °C/100 m height change, whereas in winter and spring the vertical temperature decrease is small or

Table 1 Observed and modelled temperatures (°C) at Zackenberg for winter (DJF), spring (MAM), summer (JJA), autumn (SON), and the annual mean

	DJF	MAM	JJA	SON	Year
Observed 1995–2005	−19.7	−12.8	4.3	−8.8	−9.2
Modelled 1961–1990	−19.9	−11.0	4.5	−11.2	−9.4
Modelled 2021–2050	−16.7	−6.4	3.9	−10.1	−7.3
(2021–2050) – (1961–1990)	3.2	4.6	−0.6	1.1	2.1
Modelled 2051–2080	−13.3	−3.6	4.5	−8.8	−5.3
(2051–2080) – (1961–1990)	6.6	7.4	0.0	2.4	4.1

The model data is an average of the two nearest points to Zackenberg. Changes in lapse rate and the different orography of model and station data have been taken into account.

temperature is even rising with height due to shallow inversions, while autumn lies in between. With the projected increase in windiness, such inversions will become less frequent, resulting in an additional warming of about 1 °C in winter and spring of 2051–2080 caused by changes in the stratification of the lower troposphere.

While only minor changes are projected in summer, Table 1 shows that large temperature changes can be expected in winter and spring, particularly in March and April (not shown specifically). Further evidence for these changes can be seen in Figure 9F, which elucidates that the annual number of thawing days (positive daily mean temperature) for a model grid point near Zackenberg is projected to increase from 80 for 1961–1990 to 248 for the period 2051–2080. According to these simulations, there will be more than 300 such days in individual years, so that the climate at Zackenberg hardly any longer can be characterised as "high-arctic."

VI. CONCLUSIONS

Climate and climate change over the Arctic in general and Greenland in particular have recently attained much attention. Using the most recent (IPCC AR4) models and a rather large ensemble, there is growing confidence that the climate change signal can be separated from noise in the early twenty-first century despite the large natural variability (Christensen *et al.*, 2007a). Such a statement was not possible in ACIA (2005) due to the considerably smaller ensemble size, lack in spatial resolution and model deficiencies such as flux corrections. With few exceptions, the AR4 models project modest warming on the order of 2–3 °C in the southern and larger warming (around 7 °C) for the northern part of Greenland.

We report results from a transient climate simulation at an (for Greenland) unprecedented horizontal resolution of 25 km (Stendel and Christensen, 2007), which has been forced with a rather high-resolution (T63, i.e., roughly 1.8° in latitude and longitude) coupled AOGCM. The regional model shows considerably stronger temperature increase in regions where sea ice retreats than the driving global model. An assessment of extreme events reveals considerably less cold (and warmer) days than under present-day conditions. Largest changes can be expected along the east coast, in particular in the Zackenberg region, where days with a positive average temperature are projected to become the rule rather than the exception. Most of Greenland, especially the northeast, will experience more precipitation. At lower elevations, an increasing percentage of this precipitation can be expected to fall as rain instead of as snow as under present-day conditions. On the contrary, more extreme snowfall events than at present are projected on the ice sheet.

Higher temperatures and increased precipitation will have a profound effect on the hydrological budget of the region.

Arctic sea ice is responding sensitively to global warming (Meehl *et al.*, 2007). Most AR4 models and also our regional simulation show significant, but smaller, changes in sea ice in winter (at least to the middle of the twenty-first century) than in late summer, when sea ice is projected to disappear towards the end of the twenty-first century due to the ice-albedo feedback and the advection of warmer waters into the Arctic. In the HIRHAM4 simulation, late summer sea ice around Greenland disappears everywhere except in the far north where, however, it is substantially reduced.

Changes in arctic albedo can affect lower latitudes, as recently shown (Dethloff *et al.*, 2006). Both global and regional models agree in substantial changes in sea ice cover during the twenty-first century. Via changes in storm tracks, such future arctic changes can be expected to influence large parts of the Northern Hemisphere.

ACKNOWLEDGMENTS

We thank the secretariats of the Intergovernmental Panel on Climate Change (IPCC) and the Arctic Climate Impact Assessment (ACIA) for permission to use figures from the IPCC AR4 and ACIA reports. The NAO data used in Figure 1 were obtained from Jim Hurrell's Web site. John Cappelen is thanked for comments on Greenland's climate and Jim Hurrell and Inger Hanssen-Bauer for valuable comments that helped to improve the manuscript. The transient high-resolution climate change experiment for Greenland was supported by the Danish Environmental Protection Agency, Ministry of the Environment, under Grant No. M127/001-0237 and the Danish Climate Centre.

REFERENCES

ACIA (2005) *Arctic Climate Impact Assessment*, 1042 pp. Cambridge University Press, New York.

Ambaum, M.H. and Hoskins, B.J. (2002) *J. Clim.* **15**, 1969–1978.

Ambaum, M.H., Hoskins, B.J. and Stephenson, D.B. (2001) *J. Clim.* **14**, 3495–3507.

Anisimov, O.A. and Nelson, F.E. (1997) *Clim. Change* **35**, 241–258.

Anisimov, O.A., Shiklomanov, N.I. and Nelson, F.E. (1997) *Glob. Planet. Change* **15**, 61–77.

Bamber, J., Ekholm, S. and Krabill, W.B. (2001) *J. Geophys. Res.* **106**, 33773–33780.

Barnett, T.P., Pierce, D.W., Latif, M. and Dommenget, D. (1999) *Geophys. Res. Lett.* **26**, 615–618.

Bengtsson, L., Semenov, V.A. and Johannessen, O.M. (2004) *J. Clim.* **17**, 4045–4057.

Biondi, F., Gershunov, A. and Cayan, D.R. (2001) *J. Clim.* **14**, 5–10.
Bogdanova, E.G., Ilyin, B.M. and Dragomilova, I.V. (2002) *J. Hydrometeorol.* **3**, 700–713.
Box, J.E. (2002) *Int. J. Clim.* **22**, 1829–1847.
Broecker, W.S., Bond, G. and Klas, M.A. (1990) *Paleoceanography* **5**, 469–477.
Cappelen, J. (2007) *Danish Met. Inst. Tech. Report* **07–05**.
Cappelen, J., Jørgensen, B.V., Laursen, E.V., Stannius, L.S. and Thomsen, R.S. (2001) *Danish Met. Inst. Tech. Rep.* **00–18**.
Chapman, W.L. and Walsh, J.E. (2007) *J. Clim.* **20**, 609–632.
Christensen, J.H. and Kuhry, P. (2000) *J. Geophys. Res.* **105**, 29647–29658.
Christensen, J.H., Christensen, O.B., Lopez, P., van Meijgaard, E. and Botzet, M. (1996) *Danish Met. Inst. Sci. Rep.* **96–4**.
Christensen, O.B., Christensen, J.H., Machenhauer, B. and Botzet, M. (1998) *J. Clim.* **11**, 3204–3229.
Christensen, J.H., Hewitson, B., Busuioc, A., Chen, A., Gao, X., Held, I., Jones, R., Kolli, R.K., Kwon, W.-T., Laprise, R., Magaña Rueda, V.Mearns, L., *et al.* (2007a) In: *Climate Change 2007: The Physical Science Basis* (Ed. by S. Solomon, D. Qin, M. Manning, Z. Chen, M. Marquis, K.B. Averyt, M. Tignor and H.L. Miller). Cambridge University Press, Cambridge.
Christensen, J.H., Carter, T.R., Rummukainen, M. and Amanatidis, G. (2007b) *Clim. Change* **81**, 1–6.
Comiso, J.C. (2002) *Geophys. Res. Lett.* **29**, 10.1029/2002GL015650.
Dansgaard, W., White, J.W. and Johnsen, S.J. (1989) *Nature* **339**, 532–534.
Debernard, J., Køltzow, M.Ø., Haugen, J.E. and Røed, J.E. (2003) *Reg. Clim. General Technical Report* **7**, 59–69. Norwegian Met. Inst.
Deser, C. (2000) *Geophys. Res. Lett.* **27**, 779–782.
Deser, C., Alexander, M.A. and Timlin, M.S. (2003) *J. Clim.* **16**, 57–72.
Deser, C., Phillips, A.S. and Hurrell, J.W. (2004) *J. Clim.* **17**, 3109–3124.
Dethloff, K., Abegg, C., Rinke, A., Hebestadt, I. and Romanov, V.F. (2001) *Tellus* **53A**, 1–26.
Dethloff, K., Schwager, M., Christensen, J.H., Kiilsholm, S., Rinke, A., Dorn, W., Jung-Rothenhäusler, F., Fischer, H., Kipfstuhl, S. and Miller, H. (2002) *J. Clim.* **15**, 2821–2832.
Dethloff, K., Rinke, A., Benkel, A., Køltzow, M., Sokolova, E., Kumar Saha, S., Handorf, D., Dorn, W., Rockel, B., von Storch, H., Haugen, J.E. Røed, L.P., *et al.* (2006) *Geophys. Res. Lett.* **33**, 10.1029/2005GL025245.
Dorn, W., Dethloff, K., Rinke, A. and Roeckner, E. (2003) *Clim. Dyn.* **21**, 447–458.
Evans, M.N., Cane, M.A., Schrag, D.P., Kaplan, A., Linsley, B.K., Villalba, R. and Wellington, G.M. (2001) *Geophys. Res. Lett.* **28**, 3689–3692.
Fyfe, J.C., Boer, G.J. and Flato, G.M. (1999) *Geophys. Res. Lett.* **26**, 1601–1604.
Gedalof, Z., Mantua, N.J. and Peterson, D.L. (2002) *Geophys. Res. Lett.* **29**, 10.1029/2002GL015824.
Gillett, N.P., Allen, M.R. and Williams, K.D. (2003) *Q. J. Roy. Meteor. Soc.* **129**, 947–966.
Goodison, B.E., Lottie, P.Y.T. and Yang, D. (1998), *In* WMO-TD 872, 87 pp. World Meteorological Organization, Geneva+ annexes.
Groisman, P.Ya. and Rankova, E.Ya. (2001) *Int. J. Climatol.* **21**, 657–678.
Gu, D.F. and Philander, S.G.H. (1997) *Science* **275**, 805–807.
Hanssen-Bauer, I. (2007) In: *Arctic alpine ecosystems and people in a changing environment* (Ed. by J.B. Orbæk, R. Kallenborn, I. Tombre, E.N. Hegseth, S. Falk-Petersen and A.H. Hoel), 434 pp. Springer Verlag, Berlin.

Hanssen-Bauer, I. and Førland, E.J. (1998) *Clim. Res.* **19**, 143–153.
Hanssen-Bauer, I., Førland, E.J., Tveito, O.E. and Nørdli, P.O. (1997) *Nordic Hydrol.* **28**, 21–36.
Hoerling, M.P., Hurrell, J.W. and Xu, T. (2001) *Science* **292**, 90–92.
Hoerling, M.P., Hurrell, J.W., Xu, T., Bates, G.T. and Phillips, A. (2004) *Clim. Dyn.* **23**, 391–405.
Holland, M.M., Bitz, C.M. and Tremblay, B. (2006) *Geophys. Res. Lett.* **33,** 10.1029/2006GL028024.
Hurrell, J.W. (1995) *Science* **269**, 676–679.
Hurrell, J.W., Kushnir, Y., Ottersen, G. and Visbeck, M. (2003) *Geophys. Monogr.* **134**, 1–35.
Hurrell, J.W., Hoerling, M.P., Phillips, A. and Xu, T. (2004) *Clim. Dyn.* **23**, 371–389.
IPCC (2007) *Climate Change 2007: The Physical Science Basis* (Solomon, S., Qin, D., Manning, M., Chen, Z., Marquis, M., Averyt, K.B., Tignor, M. and Miller, H.L. (eds.), (2007) p. 996. Cambridge University Press, Cambridge, UK and New York, NY, USA.
Johannessen, O.M., Bengtsson, L., Miles, M.W., Kuzmina, S.I., Semenov, V.A., Alekssev, G.V., Nagurnyi, A.P., Zakharov, V.F., Bobylev, L.P., Pettersson, L.H., Hasselmann, K. and Cattle, H.P. (2004) *Tellus* **56A**, 328.
Jones, P.D. and Moberg, A. (2003) *J. Clim.* **16**, 206–223.
Jones, P.D., Jónsson, T. and Wheeler, D. (1997) *Int. J. Climatol.* **17**, 1433–1450.
Jones, P.D., Osborn, T.J. and Briffa, K.R. (2003) *Geophys. Monogr.* **134**, 51–62.
Jungclaus, J.H., Keenlyside, N., Botzet, M., Haak, H., Luo, J.-J., Latif, M., Marotzke, J., Mikolajewicz, U. and Roeckner, E. (2006) *J. Clim.* **19**, 3952–3972.
Källén, E. (Ed.), (1996) In: *HIRLAM Documentation Manual, System 2.5.*, 126 pp. Swed. Meteorol. and Hydrol. Inst., Norrköping.
Karl, T.R., Quayle, R.G. and Groisman, P.Ya. (1993) *J. Clim.* **6**, 1481–1494.
Kiilsholm, S., Christensen, J.H., Dethloff, K. and Rinke, A. (2003) *Geophys. Res. Lett.* **30**, 10.1029/2002GL015742.
Kristjánsson, J.E. and McInnes, H. (1999) *Q. J. Roy. Meteor. Soc.* **125**, 2819–2834.
Kuzmina, S.I., Bengtsson, L., Johannessen, O.M., Drange, H., Bobylev, L.P. and Miles, M.W. (2005) *Geophys. Res. Lett.* **32**, 10.1029/2004GL021064.
Lynch, A.H., Cassano, E.N., Cassano, J.J. and Lestak, L. (2003) *Mon. Wea. Rev.* **131**, 719–732.
Manabe, S. and Wetherald, R.T. (1975) *J. Atm. Sci.* **32**, 3–15.
Mantua, N.J., Hare, S.R., Zhang, Y., Wallace, J.M. and Francis, R.C. (1997) *Bull. Am. Met. Soc.* **78**, 1069–1079.
Marsland, S.J., Haak, H., Jungclaus, J.H., Latif, M. and Roeske, F. (2003) *Ocean Model.* **5**, 91–127.
Maslanik, J.A., Lynch, A.H., Serreze, M.C. and Wu, W. (2000) *J. Clim.* **13**, 383–401.
Meehl, G.A., Stocker, T.F., Collins, W.D., Friedlingstein, P., Gaye, A.T., Gregory, J. M., Kitoh, A., Knutti, R., Murphy, J.M., Noda, A., Raper, S.C.B.Watterson, I.G., *et al.* (2007) In: *Climate Change 2007: The Physical Science Basis* (Ed. by S. Solomon, D. Qin, M. Manning, Z. Chen, M. Marquis, K.B. Averyt, M. Tignor and H.L. Miller). Cambridge University Press, Cambridge.
Mikolajewicz, U., Sein, D.V., Jacob, D., Königk, T., Podzun, R. and Semmler, T. (2005) *Meteorol. Z.* **14**, 793–800.
Miller, R.L., Schmidt, G.A. and Shindell, D.T. (2006) *J. Geophys. Res.* **111**, 10.1029/2005JD006323.
Monahan, A.H., Fyfe, J.C. and Pandolfo, L. (2003) *J. Clim.* **16**, 2005–2021.

Nakicenovic, N., Alcamo, J., Davis, G., de Vries, B., Fenhann, J., Gaffin, S., Gregory, K., Grübler, A., Jung, T.Y., Kram, T., La Rovere, E.L.Michaelis, L., *et al.* (2000) *IPCC Special Report on Emissions Scenarios*. Cambridge University Press, Cambridge.
Nelson, F.E., Anisimov, O.A. and Shiklomanov, N.I. (2001) *Nature* **410**, 889–890.
Newman, M., Compo, G. and Alexander, M.A. (2003) *J. Clim.* **23**, 3853–3857.
Oberman, N.G. and Mazhitova, G.G. (2001) *Norw. J. Geogr.* **55**, 241–244.
Oechel, W.C. and Vourlitis, G.L. (1994) *Trends Ecol. Evol.* **9**, 324–329.
Osborn, T.J. (2004) *Clim. Dyn.* **22**, 605–623.
Osterkamp, T.E. (2003) In: *Proceedings of the Eighth International Conference on Permafrost* (Ed. by M. Phillips, S.M. Springman and L.U. Arenson), 21–25 July 2003 Zürich, Switzerland, pp. 863–868. Balkema, Netherlands.
Ostermann, H. (1942) *Meddr. Grønland* **129**(2), 103 pp.
Pavlov, A.V. (1994) *Permafr. Periglac. Proc.* **5**, 101–110.
Peterson, T.C. and Vose, R.S. (1997) *Bull. Am. Met. Soc.* **78**, 2837–2849.
Peterson, T.C., Gallo, K.P., Livermore, J., Owen, T.W., Huang, A. and McKittrick, D.A. (1999) *Geophys. Res. Lett.* **26**, 329–332.
Polyakov, I.V., Alekseev, G.V., Bekryaev, R.V., Bhatt, U.S., Colony, R.L., Johnson, M. A., Karklin, V.P., Walsh, J.D. and Yulin, A.V. (2003a) *J. Clim.* **16**, 2078–2085.
Polyakov, I.V., Bekryaev, R.V., Alekseev, G.V., Bhatt, U.S., Colony, R.L., Johnson, M.A., Maskshtas, A.P. and Walsh, J.D. (2003b) *J. Clim.* **16**, 2067–2077.
Putnins, P. (1970) In: *World Survey of Climatology* (Ed. by S. Orvig), Vol. 14, pp. 3–113. Elsevier.
Räisänen, J. (2002) *J. Clim.* **15**, 2395–2411.
Randall, D.A., Wood, R.A., Bony, S., Colman, R., Fichefet, T., Fyfe, J., Kattsov, V., Pitman, A., Shukla, J., Srinivasan, J., Stouffer, R.J., Sumi, A., *et al.* (2007) In: *Climate Change 2007: The Physical Science Basis* (Ed. by S. Solomon, D. Qin, M. Manning, Z. Chen, M. Marquis, K.B. Averyt, M. Tignor and H. L. Miller). Cambridge University Press, Cambridge.
Rauthe, M., Hense, A. and Paeth, H. (2004) *Int. J. Climatol.* **24**, 643–662.
Rind, D., Perlwitz, J. and Lonergan, P. (2005a) *J. Geophys Res.* **110**, 10.1029/2004JD005103.
Rind, D., Perlwitz, J., Lonergan, P. and Lerner, J. (2005b) *J. Geophys. Res.* **110**, 10.1029/2004JD005686.
Rinke, A., Gerdes, R., Dethloff, K., Kandlbinder, T., Karcher, M., Kauker, F., Frickenhaus, S., Köberle, C. and Hiller, W. (2003) *J. Geophys. Res.* **108**, 10.1029/2002JD003146.
Rinke, A., Dethloff, K., Cassano, J.J., Christensen, J.H., Curry, J.A., Du, P., Girard, E., Haugen, J.E., Jacob, D., Jones, C.G., Køltzow, M.Laprise, R., *et al.* (2006) *Clim. Dyn.* **26**, 459–472.
Rodwell, M.J., Rowell, D.P. and Folland, C.K. (1999) *Nature* **398**, 320–323.
Roeckner, E., Oberhuber, J.M., Bacher, A., Christoph, M. and Kirchner, I. (1996) *Clim. Dyn.* **12**, 737–754.
Roeckner, E., Bäuml, G., Bonaventura, L., Brokopf, R., Esch, M., Giorgetta, M., Hagemann, S., Kirchner, I., Kornblueh, L., Manzini, E., Rhodin, A.Schlese, U., *et al.* (2003) *Max Planck Institute for Meteorology Report* **349**, 127.
Romanovsky, V.E., Burgess, M., Smith, S., Yoshikawa, K. and Brown, J. (2002) *Eos* **83**, 589–594.
Romanovsky, V.E., Christensen, J.H., Sazonova, T.S., Stendel, M., Walsh, J.E., Kiilsholm, S. and Sergueev, D.O. (2004) In: *Proceedings of the ACSYS Final Science Conference* (Ed. by C. Dick, T. Fichefet, B. Miville and T. Villinger),

11–14 November 2003 World Meteorological Organization, Geneva, WCRP-118 (CD), WMO/TD 1232.

Rothrock, D.A., Yu, Y. and Maykut, G.A. (1999) *Geophys. Res. Lett.* **26**, 3469–3472.

Schneider, N. and Cornuelle, B.D. (2005) *J. Clim.* **18**, 4355–4373.

Semmler, T., Jacob, D., Schluenzen, K.H. and Podzun, R. (2005) *J. Clim.* **18**, 2515–2530.

Serreze, M.C., Walsh, J.E., Chapin, E.S., III, Osterkamp, T., Dyurgerov, M., Romanovsky, V.E., Oechel, W.C., Morison, J., Zhang, T. and Barry, R.G. (2000) *Clim. Change* **46**, 159–207.

Shiklomanov, I.A., Shiklomanov, A.I., Lammers, R.B., Peterson, B.J. and Vörösmarty, C.J. (2000) In: *The Freshwater Budget of the Arctic Ocean* (Ed. by E.L. Lewis, E.P. Jones, P. Lemke, T.D. Prowse and P. Wadhams), pp. 281–296. Kluwer Academic Publishers, Dordrecht.

Shindell, D.T., Miller, R.L., Schmidt, G.A. and Pandolfo, L. (1999) *Nature* **399**, 452–455.

Shindell, D.T., Schmidt, G.A., Miller, R.L. and Rind, D. (2001) *J. Geophys. Res.* **106**, 7193–7210.

Sigmond, M., Siegmund, P.C., Manzini, E. and Kelder, H. (2004) *J. Clim.* **17**, 2352–2367.

Smith, S.L. and Burgess, M.M. (1999) *IAHS Publications* **256**, 71–80.

Stefan, J. (1891) *Ann. Phys.* **42**, 269–286.

Stendel, M. and Christensen, J.H. (2002) *Geophys. Res. Lett.* **29,** 10.1029/2001GL014345.

Stendel, M. and Christensen, J.H. (2007) *Danish Climate Centre Report* **07-02**. Also available on http://klimagroenland.dmi.dk.

Stendel, M., Schmith, T., Roeckner, E. and Cubasch, U. (2002) *Danish Met. Inst. Report* **02–1**, 51.

Stendel, M., Romanovsky, V.E., Christensen, J.H. and Sazonova, T.S. (2007) *Glob. Planet. Change* **56**, 203–214.

Stephenson, D.B., Wanner, H., Brönnimann, S. and Luterbacher, J. (2003) *Geophys. Monogr.* **134**, 37–50.

Stevenson, D.S., Dentener, F.J., Schultz, M.G., Ellingsen, K., van Noije, T.P.C., Wild, O., Zeng, G., Amann, M., Atherton, C.S., Bell, N., Bergmann, D.J., Bey, I., *et al.* (2006) *J. Geophys. Res.* **111**, 10.1029/2005JD006338.

Stocker, T.F., Clarke, G.K.C., Le Treut, H., Lindzen, R.S., Meleshko, V.P., Mugara, R.K., Palmer, T.N., Pierrehumbert, R.T., Sellers, P.J., Trenberth, K.E. and Willebrand, J. (2001) In: *Climate Change 2001: The Scientific Basis* (Ed. by J.T. Houghton, Y. Ding, D.J. Griggs, M. Noguer, P.J. van der Linden, X. Dai, K. Maskell and C.A. Johnson), pp. 418–470. Cambridge University Press, Cambridge.

Tjernström, M., Žagar, M., Svensson, G., Cassano, J.J., Pfeifer, S., Rinke, A., Wyser, K., Dethloff, K., Jones, C., Semmler, T. and Shaw, M. (2005) *Boundary-Layer Meteorol.* **117**, 337–381.

Thompson, D.W.J. and Wallace, J.M. (1998) *Geophys. Res. Lett.* **25**, 1297–1300.

Thompson, D.W.J. and Wallace, J.M. (2000) *J. Clim.* **13**, 1000–1016.

van Loon, H. and Rogers, J.C. (1978) *Mon. Wea. Rev.* **106**, 296–310.

Vimont, D.J., Battisti, D.S. and Hirst, A.C. (2001) *Geophys. Res. Lett.* **28**, 3923–3926.

Walsh, J.E., Chapman, W.L. and Shy, T.L. (1996) *J. Clim.* **9**, 480–488.

Walsh, J.E., Chapman, W.L., Romanovsky, V.E., Christensen, J.H. and Stendel, M. (in press). *J. Clim.*

Wei, H., Gutowski, W.J., Vorosmarty, C.J. and Fekete, B.M. (2002) *J. Clim.* **15**, 3222–3236.

Late Quaternary Environmental and Cultural Changes in the Wollaston Forland Region, Northeast Greenland

OLE BENNIKE, MIKKEL SØRENSEN, BENT FREDSKILD, BJARNE H. JACOBSEN, JENS BÖCHER, SUSANNE L. AMSINCK, ERIK JEPPESEN, CLAUS ANDREASEN, HANNE H. CHRISTIANSEN AND OLE HUMLUM

SUMMARY

This chapter provides a review of proxy data from a variety of natural archives sampled in the Wollaston Forland region, central Northeast Greenland. The data are used to describe long-term environmental and climatic changes. The focus is on reconstructing the Holocene conditions particularly in the Zackenberg area. In addition, this chapter provides an overview of the archaeological evidence for prehistoric occupation of the region.

The Zackenberg area has been covered by the Greenland Ice Sheet several times during the Quaternary. At the Last Glacial Maximum (LGM, about

ADVANCES IN ECOLOGICAL RESEARCH VOL. 40 0065-2504/08 $35.00
© 2008 Elsevier Ltd. All rights reserved
DOI: 10.1016/S0065-2504(07)00003-7

22,000 years BP), temperatures were much lower than at present, and only very hardy organisms may have survived in the region, even if ice-free areas existed. Marked warming at around 11,700 years BP led to ice recession, and the Zackenberg area was deglaciated in the early Holocene, prior to 10,100 years BP. Rapid early Holocene land emergence was replaced by a slight transgression in the late Holocene.

During the Holocene, summer solar insolation decreased in the north. Following deglaciation of the region, summer temperatures probably peaked in the early to mid-Holocene, as indicated by the occurrence of a southern beetle species. However, the timing for the onset of the Holocene thermal maximum is rather poorly constrained because of delayed immigration of key plant species. During the thermal maximum, the mean July temperature was at least 2–3°C higher than at present. Evidence for declining summer temperatures is seen at around 5500, 4500 and 3500 years BP. The cooling culminated during the Little Ice Age that peaked about 100–200 years ago.

The first plants that immigrated to the region were herbs and mosses. The first dwarf shrubs arrived in Northeast Greenland prior to 10,400 years BP, and dwarf birch arrived around 8800 years BP. The first people arrived about 4500 years BP, but the region was depopulated several times before the last people disappeared some time after 1823 AD, perhaps as a consequence of poor hunting conditions during the peak of the Little Ice Age.

I. INTRODUCTION

The monitoring programmes at Zackenberg in high-arctic Northeast Greenland (74°30′N, 20°30′W) have only been running for 10 years, and the oldest historical records of the flora and fauna of the region go back less than 200 years. To evaluate the magnitude of ongoing changes at Zackenberg, it is important to compare present conditions to long-term records of environmental changes. For this purpose, proxy data from a variety of natural archives can be used. Here we will review some of the evidence available from Northeast Greenland, with a focus on the Wollaston Forland region and in particular the area around Zackenberg.

The Zackenberg lowland is dominated by heaths, grasslands, fens, arctic willow *Salix arctica* snow-beds and sparsely vegetated windblown abrasion areas. White arctic bell-heather *Cassiope tetragona* heaths, covered by snow during winter, are frequent. Open mountain avens *Dryas*-dominated heaths are found on drier soil. Dwarf birch *Betula nana*, arctic blueberry *Vaccinium uliginosum* and crowberry *Empetrum* are rare. Common to all fens are polargrass *Arctagrostis latifolia*, arctic cotton-grass *Eriophorum scheuchzeri*, *Salix arctica* and viviparous knotweed *Polygonum viviparum*. In grasslands, *Eriophorum scheuchzeri* is replaced by tall cotton-grass *E. triste*. *Salix actica* snow-beds cover fairly large hummocky areas on level ground, whereas herb-rich snow-beds with herb-like willow *Salix herbacea* are found on some south-facing slopes. Further details are provided by Elberling *et al.* (2008, this volume).

The modern invertebrate fauna at Zackenberg is typical of high-arctic Greenland by its very low species diversity compared to southern latitudes. Nonetheless, like the flora it is relatively species rich for this latitude in East Greenland. The fauna is dominated by the order Diptera (flies), with regard to both species and individuals. Stinging mosquitoes are represented by only one species, *Aedes nigripes*, which is numerous. Lepidoptera (butterflies and moths) are frequent, and in sunny weather four species of butterflies characterise the landscape. The order Hymenoptera is represented by a high number of small, parasitic species, but the large arctic bumblebee *Bombus polaris* is also common. Only two species of beetles have been found. One specimen of the tiny *Latridius minutus* has been found in a pitfall sample. This species is synanthropic and probably introduced recently to the Zackenberg area by humans. The other, the small rove beetle *Gnypeta cavicollis*, inhabits marshes along the coast. This species is unique in being one of the very few real arctic beetle species. In Northeast Greenland, it seems to have a southern boundary at the low-arctic/high-arctic transition at Scoresby Sund.

The results of early archaeological field work were summarised by Glob (1946). Later archaeological surveys were conducted by Bandi and Meldgaard (1952), Andersen (1975) and Stenico and Woolmore (2002). These studies concentrated on remains from the Thule culture. During reconnaissance work in 2003–2005, major parts of the coastal regions were surveyed (Sørensen *et al.*, 2004; Grønnow *et al.*, 2005; C. Andreasen, unpublished data), whereas the inland areas have not yet been systematically surveyed. The late field work has focussed on palaeo-Eskimo traditions.

The chronologies in this chapter are based on published and unpublished accelerator mass spectrometry (AMS) or conventional radiocarbon age determinations (Tables 1–3), and some published luminescence age determinations. These are given in calendar years before present. Calibration is according to the INTCAL04 data-set (Reimer *et al.*, 2004), using the OxCal. v. 3.10 programme (Ramsey, 2001). Chronologies of ice cores are based on "ice core years BP."

II. THE GLACIAL HISTORY

The presence of glacial erratics at high elevations shows that the entire region has been inundated by the Greenland Ice Sheet (Bretz, 1935). Such an extensive glaciation has been referred to the early part of the last glacial stage (Hjort, 1981) or to the penultimate glacial stage (Hjort and Björck, 1984). Buried glacier ice in an ice-cored moraine at Kap Herschell on southeastern Wollaston

Table 1 Conventional radiocarbon dates of marine shells

Lab. no.	Altitude (m)	Locality	Age, ^{14}C years BP	^{14}C Calibrated age	Reference
I-9133	31	Zackenberg	8835 ± 130	9890–9530	Weidick, 1977
K-6579	30	Zackenberg	9090 ± 100	10,210–9910	Christiansen *et al.*, 2002
K-6580	8	Zackenberg	8300 ± 85	9260–9000	Christiansen *et al.*, 2002
K-6032	0.5	Zackenberg	6070 ± 100	6890–6630	Christiansen *et al.*, 2002
K-3217	13	Cardiocerasdal	7600 ± 110	8380–8170	S. Funder, unpublished data
K-3218	19	Lille Sødal	7470 ± 105	8290–8010	S. Funder, unpublished data

Forland and associated glacio-lacustrine sediments were overridden by an outlet glacier during the Last Glacial Maximum (LGM) (Houmark-Nielsen *et al.*, 1994). Pre-LGM glacio-lacustrine sediments in the valley Zackenbergdalen, documented by Christiansen and Humlum (1993) and Christiansen *et al.* (2002), must also have been overridden by ice.

Recent studies of the Quaternary deposits on the (offshore) continental shelf and slope suggest that the ice margin extended far onto the shelf and probably reached the shelf break during the LGM (Evans *et al.*, 2002; Ó Cofaigh *et al.*, 2004). This is in accordance with onshore studies conducted further north in East Greenland (Bennike and Weidick, 2001).

If the ice margin reached the shelf break during the LGM, it is possible that the mountains in the region were glaciated. However, glacial trimlines on mountain sides in the Zackenberg area may indicate that during the last glaciation only valley glaciers occupied the Zackenberg area. Also geophysical modelling indicates that the ice was 500 to 1000 m thick in the Wollaston Forland region during the LGM, based on the assumption that the ice sheet margin was situated at the outer coast (Fleming and Lambeck, 2004). Therefore, it is possible that fast moving outlet glaciers flowed out from the fjords, whereas slower moving smaller valley glaciers were found in the alpine landscape between the fjords.

The position of the Greenland Ice Sheet margin during the LGM is however debated. Previous researchers have proposed that the ice margin was situated on the inner shelf, close to the outer coast (Hjort, 1981; Funder and Hansen, 1996). One of the main arguments for this was based on weathering differences, which, however, is a difficult criterion to use because cold-based ice may leave weathered surfaces intact.

Minimum ages for the deglaciation of the Wollaston Forland region and regions further north and south are provided by radiocarbon age determinations of shells from raised marine deposits (Figure 1). It appears that some areas at the outer coast became ice-free in the earliest Holocene. The inner parts of the fjords were deglaciated 3000–4000 years later. The oldest radiocarbon date from the Zackenberg area is 10,100 years BP (Table 1), which provides a minimum date for the deglaciation of the adjacent part of Young Sund. Radiocarbon dates from the region have been published by Håkansson (1974, 1975, 1978), Weidick (1976, 1977, 1978), Hjort (1979, 1981), Funder and Petersen (1980), Björck *et al.* (1994) and Christiansen *et al.* (2002). In addition, one age determination from Store Koldewey and one from Geographical Society Ø are included (O. Bennike, unpublished data).

III. SEA-LEVEL CHANGES

As the ice receded, the sea inundated low-lying land areas. Based on geomorphological evidence, the marine limit falls from *c.* 70 m above present sea level in the Zackenberg area to *c.* 40 m a.s.l. near the present margin of the Greenland Ice Sheet. However, the highest marine fossils are reported from an elevation of only 31 m a.s.l. (Table 1). Relative sea level fell rapidly during the early Holocene, and it reached close to the present-day level around 5000 years BP (Table 1 and Figure 2). During the past *c.* 4000 years, when the region has been inhabited, only minor relative sea-level changes have occurred, since ruins from the different periods are often situated at the same sites. However, the sea has transgressed several house ruins from the Thule culture (Gelting, 1934).

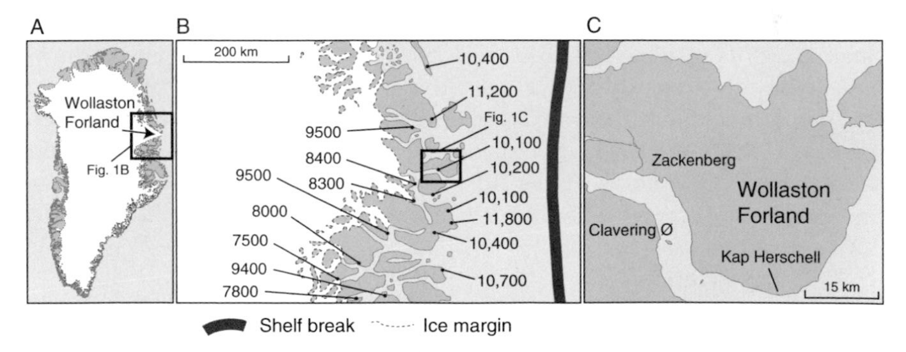

Figure 1 (A) Map of Greenland showing the location of the Wollaston Forland region. (B) Map of central Northeast Greenland showing the distribution of the oldest radiocarbon age determinations of mollusc shells from raised marine deposits. The dates are given in calendar years BP (modified from Bennike and Björck, 2002). (C) Map of the Wollaston Forland region.

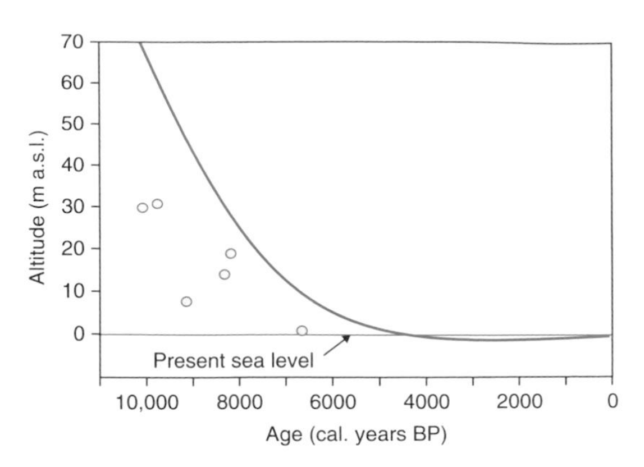

Figure 2 Provisional relative sea-level curve for the Zackenberg area, partly based on dates of shells of sub-littoral bivalves. The curve starts at 70 m a.s.l. (the marine limit) and at 10,100 years BP (the deglaciation date).

IV. HOLOCENE ENVIRONMENTAL CHANGES

A. Lake Sediment Evidence

Sediment cores were collected from Boresø in the northwestern part of Zackenbergdalen (see map in Cristoffersen *et al.*, 2008, this volume). The lake is situated above the marine limit, at about 95 m a.s.l. A 150 cm long sequence was collected at a water depth of 271 cm, at 74°30′N, 20°38′W; the record extends back to *c.* 9000 years BP.

The chronology is based on 18 radiocarbon dates, mostly of bulk sediment samples and water mosses (Table 2 and Figure 3). The sediments contain small amounts of microscopic coal (Figure 4), which means that dating of bulk sediment samples will give too old ages. However, it appears from the age–depth model that the bulk sediment dates are only a few hundred years too old, at the most. Based on pollen, macrofossils, Cladocera and sediment characteristics (Figures 4–7), the Boresø sequence has been divided into five time periods, which are described and discussed below.

1. Period 1 (c. 9000–8300 Years BP)

During Period 1 the area was characterised by very open pioneer vegetation dominated by purple saxifrage *Saxifraga oppositifolia*, but including mountain sorrel *Oxyria digyna*, moss campion *Silene acaulis* and/or sandwort *Minuartia* sp., buttercup *Ranunculus* sp. and other forbs. Grasses (Poaceae)

Table 2 Accelerator mass spectrometry radiocarbon dates from a sediment core from Boresø

Lab. no.	Depth (cm)	Material	Age, ^{14}C years BP	Calibrated age	$\delta^{13}C$ (‰)
AAR-4140	9–14	Bulk	3415 ± 50	3700–3590	−21.5
AAR-4139	19–24	Terrestrial moss	3505 ± 55	3840–3690	−28.2
AAR-3450	24–29	Bulk	3780 ± 45	4230–4085	−20.8
AAR-4138	24–29	Water moss	3685 ± 65	4090–3900	−25[a]
AAR-3449	53–59	Bulk	4850 ± 75	5650–5490	−19.9
AAR-4137	53–59	Water moss	4730 ± 60	5580–5330	−24.9
AAR-4141	69–74	Water moss	5570 ± 65	6410–6300	−24.3
AAR-4142	69–74	Terrestrial plant	5500 ± 55	6310–6220	−27.7
AAR-4135	78–85	Bulk	5720 ± 65	6630–6420	−19.3
AAR-4136	78–85	Water moss	5370 ± 75	6280–6000	−26.5
AAR-3448	104–109	Bulk	6315 ± 55	7240–7180	−19.6
AAR-4133	104–109	Water moss	6105 ± 65	7140–6880	−26.0
AAR-4134	104–109	Terrestrial plant	6320 ± 70	7260–7170	−26.4
AAR-3447	129–133	Bulk	7475 ± 65	8330–8140	−20.1
AAR-3446	138–144	Bulk	8020 ± 65	8990–8680	−19.7
AAR-4132	138–144	Water moss	8050 ± 90	9000–8720	−25.3
AAR-3226	157–159	Water moss	5855 ± 45	6735–6645	−25.4
AAR-4144	157–159	Water moss	5820 ± 60	6730–6540	−26.3

[a]Assumed value.

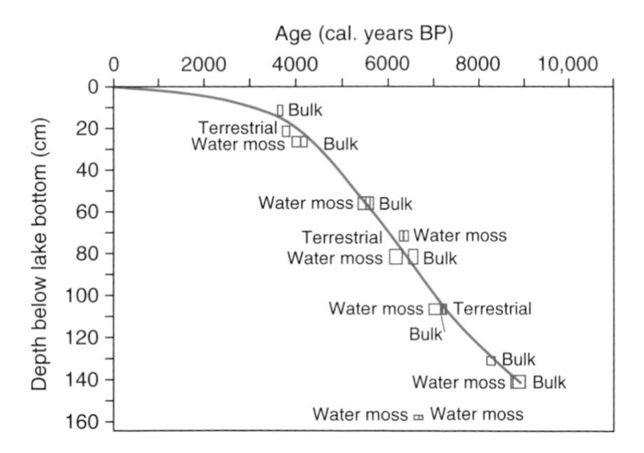

Figure 3 Age–depth model for the Boresø sequence.

reached a maximum in the middle of the period, and Cyperaceae were spreading. The ericaceous dwarf shrubs *Empetrum* and *Cassiope tetragona* were rare, and *Salix* and *Dryas* were missing. Like in all other studied sequences from high-arctic Greenland, the resistant spores of Pteridophytes (ferns etc.) are frequent in the oldest lake sediments. We suggest that virtually

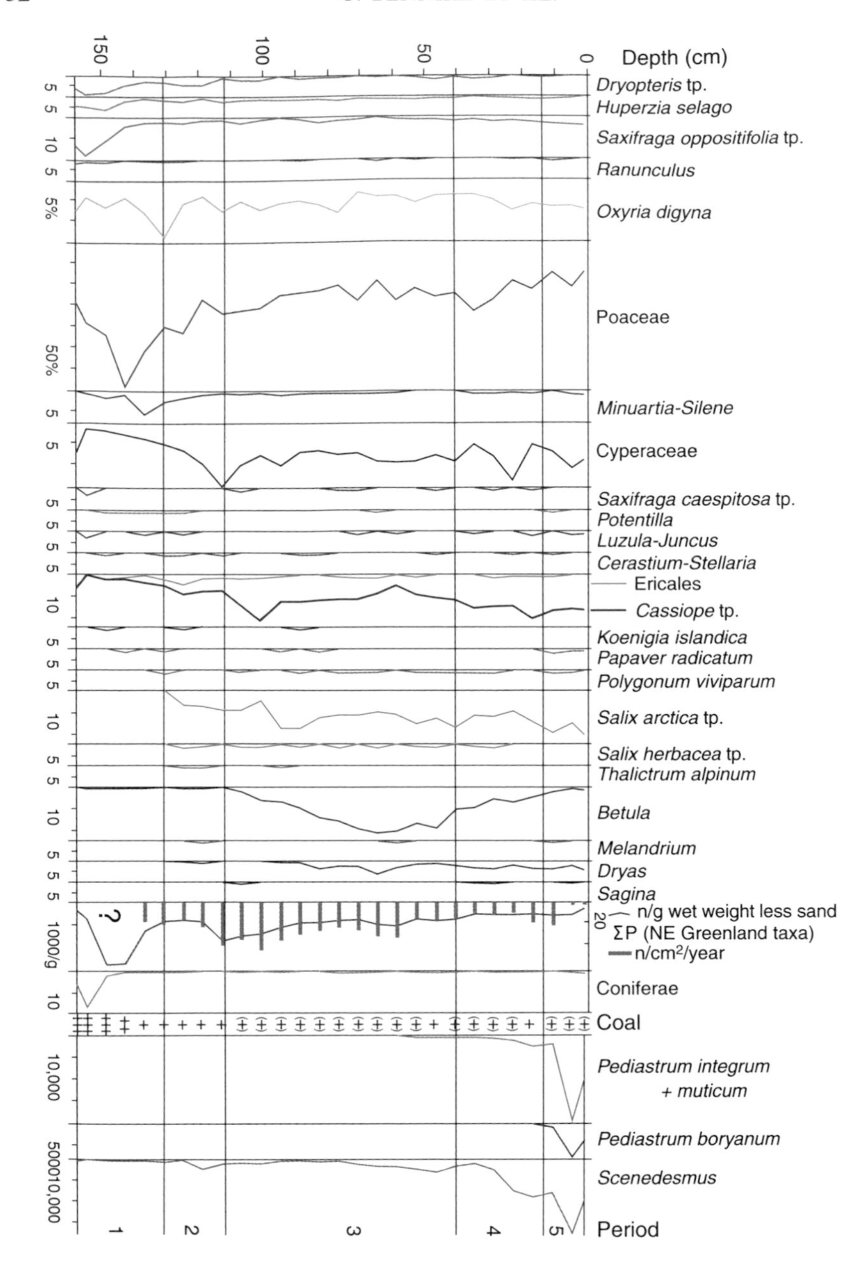

Figure 4 Simplified pollen percentage diagram from Boresø. The full diagram was published by Fredskild (2006). Coniferae is kept outside the pollen sum. The occurrence of some green algae is also indicated (n/g sediment). (+): very rare, +: rare, ++: common, +++: very common.

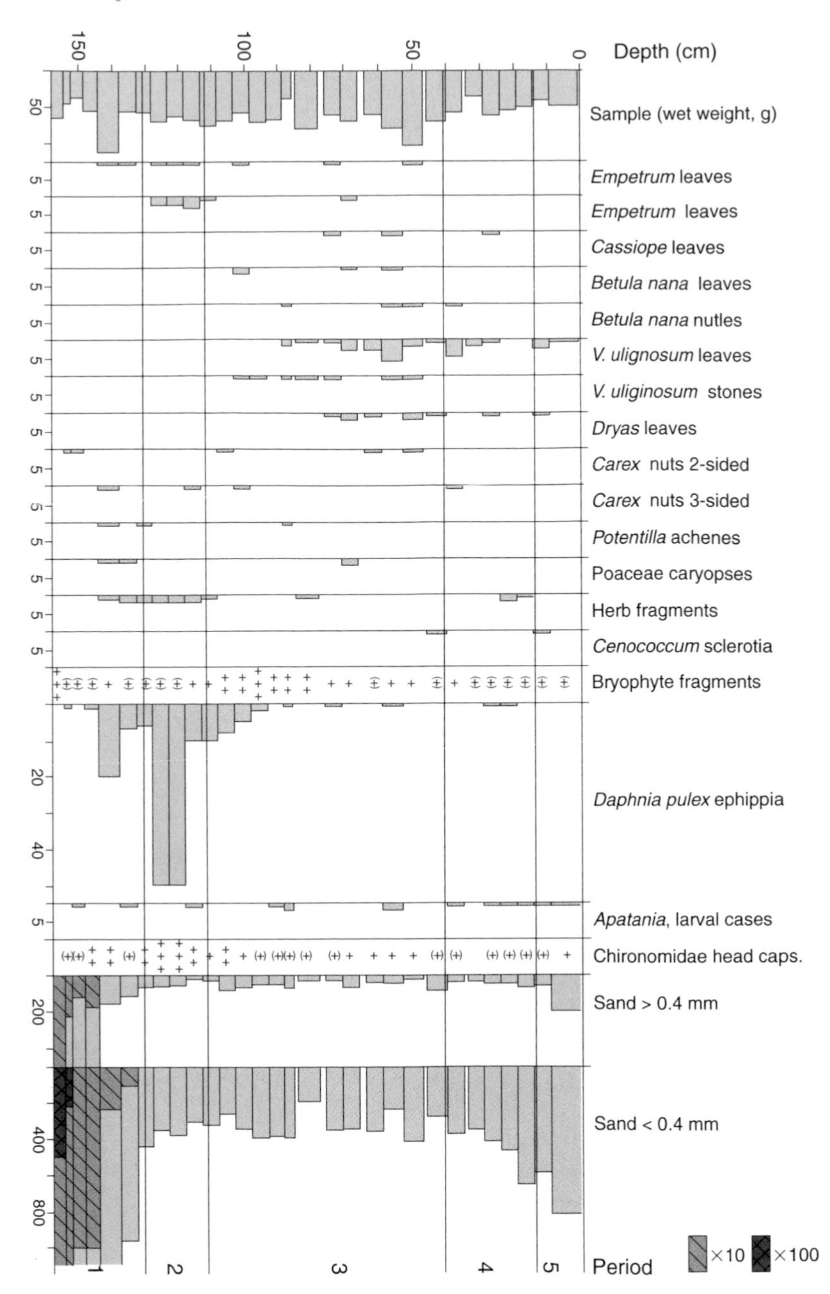

Figure 5 Macrofossil diagram from Boresø with numbers per sample. Samples were wet sieved on a 0.4-mm sieve. Sand was measured in milligram per 100 gram (mg/100 g) wet sediment.

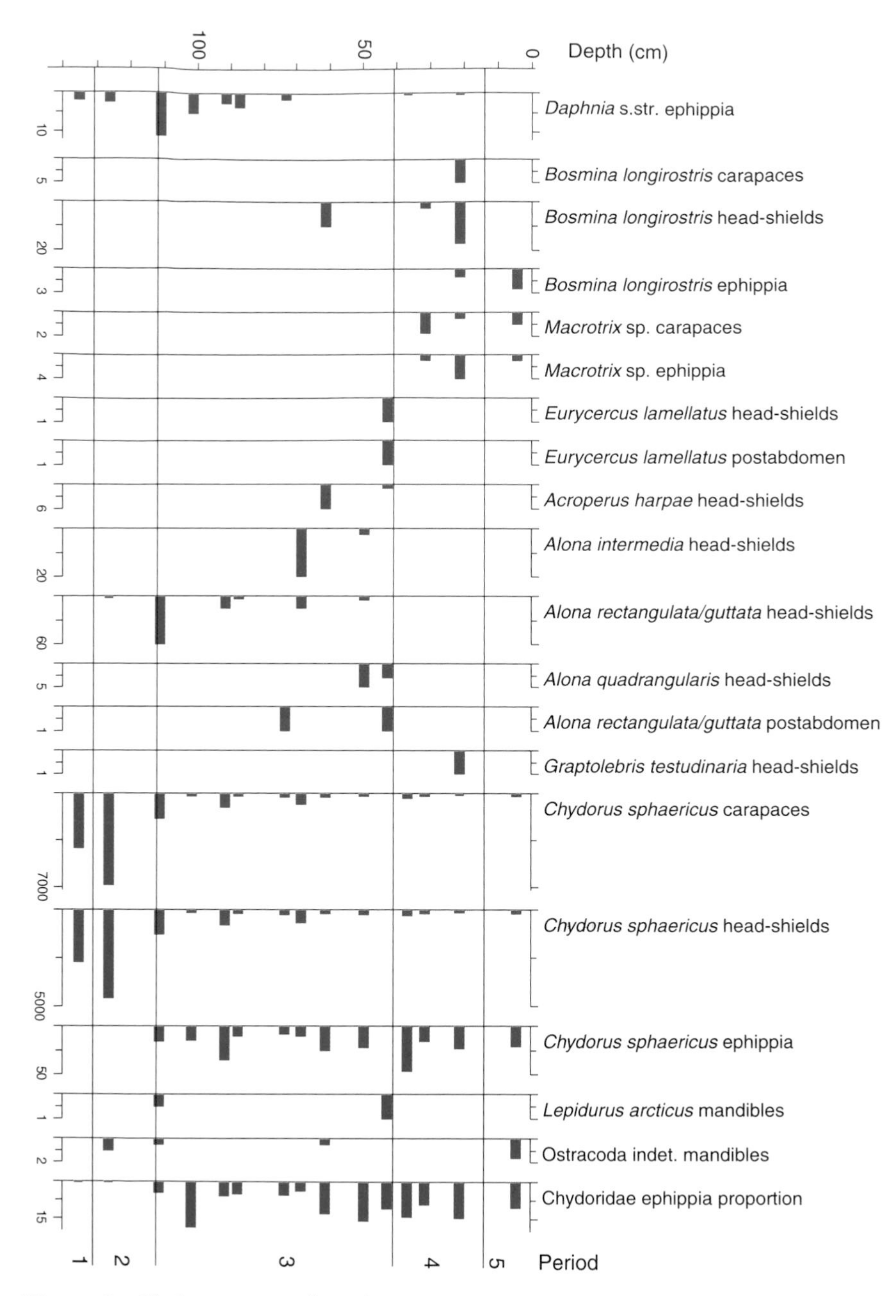

Figure 6 Cladoceran stratigraphy at Boresø, showing number of fragments per gram dry weight. The Chydoridae ephippia proportion shows the ratio of chydorid ephippia to head-shields + ephippia (in %). The ratio is close to zero in the two lower samples.

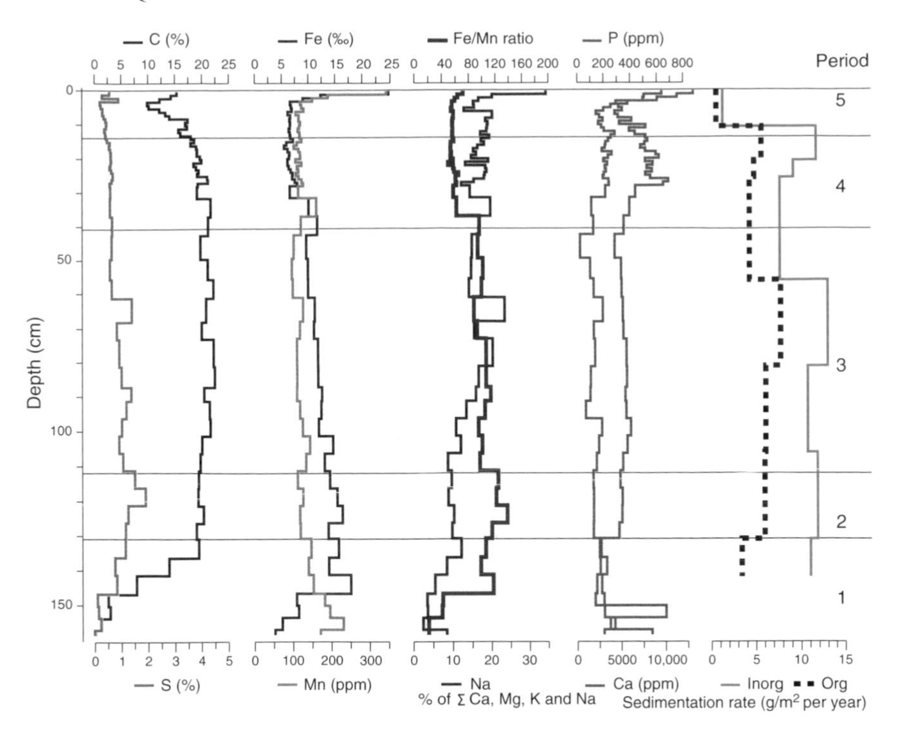

Figure 7 Geochemical parameters for the Boresø sequence.

all Pteridophyte spores in minerogenic sediments in high-arctic lakes are rebedded. The algal productivity in the lake was initially low, presumably a result of a large amount of suspended clay in the lake water. Remains of *Daphnia pulex*, *Alona*, chironomids and *Apatania zonella* are found, and *Chydorus sphaericus* shows a pronounced peak. The common presence of *Daphnia* suggests that fish were absent (Jeppesen *et al.*, 2001, 2003a; Lauridsen *et al.*, 2001).

The oldest recovered sediments are fine-grained sediments low in organic material, in the earliest part with some washed-in and blown-in fine sand, reflecting the recently deglaciated and sparsely vegetated catchment. Because of the low initial biological productivity in the lake and the unweathered soils in the surroundings of the lake, the earliest sediments show low iron content and relatively high content of manganese and inorganic phosphorous. In this period the productivity in the lake increased and the average rate of accumulated organic material was 3.4 g/m^2 per year, whereas the average rate of accumulated inorganic materials was around 11 g/m^2 per year.

Vegetation spread in the catchment, enhancing weathering processes and giving rise to higher Fe/Mn ratios under more reducing conditions at the lake bottom. Inferred from the Na percentage value, the area probably became more influenced by spray from less ice-covered waters outside Wollaston Forland.

During Period 1 delta deposits began to form at the mouth of Zackenbergelven (Christiansen *et al.*, 2002). The early to mid-Holocene delta deposits were elevated above sea level because of isostatic rebound.

2. *Period 2 (c. 8300–7250 Years BP)*

The transition to Period 2 is placed at the first appearance of *Salix arctica* and *S. herbacea* pollen. New to the period are also alpine meadow-rue *Thalictrum alpinum*, asphodel *Tofieldia* sp. and harebell *Campanula* sp. The present-day northern range limit of the wind-pollinated *Thalictrum alpinum* is located on the southern part of Clavering Ø (74°06′N), but its pollen have been found in lakes up to 400 km north of its present distribution. Thus, the presence of its pollen in Boresø is ascribed to long-distance pollen dispersal. Ericaceous dwarf shrubs, especially *Cassiope*, were spreading. The only ericaceous species proved by macroremains is *Empetrum*, the stones of which occur in all samples in the zone, indicating relatively high summer temperatures.

Saxifraga oppositifolia decreases during Period 2. Most of the surroundings of the lake became covered by vegetation, but fell-field and pioneer vegetation types were still important. Judging from the regular occurrence of *Empetrum* stones, summer temperatures were higher than at present. The production of *Pediastrum integrum* + *muticum* and *Scenedesmus* increases slightly, and the numbers of *Daphnia* ephippia, *Chydorus sphaericus* remains and Chironomidae head capsules reached their maxima. The cladoceran fauna was species poor in this period. The sediments show rather low Na percentage values and the highest Fe/Mn ratios observed in the sequence, indicating rather reducing conditions at the lake bottom and a continental type of climate.

The delta deposits from Period 2 are rich in leaves of *Salix herbacea*, whereas no leaves of *Salix arctica* were found. A sample of *Salix herbacea* leaves has been dated to about 7900 years BP (Christiansen *et al.*, 2002). The arthropod fauna in two samples was poor. Nonetheless, the finds are remarkable. At present, leaf beetles are totally absent from Greenland (Böcher, 1988). However, well-preserved fragments from a leaf beetle identified as *Phratora polaris* were found. Today the species lives in Iceland, sub-arctic Fennoscandia and northern Russia. It is mainly found in the outskirts of forests and along water courses, where it feeds on leaves of different species of willow. Remains of *Phratora polaris* have also been recovered from two localities in Jameson Land 400 km further south in Northeast Greenland, and these finds date to between 8700 and 7900 years BP (Böcher and Bennike, 1996; Bennike *et al.*, 2000).

The former occurrence of the species in the Zackenberg area demonstrates the previous surprisingly wide Greenland distribution during the Holocene thermal maximum. The species may have succumbed when the climate deteriorated in the mid- to late Holocene. However, it possibly survived much longer in warmer West Greenland, where a fragment of probably the same species was found during excavations of Norse layers in the Godthåbsfjord area (P. Buckland, personal communication).

Also a few remains of the Greenland seed bug *Nysius groenlandicus* were found in the delta deposits. At present the species has a wide distribution in Greenland (Böcher, 2001), but the finds from the Zackenberg delta together with an earlier record from Jameson Land (Böcher and Bennike, 1996) indicate an amazingly early and northerly colonisation of East Greenland following the last deglaciation.

In addition to a low number of unidentified fragments of Lepidoptera, parasitic Hymenoptera and oribatid mites, one abdominal tergum of the syrphid fly *Syrphus torvus* was identified from the delta deposits. The species is at present fairly common in Greenland north to about 70°N, about 400 km south of Zackenberg.

3. Period 3 (c. 7250–5500 Years BP)

At the beginning of this period *Betula nana* was spreading to the surroundings of the lake, as shown by leaves and fruits in the sediment and by increasing pollen percentage values, reaching a maximum of 22% at the end of the period. For comparison, *B. nana* makes up 24% of the pollen sum in recent gyttja from the interior part of Scoresby Sund *c.* 500 km to the south (Funder, 1978). Another important dwarf shrub is *Vaccinium uliginosum*, the seeds and/or leaves of which were found in 9 of the 11 macrofossil samples from this period. In the middle of the period *Dryas* was spreading. *Cassiope* and *Salix arctica* were more frequent than in the preceding periods. Judging from the low concentration of algae, *Daphnia*, remains of aquatic insect larvae and chydorids, the lake was probably nutrient poor. However, the C content shows the highest values during this period and also the rates of inorganic and organic sedimentation show maximum values of 13.0 and 7.7 g/m^2 per year, respectively.

Increasing Na percentage values and a decreasing Fe/Mn ratio indicate significant atmospheric transport from the south, increasing precipitation and consequently increasing drainage and run off from the landscape during the summer. The almost complete disappearance of *Daphnia* ephippia and the presence of remains of *Bosmina*, though in low abundance, suggest that fish entered the lake, likely triggered by contact to the river systems. The simultaneous major reduction in chironomids points in the same direction

(E. Jeppesen, unpublished data). Today, a small landlocked population of dwarf arctic char *Salvelinius alpinus* is present in the lake, and *Daphnia* is absent in most years but present in a nearby lake without fish (Christoffersen *et al.*, 2008, this volume). More species of cladocerans are present in this period and in the beginning of the next period.

Palaeosoils in the area, dated to the mid-Holocene, and at many places covered by late Holocene aeolian sediments, show weathering and leaching, features characteristic of sub-arctic podzols (Elberling *et al.*, 2008, this volume), also suggesting higher precipitation and higher temperatures than at present. The *Betula nana–Vaccinium uliginosum* heaths characteristic of this period imply higher summer temperatures than today. The rising *Betula* curve may indicate that the thermal maximum was reached at the end of the period, around 5500 years BP. However, the late arrival of *B. nana* hampers the interpretation of palaeotemperatures. The amount of washed-in or blown-in sand shows a minimum, probably suggesting a decrease in fell-fields and denser vegetation cover. The spreading of *Dryas* is probably indicative of larger areas with dry heaths on well-drained sites in the Zackenberg lowlands.

4. *Period 4 (c. 5500–4500 Years BP)*

The lower boundary of this period is defined by the decrease in *Betula* pollen. This indicates a decrease in summer temperature, supported by the fact that leaves but no fruits of *Vaccinium uliginosum* were found, and no macroscopic remains of *Empetrum* occur. *Cassiope* was spreading. The *Sagina* pollen most likely are from snow pearlwort *S. intermedia*, which is characteristic of *Salix arctica* snow-beds in this area. The chydorid concentration was low and the proportion of chydorid ephippia to head-shields is in good accordance with the suggested decline in temperature (Jeppesen *et al.*, 2003b; Sarmaja-Korjonen, 2003). Also a lower species richness of cladocerans may reflect a cooling trend.

The C content decreases slightly, but the average sedimentation rates of both organic and inorganic material in Period 4 were 40–50% lower than in Period 3, indicating lower temperatures and lower precipitation. A generally lower Na content suggests a weakening influence of spray, probably due to a change in the wind regime or enhanced sea ice cover. Slightly increasing values of Ca and P and fine sand in the latest part of this period may reflect changes in the wind regime, with blowing in fine inorganic material from the sedimentary mountains to the north. Probably due to this wind born contribution of nutrients to the lake, a slight increase in algae production is observed.

5. *Period 5 (c. 4500 Years BP to Present Day)*

During Period 5, the decrease in *Betula* continued, whereas *Cassiope*, *Salix arctica*, *Saxifraga oppositifolia* and *Oxyria digyna* increased, indicating more extensive snow-beds and snow-protected heaths. Arctic poppy *Papaver radicatum*, increasing in the later part of the period, grew in different heath types, often with *Cassiope*, and on abrasion ridges. The chydorid density is also low in this period and the proportion of chydorid ephippia to head-shield is very high, supporting the evidence for a colder climate during this period. Towards the end of the period, the algae production first increased abruptly, and then declined. The decrease in the recent sample was caused by a very high content of water. The increase was partly a result of the change in sedimentation rate, but this cannot explain the *Pediastrum* blooming, which was probably caused by a larger supply of nutrients to the lake (Christoffersen *et al.*, 2008, this volume).

The C content decreases from above 20% to a minimum of less than 10%, indicating decreasing summer temperatures. At this period border, wind born fine sand and Ca- and P-rich material blown in by northerly winds increase, enhancing the production of algae. For the period *c.* 4500–3500 years BP sedimentation rates of especially inorganic material increase slightly, indicating enhanced snow coverage and decreasing temperatures. From about 3800 to 3500 years BP, low temperatures, high precipitation values and snow coverage characterised the area. After 3500 years BP, temperatures decreased further, presumably due to a weakening of the influence from warm and moist air masses from the south, and a strengthening of northerly winds, primarily during the winter period. Precipitation in the area decreased, and the limited amount of snow was redistributed in the landscape, giving rise to nivation activity, which increased strongly after 3500 years BP (Christiansen *et al.*, 2008, this volume). The very low sedimentation rate during the past 3500 years was presumably due to a marked shortening of the ice-free period of the lake. After the Little Ice Age, increasing temperatures gave rise to more organic-rich sediments.

6. *Further Discussion of Lake Sediment Evidence*

In addition to organic production in the lake, organic and inorganic material, including pollen, are brought into the lake from the catchment. Three different ways can be envisaged for the latter: (1) washed in by streams and seeping water from melting snow-patches during spring and summer, (2) melting out from ice and snow on the lake and (3) blown in during the ice-free period. The complex climatic control on these different processes combined with the timing of plant immigration, especially warmth indicators like *Betula nana*, hampers the reconstruction of climatic changes.

Fairly warm conditions were established at the latest when the pioneer vegetation (Period 1) changed to heaths, fens, snow-beds and other plant communities. Periods 2 and 3 were significantly warmer than today. Whether Period 2, as indicated by the occurrence of *Empetrum* stones, was warmer than Period 3, when *Vaccinium* flourished, cannot be deduced. However, the presence of *Phratora polaris* at 7900 years BP indicates subarctic conditions locally, similar to those found in southernmost Greenland today. Fish likely entered the lake during Period 3 as a result of contact to the river/marine system.

The gradual increase of *Betula nana* in Period 3 may partly be a result of its competitiveness as compared with other dwarf shrubs after its late immigration, but the decrease at the transition to Period 4 must reflect declining summer temperatures starting at about 5500 years BP. Cooling is also indicated by the lack of *Empetrum* stones in Periods 4 and 5. The cooling continued, as illustrated by rates of sedimentation and other sediment characteristics, by a reduction in cladoceran remains in the sediment, and an increased ratio of ephippia to head-shields of chydorids. Continued cooling and longer lasting snow-cover can also explain the increase in *Cassiope* and *Salix arctica*, and the decrease in *Salix herbacea*, characteristic of early snow-free "herb-slope like" spots together with *Vaccinium* and other ericaceous dwarf shrubs from early snow-free heaths. Throughout the Holocene, *Cassiope* became more important, whereas the number, or at least the flowering, of the other ericales decreased. The decrease in *Betula nana* and "other ericales" after 5500 years BP and the disappearance of *Salix herbacea* pollen in Period 5 indicate colder summers. Another indication of late Holocene cooling is given by the decrease in pollen influx, seen in most Greenlandic and some Canadian pollen diagrams, especially from the High Arctic (Fredskild, 1985).

The overall sediment characteristics, the organic carbon and the sulphur content, rates of sedimentation of both organic and inorganic material, the content of authigenic and adsorbed metals and phosphorus document changing conditions in the lake. These changes are generally controlled by regional climatic changes, but are also influenced by local landscape feedbacks. These feedbacks comprise mainly the effect of changing ice- and snow-cover on the lake and in the catchment together with the effects of pronounced changes in the intensity and the direction of winds.

During Periods 2–4 organic sedimentation rates vary from 4.3 to 7.7 g/m^2 per year and inorganic sedimentation rates from 7.6 to 13 g/m^2 per year. At around 3500 years BP sedimentary conditions changed dramatically. The sedimentation rate dropped by a factor of 10, to average values of 0.6 and 1.3 g/m^2 per year for the sedimentation of organic and inorganic material for the past 3500 years. This change probably reflects lower temperatures, lower levels of erosion and lower productivity in the lake.

B. Periglacial Evidence

Since the last deglaciation, the Zackenberg landscape has been underlain by permafrost with significant periglacial landscape activity. Active periglacial landforms are described by Christiansen *et al.* (2008, this volume), but palaeoenvironmental reconstructions have also been performed. Nivation forms and sediments represent important palaeoenvironmental archives in periglacial sedimentary landscapes in Zackenbergdalen (Christiansen, 1998). In particular during the Little Ice Age, significantly increased nivation activity with rapid filling of nival basins, deposition of large nival fans and nival sedimentation modifying pre-existing valleys took place (Christiansen, 1998). Climatically, this increase was primarily caused by increased wind activity during winter with more drifting snow. This led to accumulation of larger than the modern average snow-patches. The dominating winter wind direction was from the north, as it has been over the past 6500 years (Christiansen and Humlum, 1993). In the period from AD 1420 to AD 1690 increased niveo–aeolian and niveo–fluvial sedimentation with sedimentation rates of 0.1–0.6 cm/year took place in the Zackenberg lowland (Christiansen, 1998). The average Holocene nival sedimentation rate since around 9000 years ago was only 0.03 mm/year in the bedrock-dominated western side of Zackenbergdalen (Christiansen, 1994). Also one detailed nival basin study showed increased erosion of organic material in the period from AD 1420 to AD 1500–1580, at the time of highest nivation activity. After this period the source areas for nival erosion of organic material became partly depleted, indicating that the vegetation cover was then significantly reduced (Christiansen, 1998). However, niveo–fluvial re-sedimentation in nivation hollows has been found to have started some time before 2900 ± 300 years BP, but with a lower mean sedimentation rate of 0.2–0.3 mm/year (Christiansen *et al.*, 2002).

Also the avalanche activity in the Zackenberg area has been studied and dated by combined geomorphological studies, Schmidt Hammer rebound values and lichenometry (Christiansen *et al.*, 2002). Avalanche activity started around 2100 years BP and was particularly significant from AD 1070 to AD 1750 including the Little Ice Age period.

C. Pedological Evidence

A well-developed podsol exists on several parts of the fluvial bars in Zackenbergdalen, buried below 10–40 cm of aeolian sand (Jakobsen, 1992; Christiansen *et al.*, 2002). Podsolisation here started around 7400 years BP and continued throughout the Holocene thermal maximum, with activity until around 3000–2400 years BP (Christiansen *et al.*, 2002). In the coldest part of the late Holocene, no podsol formation took place in the Zackenberg area. The aeolian sand that bury the podsols can have been deposited at least since 6500 years BP, when the present northerly dominating winds already had started to cause aeolian sedimentation in the Zackenberg area (Christiansen and Humlum, 1993).

D. Comparisons with Other Climate Records

Decreasing summer solar insolation has occurred throughout the Holocene (Figure 8A), which is one of the factors that govern the long-term climatic development. However, in the earliest Holocene, a major part of the energy was used to melt the low-lying parts of the Greenland Ice Sheet, which was therefore subject to major recession.

Some palaeoclimate records from central East and Northeast Greenland are presented in Figure 9. On the summit of the Greenland Ice Sheet, maximum melting took place during the period from 8000 to 6000 years BP (B). On Renland, the thermal maximum lasted from *c.* 8000 to *c.* 5000 years BP, as inferred from δ^{18}O values from an ice core (C), and on Geographical Society Ø it lasted from *c.* 9000 to *c.* 5000 years BP, as deduced from the content of biogenic opal in a lake sediment sequence (E). The youngest dates of the southern blue mussel *Mytilus edulis* in Northeast Greenland are around 6500 years BP (F). Further north in Northeast Greenland, Nioghalvfjerdsfjorden was without glacier cover in the time period 7700–4500 years BP (G). Raffles Sø became perennially ice covered at *c.* 1800 years BP (H). Based on present-day borehole temperatures, a pronounced Medieval warm period and a pronounced Little Ice Age have been modelled (D). Major late Holocene changes are also seen in the Zackenberg area, based on analyses of lake sediments and peat cores.

An early to mid-Holocene maximum in macro-remains of *Empetrum* and *Betula nana* has been recognized in a lake on Jameson Land (Bennike and Funder, 1997), similar to that seen at Zackenberg. At Zackenberg the dating of the *Betula* decline is around 5500–5000 years BP, which is in accordance with the Scoresby Sund data. Here the first temperature decrease is registered in the coastal area by a decrease in *Betula nana* heaths and increase in pollen of *Salix arctica* and *Cassiope* around 5100 years BP, and in the continental inland areas around 4500 years BP (Funder, 1978).

V. IMMIGRATION HISTORY OF PLANTS AND ANIMALS

It has been much debated if the Greenland plants and animals survived the last ice age, or if they immigrated after the LGM (Funder, 1979; Bennike, 1999). However, results from the ice core studies indicate that the mean

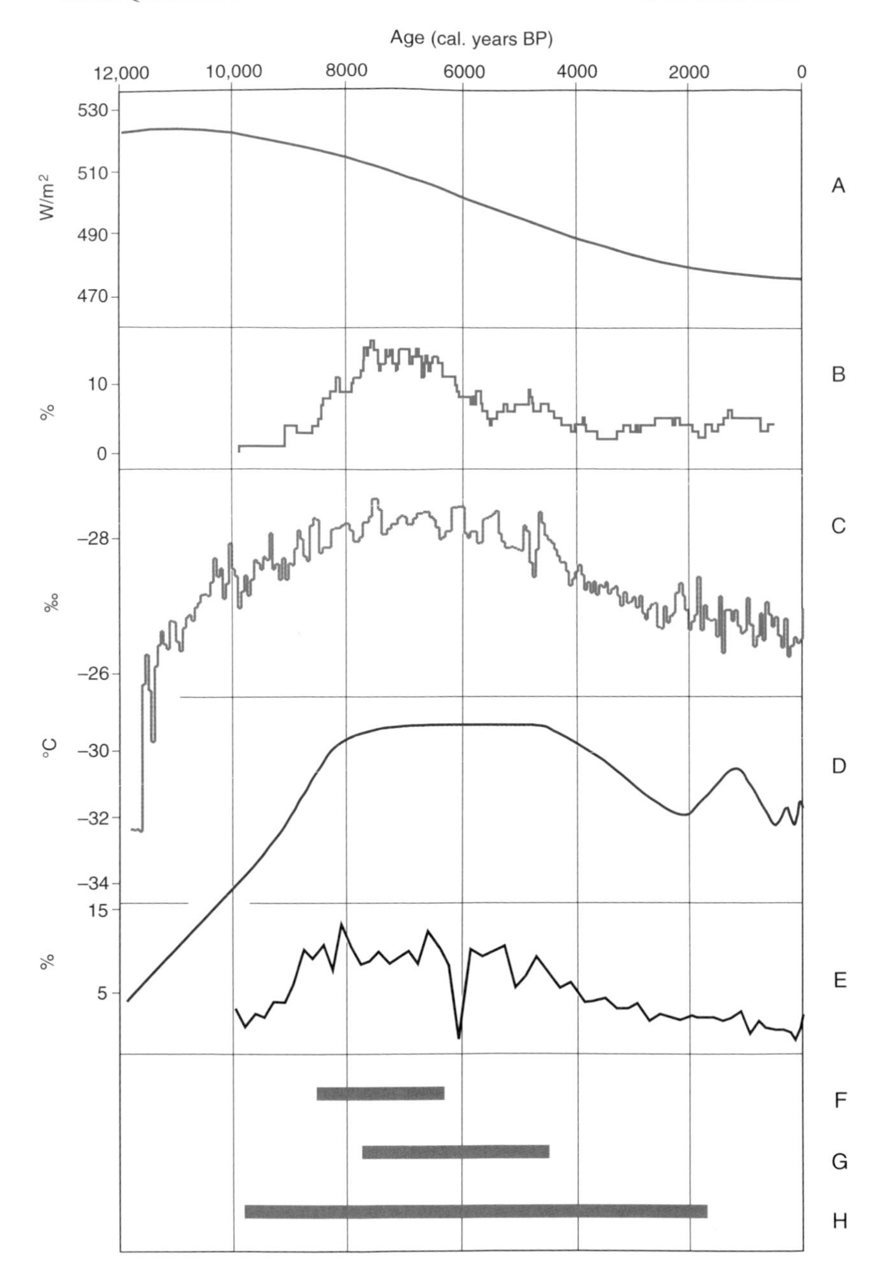

Figure 8 (A) June solar insolation at 60°N (Berger and Loutre, 1991). (B) Proportion of years with meltwater layers from the GISP2 ice core (running 1000 years means) (Alley and Anandakrishnan, 1995). (C) Oxygen isotope ratios in the Renland ice core (δ^{18}O, Johnsen *et al.*, 1992). (D) Palaeotemperatures at GRIP, modelled from the present-day temperatures in the borehole (°C, Dahl-Jensen *et al.*, 1998). (E) Occurrence of biogenic opal in a lake deposit from Geographical Society Ø in

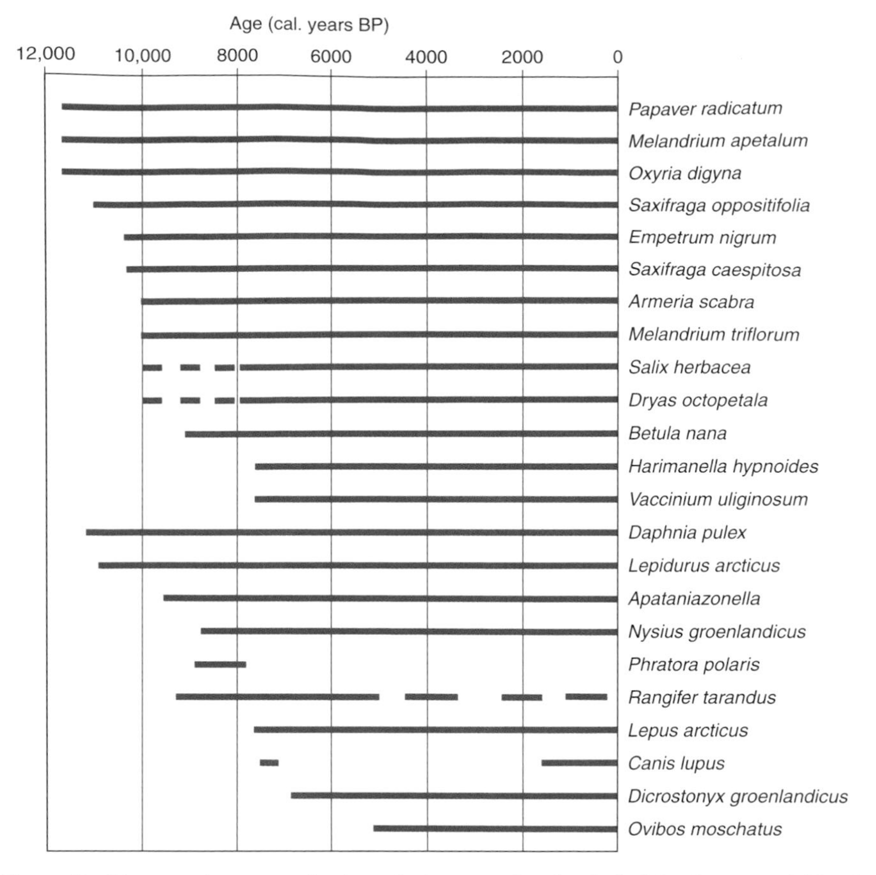

Figure 9 Temporal ranges of selected plants and animals in Northeast and North Greenland. The transition from the Holocene to the last glacial stage is dated to around 11,650 years BP (data from Meldgaard, 1991; Funder, 1978; Bennike, 1997, 1999, 2000; Bennike and Andreasen, 2005; O. Bennike, unpublished data).

annual temperature in central Greenland was around 25°C lower than at present during a short time period at 21,000 years BP (Dahl-Jensen *et al.*, 1998). Even though these results cannot be directly transferred to the ice-free parts of Greenland, they indicate that only species adapted to extreme cold

---

East Greenland (%, Wagner *et al.*, 2000). (F) Radiocarbon dates of blue mussel *Mytilus edulis* shells from raised marine deposits in central East Greenland (mainly from Hjort and Funder, 1974). (G) Range of radiocarbon dates of remains of marine animals and drift wood from Nioghalvfjerdsfjorden in northern East Greenland (79°N). The dates show when the fjord was without glacier cover (Bennike and Weidick, 2001). (H) Relatively warm period without perennial lake ice cover on Raffles Sø in East Greenland (Cremer *et al.*, 2001).

could survive. With respect to mammals and birds, we consider it unlikely that any species could survive in Northeast Greenland.

A single sample with plant remains from Northeast Greenland has been dated to around 11,800 years BP, and several samples have been dated to around 11,000 years BP (Bennike *et al.*, 1999). These samples contain remains of mosses and herbs only. The early herbs comprise *Papaver radicatum*, nodding lychnis *Melandrium apetalum*, *Oxyria digyna*, chickweed *Cerastium* sp., *Minuartia* sp., whitlow-grass *Draba* tp., cinquefoil *Potentilla* sp., sedge *Carex* sp., *Eriophorum* sp. and Poaceae indet. (Figure 9). Most of these taxa are widely distributed in Greenland at present, but *Melandrium apetalum* is found only in the northern part, and several of them are more frequent in the north than further south. *Empetrum* may have been the first woody plant to arrive in Northeast Greenland; its stones have been recorded from layers dated to 10,400 years BP. The fruits are adapted to dispersal by birds, which in combination with wind were probably one of the key factors for long-distance chance dispersal of plants to Greenland.

The appearance of *Salix arctica* and *S. herbacea* in Boresø can be dated to around 8300 years BP. In the Thule area, the high-arctic *Salix arctica* was present at around 7600 years BP (Fredskild, 1995), and near Nioghalvfjerdsfjorden in Northeast Greenland it was present at 8800 years BP (Bennike and Weidick, 2001), whereas it did not reach southwards to Scoresby Sund until around 6700 years BP (Funder, 1978). *Salix herbacea* immigrated earlier to Northeast Greenland, perhaps already around 10,000 years BP (Funder, 1978). In the Scoresby Sund area, *Betula nana* arrived *c.* 8800 years BP (Funder, 1978). At Zackenberg its arrival is dated to around 7250 years BP. The timing of the arrival of *Dryas octopetala* in Northeast Greenland is poorly determined, but it may have arrived at around 10,000 years BP (Funder, 1978). Overall, palaeoecological studies show that many species have high migration rates, even across the North Atlantic Ocean (Bennike, 1999). This agrees with results of molecular studies, which demonstrate an "extreme dispersal ability" of some plant species (Brochmann *et al.*, 2003).

The raised delta layers at Zackenberg also contain droppings of arctic hare *Lepus arcticus*, dated to around 7900 years BP, which is the oldest record of this species from Greenland (Christiansen *et al.*, 2002). Knowledge about the history of mammals and birds is scarce (Bennike, 1997). The oldest reindeer *Rangifer tarandus* remains are dated to 7100 years BP, the oldest lemming *Dicrostonyx groenlandicus* remains to 6700 years BP (O. Bennike, unpublished data) and the oldest musk ox *Ovibos moschatus* remains to 5000 years BP (Bennike and Andreasen, 2005). Evidence from lake sediments suggests that the little auk *Alle alle* was present at 7500 years BP (Wagner and Melles, 2001). At around 4500 years BP, when the first humans arrived in Greenland, the present-day vertebrate fauna appears to have been established (Darwent, 2003).

VI. ARCHAEOLOGY

The overall present landscape of Northeast Greenland has not changed much since the first inhabitants arrived to the area about 4500 years BP and during surveys we travel in the same landscape as our prehistoric human ancestors. Because of the low organic production and the low sedimentation rates, artefacts from the former inhabitants can be found on the terrain surface. Because of these two circumstances, substantial prehistoric and historic cultural information can be gained with a minimal effort, compared to other parts of the world.

However, human settlements from different time periods are often located on the same raised beaches, and younger inhabitants have often reused and destroyed older ruins. This problem goes especially for ecological "hot spots" where aggregation camps have been situated. Another problem concerns preservation of organic material. The climatic conditions and lack of bedrock carbonates mean that the only organic materials normally preserved at pre-Thule sites are charcoal and teeth (enamel).

A. Independence I

The first humans who entered the Wollaston Forland region belonged to an early eastern part of the Arctic Small Tool traditions, named the "Independence I culture" (Andreasen, 2004a). This group of people entered the region around 4500 years BP, probably within few generations from Beringia, thus travelling more than 5000 km. The Independence I can be distinguished by their tent rings, often of midpassage type with a central "box hearth" (Andreasen, 2004a). The tradition can be separated from other palaeo-Eskimo traditions by a characteristic lithic technology and artefact design (Figure 10). Mainly due to the work of Knuth (1967), we know that the Independence I came through North Greenland down the east coast to Scoresby Sund, which is their southernmost habitation area (Grønnow and Jensen, 2003). Former Independence I presence in the Wollaston Forland region is suggested from radiocarbon dated charcoal from local *Salix* wood from a fireplace at the Independence I site Røde Hytte in Jameson Land (Tuborg and Sandell, 1999). The two early radiocarbon dates from Røde Hytte (Table 3) strongly indicate that the Independence I came through the Wollaston Forland region already around 4500 years BP, on their way from North Greenland to Jameson Land. From their remains it seems that the Independence I used the Wollaston Forland region over a long time. Dates from North Greenland suggest that the Independence I disappeared

Figure 10 Lithic artefacts typical of Independence I (4500–3900 years BP). (A) Burin, (B) burin spall, (C, D) two fragments of broad microblades, (E) burin with notches at its base and a grounded distal end and (F) a distal part of large oval knife blade. The burin "e" is because of its grinding atypical of Independence I. This burin may be interpreted as evidence of contact between Saqqaq and Independence I in central East Greenland.

around 3900 years BP from Greenland, and it is thus possible that the Independence I travelled through and used the Wollaston Forland region for up to 600 years.

Table 3 Accelerator mass spectrometry radiocarbon dates from archaeological sites

Lab. no.	Material	Site	Age, ^{14}C years BP	Calibrated age	Culture
IA-22759	*R. tarandus*	Tyroler Fjord	174 ± 30	AD 1660–1950	Thule
KIA-22761	*R. tarandus*	Schumacher Ø	454 ± 28	AD 1425–1450	Thule
AAR-4689[a]	*P. hispida*	Zackenberg	70 ± 35	AD 1690–1920	Thule
AAR-1182[b]	*Salix* sp.	Røde Hytte	4000 ± 75	BC 2660–2350	Independence I
AAR-1184[b]	*Salix* sp.	Røde Hytte	4030 ± 90	BC 2860–2460	Independence I

[a]Published by Christiansen *et al.* (2002).
[b]Published by Tuborg and Sandell (1999).

B. Saqqaq

The material remains from Saqqaq are few and sparse. Like Independence I, Saqqaq had a specific lithic technology and artefact design, which makes it possible to distinguish between the two traditions (Sørensen, 2005). In the Wollaston Forland region the distinction between Independence I and Saqqaq can be made because of differences in lithic material choice, the methods in which large lithic preforms are worked out and in the burin technology. At Kap Berghaus, a typical Saqqaq burin made from local basalt with a ground distal end was collected as a stray find (Andersen, 1975) (Figure 11). At the same site and at another site a little further north, "Grønlænderhuse," some large bifacial cores made from basalt, typical to Saqqaq, were found (Sørensen *et al.*, 2004). These material remains are today the most northern evidence of Saqqaq on the east coast of Greenland. No structures from Saqqaq have yet been located in the region.

Saqqaq came from South Greenland, migrating northwards into the Wollaston Forland region. The Saqqaq is formerly known from the Scoresby Sund and Ammassalik regions on the east coast, but seems to have had its main habitation in the Disko Bugt region of central West Greenland (Grønnow, 2004). Because of the lack of radiocarbon dates in eastern and southern Greenland, we are still uncertain about the chronology of Saqqaq in East Greenland, but it seems that Saqqaq came later than Independence I.

Figure 11 Burin from the Saqqaq culture (4500–2800 years BP), found at Kap Berghaus in Young Sund. The burin is the northernmost evidence of Saqqaq on the east coast of Greenland. It is made of basalt and has a grounded distal end. This particular type of basalt is known from Basaltø in Young Sund, *c.* 6 km from the finding place. The burin is 3.8 cm long.

C. Greenlandic Dorset

The latest reviews of what was formerly known as the Dorset I and the Independence II suggest that these two "cultures" have to be considered as one (Elling, 1992; Jensen, 2004; Sørensen, 2005), termed the Greenlandic Dorset (Grønnow and Sørensen, 2006).

Evidence of Greenlandic Dorset is, like Independence I, substantial within the Wollaston Forland region. Architecturally, Greenlandic Dorset sites are characterised by tent rings most often of midpassage type, but different from the Independence I structures, often by the use of lots of flat stone slabs in the floor construction (Andreasen, 2004b; Sørensen *et al.*, 2004). Moreover, a special triangular 1-m wide structure located on the outer Wollaston Forland coasts is related to the Greenlandic Dorset. These structures consist of a pavement made of flat stone slabs and could be a remnant from a special light tent construction (Andreasen, 2004b; Sørensen *et al.*, 2004).

At several ruin sites in the Wollaston Forland region Greenlandic Dorset lithic inventories have been observed. A typical material process used by the Dorset is heat treatment, which was used to improve the quality of microcrystalline quartz for knapping. The heat-treated lithic material becomes shiny and transparent and often red. Artefacts unique to Greenlandic Dorset found in the Wollaston Forland region are burin with grounded edges, narrow blades, end scrapers with widening edges, and stemmed end blades (Figure 12).

Figure 12 Lithic artefacts from Greenlandic Dorset (2800–2000 years BP). (A) Knife blade with notches, (B–D) burins made from basalt with grounded cutting edges and grounded distal ends, (F) microblade core, (G–I) small knives made from microblades (there is small retouch on the proximal end made to haft the knives) and (E, J, K) end scrapers with widening edges.

The migration routes to Wollaston Forland taken by Greenlandic Dorset are difficult to establish, since it has a circum-Greenlandic distribution. They could therefore have come from both the north and the south. However, the use of basalt for burins indicates an affinity to West Greenland and may therefore be an argument for migration around South Greenland and into the Wollaston Forland region (Sørensen, 2005). Greenlandic Dorset is known to have existed in Greenland from 2800 to 2000 years BP. Yet no radiocarbon dates from Greenlandic Dorset are available from the Wollaston Forland region due to lack of organic material. Age determinations from Jøkelbugt and Dove Bugt to the north suggest that the northeastern part of Greenland could have been used during the whole period (Andreasen, 2004b; Bennike and Andreasen, 2005). However, because of plateaus in the calibration curve between 2800 and 2400 years BP, radiocarbon dating of Greenlandic Dorset is problematic.

D. Thule

The Thule culture is well known from the Wollaston Forland region due to the use of conspicuous habitation structures, such as winter houses and tent rings, but only a few age determinations are available (Table 3). The latest evidence of the Thule in the region comes from the southern tip of Clavering Ø, where captain Clavering in August 1823 met a small group of Inuit (Clavering, 1830). When Koldewey visited the same regions in 1869–1870, the Thule people had apparently disappeared (Koldewey, 1874).

It appears that the Thule culture was established in the region at the end of the Medieval warm period, and it may have succumbed during the Little Ice Age, perhaps due to poor hunting conditions. At Zackenberg a seal bone from a Thule winter house ruin has been dated (Christiansen *et al.*, 2002). The calibrated date suggests that the bone is younger than 1650 AD, and it probably derives from the period shortly after 1800 AD. A reindeer tooth from a Thule tent ring on a small island close to Kap Schumacher was dated to 1425–1450 AD (Table 3). A reindeer antler related to a Thule tent ring in Tyrolerfjord was dated to 1660–1950 AD (Table 3). No certain radiocarbon dates from the Thule culture are earlier than from the beginning of the fifteenth century AD from central East Greenland, and it thus seems that the Thule culture used and travelled in the Wollaston Forland region regularly between 1400 and 1850 AD. Artefacts from the Thule culture are characteristic and plenty because organic material is often preserved in middens at the winter settlements. Lithic artefacts are most often shaped by grinding and only seldom by knapping. Organic artefacts are characteristic by being worked mainly by drilling in order to split and shape the material (Figure 13).

E. Landscape Use and Settlement Patterns

The investigation of the use of the landscape in the Wollaston Forland region is based on information from the GeoArk 2003 and the GeoArk 2005 expedition surveys and on literature about the trapping period (Mikkelsen, 2001) (Figure 14). An attempt is made to separate the habitations into settlements of short duration and longer duration. It is assumed that fireplaces, stray finds, or small light structures mainly indicate summer dwellings from the palaeo-Eskimo periods. From the Thule period we have substantial ethnographic evidence for the relation between season and settlement type (e.g., tents are used in the summer and semi sub-terrain houses in winter). For the Danish and Norwegian twentieth-century trapping period historical documentation is available (i.e., occasional use of cabins in the winter and the larger stations during summer and winter).

When information on all located human activity is plotted and separated on a map by colour and size according to tradition and type of structure, it becomes clear that some particular spots in the region have been favoured through all periods and during all seasons: the mouth of Young Sund, the strait between Sabine Ø and Wollaston Forland, Kap Schumacher and the inner part of Tyrolerfjord. This settlement pattern is understood when

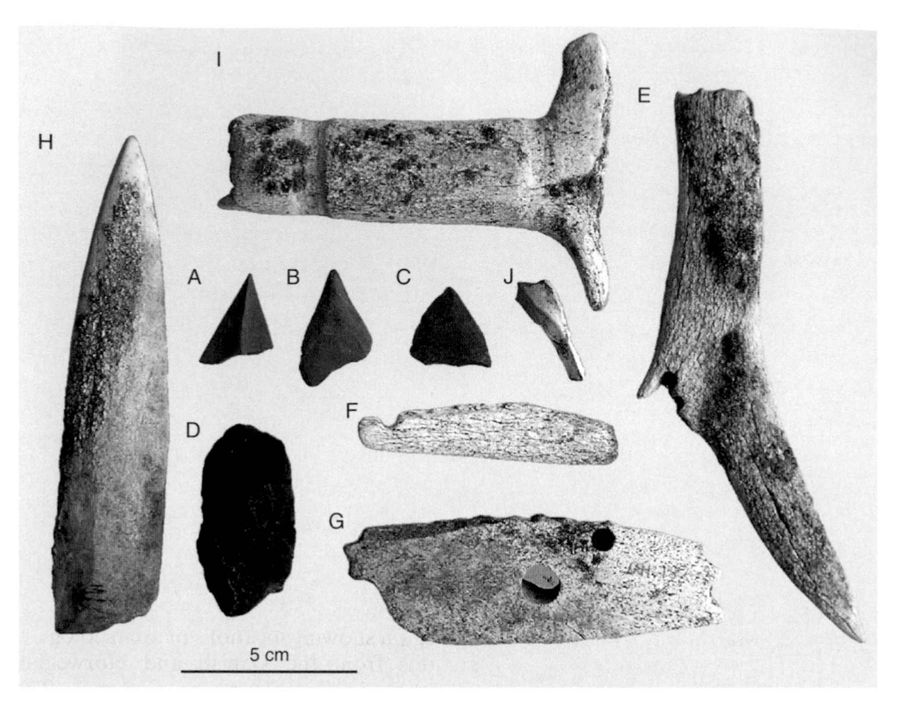

Figure 13 Artefacts from the Thule culture (1400–1850 AD). (A–C) Fragments of harpoon points made from slate by grinding, (D) axe pre-form made from basalt, (E, F) pieces of worked reindeer antler, (G) fragment of sledge runner made from whalebone, (H) knife or point made from antler, (I) shaft of unknown use made from antler, and (J) piece of walrus tusk.

satellite images showing the modern extent of sea ice during autumn and spring in the region are studied. Because of surface currents in the outer fjords and straits, strong winds, and freshwater runoff from rivers in the inner fjords, these places are the first to become ice-free in the spring and the last to be closed by sea ice during the autumn. The settlement at Kap Schumacher can be explained by a polynia, which evolves during spring close to the east coast of Kuhn Ø. This polynia is caused by early melting of the sea ice due to wind-transported black silt and sand being deposited on it.

Thus, polynias with concentrations of sea mammals develop especially in the outer fjords. Larger settlements of the prehistoric hunters (Independence I, Saqqaq, Greenlandic Dorset, Thule) were most often placed near these polynias, where sea mammals were probably hunted from the ice edge. However, winter settlements from the Thule culture are sometimes situated in the inner fjords.

On Hvalros Ø a very large aggregation camp, with several hundred structures, is found. The site is characterised by Thule structures, such as tent

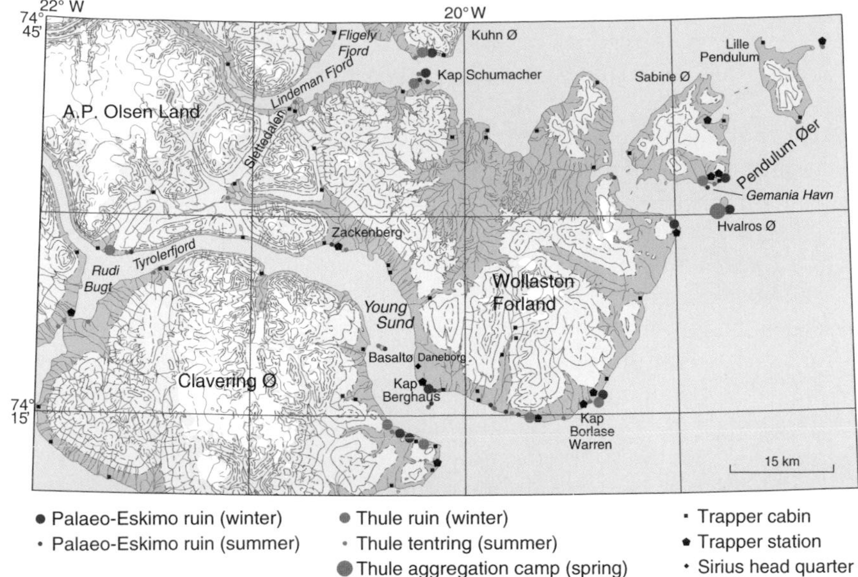

Figure 14 Map of the Wollaston Forland region showing locations of archaeological localities together with cabins and stations from the Danish and Norwegian twentieth-century trapper period. For location in Greenland, see Figure 1. The location of trapper cabins and trapper stations is partly from Mikkelsen (2001).

houses, shooting blinds and wall shelters. However, the same site is also rich in finds from palaeo-Eskimo periods, unfortunately usually disturbed by the Thule. The site also contains the region's only known Thule "festival house," a 10 m wide round stone build structure. It is ethnographically known that such structures were used when many people from the Thule culture gathered while they were hunting whales and used it for festivals (Thalbitzer, 1923). Thus, the Thule aggregation site can be interpreted as a key site in East Greenland, where Inuit from large areas met during early spring to hunt whales.

Attention must also be given to the prehistoric habitations at Zackenberg since these are atypically situated in a middle fjord area. However, the river Zackenbergelven is rich in arctic char from July to September and seals rest on the fjord ice in spring. These resources were probably also present, known and exploited by prehistoric people. The summer habitation in Lindeman Fjord (Slettedalen) and on the south side of A. P. Olsen Land (Tyrolerfjord) can be related to inland hunting of reindeer and musk ox during summer. These coastal sites are situated at gateways to the inland, through the valleys. The site in A. P. Olsen Land supports this viewpoint by structures such as "hunting beds" typical for Thule inland hunting sites. Not many hunting sites have been located in the inland, but a few from the Thule period document that the inland was used.

Little is known about the prehistoric subsistence. However, from Dødemandsbugten on southwest Clavering Ø, excavations were undertaken in the early 1930s of Thule sub-terrain house ruins by Larsen (1934) and later by Bandi and Meldgaard (1952). Bones from reindeer, musk ox, dog *Canis familiaris*, arctic fox *Alopex lagopus*, arctic hare, polar bear *Ursus maritimus*, walrus *Odobenus rosmarus*, narwhale *Monodon monoceros*, harp seal *Phoca groenlandica*, ringed seal *Phoca hispida* and bearded seal *Erignathus barbatus* were identified from the site (Degerbøl, 1934).

We suppose that this range of species reflects the animals hunted and used by the Thule people on the site, and maybe more generally in the region. Scattered finds of bones from Thule sites in the Wollaston Forland region support this hypothesis. It does, however, seem that musk ox bones are very rare at Thule sites (Darwent and Darwent, 2004). Tools made from bones or teeth by the Thule people in East Greenland come from rare musk ox bones, numerous caribou antlers and bones, walrus and narwhale tusks and bones, seal and whalebones and possibly some bird bones.

No bones have yet been excavated from palaeo-Eskimo sites in the Wollaston Forland region, and we consider it doubtful if palaeo-Eskimo sites with preserved bones can be located because of poor preservation conditions.

Both the "Independence I culture" and the Greenlandic Dorset were adapted to a cold and dry high-arctic environment, with marine mammals accessible from polynias in ice-covered coastal waters and with land mammals that benefited from a thin winter snow-cover and long snow-free summers. The "Independence I culture" appears to have abandoned the region around 3900 years BP, perhaps during a time period with a somewhat warmer climate with more snow on land and more open water. Such climatic conditions may have led to declining populations of mammals, and marine mammals may have been more inaccessible. Shortly after 2000 years BP the later palaeo-Eskimo culture, the Greenlandic Dorset, presumably disappeared from the region.

The Thule culture entered the region during a time of cooling at the beginning of the Little Ice Age, and disappeared from the area some hundred years later, at a time characterised by some of the coldest conditions during the entire Holocene period. There is no simple relation between climate, biological productivity and the success of human cultures. We assume that a delicate balance existed between livelihood strategies and challenges posed by changing climates, undermining the mobility of humans and their access to resources.

In all periods the human settlement pattern in the Wollaston Forland region was surprisingly similar. This is seen on a regional scale, but is certainly also true for the very specific locations. Often Independence I, Greenlandic Dorset, and Thule have chosen exactly the same spots for their settlements, and the ruins are today often merged together. This pattern reflects several similar factors that have structured the relationship between humans and the landscape, through time.

First of all, the available resources in the region are restricted and often extremely seasonal, which means that the knowledge needed to survive as a hunter in the region had to be specific and somehow the same for all the traditions with a hunting economy. This explains to some extent the similar use of the landscape during prehistory. However, it is difficult to understand why the settlements are not only located in the same areas but also usually on exactly the same spots. Factors like well-drained substrates, favourable wind conditions, freshwater availability in summer and good "harbours" must have been crucial. But when one stands in the huge landscapes of Wollaston Forland one wonders if not a common intuition and interpretation of the landscape, developed during the evolution of mankind, could play a role in selecting the same spots for settlements over millennia.

VII. CONCLUSIONS

Most likely the Wollaston Forland region has been completely covered by the Greenland Ice Sheet several times during the Quaternary. During the LGM, around 22,000 years ago, the margin of the Inland Ice may have extended to the shelf break off Northeast Greenland. Deglaciation of the Wollaston Forland commenced at the outer coast between 11,000 and 12,000 years BP, whereas the Zackenberg area was deglaciated somewhat later, before 10,100 years BP, and it took another 3000–4000 years before the inner fjords of the region were deglaciated. Isostatic rebound led to rapid emergence in the early Holocene, but in the late Holocene, low-lying coastal areas were transgressed by the sea.

Mosses and herbs spread over the deglaciated terrain, soon followed by invertebrates, dwarf shrubs and vertebrates. The Holocene development in lake Boresø can be divided into five periods. During the early Holocene Period 1 vegetation spread over the recently deglaciated terrain. During the early to mid-Holocene Periods 2 and 3 plant life peaked, and these periods correspond to the Holocene thermal maximum. A southern beetle species colonised the region, and peak production of fruits and pollen of several dwarf shrub species is seen. The mean July temperature then was at least 2–3°C higher than at present. During the mid- to late Holocene Periods 4 and 5 summer temperatures declined, especially at around 5500, 4500 and 3500 years BP. Significant periglacial activity affected large parts of the sedimentary lowlands during the Little Ice Age.

Prehistoric people periodically entered the region after the Holocene thermal maximum, at a time when the modern day vertebrate fauna had become established. The first people arrived at around 4500 years BP. Traces of Independence I, Saqqaq, Greenlandic Dorset and Thule have been found in the region, but it was depopulated at around 3900 years BP, at around 2000 years BP and between 1823 and 1870 AD. The prehistoric people often settled on the same spots, and this also applies to the trappers that inhabited the region in the last century.

The last disappearance of prehistoric man from the region coincides with the culmination of the Little Ice Age. The prehistoric hunting cultures lived in a high-arctic landscape similar to that of today, characterised by coastal waters covered by sea ice for most of the year, by scattered polynias and by land areas with limited winter snow-cover, which was redistributed by strong northerly winds.

ACKNOWLEDGMENTS

Parts of the data presented in this chapter were collected during two expeditions carried out in 2003 and 2005 by archaeologists and geographers. The GeoArk 2003 and GeoArk 2005 expeditions were funded and organised by the Royal Danish Geographical Society and included participants from the Geographical Institute, University of Copenhagen, the Greenland National Museum and Archive, and SILA—The Greenland Research Centre at the National Museum of Denmark. We are grateful to Christian Hjort and Bernd Wagner who reviewed the manuscript.

REFERENCES

Alley, R.B. and Anandakrishnan, S. (1995) *Ann. Glaciol.* **21**, 64–70.
Andersen, S.H. (1975) *Rapport over en rejse i Østgrønland 1975*. Århus.
Andreasen, C. (2004a) In: *Grønlands Forhistorie* (Ed. by H.C. Gulløv), pp. 37–65. Gyldendal, Copenhagen.
Andreasen, C. (2004b) In: *Grønlands Forhistorie* (Ed. by H.C. Gulløv), pp. 113–141. Gyldendal, Copenhagen.
Bandi, H.-G. and Meldgaard, J. (1952) *Meddr. Grønland.* **126**(4), 85.
Bennike, O. (1997) *Quatern. Sci. Rev.* **16**, 899–909.
Bennike, O. (1999) *Polar Rec.* **35**, 323–336.
Bennike, O. and Andreasen, C. (2005) *Polar Rec.* **41**, 305–310.
Bennike, O. and Björck, S. (2002) *J. Quatern. Sci.* **17**, 211–219.
Bennike, O. and Funder, S. (1997) *Geol. Greenland Survey Bull.* **176**, 80–83.
Bennike, O. and Weidick, A. (2001) *Boreas* **30**, 205–227.
Bennike, O., Björck, S., Böcher, J., Hansen, L., Heinemeier, J. and Wohlfarth, B. (1999) *J. Biogeogr.* **26**, 667–677.

Bennike, O., Björck, S., Böcher, J. and Walker, I. (2000) *Bull. Geol. Soc. Denmark* **47**, 111–134.
Berger, A. and Loutre, M.F. (1991) *Quatern. Sci. Rev.* **10**, 297–317.
Björck, S., Wohlfahrt, B., Bennike, O., Hjort, C. and Persson, T. (1994) *Boreas* **23**, 513–523.
Bretz, J.H. (1935) *Am. Geographical Soc. Spec. Publ.* **18**, 159–266.
Brochmann, C., Gabrielsen, T.M., Nordal, I., Landvik, J.Y. and Elven, R. (2003) *Taxon* **52**, 417–450.
Böcher, J. (1988) *Meddr. Grønland, Bioscience* **26**, 1–100.
Böcher, J. (2001) *Insekter og andre smådyr – i Grønlands fjeld og ferskvand*. Nuuk, Atuagkat.
Böcher, J. and Bennike, O. (1996) *Boreas* **25**, 187–193.
Clavering, D.C. (1830) *Edinburgh New Philosophical J.* **9**, 1–30.
Christiansen, H.H. (1994) *Quatern. Sci. Rev.* **13**, 491–496.
Christiansen, H.H. (1998) *Holocene* **8**, 719–728.
Christiansen, H.H. and Humlum, O. (1993) *Geografisk Tidsskr.* **93**, 19–29.
Christiansen, H.H., Bennike, O., Böcher, J., Elberling, B., Humlum, O. and Jakobsen, B.H. (2002) *J. Quatern. Sci.* **17**, 145–160.
Cremer, H., Wagner, B., Melles, M. and Hubberten, H.-W. (2001) *J. Paleolim.* **26**, 67–87.
Dahl-Jensen, D., Mosegaard, K., Gundestrup, N., Clow, G.D., Johnsen, S.J., Hansen, A.W. and Balling, N. (1998) *Science* **282**, 268–271.
Darwent, C.M. (2003) *Meddr. Grønland, Man & Society* **29**, 348–395.
Darwent, C.M. and Darwent, J. (2004) In: *Zooarchaeology and Conservation Biology* (Ed. by R.L. Lyman and K.P. Cannon), pp. 61–87. The University of Utah Press, Salt Lake City.
Degerbøl, M. (1934) *Meddr. Grønland* **102**, 173–180.
Elling, H. (1992), *In* Grønlandsk Kultur og Samfundsforskning 1992 pp. 50–70 Nuuk.
Evans, J., Dowdeswell, J.A., Grobe, H., Niessen, F., Stein, R., Hubberten, H.-W. and Whittington, R.J. (2002) In: *Glacier-Influenced Sedimentation on High-Latitude Continental Margins* (Ed. by J.A. Dowdeswell and C. Ó Cofaigh), *Geol. Soc.* London, Spec. Publ. **203**, pp. 149–179.
Fleming, K. and Lambeck, K. (2004) *Quatern. Sci. Rev.* **23**, 1053–1077.
Fredskild, B. (1985) In: *Quaternary Environments, Eastern Canadian Arctic, Baffin Bay and western Greenland* (Ed. by J.T. Andrews), pp. 643–681. Allen & Unwin, Boston.
Fredskild, B. (1995) *Meddr. Grønland, Geoscience* **14**, 20 pp.
Fredskild, B. (2006) In: *Zackenberg Ecological Research Operations, 11th Annual Report, 2005* (Ed. by A.B. Klitgaard, M. Rasch and K. Caning), pp. 92–94. Danish Polar Center, Copenhagen.
Funder, S. (1978) *Grønlands Geol. Unders. Bull.* **129**, 66.
Funder, S. (1979) *Palaeogeogr. Palaeoclimatol. Palaeoecol.* **28**, 279–295.
Funder, S. and Hansen, L. (1996) *Bull. Geol. Soc. Denmark* **42**, 137–152.
Funder, S. and Petersen, K.S. (1980) *Bull. Geol. Soc. Denmark* **28**, 115–122.
Gelting, P. (1934) *Meddr. Grønland* **101**(2), 340.
Glob, P.V. (1946) *Meddr. Grønland* **144**(6), 40.
Grønnow, B. (2004) In: *Grønlands Forhistorie* (Ed. by H.C. Gulløv), pp. 66–108. Gyldendal, Copenhagen.
Grønnow, B. and Jensen, J.F. (2003) *Meddr. Grønland, Man & Society* **29**, 1–403.

Grønnow, B. and Sørensen, M. (2006) In: *Dynamics of Northern Societies* (Ed. by B. Grønnow, H.C. Gulløv and J. Arneborg), pp. 59–74. SILA/NABO conference, Copenhagen, May 2004, The National Museum, Copenhagen.
Grønnow, B., Jakobsen, B.H. and Møller, H.S. (2005) In: *Weekendavisen*, 17 November 2001, pp. 6–7, Copenhagen.
Håkansson, S. (1974) *Radiocarbon* **16**, 307–330.
Håkansson, S. (1975) *Radiocarbon* **17**, 174–195.
Håkansson, S. (1978) *Radiocarbon* **20**, 416–435.
Hjort, C. (1979) *Boreas* **8**, 281–296.
Hjort, C. (1981) *Boreas* **10**, 259–274.
Hjort, C. and Björck, S. (1984) *Geol. Fören. Stockholm Förh.* **105**, 235–243.
Hjort, C. and Funder, S. (1974) *Boreas* **3**, 23–33.
Houmark-Nielsen, M., Hansen, L., Jörgensen, M.E. and Kronborg, C. (1994) *Boreas* **23**, 505–512.
Jakobsen, B.H. (1992) *Danish Geogr. J.* **92**, 111–115.
Jensen, J.F. (2004) Tent rings and stone tools. Unpublished Ph.D. thesis. The Faculty of Humanities. University of Copenhagen, Denmark.
Jeppesen, E., Landkildehus, F., Lauridsen, T. and Amsinck, S. (2001) *Hydrobiologia* **442**, 329–337.
Jeppesen, E., Jensen, J.P., Jensen, C., Faafeng, B., Brettum, P., Hessen, D., Søndergaard, M., Lauridsen, T. and Christoffersen, K. (2003a) *Ecosystems* **6**, 313–325.
Jeppesen, E., Jensen, J.P., Lauridsen, T.L., Amsinck, S.L., Christoffersen, K. and Mitchell, S.F. (2003b) *Hydrobiologia* **491**, 321–330.
Johnsen, S.J., Clausen, H.B., Dansgaard, W., Gundestrup, N.S., Hansson, M., Jonsson, P., Steffensen, J.P. and Sveinbjørnsdottir, A.E. (1992) *Meddr. Grønland, Geoscience* **29**, 22.
Knuth, E. (1967) *École Pratique Hautes Études, Cont. Centre d'Études Arctiques Finno-Scandinaves* **5**, 70.
Koldewey, K. (1874) *Die zweite deutsche Nordpolarfahrt in dem Jahren 1869 und 1870 unter Führung des Kapitän Karl Koldewey*. F.A. Brockhaus, Leipzig.
Larsen, H. (1934) *Meddr Grønland* **102**, 4, 185 pp.
Lauridsen, T., Jeppesen, E., Landkildehus, F., Christoffersen, K. and Søndergaard, M. (2001) *Hydrobiologia* **442**, 107–116.
Meldgaard, M. (1991) In: *4th North American Caribou Workshop* (Ed. by C. Butler and S.P. Mahoney), pp. 37–63.
Mikkelsen, P.S. (2001) *Nordøstgrønland 1908–60. Fangstmandsperioden*. Aschehoug, Copenhagen.
Ó Cofaigh, C., Dowdeswell, J.A., Evans, J., Kenyon, N.H., Taylor, J., Mienert, J. and Wilken, M. (2004) *Mar. Geol.* **207**, 39–54.
Ramsey, C.B. (2001) *Radiocarbon* **43**, 355–363.
Reimer, P.J., Baillie, M.G.L., Bard, E., Bayliss, A., Beck, J.W., Bertrand, C.J.H., Blackwell, P.G., Buck, C.E., Burr, G.S., Cutler, K.B., Damon, P.E., Edwards, R.L., *et al.* (2004) *Radiocarbon* **46**, 1029–1058.
Sarmaja-Korjonen, K. (2003) *Holocene* **13**, 691–700.
Stenico, J. and Woolmore, R. (2002) *The Arcturus Wollaston Forland expedition, North-East Greenland, July/August 2002*. Cheltenham, Bruxelles.
Sørensen, M. (2005) *Teknologi og tradition i Østarktis 2500 BC–1200 AD. En dynamisk teknologisk undersøgelse af de litiske inventarer i de palæoeskimoiske traditioner*. SILA—The Greenland Research Centre at the National Museum of Denmark.

Unpublished Ph.D. thesis. Institute of Archaeology, University of Copenhagen, Denmark.
Sørensen, M., Andreasen, C., Jakobsen, B.H. and Møller, H.S. (2004) *Ekspedition "Geo-Ark 2003". Arkæologisk-geografisk rekognoscering af området Wollaston Forland.* The Royal Danish Geographical Society, The National Museum, Copenhagen and Greenland National Museum & Archives, Nuuk, 102 pp.
Thalbitzer, W. (1923) *Meddr. Grønland* **40**, 739.
Tuborg, H. and Sandell, B. (1999) *Danish Polar Center Publication* **6**, pp. 149.
Wagner, B. and Melles, M. (2001) *Boreas* **30**, 228–239.
Wagner, B., Melles, M., Hahne, J., Niessen, F. and Hubberten, H.-W. (2000) *Palaeogeogr., Palaeoclimatol., Palaeoecol.* **160**, 45–68.
Weidick, A. (1976) *Rapp. Grønlands Geol. Unders.* **80**, 136–144.
Weidick, A. (1977) *Rapp. Grønlands Geol. Unders.* **85**, 127–129.
Weidick, A. (1978) *Rapp. Grønlands Geol. Unders.* **90**, 119–124.

Climate Change Influences on Species Interrelationships and Distributions in High-Arctic Greenland

DAVID R. KLEIN, HANS HENRIK BRUUN,
REBEKKA LUNDGREN AND MARIANNE PHILIPP

SUMMARY

Biotic communities in Northeast Greenland have an insular character as a consequence of the complex geomorphological nature of the ice-free land and its interdigitation with glacial ice and the sea. Post Pleistocene movements of most plants and animals into the region have generally followed East and North Greenland coastal routes, and the majority of the plants have North American affinities. Climatic change, bringing about reduction in the extent

ADVANCES IN ECOLOGICAL RESEARCH VOL. 40
0065-2504/08 $35.00
DOI: 10.1016/S0065-2504(07)00004-9
© 2008 Elsevier Ltd. All rights reserved

of sea ice adjacent to the coast and changes in seasonality and associated precipitation and air movements, influences patterns of activity, growth, reproduction, and dispersal of all life forms present. Climate-associated changes in the biotic communities of the region are altering inter-species interactions, notably pollination, seed dispersal and plant–herbivore relations.

Sexual reproduction and dispersal of propagules, primarily seeds, are essential processes underlying maintenance of genetic diversity in plant communities in Northeast Greenland. Wind and water transport of seeds are primary methods by which plants disperse and become established in the High Arctic, particularly at shorter distances. Birds and mammals are also involved and may be of particular significance to long-distance seed dispersal. In Northeast Greenland, dispersal of viable seeds may frequently occur by passage through the guts of geese and musk oxen.

Research at Zackenberg on the role of insects in pollination of flowering plants has shown that Diptera species, primarily flies, dominate among the insect species visiting flowers each summer. Diptera, Lepidoptera (butterflies and moths), Hymenoptera (bumble bees and small wasps), and one Hemiptera (true bugs) species have constituted the primary pollinators at Zackenberg. Arctic willow *Salix arctica*, white arctic bell heather *Cassiope tetragona*, and mountain avens *Dryas octopetala* are the primary species represented in the pollen present on pollinating insects at Zackenberg. The effects of climate warming that may enhance environmental conditions for plant growth in Northeast Greenland and accelerate invasion of new species will also be tied to the relationship of specific plant species to their insect pollinators. Those plants that are self-pollinated may have an initial advantage in an environment where insects and their plant relationships are being altered by the changing climate.

An increase in growth and dispersal of shrubs in the Arctic is occurring as a consequence of climate warming. Increases in shrubs with more upright growth form, especially willows, will generate microhabitats not previously present in the High Arctic. The new habitats will make possible the invasion of new insect, mammal, and bird herbivores, as well as their parasites and predators.

I. INTRODUCTION

The High Arctic of Northeast Greenland is unique among land areas of the Arctic. It is cooled by the southward- flowing Arctic Ocean waters of the East Greenland Current and is characterized by an extensive band of adjacent sea ice. Northeast Greenland includes a broad band of largely ice-free land between the coast and the Greenland Ice Sheet that decreases in width as it extends northward from Scoresby Sund at 70 °N to nearly 82 °N (see map in

Meltofte and Rasch, 2008, this volume). This relatively narrow belt of land includes rivers, streams, and lakes of varying sizes and is dissected by fjords, bays, and glaciers that extend from the Greenland Ice Sheet to the sea.

Biotic communities are the product of the local climatic and geomorphic characteristics of the environment that provide both options and constraints to their development. Interspecies relationships have also shaped the processes and dynamics of these developing ecosystems. The present biotic communities in the High Arctic of Greenland are largely the product of their development during the Holocene, and they exhibit a corresponding decline in complexity with increasing latitude. Although most life forms present in the High Arctic of Northeast Greenland today found their way there during the last 8–10 thousand years, some few species in the High Arctic may have derived from refugia that were present there during the last glacial epoch (Funder, 1979; Bennike, 1999).

Global climate changes during the past several decades have been pronounced in the Arctic, accounting for warming in many regions of the Arctic, increased seasonal absence of sea ice adjacent to land areas, and associated changes in precipitation and wind (ACIA, 2005). In Northeast Greenland, a warming climate, leading to decrease in seasonal extent and duration of sea ice near the coast, brings associated changes in precipitation, duration and depth of snow-cover, amount of soil moisture, and extent of cloud-cover (Stendel *et al.*, 2008, this volume), all of which influence plant growth and community structure, invertebrate activity, and underlay the timing and success of vertebrate reproduction in the high latitudes of the Arctic.

With climate warming, ecosystem interactions in the Arctic are generally accelerated, and to a large extent through increases in soil processes and associated increased nutrient availability for plant growth (Chapin *et al.*, 1995). Availability of soil moisture has been and will continue to be a major factor limiting plant growth in land areas of the High Arctic. If existing plant communities are to be able to expand spatially and in elevation during longer and warmer summers, expected increased precipitation as sea ice decreases must more than compensate for increased losses of soil moisture through evaporation, deeper annual thaw depths that accelerate soil drainage, and water loss through plant transpiration. Plant and invertebrate seasonal dormancy is expected to shorten as the climate warms. However, available soil moisture will continue to be a major factor influencing the timing of senescence and initiation of the onset of dormancy in terrestrial plants and insect herbivores. Interspecies interactions will obviously be dynamically altered, inclusive of parasitism, mutualism, symbiosis, competition, and predation. Reproduction in some plants may be dependent on their pollination by adult stages of insect species, some of which may have been defoliators as larvae. Consequences of climate change for high-arctic biota will be profound, affecting changes in the structure, function, and dynamics of ecosystems.

In the interactions between plants and animals, short-term costs are generally offset by longer-term benefits to the species involved. For example, plant tissue losses to vertebrate herbivores may be offset or compensated by acceleration of nutrient recycling and increased nutrient availability to the plant, stimulated compensatory growth, accumulation of chemical or other defences against further tissue loss to herbivores and facilitation of seed production and dispersal. These plant responses can be thought of as phenotypic adaptations to a changing environment at the individual organism level that are dependent on flexibility within the genome of the individual plant. Longer-term changes at the species level may come about through natural selection stimulated by climate change, and resulting in genetic change. This is evolutionary adaptation.

Co-relationships between plants and animals, though not unique to high-arctic ecosystems, are constrained in the High Arctic by the narrow balance in primary productivity by plants during the brief seasonal window available for plant growth, flowering and reproduction. Animals at higher trophic levels are dependent on plant productivity, either directly as herbivores or indirectly through predation on the herbivores, for their sustenance and reproduction. Invertebrate species in terrestrial systems pass the winter in dormant or quiescent stage, which restricts growth and reproduction to summer. In aquatic systems, where liquid water remains available throughout winter, growth is quiescent in the absence of light, but seasonal growth and reproduction begins when returning light can reach the water column through the ice-cover. Vertebrate herbivores that remain active throughout the entire year in the Arctic, the musk ox *Ovibos moschatus*, arctic hare *Lepus arcticus* and collared lemming *Dicrostonyx groenlandicus*, in contrast to those that migrate to lower latitudes, the rock ptarmigan *Lagopus mutus* and geese *Anser* and *Branta* spp., are most likely to show quick responses to changes in the climate of the High Arctic. Specific responses may be positive as well as negative for the individual species involved.

II. PLANT DISPERSAL AND INFLUENCES ON SPECIES RICHNESS AND GENETIC DIVERSITY

A. Sexual Reproduction and Dispersal

Sexual reproduction and dispersal of propagules away from the parents is an important part of any plant life cycle. In the High Arctic, however, flowering, pollination and ripening of seeds may be severely limited by the short summer.

Sexual reproduction may fail completely except in infrequent benign years (Savile, 1972). Moreover, recruitment of new individuals into plant populations from seed appears in general to be less frequent the harsher the environment. Thus, clonal reproduction dominates over sexual reproduction in most plant species in the High Arctic, rendering plant individuals (genets) extremely long lived once established. Despite this general picture of stasis, seed dispersal may be an important process in community dynamics, especially at slightly longer timescales. Glacier forelands and land-slides become re-vegetated, albeit slowly and plant migration as a consequence of climatic change must be brought about primarily through dispersal of seeds. Despite this fact, dispersal processes have rarely been studied directly in the Arctic. In the face of present and future climate change, quantification of the efficacy of plant dispersal and success of establishment is badly needed.

B. Sources of Origin for Northeast Greenland Flora

The most dramatic recent climate change in the High Arctic was at the onset of the Holocene. Plants colonized the areas newly freed from ice in Northeast Greenland. For most species, colonizing individuals came from refugia far from Greenland by long-distance migration. Böcher *et al.* (1959) and Bay (1992) cited existing plant species distribution in Northeast Greenland and their derivation to be from either western, that is, the Canadian Arctic, or eastern flora, that is, Southeast Greenland and Svalbard. Their rankings indicated that about twice as many of the Northeast Greenland flora derived from the western flora across North Greenland as had derived from the south via the eastern Greenland coast.

The means by which seeds have been transported over vast distances were much debated by naturalists for more than a hundred years. In fact, the viewpoints were either that migration of terrestrial biota was impossible (Löve, 1963), which was then taken as evidence for *in situ* survival (nunatak hypothesis: Warming, 1888; Böcher, 1951; Dahl, 1963), or that survival of biota during full glacial conditions was impossible, which was then taken as evidence that long-distance dispersal had taken place (tabula rasa hypothesis: Nathorst, 1892; Ostenfeld, 1926; Nordal, 1987). However, in the absence of methods for quantification of long-distance dispersal, the debate was inconclusive.

The recent development of molecular methods has enabled evaluation of the value of these hypotheses. Abbott and Brochmann (2003) reviewed the molecular evidence for post-glacial (or earlier) dispersal of plants across the North Atlantic. They found, for a host of species, that present-day genetic patterns strongly suggested recent dispersal events. Present-day genetic patterns cannot rule out glacial survival, they only make strong evidence that genetic patterns that might have arisen in isolated refugia have been completely swamped by post-glacial gene flow. Investigations of mitochondrial DNA haplotype distribution in purple saxifrage *Saxifraga oppositifolia* has presented unequivocal evidence for seed dispersal across the North Atlantic (Abbott *et al.*, 2000), and it is even possible to determine the direction of migration, namely, from east to west (Siberia, Franz Josef Land, Svalbard, North Greenland). However, it has arguably taken place in an earlier interglacial.

C. Dispersal by Physical Processes

By what means could such long-distance dispersal have taken place? While considering trans-Atlantic dispersal impossible for most plant species, Löve (1963) enumerated species for which she considered long-distance dispersal possible, although highly improbable, namely, 10 species of marine hydrophytes and sea shore plants potentially dispersed by sea currents, and a slightly larger group of plants having diaspores lighter than 0.01 mg and no longer than 0.2 mm (37 species of cryptogams, 14 species of seed plants) potentially dispersed by wind. Nordal (1987) suggested drift ice, icebergs and birds as the most probable dispersal vectors. Hultén (1962) substantiated that contention by reporting living individuals of three vascular plant species, snow grass *Phippsia algida*, chickweed *Stellaria laeta* and *S. oppositifolia*, rafting in the Beaufort Sea on an "ice island" probably originating on the shelf of Ellesmere Island. Johansen and Hytteborn (2001) reviewed the possibilities for intercontinental dispersal of diaspores via driftwood and drift ice in light of recent developments in palaeo-oceanography and studies on ice-rafted debris and driftwood. They concluded that ice-rafting of diaspores is likely to have brought about effective long-distance dispersal from northern Siberia to East Greenland and northern Scandinavia, and mentioned four plant species with isolated occurrences in these areas as likely examples. Seed dispersal with ice-rafted debris along the Transpolar Drift Stream could also explain the distribution of one particular haplotype of *S. oppositifolia* chloroplast DNA in two geographically distant locations, the Taimyr Peninsula and North Greenland (Abbott *et al.*, 2000).

D. Animals as Seed Dispersers

A further possibility is that plant seeds have been transported by sticking to the feathers or feet of birds or through bird guts. On the basis of fossil pollen of seed plants, Iversen (1952) distinguished a group of probable glacial survivors and a group of early immigrants, both consisting of 16 plant species, the latter group primarily hydrophytes. Birds, like the barnacle goose *Branta leucopsis*, migrating from their winter quarters in western Europe, were considered the most likely dispersal vector (Figure 1A). Geese have been shown to disperse seeds of terrestrial plants elsewhere (Willson *et al.*, 1997), and other birds may have contributed as well. For example, the snow bunting *Plectrophenax nivalis* was shown to excrete viable seeds on the new volcanic island Surtsey (Friðriksson and Sigurðsson, 1969), and tens of thousands of snow buntings migrate from northern Norway to high-arctic Greenland each year (Meltofte, 1983). The rock ptarmigan does not migrate across the North Atlantic, but it does make annual migrations between the Low and the High Arctic of Greenland, and it is known to ingest and defecate viable seeds (Ekstam, 1897; Gelting, 1937) (Figure 1B). Humans have apparently accounted for the presence of a few rare plant species in Northeast Greenland as is evident from their primary association with former Inuit habitation sites (Bay, 1992). With continuing and expected increased human presence and visitation in Northeast Greenland associated with arctic research, weather station operation, military patrols, mineral exploration and tourism, there is considerable potential for humans to be the agents for establishment of invasive plant and invertebrate species.

Seed dispersal by migratory birds is perhaps the most likely, but a unique contribution by more sedentary birds and terrestrial mammals during their post-glacial re-colonisation of Greenland is possible, although it has been completely neglected by past reviews. Pakeman (2001) presented evidence for mammal endozoochorous seed dispersal in the Holocene migration of angiosperms in Europe. From a modelling study, he concluded that large mammalian herbivores could have been the main dispersal vector of herbaceous species, even at relatively low probabilities for ingestion and gut

Figure 1 Birds can play an important role in seed dispersal. (A) Tracks and faeces have been left by a goose that visited and fed on individual plants of *Puccinellia* sp. on the exposed lake bed of glacier-damned Blåsø in Kronprins Christian Land (79°38′N). (B) Rock ptarmigan that breed in North and Northeast Greenland migrate to south in winter, and thus have the potential to accelerate the northward movement of plants when they return to their breeding grounds. Photos: D.R. Klein.

survival, especially animal species with relatively large home ranges. Terrestrial mammals, such as reindeer *Rangifer tarandus*, wolf *Canis lupus* and arctic fox *Alopex lagopus*, are known to occasionally travel over vast stretches of sea ice and, thus, have the potential to disperse seeds at similar distances. Even the polar bear *Ursus maritimus*, nominally a carnivore, has been observed grazing in sedge-dominated meadows in both Northeast Greenland and other areas of their distribution (Elliott, 1882; C. Bay, personal communication). Understanding the processes driving the Holocene re-immigration will help us predict species migration potential under present and future climate change.

E. Problems of Predictability

Events of long-distance seed dispersal are often claimed to be inherently stochastic and barely predictable. This is because a decreasing number of seeds are dispersed with increasing distance from the source plant, with a very small amount dispersed to really long distances (Harper, 1977). The claim of unpredictability has, however, rarely been put to test. Recent modelling of long-distance wind dispersal has increased predictability to a considerable extent. Relatively few parameters are needed, inclusive of a few plant traits (height of seed release, seed falling velocity), real-landscape anemometry and both horizontal and vertical wind speed (Tackenberg *et al.*, 2003). Similarly, the potential for prediction of zoochorous seed dispersal needs to be investigated. Among the parameters needed for modelling of zoochorous dispersal are attachment and detachment probability (for epizoochory), ingestion probability and passage rate (for endozoochory) and animal movement behaviour, which is probably in itself more complex than wind "behaviour." Assessment of these parameters has been the focus of research conducted at Zackenberg (H.H. Bruun, unpublished data).

Seed dispersal influences composition and structure of populations both at the species level and at the genetic level. Most often we look at dispersal as a positive feature. Dispersal is seen as a mechanism for individuals to colonize new localities and to increase diversity locally as well as in distinct communities. Seed dispersal is, however, a risky business, as many seeds are deposited at sites unsuitable for germination and establishment.

F. Long-Distance versus Short-Distance Dispersal

Common to all seed dispersal studies is the difficulty in obtaining knowledge about seeds dispersed over large distances. It is extremely difficult to follow the fate of seeds after they have left their parent (Wang and Smith, 2002). Examination of genetic variation among populations may, however, provide an index of the degree of genetic contribution from seed dispersal. If the genetic variation is the same in two populations, the chance of a good genetic exchange among the populations is high. If no genetic communication occurs, genetic differentiation will most probably happen over time due to local adaptations or genetic drift. Yet another indirect method is by analyzing genetic markers in seeds or young individuals to try to find out from which adult individuals or populations the seed most likely originated.

Seeds dispersed over short distances will most likely fall within the environment of the parent plant, which has proved to be suitable for establishment, growth and reproduction of the species. The chance for heavy competition from conspecifics, however, is considerable in close vicinity of the parent plant. Seeds dispersed over longer distances have a chance to colonize areas where competition with conspecifics is low, but there is a high chance of being deposited at sites where germination and establishment is not possible. Different species possess different seed dispersal strategies corresponding to other life history traits and to the environment of each species. Arctic plant species are mostly dispersed by wind and water. Many species have no special equipment for dispersal, and seeds are merely dropped from the parent plant to the ground. Seeds may be eaten more or less accidentally by vertebrates or become attached to them and thereby can be dispersed over considerable distances.

For some species, mature seeds can either germinate immediately or in the following spring, or they can be incorporated in seed banks in the soil. Seeds germinating directly contribute to renewal of the population and to instantaneous continuation of the alleles of the parent plant in contrast to a questionable future for those alleles from seeds that enter a seed bank. Upon germination, the seedling may become exposed to poor conditions for establishment, whereas seeds in seed banks have the chance for activation when conditions may be more favourable. Seeds from several arctic species tend to accumulate in seed banks. Since not all growing seasons in the Arctic result in seed production but may be suitable for seed germination, inclusion of some seeds in a seed bank may enhance long-term reproductive success. Arctic seed banks are probably rather long lived due to the low soil temperature. Therefore, the complex of species and genes represented in a seed bank may differ from species growing there or to the alleles recently found there. Seed banks or seed dispersal in time have thus the potential to influence the vegetation above ground both at the species level and at the genetic level.

G. Seed Dispersal and Community Population Structure

Seed dispersal in space and time influences the population structure. In some populations, new individuals are established only after disturbance. If the total population of a species is erased or un-vegetated areas are appearing, as in front of receding glaciers, seed dispersal is essentially the only way of plant establishment. If seeds are dispersed into existing vegetation, this might increase the diversity in a population or possibly reduce it through competition.

The total amount of gene flow (pollen and seed flow) determines differentiation among populations. At the genetic level, the exchange of only one allele per generation between populations is necessary to homogenize allele frequencies (Ellstrand and Elam, 1993). Arctic plant species often have very long generation time, in some cases requiring several years before reproduction can occur (Savile, 1972). Consequently, only a small amount of gene flow among populations would seem to be necessary to even out differences. However, establishment of new individuals is in most communities very rare, thus continuous gene flow is needed to secure availability of seeds at the appropriate moment where establishment is possible. The genetic population structure has been investigated for some arctic species (Abbott *et al.*, 2000; Abbott and Brochmann, 2003; Philipp and Siegismund, 2003). For the *Dryas integrifolia–Dryas octopetala* complex in East Greenland, it has been demonstrated that populations separated by increasing geographic distance show increasing genetic differentiation in terms of different frequencies of alleles (Høye *et al.*, 2007). Genetic differentiation was therefore found in this species, which is both insect pollinated and possesses wind-dispersed seeds, both of which promote gene flow. As found in many species, local differentiation most likely occurs because of limited communication among populations separated by a distance shorter than the most frequent pollen and seed dispersal distances.

H. Propagule Dispersal via Vertebrate Guts and Influences on Species Richness and Genetic Diversity at the Community Level

The correspondence between composition of the endozoochorous seed rain and of the receiving community has been investigated in the Zackenberg area. Dung of four vertebrate species has been investigated, namely, musk ox, arctic fox, arctic hare and pink-footed and barnacle geese (Figure 2). The question is whether immigration, that is, seed dispersal (potentially), contributes to the generation and maintenance of species richness within plant communities and genetic diversity in populations of constituent species. Musk oxen and geese appear to disperse far more seeds in their faeces than foxes and hares. Goose droppings collected during autumn have high content of viable seeds (80% of samples contain seed, with an average of more than 400 seeds per 100 g dung). In total, 12 species of vascular plants were encountered as viable seeds in goose droppings. Geese are relatively mobile

Figure 2 Musk oxen ingest seeds along with other plant tissues (A), some of which pass through their guts and remain viable in their faeces (B). The faeces provide fertilization for plants that may be adjacent to them as well as for any germinating seeds present in the faeces. Photos: D.R. Klein.

at the landscape scale, even during molt, and may disperse seeds between similar habitats. In addition, geese stopping over during the southbound autumn flight may transport seeds between different valleys in the island-like ice-free fringe in Northeast Greenland. The northbound dispersal of seeds during spring migration has not been investigated. As a general rule, bird intestines are emptied before or early into long-range flights. They often, however, alight and feed for a few hours where possible during their northward flights. It is highly likely therefore that geese have contributed significantly to the dispersal of a number of plant species from Iceland to East Greenland (Halliday *et al.*, 1974; Olesen, 1987; Bennike and Anderson, 1998).

Musk ox dung contains fewer viable seeds per unit weight (on average 30 seeds per 100 g dung), but this is more than compensated for by the larger size of dung patches and the less selective ingestion behaviour of musk oxen compared to geese. Seeds of more than 20 plant species have been found in musk ox dung, on average more than four per dung sample. Mobility of musk oxen over the landscape is high. Thus, this mammal is likely to be important in population and community dynamics of many plant species. Musk oxen may have brought plants species along its Holocene immigration route from Arctic Canada to North and East Greenland, but is less likely to contribute significantly to present-day long-distance plant dispersal.

Dung pellets of arctic hare and scats of arctic fox contain little viable plant seeds. For the hare, this means a limited potential as a seed dispersal vector. The potential of the fox is probably less easy to assess. The fox's diet is probably highly variable due to its opportunistic feeding behaviour (Kapel, 1999). Foxes may have a seed-free diet in some periods, but ingest berries (intentionally), soil and vegetation (unintentionally) in other periods. Feeding experiments have shown that seeds of many arctic plants survive fox gut passage (Graae *et al.*, 2004). In addition, arctic foxes are highly mobile during some periods of the year (Anthony, 1997), and are reported to sometimes cross vast stretches of sea ice, for example, between Svalbard and Greenland. Polar bears presumably offer a potential for seed dispersal similar to arctic foxes. Although polar bears only occasionally feed on vegetation, this is most likely to occur when they may become stranded on land if pack ice, normally adjacent to the coast, melts or is blown far offshore (Derocher *et al.*, 2004), a condition that may become increasingly common in the future with continued climate warming. Individual polar bears are known to make extensive journeys overland under such circumstances (A. Balser, personal communication). Polar bears are genetically close to brown bears *Ursus arctos* (Talbot and Shields, 1996), and in Alaska the latter have been shown to pass large amounts of viable seed through their guts (Willson, 1993; D.R. Klein, personal observation).

In addition to their role as seed dispersers, all mammals and birds also recycle nutrients to the soil through their excreta and decomposition of their own bodies. This may be important where accumulation of excreta and decomposing prey remains at raptor nesting sites, carnivore dens and bird perches may create optimal conditions for both deposition and germination of seed, and especially the establishment of new species invading from more nutrient-rich soil areas at lower latitudes (Figure 3A, B). Klein and Bay (1991, 1994), in their investigations of herbivore–plant community relationships in North Greenland, made several first regional records of plant species. The sites within the area of the survey where these plant species new to

Figure 3 (A) Around fox dens and raptor nest sites, accumulating excreta and decomposing prey remains provide fertilization for dispersing and invading plants. It was at this fox den in Nansen Land (83 °N) that the northern most occurrence of arctic harebell *Campanula uniflora* was found (Fredskild *et al.*, 1992). (B) This decomposing carcass of a musk ox has created a favourable site for plant establishment through release of nutrients to the soil. It also has generated a micro environment that captures wind-blown snow (and possibly seeds). The melting snow provides moisture for germinating and growing plants and provides an additional possible seed source from the gut contents. Photos: D.R. Klein.

the region were found were most often at fox dens, raptor aeries or bird perches fertilized by the occupants and decomposing musk ox carcasses. All of these sites were likely also fertilized through visits and excreta marking by foxes and wolves.

At the level of genetic diversity in plant populations, investigations have been made of the endozoochorous dispersal of viviparous knotweed *Polygonum viviparum* bulbils through the musk ox and barnacle goose. Amplified Fragment Length Polymorphism (AFLP) markers were used to fingerprint individual bulbils in the faeces and in established individuals in the vegetation surrounding each dung patch. Unexpectedly high levels of clonal diversity were found in populations of *P. viviparum*, a species which relies almost exclusively on vegetative reproduction. There also was considerable variation in clonal diversity among populations. Assignments tests have shown that the genetic composition in bulbils retrieved from faeces was different from the genetic composition in the populations within which the faeces were deposited. The high genetic diversity found in existing populations of *P. viviparum* in the Zackenberg area is likely to have been brought about to a large extent through grazing and bulbil dispersal by geese and musk oxen.

III. POLLINATION AND INVADING SPECIES

A. Pollinator–Plant Relationships at Zackenberg

The severe and inter-annually variable, high-arctic environment of Northeast Greenland supports a flora and associated fauna of low complexity and species richness (Downes, 1964; Bay, 1992). With climate warming, species richness in high-arctic ecosystems is expected to increase as new species invade, largely from lower latitudes. Among invertebrates, many insect species living at lower latitudes, especially pollinators, may be pre-adapted to invade high-arctic ecosystems. In the Zackenberg region, Elberling and Olesen (1999) found that somewhat more than 76 insect species are associated with pollination of flowering plants through their visitation to 31 plant species (Figure 4). Plant species appeared to have a random relationship to the complex of species of insects that visited them, whereas there was a more frequent association with specific plant species by the insect species. Diptera dominate the insect species involved in pollination. Among the more than 50 dipteran species involved, the greatest numbers and most species are from the families Muscidae (muscid flies), Chironomidae (midges) and Syrphidae (syrphid flies), in that order. Hymenoptera and Lepidoptera species are also important pollinators in this high-arctic region with about 10 species of each order participating. Only one hemipteran species has been observed to be involved in pollination. In comparison to other high-latitude systems,

Figure 4 Diptera were the most common insect group found to be involved in pollination in the Zackenberg area. (A) A fly and a polar fritillary butterfly visit a *Taraxicum* spp. flower. (B) The few Lepidoptera species present in Northeast Greenland have a multi-year development stage as herbivores, but during their brief adult stage they are important as pollinators. (C) Bumble bees are effective pollinators in the High Arctic and are able to be active over a broader temperature range than other insects. Photos: D.R. Klein.

proportions of species in the orders Diptera, Hymenoptera and Lepidoptera associated with flower visitation are similar (Mosquin and Martin, 1967; Hocking, 1968; Kevan, 1970, 1972; Elberling and Olesen, 1999; Lundgren and Olesen, 2005). Remarkable similarity exists between insect pollinator species at Zackenberg and those at Ellesmere Island in the Canadian High Arctic (Kevan, 1973; Elberling and Olesen, 1999). Among the Diptera at Zackenberg, visiting plant inflorescences are nectar gatherers, blood suckers and predators, whereas most hymenopterans are parasitoids and their role in pollination is not understood (Elberling and Olesen, 1999). However, insects, whatever is their primary role in ecosystem relationships, can be attracted to flowers of many high-arctic plant species because of the heat focusing characteristic of the flowers. Thus, they can become pollinators.

Identification of the sources of pollen carried by the insect visitors to flowers by Elberling and Olesen (1999) showed that the richest plant species represented by the pollen on the insect bodies were *Salix arctica* (pollen present on 40 visitor species), *Cassiope tetragona* (32) and *D. octopetala* (31). Plant species most frequently visited by pollinators in their study were *S. arctica* (46 visits), *D. octopetala* (40), *C. tetragona* (38), *S. oppositifolia* (34) and moss campion *Silene acaulis* (29).

B. Climate Change Influences on Pollination and Invading Species

Climate warming and associated changes in regional and seasonal precipitation patterns may have resulted, and are likely to result, in the movement of new plants and insects into the Zackenberg region. The most likely sources of these potential new ecosystem components are from existing and more species-rich arctic plant communities at lower latitudes in East Greenland as well as from arctic Canada via a North Greenland route. Details of the nature of arrival of plants and insect pollinators, however, are hard to predict. They will be entering already occupied plant communities with established pollinator associations. A basic question is, how quickly and to what extent these high-arctic plant communities may become altered to make them less competitive to the invaders.

Most flowering plants in the Arctic are insect pollinated, the major exceptions being grasses, sedges and rushes. Insect pollinators in the Arctic are largely generalists, although some plant groups are clearly favoured by certain insect species. Thus, arrival of new plants and insects that are pre-adapted to arctic conditions are not constrained by species-specific mutual relationships. The flowering phenology of invading plant species may play a role in synchronization with pollinating species. The low density that characterizes invading plant species, however, will provide an advantage for their establishment to those species that are self-pollinated. A major obstacle to be overcome by invading plants and insect pollinators will be competition with species that are already established components of existing ecosystems. Additionally, both plants and their pollinators, whether resident species or invading ones, will be subject to the influences, both positive and negative, imposed on them by other invading species. For plants, this could include invertebrate and vertebrate herbivores, seed dispersers, parasites and diseases and for insects, this could include predators, parasitoids and diseases. Invasion and establishment of new plant and insect pollinators into the Zackenberg plant communities, although presumably facilitated by the changing climate, will nevertheless be a slow process.

C. Changing Plant Community Structure and Invasion of New Species

An expected major change in the flora of high-arctic regions associated with a warming climate is that of movement of shrub species northward. The northward movements of shrubs in arctic tundra regions have been associated with previous periods of climatic warming during the Holocene (Sturm *et al.*, 2005), and specifically at Zackenberg (Christiansen *et al.*, 2002). As a consequence of recent climate warming, alder *Alnus* spp. and birch *Betula* spp. have both increased in density and expanded on the landscape in regions of the Low Arctic (Chapin *et al.*, 2000). In Northeast Greenland, where few shrub species are currently present, an expansion on the landscape, and especially in elevation, of those shrub species there now, inclusive of *S. arctica* and *Betula nana*, dwarf birch, can be expected. Species, such as

Salix arctophila, *Salix glauca* and *Betula glandulosa*, shrub birch, which are already present at lower latitudes in Greenland can be expected to reach Northeast Greenland in the not too distant future with continued climate warming. Climate warming, leading to an extended summer growth period, may alter reproductive strategies of some arctic plant species (Elmquist *et al.*, 1988; Crawford and Balfour, 1990). The altered sex ratio of *S. arctica* at high latitudes may change as relationships to both herbivores and insect pollinators become altered by the changing climate (Klein and Bay, 1991, 1994; Hjältén, 1992; Klein *et al.*, 1998; Nyman and Julkunen-Tiitto, 2000; Ikonen, 2001).

In a meta-analysis of field studies from a wide range of ecosystems, Parker *et al.* (2006) challenge the hypothesis that invasive exotic plants become a problem in their adoptive lands because they leave their coevolved herbivores behind. In contrast, herbivores already present in the invaded communities are better able to suppress or inhibit establishment of invaders than are the herbivores that have evolved with those plants in their original home. In a reverse relationship, introduced or invading herbivores are harder on native plants in the lands they invade than on introduced plants, including those with which they coevolved. Thus, the replacement of native with exotic herbivores may trigger an ecological "meltdown," or ecosystem regime shift, whereby one exotic species facilitates invasions by others.

The collared lemming, among the vertebrate herbivores of Northeast Greenland, may play the most important role in inhibiting the establishment of new vascular plant species in Northeast Greenland. During the peaks of its 3–4-year cycles, the numbers of lemmings can reach such high levels that their impact on aboveground plant tissues can greatly exceed that of other herbivores. Collared lemmings, however, are dietary specialists that select for dicotyledenous plant species, and their greatest impact on vegetation occurs beneath the snow-cover as their populations build to peak levels in late winter (Berg, 2003; Berg *et al.*, 2008, this volume). In North and Northeast Greenland, willow dominates their winter diet when they are largely restricted to plant communities with sufficient snow-cover to offer protection from predation (Klein and Bay, 1991). Since most willow species present in the Arctic require snow-cover in winter, dispersal of new willow species to Northeast Greenland may be inhibited in their establishment by lemmings.

IV. CONCLUSIONS AND FUTURE PERSPECTIVES

Biotic communities in Northeast Greenland largely result from development during the past 8–10 thousand years of the Holocene. These biotic communities have an insular character as a consequence of the complex geomorphological nature of the ice-free land, interrupted where glaciers connect the

Greenland Ice Sheet to the sea. With the exception of a few possible plant and invertebrate animal species that may have been present there during the last glacial epoch, most existing species have derived from lower latitudes by following coastal routes in East and North Greenland. A few plants, however, appear to have arrived from the east across the North Atlantic and Arctic Ocean.

Climatic warming globally, most pronounced in regions of the Arctic, is resulting in reduction in the extent of sea ice adjacent to the coast in Northeast Greenland, which is a driver of associated changes in seasonal patterns of precipitation. Patterns of activity, growth, reproduction and dispersal of all life forms present are changing in response to the changing climate. Inter-species interactions, notably pollination, seed dispersal and herbivory, are also being altered by the changing climate.

Sexual reproduction and dispersal of propagules, primarily seeds, are essential mechanisms underlying maintenance of biodiversity in plant populations and communities in Northeast Greenland. They also are necessary for plant species dispersal in these high-arctic plant communities that decrease in species richness with increasing latitude. Wind and water transport of seeds are primary methods by which plants disperse and become established in new areas, at least at shorter distances. Birds and mammals also are important in transport of seeds and other plant propagules, and these vectors may be particularly important to dispersal at longer distances. In Northeast Greenland, seed dispersal may occur by passage through the gut of viable seeds that have been consumed, primarily by geese and musk oxen. Attachment of seeds to the feet and plumage of migrating birds is presumably important for long-distance transport, especially by waterfowl. Locally, attachment of seeds to feet and hair of mammals may also assist in plant dispersal.

The role of insects in pollination of flowering plants is of major importance for the establishment and maintenance of plants and their associated plant community structure and dynamics in the High Arctic of Northeast Greenland. Research at Zackenberg has demonstrated that Diptera species, primarily flies, dominate among the insect species visiting flowers each summer. Diptera, Lepidoptera, Hymenoptera and one Hemiptera species have constituted the primary pollinators at Zackenberg. *S. arctica, C. tetragona* and *D. octopetala* are the primary species represented in the pollen present on pollinating insects at Zackenberg.

In Northeast Greenland, as in other regions of the Arctic, most flowering plants are insect pollinated. Insect pollinators, like flowering plants, decrease in species richness with increasing latitude throughout the Arctic. Thus, the effects of climate warming in enhancing environmental conditions for plant growth in Northeast Greenland and invasion of new species will also be tied to the relationship of specific plant species to their insect pollinators. These plant–insect relationships may be complicated by the differential phenology of plants and timing of insect emergence from larval stages. Those plants that are self-pollinated may have an initial advantage in an environment where insects and their plant relationships are being altered by the changing climate.

A major change occurring in plant community structure in the Arctic as a consequence of climate warming is the increase in growth and dispersal of shrubs. This will have a much greater influence on ecosystem structure and dynamics in the High Arctic than in the Low Arctic, where shrubs are already a dominant component of many plant communities. Increases in shrubs with more upright growth form, especially willows, will have a major effect by generating microhabitats not previously present in the High Arctic. The new habitats will make possible the invasion of new insect, mammal and bird herbivores, as well as their parasites and predators.

ACKNOWLEDGMENTS

We appreciate the assistance and contributions provided by Christian Bay, Per Mølgaard, Fiona Danks, David Murray, Mads C. Forchhammer, Henning Thing, Kirsten Christoffersen, Torben Christensen, Hans Meltofte and an anonymous reviewer. The Danish Polar Center provided logistic services and support.

REFERENCES

Abbott, R.J. and Brochmann, C. (2003) *Mol. Ecol.* **12**, 299–313.

Abbott, R.J., Smith, L.C., Milne, R.I., Crawford, R.M.M., Wolff, K. and Balfour, J. (2000) *Science* **289**, 1343–1346.

ACIA (2005) In: *Arctic Climate Impact Assessment,* Cambridge University Press, New York, 1042pp.

Anthony, R.M. (1997) *Arctic* **50**, 147–157.

Bay, C. (1992) *Meddr. Grønland, Biosci* **36**, 1–102.

Bennike, O. (1999) *Polar Rec.* **35**, 323–336.

Bennike, O. and Anderson, J.N. (1998) *Nord. J. Bot.* **18**, 499–501.

Berg, T.B.G. (2003) The collared lemming (Dicrostonyx groenlandicus) in Greenland: Population dynamics and habitat selection in relation to food quality. PhD thesis, National Environmental Research Institute, Ministry of the Environment, Denmark.

Böcher, T.W. (1951) *J. Ecol.* **39**, 376–395.

Böcher, T.W., Holmen, K. and Jakobsen, K. (1959) *Meddr. Grønland* **163**(1), 32.

Chapin, F.S., III, Shaver, G.R., Giblin, A.E., Nadelhoffer, K.G. and Laundre, J.A. (1995) *Ecology* **76**, 694–711.

Chapin, F.S., III, McGuire, A.D., Randerson, J., Pielke, R., Sr, Baldocchi, D., Hobbie, S.E., Roulet, N., Eugster, W., Kasischke, E., Rastetter, E.B., Zimov, S.A. Oechel, W.C., *et al.* (2000) *Global Change Biol.* **6**, 211–223.

Christiansen, H.H., Bennike, O., Böcher, J., Elberling, B., Humlum, O. and Jackobsen, B.H. (2002) *J. Quat. Sci.* **17**, 145–160.
Crawford, R.M.M. and Balfour, J. (1990) *Flora* **184**, 291–302.
Dahl, E. (1963) In: *North Atlantic Biota and Their History: A Symposium Held at University of Iceland, Reykjavík July 1965* (Ed. by Á. Löve and D. Löve), pp. 173–188. Pergamon Press, London.
Derocher, A.E., Lunn, N.J. and Stirling, I. (2004) *Integr. Comp. Biol.* **44**, 163–176.
Downes, J.A. (1964) *Can. Entomol.* **96**, 279–307.
Ekstam, O. (1897) *Tromsø Museum Aarshefte* **20**, 1–66.
Elberling, H. and Olesen, J.M. (1999) *Ecography* **22**, 314–323.
Elliott, H.W. (1882) *Monograph of the Seal Islands of Alaska.* Fisheries of the United States, U.S. Department of the Interior, Washington, DC.
Ellstrand, N.C. and Elam, D.R. (1993) *Annu. Rev. Ecol. Syst.* **24**, 217–242.
Elmquist, T., Ericson, L., Danell, K. and Salomonson, A. (1988) *Oikos* **51**, 259–266.
Fredskild, B., Bay, C., Feilberg, J. and Morgensen, G.S. (1992) In: *Grønlands Botaniske Undersøgelse 1991,* Botanisk Museum, University of Copenhagen. pp. 18–25.
Friðriksson, S. and Sigurðsson, H. (1969) *Náttúrufræðingurinn* **39**, 32–40.
Funder, S. (1979) *Palaeogeogr. Palaeocl.* **28**, 279–295.
Gelting, P. (1937) *Meddr. Grønland* **116**(3), 196.
Graae, B.J., Pagh, S. and Bruun, H.H. (2004) *Arct. Antarct. Alp. Res.* **36**, 468–473.
Halliday, G., Kliim-Nielsen, L. and Smart, I.H.M. (1974) *Meddr. Grønland* **199**(2), 47.
Harper, J.L. (1977) *Popul. Biol. Plants.* Academic Press, London.
Hjältén, J. (1992) *Oecologia* **89**, 253–256.
Hocking, B. (1968) *Oikos* **19**, 359–388.
Hultén, E. (1962) *Svensk Bot. Tidskr.* **56**, 362–364.
Høye, T.T, Ellebjerg, S.M. and Philipp, M. (2007) *Arct. Antarct. Alp. Res.* **39**, 412–421.
Ikonen, A. (2001) Leaf beetle feeding patterns on, and variable plant quality in, Betulaceous and Salicaceous hosts. PhD thesis, Department Biology, University of Joensuu, Finland.
Iversen, J. (1952) *Oikos* **4**, 85–103.
Johansen, S. and Hytteborn, H. (2001) *J. Biogeogr.* **28**, 105–115.
Kapel, C.M.O. (1999) *Arctic* **52**, 289–293.
Kevan, P.G. (1970) High Arctic insect-flower relations: The interrelationships of arthropods and flowers at Lake Hazen, Ellesmere Island, Northwest Territories, Canada. PhD thesis, University of Alberta, Edmonton, Alberta, Canada.
Kevan, P.G. (1972) *J. Ecol.* **60**, 831–847.
Kevan, P.G. (1973) *Anz. Schädlingskd. Pfl.* **46**, 3–7.
Klein, D.R. and Bay, C. (1991) *Holarctic Ecol.* **14**, 152–155.
Klein, D.R. and Bay, C. (1994) *Oecologia* **97**, 439–450.
Klein, D.R., Bay, C. and Danks, F.S. (1998) *Herbivore influences on reproductive strategies of arctic willow, Salix arctica.* Report to National Geographic Society.
Lundgren, R. and Olesen, J.M. (2005) *Arct. Antarct. Alp. Res.* **37**, 514–520.
Löve, D. (1963) In: *North Atlantic Biota and their History: A symposium held at University of Iceland, Reykjavík July 1962* (Ed. by Á. Löve and D. Löve), pp. 189–205. Pergamon Press, London.
Meltofte, H. (1983) *Polar Res,* **1** (n.s.), 185–198.
Mosquin, T. and Martin, J.E. (1967) *Can. Field Nat.* **82**, 201–205.
Nathorst, A.G. (1892) *Englers Botanische Jahrbücher* **14**, 183–221.
Nordal, I. (1987) Tabula rasa after all? *J. Biogeogr.* **14**, 377–388.
Nyman, T. and Julkunen-Tiitto, R. (2000) *Proc. Natl. Acad. Sci. USA* **97**, 13184–13187.

Olesen, J.M. (1987) *Can. J. Bot.* **65**, 1509–1513.
Ostenfeld, C.H. (1926) *Biol. Med.* **6**, 1–71.
Pakeman, R.J. (2001) *J. Biogeogr.* **28**, 795–800.
Parker, J.D., Burkepile, D.E. and Hay, M.E. (2006) *Science* **311**, 1459–1461.
Philipp, M. and Siegismund, H.R. (2003) *Mol. Ecol.* **12**, 2231–2242.
Savile, D.B.O. (1972) *Res. Branch Monogr.* **6**, 1–81.
Sturm, M., Schimel, J., Michaelson, G., Welker, J.M., Oberbauer, S.F., Liston, G.E., Fahnestock, J. and Romanovsky, V.E. (2005) *Bioscience* **55**, 17–26.
Tackenberg, O., Poschlod, P. and Bonn, S. (2003) *Ecol. Monogr.* **73**, 191–205.
Talbot, S. and Shields, G.F. (1996) *Mol. Phylogenet. Evol.* **5**, 477–494.
Wang, B.C. and Smith, T.B. (2002) *Trends Ecol. Evol* **17**, 379–386.
Warming, E. (1888) *Meddr. Grønland* **12**, 223.
Willson, M.F. (1993) *Oikos* **67**, 159–176.
Willson, M.F., Traveset, A. and Sabag, C. (1997) *J. Field Ornithol.* **68**, 144–146.

The Study Area at Zackenberg

HANS MELTOFTE AND MORTEN RASCH

SUMMARY

The Zackenberg Research Station is situated in central Northeast Greenland (74°30′N, 20°30′W) and is open during June–August. The climate is high arctic, and the study area is mountainous with deep valleys and fjords. The site was chosen because of its position close to the northern limit of extensive vegetation cover in the lowlands and "midway" between the cool and foggy outer coast and the arid inland. Furthermore, the area holds a wide variety of physical landscape features and biota.

The main study area is a 2–3 km wide valley, Zackenbergdalen, where the research station is situated close to a small runway. The management of logistics, monitoring and research is organised under the umbrella Zackenberg Ecological Research Operations (ZERO), housed at the Danish Polar Center. The long-term monitoring programmes are run by governmental research departments and institutes.

I. ZACKENBERG IN THE HIGH ARCTIC

Greenland is the largest island in the world. It extends 2670 km from Kap Farvel in the south to Peary Land in the north as the northernmost land area in the world. About 82% of Greenland is covered by the up to 3500 m thick ice sheet, "The Greenland Ice Sheet" or "Indlandsisen" in Danish. Glacial processes during several Quaternary glaciations have formed the large-scale

ADVANCES IN ECOLOGICAL RESEARCH VOL. 40
© 2008 Elsevier Ltd. All rights reserved
0065-2504/08 $35.00
DOI: 10.1016/S0065-2504(07)00005-0

landscape in the ice-free part of the country surrounding the ice sheet. Mountainous landscapes with peaks of up to 3693 m together with valleys and deep fjords are found in all the ice-free parts of Greenland, while extensive lowland areas are few.

Greenland can largely be divided into a northern high-arctic area and a southern low-arctic area each covering about half the ice-free part of the country. A small sub-arctic area is found near the southernmost tip (Figure 1 and Box 1).

Because of the often several hundred kilometre wide belt of polar pack ice, "Storisen," drifting down along the east coast of Greenland with the East Greenland Current, the high-arctic zone extends much further south in

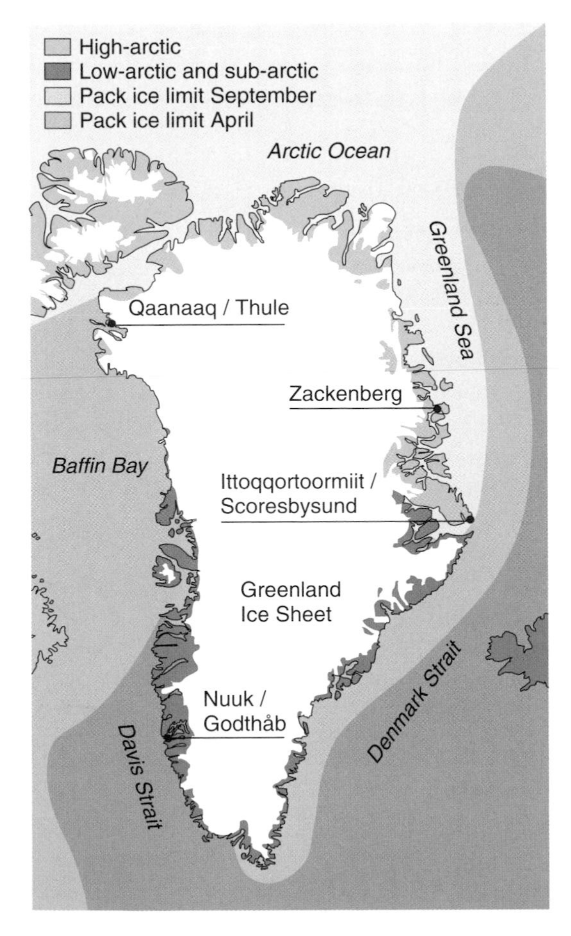

Figure 1 Map of Greenland with climate zones and extent of polar pack ice in spring (April) and autumn (September).

Box 1

What is the Arctic?

The name Arctic derives from ancient Greek *Arktikós*, the land of the Great Bear, which is the star constellation close to the North Star *Arcturus*.

There are several meanings of the Arctic. From a geophysical point of view, the Arctic may be defined as the land and sea north of the Arctic Circle, where there is midnight sun in the summer and winter darkness. But from an ecological point of view, it is more meaningful to use the name for the land north of the tree line, which generally has a mean temperature below + 10–12 °C for the warmest month, that is, July (Jonasson *et al.*, 2000). With this definition, the arctic land area comprises about 7.5 million km^2, or some 5.5% of the land surface on Earth. Similarly, the arctic sea is the area with annual ice cover at least in parts of the year, comprising the Arctic Ocean with adjacent seas.

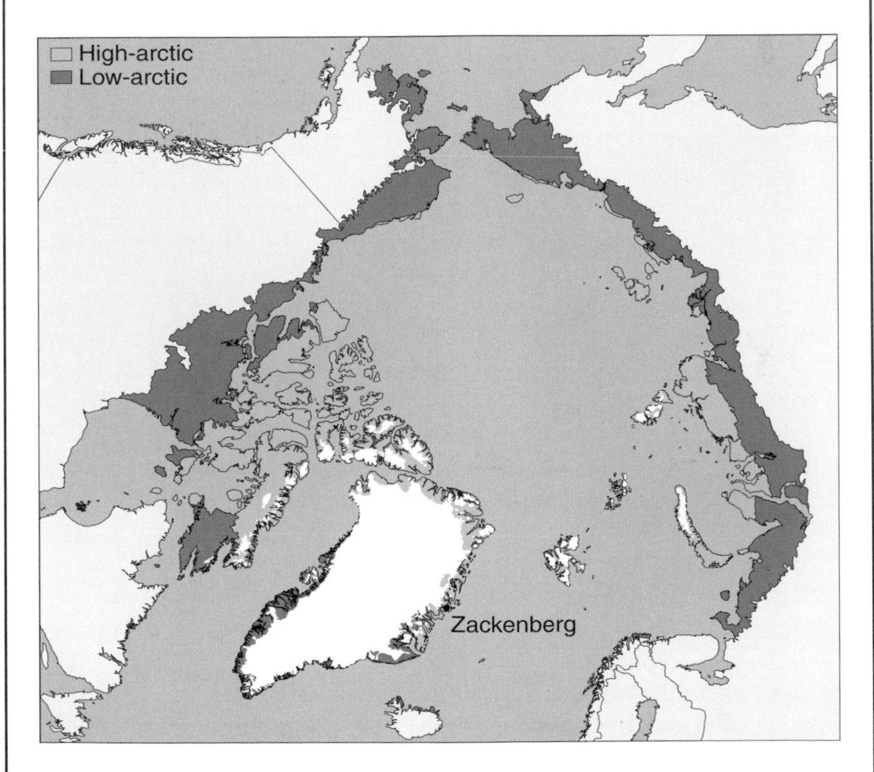

Box Figure 1 The low- and high-arctic zones with the position of Zackenberg.

(continued)

Box 1 *(continued)*

The Arctic may be divided into a number of subzones based on floristic types (CAVM Team, 2003). Here, the division between the High Arctic and the Low Arctic is most relevant, and we use the separation given by Bliss (1997), meaning that the divide is close to a mean of + 6 °C for the warmest month. According to these criteria, the extension of the two arctic zones is shown in Box Figure 1.

The vegetated lowland of the Arctic is often named tundra, which originates from Finnish *tunturi*, meaning land without forest. In general, the Low Arctic has much more lush vegetation than the High Arctic, where large lowland areas may be almost devoid of vegetation, like the arctic deserts of the northernmost lands in the world.

The subarctic is the northernmost part of the Boreal zone, that is, the area between the timberline and the tree line. Hence, the subarctic is not part of the Arctic, like the subtropics are not part of the tropics. Like the Arctic, the Boreal has its name from Greek, in that Boreas was the god of the cold northern winds and bringer of winter.

These zones often continue to the south—out of the Arctic—as subalpine, low-alpine, and high-alpine zones in mountainous area.

East Greenland than on the west coast. Furthermore, this wide and dense belt of polar pack ice affects the climate in Northeast and North Greenland by making it more continental with very cold winters, little precipitation and sunny summers. Hence, precipitation in the interior arctic deserts of Peary Land in northernmost Greenland is only about 1% (25 mm per year) of the precipitation on the southernmost tip of Greenland (2500 mm per year). The precipitation at Zackenberg is 250 mm per year (Hansen *et al.*, 2008, this volume). Hence, exposed areas without continuous snow-cover during winter become increasingly frequent from south to north.

High-arctic Greenland is largely unpopulated. Only a few weather stations and military facilities are manned year round, while Inuit settlements are found in the southernmost parts, close to the transition to the low-arctic zone both on the east and on the west coast. The vast majority of the Greenland population of about 57,000 people live in towns and settlements on the southwest coast, the so-called "open water area," where harbours may be navigated most of the year.

In contrast to the often more than half meter deep shrubby vegetation in low-arctic Greenland, vegetation in the high-arctic part of the country is normally only a few centimetres and never more than a couple of decimetres high. In accordance with the occurrence of snow-free expanses, large areas of barren land occur in the north (Bay, 1992). Similarly, precipitation and vegetation cover decreases from the outer coast towards the ice sheet. In central Northeast Greenland this distance is 100–250 km.

The Zackenberg study area (74°30′N, 20°30′W) is situated centrally on these north–south and east–west gradients. This involves that the area has about 80% snow-cover in most springs and about 83% vegetation cover at altitudes below 300 m a.s.l. The area is close to the northernmost areas with extensive vegetation cover in lowland (Figure 2), and it is situated "midway" between the outer coast (Zackenberg—outer coast: *c.* 40 km) and the ice sheet (Zackenberg—Greenland Ice Sheet: *c.* 70 km). Hence, most summer days are sunny, but fog occurs regularly, penetrating from the outer coast (Hansen *et al.*, 2008, this volume).

II. THE STUDY AREA

The study area for the terrestrial monitoring and research within the framework of Zackenberg Ecological Research Operations (ZERO) comprises the drainage basin of the river Zackenbergelven. Like most of the ice-free land of Greenland, the Zackenberg study area is mountainous. Several peaks reach 1000–1400 m, and many areas above *c.* 1300 m have permanent ice cover mainly as small local ice caps (see map in Hasholt *et al.*, 2008, this volume). The river Zackenbergelven has a catchment area of *c.* 514 km^2 of which *c.* 20% is glacier covered. The lowland around Zackenberg was deglaciated about 10,000 years ago, and since then the relative land rise has been *c.* 70 m probably with emergence in early and middle Holocene followed by submergence in Late Holocene (Bennike *et al.*, 2008, this volume).

The permafrost reaches an estimated depth of about 300 m, and the average annual maximum depth of the active layer below varies between 40 cm and 70 cm in most years (Christiansen *et al.*, 2008, this volume).

The Zackenberg study area is divided by a fault zone separating areas with Cretaceous and Tertiary sandstones topped by basalts above *c.* 600 m a.s.l. to the east (covering 90 km^2 of the drainage basin to Zackenbergelven) from Caledonian gneissic and granite bedrock to the west (covering 422 km^2 of the drainage basin to Zackenbergelven) (Koch and Haller, 1971).

Besides the rather barren mountains, vegetated valleys and lower mountain slopes make up a significant part of the study area. Most investigations are carried out within the 2–3 km wide valley, Zackenbergdalen, where the research station is situated (Figure 3). From here, the valley Lindemansdalen continues to the north, and the much longer valley, Store Sødal, penetrate A.P. Olsen Land to the west, where it widens up and divides into four tributary valleys headed by glaciers. Situated centrally in the lower part of Store Sødal is the 27 m deep lake Store Sø. From Zackenbergdalen, coastal lowland continues along the south coast of Wollaston Forland towards the

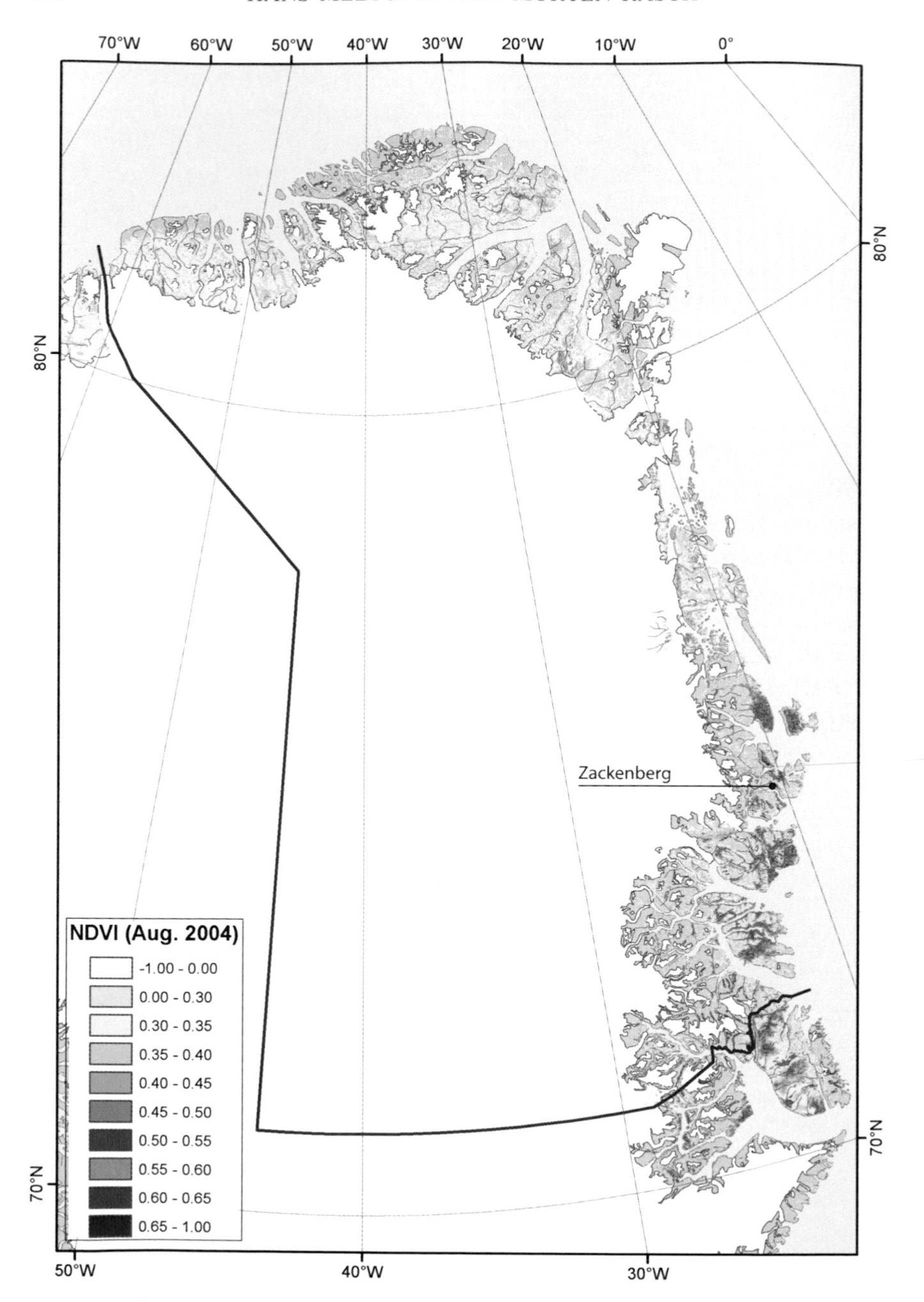

Figure 2 Vegetation greening in high-arctic Greenland (moderated from Aastrup *et al.*, 2005). The red line demarcates the National Park of North and East Greenland.

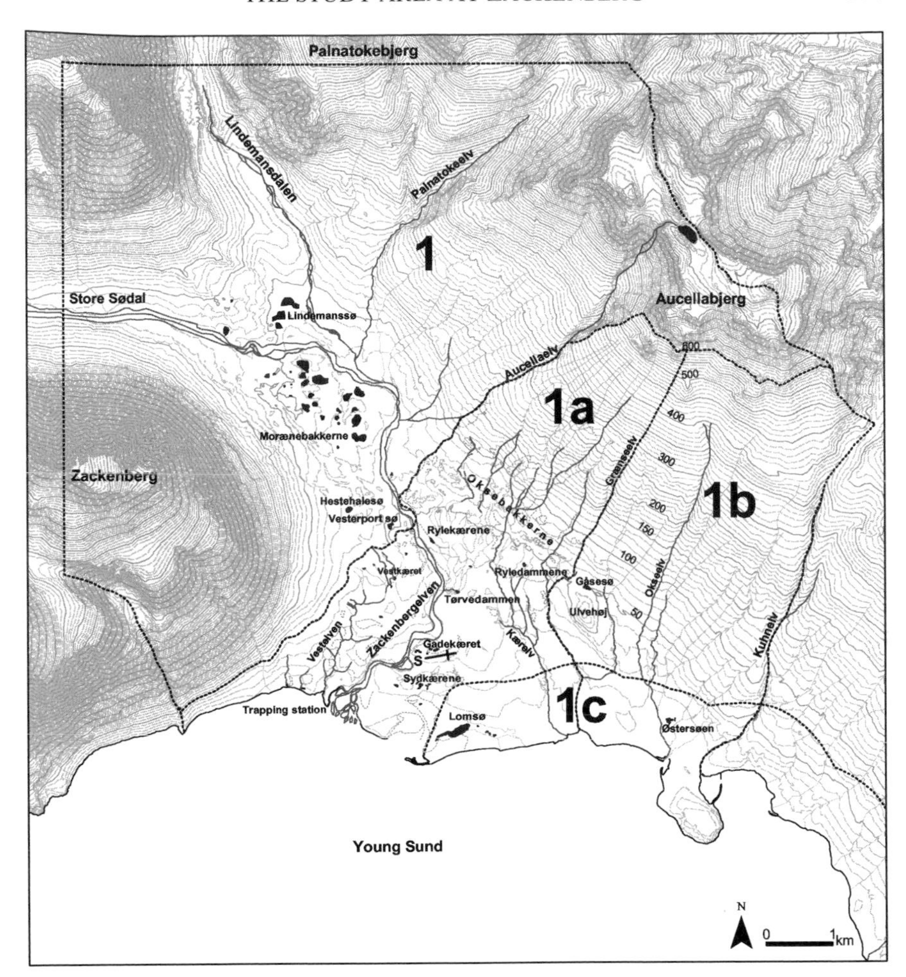

Figure 3 Research zones within the Zackenberg study area with different regulations for traffic and research. Zone 1 comprises the entire valley with adjacent mountain sides, where you are allowed to work alone and without weapons. Zone 1a is the main study area, where most monitoring takes place. Zone 1b is a low-impact study area devoted to research in virtually undisturbed habitats. 1c is a goose protection zone 1 km landward and seaward of the coast east of the old delta of Zackenbergelven, where traffic is minimized between 20 June and 10 August. "S" denotes the position of the research station.

military outpost Daneborg and the outer coast. To the south, the study area is limited by the 1–4 km wide and up to 360 m deep fjords Young Sund and Tyrolerfjord.

Zackenbergdalen itself was selected as the main study area because of its great diversity in physical landscape features, plant communities and other biota. Almost all Greenland high-arctic landforms and biodiversity is represented in the valley, including moraines, scree slopes, rock glaciers, gently sloping rock faces, river beds, alluvial fans, a raised delta, beach terraces, permanent and perennial snow beds and glaciers, several types of ponds and lakes, fens, heath and barren lands (see vegetation maps in Elberling *et al.*, 2008, this volume, and Bay, 1998), together with most known species of plants and animals of Northeast Greenland.

The growing season at Zackenberg is from late May in early snow-free areas, while extensive snow-cover may prevail into mid-June–early July in other areas depending on the characteristics of the season and the snow thickness. Greening peaks between mid-July and early August, and from early September frost and snow may predominate (Ellebjerg *et al.*, 2008, this volume).

Shallow ponds may thaw from late May, while lakes have up to about 1.5 m thick lake ice and may be ice covered well into July. Similarly, the fjord ice may reach a thickness of 1.5 m and does not break up until early–late July (Rysgaard *et al.*, 2003; Sigsgaard *et al.*, 2006). Ponds and lakes often freeze over during September, while the fjords may stay open until late autumn.

III. THE MANAGEMENT OF MONITORING AND RESEARCH WITHIN THE STUDY AREA

Research, monitoring and logistics at Zackenberg are coordinated by the research programme ZERO. Research and monitoring within the Zackenberg research area is managed to secure an appropriate coordination both between individual research projects and between research and monitoring. Further, the management of ZERO prevents mutual disturbance between individual research projects and between logistics and research/monitoring. All regulations are described in the Zackenberg Site Manual, which is available on the Internet (www.zackenberg.dk). These regulations are intended only as a guide for station personnel and have no legal status.

The research area at Zackenberg has been divided into several different zones mainly to allow for a different protection level in relation to disturbances within the different areas of interest (Figure 3). Most research and monitoring takes place in research zone 1a (20 km^2), where a large number of different plots run by the BioBasis monitoring programme have been established within a radius of 1 km around the climate station situated *c.* 600 m east of the research station. Similarly, most installations and plots relating to the GeoBasis programme are situated close to the climate station. Particularly the areas south of the runway are open to manipulative research, since these areas are already disturbed by blowing dust from the runway during winter, changing the albedo and hence the snowmelt in spring.

Research zone 1b is a "low-impact study area" of 16 km^2 reserved for studies, which rely on virtually undisturbed areas or periods, while research zone 1c is a goose moulting and breeding reserve of *c.* 5 km^2 with almost no traffic during 20 June–10 August to avoid disturbance of moulting flocks and family groups of barnacle geese. During the first 3 years of our stay at Zackenberg, hundreds of pink-footed geese moulted in this zone, but they have disappeared probably because of disturbances relating to our presence at the research station. No other serious effects of our activities have been recorded.

Camping is not allowed within research zone 1, while zone 2 is open to camping. Motorised traffic is limited to snow-cover in spring and two tracks close to the research station during the rest of the season.

Each year, the number of man-days in each research zone is reported in the ZERO Annual Reports, and so is sampling of specimens together with character and location of manipulative research projects. Man-days in the terrain in Zackenbergdalen (zone 1a) have ranged between 400 per season and 1200 per season, while visits in the low-impact study area and the goose reserve have amounted to 30–90 and 1–14, respectively, during 1996–2005.

Sewage from kitchen and bathroom facilities from 800 man-days per season to 1450 man-days per season is drained into river Zackenbergelven below the research station, and so are biologically degradable leftovers from the kitchen. Solid and chemical waste is flown out of the area, while combustible waste (paper etc.) is burnt at the station.

Sampling of specimens other than plants and invertebrates has only occurred on a very limited scale, and all such sampling together with manipulative projects has to be approved and registered by the ZERO Working Group before the field season.

IV. THE RESEARCH STATION

Zackenberg Research Station was established in 1995. It is owned and maintained by the Danish Polar Center and consists of six wooden houses with a total area of 275 m^2 under roof. The houses hold mess and kitchen; different laboratories; offices for visiting scientists, for monitoring programme staff and for logisticians; bathing facilities; storage room for provision and bedrooms for the staff staying at the station throughout the summer. Visiting scientists, monitoring programme staff and logisticians staying at the station for shorter periods are accommodated in large insulated tents, the so-called Weatherhaven shelters. A workshop is also situated in a large Weatherhaven shelter. In 2006, a new accommodation building and a house with garage, workshop and room for generators were erected.

The station provides electrical power (220 V, 50–60 Hz) and different qualities of water both for the running of the station and for research purposes. Zackenberg Research Station is normally open for scientists during June–August. In this period the station is staffed by two to three logisticians and a cook, all employed by the Danish Polar Center.

The Danish Polar Center provides research travels to Zackenberg as package solutions including all relevant transport from Denmark to/from Zackenberg and accommodation including full board at the station. Transport to/from Zackenberg is carried out from Iceland with chartered Short Take-Off and Landing (STOL) aircrafts, which land on the station's own 450 m runway situated only *c.* 100 m from the other station facilities.

The Danish Polar Center also provides relevant field and laboratory equipment, equipment for transport of scientists and heavy research equipment within the study area (rubber boats and an all-terrain-vehicle) and different equipment necessary to secure the safety of the scientists visiting the station (including rifles, radios, different types of satellite telephones, etc.).

ACKNOWLEDGMENT

We are grateful to Mikkel P. Tamstorf for providing the vegetation greening map.

REFERENCES

Aastrup, P., Egevang, C., Lyberth, B. and Tamstorf, P. (2005) *Naturbeskyttelse og turisme i Nord- og Østgrønland.* Danmarks Miljøundersøgelser, Faglig rapport fra DMU Nr. 545, 131 pp.

Bay, C. (1992) *Meddr Grønland, Bioscience* **36**, 102 pp.

Bay, C. (1998) *Vegetation Mapping of Zackenberg Valley Northeast Greenland.* Danish Polar Center and Botanical Museum, University of Copenhagen, 29 pp.

Bliss, L.C. (1997) In: *Polar and Alpine Tundra* (Ed. by F.E. Wielgolaski), Vol. 3, pp. 551–683. Elsevier, Amsterdam.

CAVM Team (2003) *Circumpolar Arctic Vegetation Map. Scale 1:7,500,000.* Conservation of Arctic Flora and Fauna (CAFF) Map No. 1, U.S. Fish and Wildlife Service, Anchorage, Alaska.

Jonasson, S., Callaghan, T.V., Shaver, G.R. and Nielsen, L.A. (2000) In: *The Arctic, Environment, People, Policy* (Ed. by M. Nuttall and T.V. Callaghan), pp. 275–313. Harwood Academic Publishers, Newark, New Jersy.

Koch, L. and Haller, J. (1971) *Meddr Grønland* **183,** Plate 2.

Rysgaard, S., Vang, T., Stjernholm, M., Rasmussen, B., Windelin, A. and Kiilsholm, S. (2003) *Arctic, Antarctic, and Alpine Research* **35**, 301–312.

Sigsgaard, C., Petersen, D., Grøndahl, L., Thorsøe, K., Meltofte, H., Tamstorf, M.P. and Hansen, B.U. (2006) In: *Zackenberg Ecological Research Operations, 11th Annual Report, 2005* (Ed. by A.B. Klitgaard, M. Rasch and K. Caning), pp. 11–35. Danish Polar Center, Ministry of Science, Technology and Innovation, Copenhagen.

Present-Day Climate at Zackenberg

BIRGER ULF HANSEN, CHARLOTTE SIGSGAARD,
LEIF RASMUSSEN, JOHN CAPPELEN, JØRGEN HINKLER,
SEBASTIAN H. MERNILD, DORTHE PETERSEN,
MIKKEL P. TAMSTORF, MORTEN RASCH AND BENT HASHOLT

SUMMARY

At Zackenberg (74°30′N, 20°30′W), the polar night lasts 89 days and the polar day 106 days. During the polar night, when solar energy is low or equal to zero, radiative cooling of the snow-covered surface leads to monthly mean air temperatures below −20 °C and daily minimum temperatures well below −30 °C often occur. Calm and weak winds (<2 m/s) combined with strong low-level temperature inversions are present 55–79% of the time during the winter months, when northerly winds dominate. Cyclone activity over the Greenland Sea or over the Greenland Ice Sheet often takes place during the winter season and can generate pronounced changes in the weather, which is the main factor producing the large temporal variability during the cold period. Within a few hours, the air pressure can drop drastically and the wind can increase well above 20 m/s. The temperature rises rapidly because of destruction of the cold bottom layer. Once or twice during the snow-covered period, strong foehn events can make the temperature reach the melting point and can stay well above it for several hours even in the core of the winter.

ADVANCES IN ECOLOGICAL RESEARCH VOL. 40 0065-2504/08 $35.00
© 2008 Elsevier Ltd. All rights reserved
DOI: 10.1016/S0065-2504(07)00006-2

In the snow-free summer period with continuous daylight all around the clock, an almost constant weak sea breeze from south to southeast dominates the daytime weather at Zackenberg. The mean monthly air temperatures (MMATs) vary between 3 °C and 7 °C in July and August, and the air temperature rarely gets below zero for 4–6 weeks during the warmest part of the summer season. Foehn events can occur and the wind speed can suddenly increase up to 20 m/s, when the air temperature remains above 20 °C for several hours, and the relative humidity drops well below 40%.

Precipitation at Zackenberg is strongly influenced by the cyclone activity. For the hydrological year (October 1 to September 30), the average annual accumulated precipitation was 261 mm w.eq. (water equivalent) for the years 1996–2005, of which mixed precipitation or sleet accounted for 17 mm w.eq. (7%) primarily from May until November, while liquid precipitation was 27 mm w.eq. (10% of total) and occurred only from June to September.

The observational time series have been used to downscale the datasets from the National Centers for Environmental Prediction/National Center for Atmospheric Research (NCEP/NCAR) reanalysis in order to obtain longer time series and to analyse these time series for both spatial and temporal variability at various scales. For the period 1958–2005, the NCEP/NCAR datasets show a significant increase of 1.9 mm/year in the annual accumulated precipitation. The average for the whole period 1958–2005 was 178, 15 and 22 mm w.eq. for solid, mixed and liquid precipitation, respectively. The spatial variability reveals a strong overall precipitation gradient along the east coast with nearly 2000 mm w.eq./year in the southern part (61°N) decreasing to less than 100 mm w.eq. in the northern part (83°N) (Cappelen *et al.*, 2001).

A downscaled time series of air temperatures from the Climate Research Unit (CRU) at Tyndall Centre for Climate Change Research shows a strong annual variability in the seasonal temperatures for the period 1901–2005 and a significant annual warming of 2.25 °C during the latest 15 years period (1991–2005), with the five warmest years (2005, 2004, 2002, 1996 and 2003) during the last century, all occurring within the last 10 years. The air temperature also reveals a strong latitudinal gradient along the east coast of Greenland with −1.41 °C/latitude in the coldest month (February) and a very weak gradient of only −0.12 °C/latitude in the warmest month (July).

Two external factors have a great influence on the spatiotemporal variability of the climate at Zackenberg. The annual variability in the export of Arctic Ocean sea ice through the Greenland Sea and the annual variability in the North Atlantic Oscillation (NAO) both have a great influence on the frequency of cyclones passing Zackenberg, especially during the snow-covered wintertime. Cold, dry winters (<100 mm w.eq.) seem to be caused by an above normal sea-ice export along the coast and by tree or more negative NAO months, while warm/wet winters (>200 mm w.eq.) seem to be caused by below normal sea-ice export in the Greenland Sea and tree or more positive NAO months.

I. INTRODUCTION

The arctic climate is one of the most extreme and important components within the global climate system, and like all other climate systems it exhibits variability over a wide spectrum of timescales. A fundamental problem in understanding climate variability and change is that the different system components have different response time and interact through various feedback processes. The components are never in equilibrium.

During the last 10–15 years, there has been an increased scientific consensus that the climate is likely to change within our lifetime because of anthropogenic forcing (ACIA, 2005; IPCC, 2007). Such change will take the form of a significant global warming, which is expected to be most pronounced at polar latitudes (IPCC, 2007). To monitor the expected warming and the influence on the ecosystems, numerous long-term research programs have been established all over the northern polar region within the last 10–15 years. In such a short period, data from a small network of modern automatic meteorological stations can offer timescales of daily, seasonal, and yearly variations within the study areas, but timescales of multiple decades are quite more limited in the climate databases of the Arctic (Serreze and Barry, 2005).

The modern basis of arctic science, and meteorology in particular, was established in connection with the first International Polar Year (IPY) in 1882–1883. An outcome of the IPY was the first circumpolar network of 12 principal meteorological stations, among which the newly established meteorological station at Nuuk/Goodthåb, West Greenland, was included. In connection with the International Geophysical Year in 1957–1958, the circumpolar network of meteorological stations was extended considerably, and newly established meteorological stations along the northeastern coast of Greenland (Station Nord, Danmarkshavn and Daneborg) were included in the network. During the 1970s, several manned meteorological stations (e.g., Daneborg) were replaced by automatic meteorological stations with the risk of breaks in the time series due to power failure, icing up of sensors, or damages caused by strong winds. In some respects, the losses have been balanced by improved satellite remote sensing capabilities. Routine arctic satellite coverage began in the 1970s with the first of an ongoing series of the National Oceanic and Atmospheric Administration (NOAA) Television Infrared Observation Satellites (TIROS) carrying Advanced Very High Resolution Radiometers (AVHRR) and TIROS Operational Vertical Spectrometers. The visible and infrared channels of these platforms provided for the first time useful satellite-based assessments of sea ice, snow and cloud extent and variability within the entire arctic region.

In the following years, numerous satellites with both passive and active radar systems have provided higher-resolution coverage of the arctic region, and new and more powerful computer systems offer the opportunity of a constantly updated database such as the National Centers for Environmental Prediction/National Center for Atmospheric Research (NCEP/NCAR). Outputs based on this numerical weather prediction/data assimilation system are now available back to 1958. Although the temporal resolution of the NCEP/NCAR database now is 6 hour, the spatial resolution of 2.5° is too coarse for most ecosystem studies such as the Zackenberg Ecological Research Operations (ZERO). To utilize the potential of such databases or to incorporate longer time series from neighbouring meteorological stations into ecosystem studies in a much smaller geographical scale in mountainous regions like at Zackenberg, it is vital to know the spatiotemporal variations of the most prominent climate parameters, since it is well known that the Arctic encompasses extreme climatic differences, which vary greatly by location and season. Knowing these daily, seasonal and yearly variations of the climate at a specific location such as Zackenberg is of great importance, since these variations can be much larger and can happen more rapidly than global warming trends and thus can be much more crucial for the understanding of the arctic ecosystems' ability of survival in the future. This chapter outlines the most prominent parameters of the present climate at Zackenberg and focuses on the short-term spatiotemporal variations of these parameters within the valley Zackenbergdalen and along the east coast of Greenland.

II. THE STUDY AREA AND THE METEOROLOGICAL STATIONS

Climate variables have been recorded at Zackenberg ever since the start of the research station in August 1995 in a cooperation between Asiaq, Greenland Survey and the Institute of Geography, University of Copenhagen, as part of the ClimateBasis and GeoBasis programs. Before that, short and sporadic climatic measurements were performed by Danish and Norwegian research teams shortly before World War II, and during the war, German meteorological stations were operating northeast of the area for a short period. In 1946, a meteorological station was established at Daneborg (74°18′N; 20°13′W, 12 m a.s.l.), situated in the outer part of Young Sund (Cappelen *et al.*, 2001), 23 km south-southeast of Zackenberg. However, continuous and long-term climatic records were not initiated until 1958 from where the Danish Meteorological Institute has digitised data. In August

1975, the meteorological station was closed, whereupon an automatic meteorological station was establish at Daneborg in January 1979. This station has been operating continuously ever since.

The meteorological station at Zackenberg is located on a remnant of a meltwater plain covered by homogeneous white arctic bell-heather *Cassiope tetragona* heath 38 m a.s.l. and 2 km from the coast of Young Sund (Meltofte and Thing, 1996). This location was chosen because it is representative for large parts of the landscape and vegetation of the valley floor, and because it is in the central part of the study area. In order to achieve and secure the best and most complete set of meteorological data, the station consists of two nearly identical masts. Each has separate power supply and sensors recording air temperature, ground temperature, relative humidity, precipitation, air pressure, global radiation and net radiation (all logged once every hour) together with wind direction and wind speed (logged once every 10 min) (Figure 1). The variation in snow depth has been measured every third hour since 1997.

In order to monitor the spatial variations in snow-cover and local microclimate, two micrometeorological stations (M2 and M3) were established in the summer of 2003. Station M2 was placed 17 m a.s.l. on a south-facing slope on the border between an upper zone of white arctic bell-heather and a lower zone of arctic willow *Salix arctica* snow-bed vegetation. The location was chosen as one of the first places where snow accumulates in early autumn. The snow accumulation can reach 2–3 m, and the snow can persist until mid- or late summer. Station M3 was placed 420 m a.s.l. on the southwest-facing slope of Aucellabjerg, just above the mean altitude of the lower part of the inversion layer. The station is placed on the dry part of a solifluction area, fairly densely vegetated and dominated by arctic willow and nard sedge *Carex nardina*. To monitor temperature gradients within the catchment, one temperature sensor was established in Store Sødal in August 2000 and another temperature sensor 1278 m a.s.l. on Dombjerg in August 2004 (Figure 1).

III. SOLAR RADIATION, CLOUD COVER AND RADIATION BALANCE

Geographical latitude is one of the main factors determining the weather and climate anywhere, especially in the Arctic. No matter how the arctic region is defined, its location in high latitudes limits significantly the magnitude of energy received from the sun. In regions lying north of the Arctic Circle, the most unique feature is the occurrence of seasonal day and night. At Zackenberg, the polar night lasts 89 days from November 7 to February 3, while the polar day lasts 106 days from April 30 to August 13.

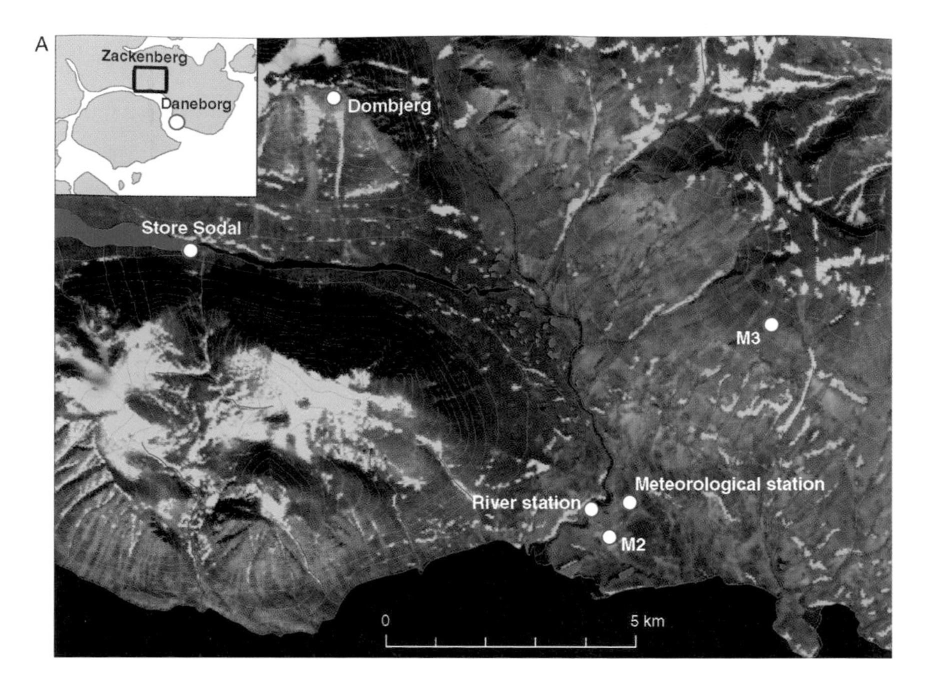

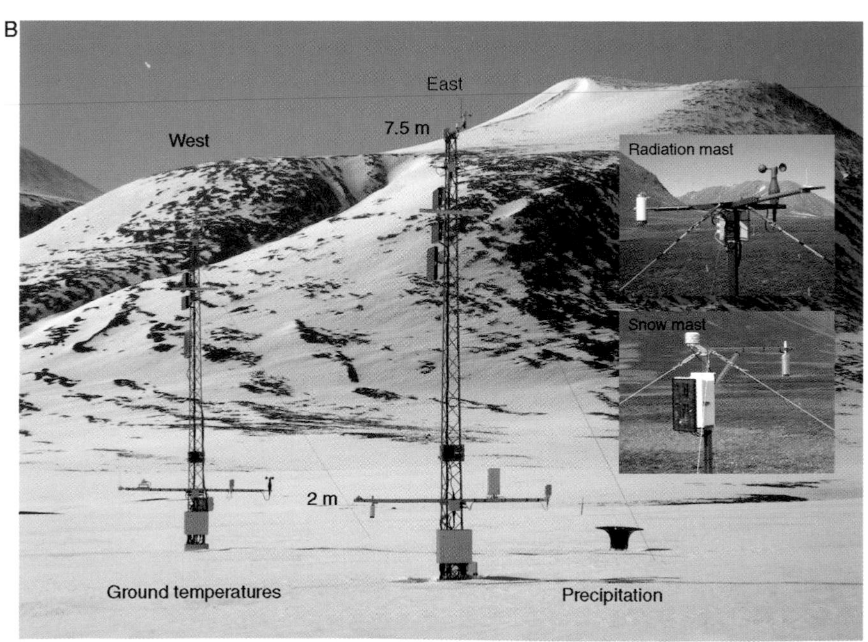

Figure 1 The automatic climate station (Climate St.) was established in August 1995 and consists of two 7.5 m high masts (west and east) with an identical sensor configuration to secure unbroken time series. The Belfort precipitation gauge with its Nipher

The surface radiation balance can be expressed as

$$Q^* = S_n + L_n = S\downarrow - S\uparrow + L\uparrow - L\uparrow = S\downarrow(1-\alpha) + L_n$$

where $S\downarrow$ is incoming solar radiation or global radiation, $S\uparrow$ the reflected solar radiation, the albedo (α) is the ratio of reflected to incoming solar radiation ($S\uparrow/S\downarrow$), while $L\downarrow$ and $L\uparrow$ are the incoming and outgoing terrestrial long wave fluxes, respectively. Net radiation, Q^*, is therefore the surplus or deficit of energy at the surface resulting from net shortwave, S_n and net long wave, L_n, radiative fluxes. The global radiation at any surface can be calculated through considerations of solar altitude and day length as well as the climatological variables of air mass transmissivity and attenuation by cloud. During the period 1996–2005, the global radiation at Zackenberg showed large daily variations (Figure 2) from 0 W/m^2 in the polar night period to a maximum of 390 W/m^2 on June 10, 1996. The extraterrestrial solar irradiance, S_{ex}, at the top of the atmosphere can easily be calculated on the basis of latitude and time of year, while $S\downarrow_{max}$ (Figure 2) might represent days with close to 0% cloud cover, $S\downarrow_{min}$ might represent days with close to 100% cloud cover and the transmissivity, τ, is the ratio of $S\downarrow_{max}/S_{ex}$.

Although the optical path length for direct solar radiation is longer for winter months, clear sky transmissivity during the cold and dry winter period (0.8–0.9) was higher at Zackenberg than during the warm/wet summer period (0.6–0.7) (Figure 3). These ranges of transmissivity are comparable to previous measurements of τ at arctic sites in Alaska and Canada (Lafleur *et al.*, 1987; Bailey *et al.*, 1989; Wendler and Eaton, 1990).

Cloud cover has first-order impact on the arctic surface radiation balance, and numerous satellite-derived gridded sets have been produced for the purpose of climate change modelling, but there are fundamental problems with satellite retrievals in high latitudes. In the visible wavelengths, clouds and the snow/ice surfaces have essentially the same reflectance, making cloud discrimination very difficult. Infrared measurements are limited by the fact that temperature differences between clouds and the surface are usually small, or the cloud tops may be warmer due to low-level temperature inversion, so cloud cover estimated from either observations or measurements is still the most reliable (Serreze and Barry, 2005). The annual variations in cloud cover at Zackenberg and Daneborg are shown in Figure 3, where three stages can be distinguished: winter, summer and transitional

---

shield is seen north of the two masts. The 1372 m high Zackenberg mountain is seen in the background. The top right inserted picture shows the radiation mast situated to the south of the main masts, while the bottom right inserted picture shows the snow mast situated to the north of the main masts. The location of the automatic weather station at Daneborg and the micrometeorological stations at M2, M3, Dombjerg and Store Sødal are found on the map at the top of the figure.

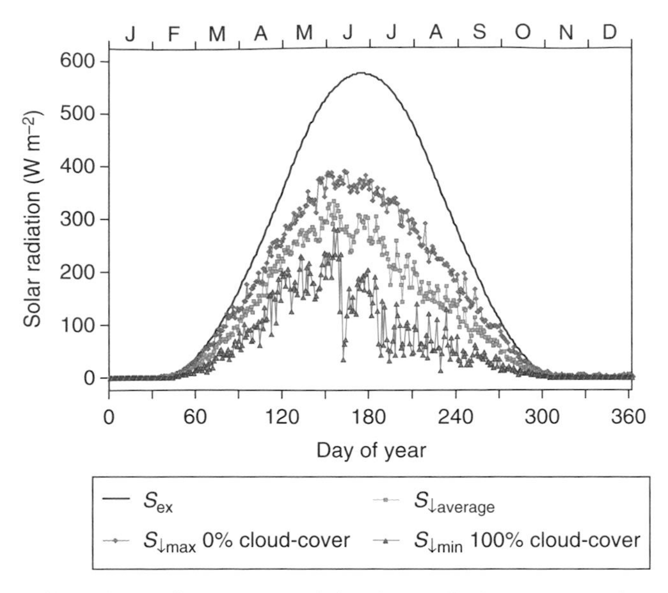

Figure 2 Daily values of extraterrestrial solar radiation, S_{ex}, at the top of the atmosphere and maximum, average and minimum values of global radiation, $S\downarrow$, for the period 1996–2005 where $S\downarrow_{max}$ may represent days with 0% cloud cover and $S\downarrow_{min}$ may represent days with nearly 100% cloud cover.

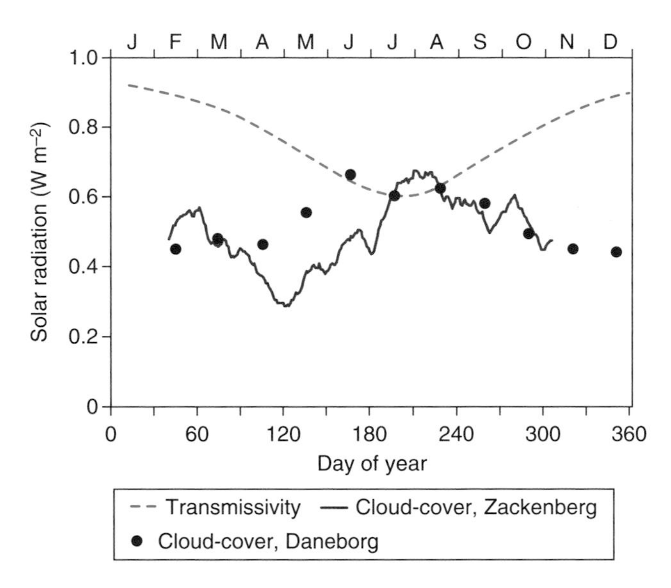

Figure 3 Modelled atmospheric transmissivity, τ, as a function of day of year, DOY, where $\tau = [1.01 - 0.45/\{1 + (\mathrm{DOY} - 199)/100\}^2]$, $r = 0.88$, $p < 0.001$, df $= 270$ and average daily cloud cover for Zackenberg and average monthly cloud cover at Daneborg. In the wintertime, the cloud cover at Zackenberg may correlate with Daneborg and may be ~45%.

(spring and autumn). In the cold half year (from November to April), the mean total cloud cover is clearly at its lowest and oscillates between 40% and 60% dominated by high cirrus clouds. In May, an abrupt increase in cloud cover is observed at both stations, and the highest cloud cover of 60–70% occurs from June to September. Normally, the amount of high cloud cover decreases in spring and summer (Przybylak, 2003), so the spring transition and high summer cloud cover are entirely accounted for by low clouds such as stratus or advective sea fog, which occur when relatively warm, moist air flows in over a cold surface (see Figure 4).

Cloud cover versus latitude along the east coast of Greenland (Figure 5) shows a significant decrease towards north from about 68% at 65°N to 50% at 77°N (Cappelen *et al.*, 2001). The decrease is mainly caused by the poleward decrease in atmospheric temperature and humidity, but also in the cyclone activity along the coast. In the summertime, the average cloud

Figure 4 Terra satellite image from August 21, 2003, at 1340 hrs shows that fog forms when warm humid air is cooled below its dew point off the Northeast Greenland coast. Warm air masses moving north are cooled when they encounter the cold Greenland Sea current, producing dense fog along the outer coast. As the air mass is cooled from below, a strong inversion is created (colder air is trapped below the warmer air) causing the fog to persist. On warm cloud-free days, the prevailing see breeze at Zackenberg (red dot) drags the fog into Young Sund. If the fog reaches Zackenberg (red dot), the sea breeze is weakened or stops and the sun radiation gradually dissolves the fog. (Image courtesy of MODIS Rapid Response Project at NASA/GSFC.)

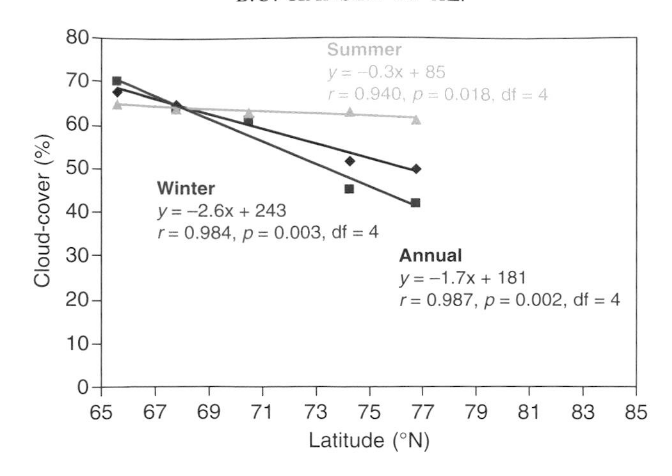

Figure 5 Annual, winter (December–February) and summer (June–August) cloud cover versus latitude along the east coast of Greenland (Cappelen *et al.*, 2001).

cover is almost constant at 63% along the coast, while the variation is greatest in the cold wintertime with a decrease of 2.6%/°lat.

Albedo or reflected solar radiation is another important parameter in the radiation balance. At Zackenberg, the albedo averages 80–90% of the total solar radiation over frozen snow-covered surfaces, and a significant and fast drop in albedo from about 80% to 10% is observed, when the snowmelt begins and the wet/dark tundra surface emerges (Figure 6). During the summer, most of the tundra surface becomes dry, and the albedo slowly increases to 15%. The surface albedo also has a significant effect on the diffuse fraction of global radiation. In the snow-covered period, the cloud cover is low and the air is cold and dry so that the scattering in the atmosphere is relatively low. But the high reflectance from the snow-cover causes the reflected radiance to be scattered once more in the atmosphere, and that is why the diffuse component of the global radiation is nearly 40% in the snow-covered period and only 25% in the snow-free period, which explain the higher $S\downarrow_{min}$ values in the first half of the year in Figure 2.

The effect of the duration and magnitude of the snow-cover on the radiation balance is clearly seen in Figure 6. The snow-cover in 1999 was more than 1 m during the first 5 months of the year, and snowmelt lasted until the first week of July, while the snow-cover in 2004 was half the thickness of 1999, and the snowmelt was complete 3 weeks earlier. Consequently, the tundra surface in 2004 received 370 MJ/m^2 more in the first 7 months than in 1999. The contrast is additionally strengthened by an early snowfall in the autumn of 1999, but the low sun angle in the autumn resulted in only 15 MJ/m^2 more in the last 5 months in 2004 than in 1999.

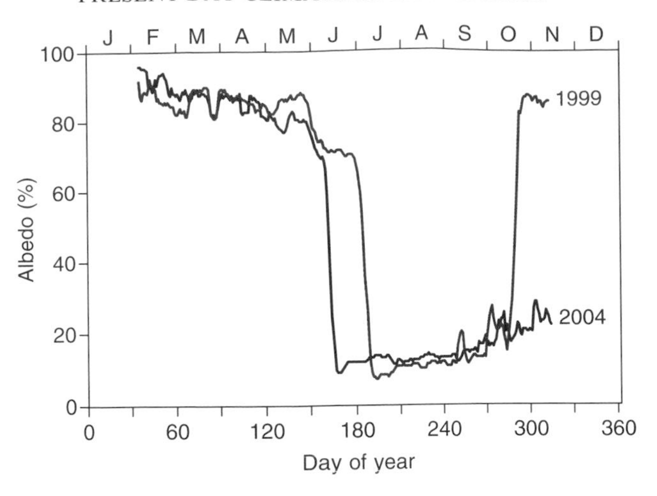

Figure 6 Annual course of the albedo at Zackenberg in two contrasting years. A cold year with a thick snow-cover with a late snowmelt (1999—blue line) and a warm year with a thin snow-cover and an early snowmelt (2004—red line).

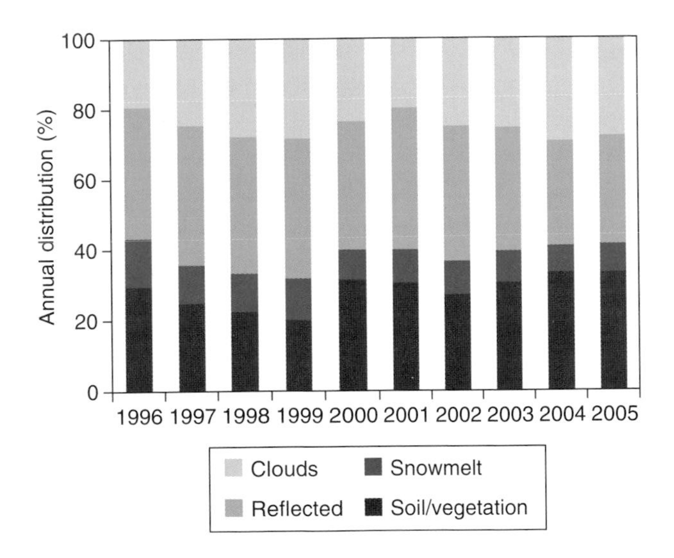

Figure 7 Annual distribution (%) of incoming solar radiation at Zackenberg. The theoretical annual solar radiation from a cloud-free sky is on average 4415 MJ/m^2.

The importance of clouds and snow for the annual radiation balance during the years 1996–2005 is shown in Figure 7. Because of the low cloud cover in spring, only 20–30% of the global radiation was blocked by clouds, while 25–37% was reflected from the surface mainly by the snow-cover, and 8–14% of the incoming solar radiation was used for snowmelt. In years with many clouds and a long period of snow (e.g., the year 1999), only 20% of the incoming solar radiation was used to heat the tundra surface. This percentage may be as high as 33% as observed in 2004 with few clouds and an early snowmelt.

The average annual patterns of net shortwave radiation (S_n), net long wave radiation (L_n) and net radiation (R_n) are presented in Figure 8. As usual, L_n is negative through the year, while R_n is negative only in winter months (October–April) and reaches zero in May, when the cloud cover increases. As for all arctic locations (Bliss, 1997), the net shortwave radiation is the controlling component in the radiation balance, and net radiation does not become positive until the snow melts in June. For most arctic sites, about 50% of the annual solar radiation is received and dissipated before snowmelt. It is no wonder that the growing season starts with such intensity, since the capture of energy by plants starts when the amount of incoming solar energy is at its maximum.

On an annual basis, $S\downarrow = 104$ W/m^2, $S_n = 53.7$ W/m^2, $L_n = -43.7$ W/m^2 and $R_n = 10.0$ W/m^2, which is quite similar to other arctic locations at the same latitude (Bliss, 1997). Only few measurements of the individual parameters in the radiation balance along the east coast of Greenland have been carried out (Cappelen *et al.*, 2001; Mernild *et al.*, 2005, 2007), but they indicate that all parameters exhibit a linear rate of decrease with latitude.

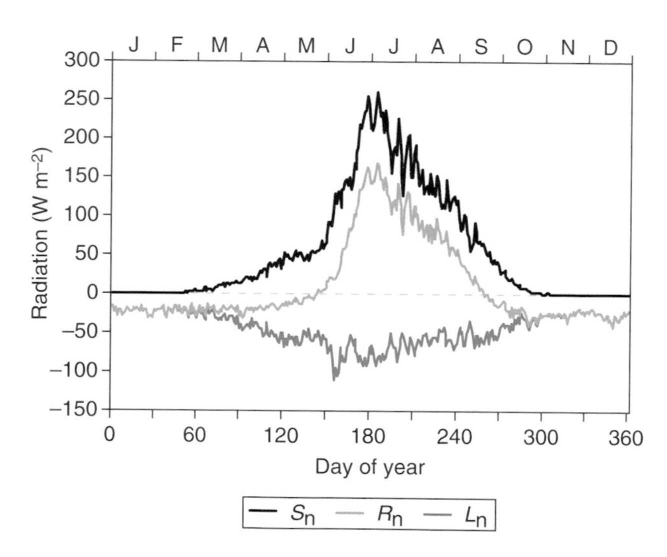

Figure 8 Daily average values of net shortwave radiation, S_n, net long wave radiation, L_n and net radiation, R_n, for Zackenberg during 1996–2005, where $R_n = S_n + L_n$.

For $S\downarrow$ this averages -1.3 W/m²/°lat., whereas for S_n and R_n, it averages, respectively, -1.7 W/m²/°lat. and -1.9 W/m²/°lat. The more rapid decrease in S_n and R_n is due to the longer snow-covered winter and spring seasons to the north and its impact through high-surface albedo. The net long wave radiation loss, L_n, also shows a small decrease northward at a rate of -0.1 W/m²/°lat., which indicates that the influence from the poleward decrease in cloudiness (Figure 5) is almost neutralized by the poleward decrease in radiative surface temperatures. Almost similar poleward gradients have been found in North American studies (Rouse, 2000).

IV. ATMOSPHERIC PRESSURE

There is a pronounced difference between the mean surface pressure patterns in winter and in summer. In the wintry atmosphere, there, typically, is a strong temperature gradient between the warm troposphere over the ocean to the south and east of Greenland and the extremely cold troposphere over the northernmost part of the Canadian Archipelago (the Canadian Cold Pole). A corresponding opposite-directed gradient of the surface pressure causes a strong prevalence of winds from northerly directions on both sides of Greenland, a "winter monsoon." In between, there is a local anticyclone over the Greenland Icecap and the adjacent coastland, being a result of the radiative cooling of the ice/snow surface. Although rather persistent, this pressure system usually is shallow and is limited to the cold surface layer below the inversion, and it can be destroyed very easily (Putnins, 1970). Off West Greenland, it tends to weaken the northerly airflow in the vicinity of the coast, while it strengthens the northerly winds along the east coast, particularly in the marginal zone of the sea-ice belt, where the temperature gradient and with that the pressure gradient is enhanced.

In the summertime, temperatures and even the surface pressure are relatively uniform around most of Greenland. Over the snow-free coastland, a local pressure gradient will arise between the outer coast, cooled by the sea ice, and the inland, warmed by the sun.

All along the coast of Greenland, rapid changes in the atmospheric pressure can occur (Putnins, 1970). They are connected with the passage of migrating high- or low-pressure systems. The greatest daily barometric ranges are found during the winter period because of the very high temperature differences in the atmosphere (Cappelen *et al.*, 2001). The greatest daily barometric range, 43.4 hPa, at Zackenberg in the period 1995–2005 occurred on January 25, 2005, when the air pressure increased from 971.3 hPa to 1014.7 hPa within 22 hours (since all pressures are measured 39 m a.s.l. about 5 hPa should be added to have the 0 m a.s.l. pressure). The greatest daily barometric ranges in the summer period are more moderate and lie just below 16 hPa.

Most lows affecting Greenland arrive from directions between south and west, steered by an upper-level cyclone, the "polar vortex," in winter centred over the Canadian Cold Pole, in summer less pronounced, and now situated over the Arctic Ocean. These patterns are closely related to the North Atlantic Oscillation (NAO) and consequently subject to considerable changes. The track of a pressure system is utmost decisive for the weather conditions at a given position (see examples in Section VI below).

The average air pressure at Zackenberg was 1008.2 hPa for the period 1996–2005. The average monthly atmospheric pressure (Figure 9) was highest in the springtime (April: 1013.9 hPa), while it was lowest around midwinter (February: 1003.8 hPa). This annual variation is of the so-called polar type (Hovmøller, 1947). The spring maximum may be seen as a compensation for a pronounced pressure fall over the rapidly warming continents, with snow and ice remaining all over the Arctic. Also in spring, the intensity and influence of the Atlantic storm systems decrease considerably, and the weather at high latitudes becomes more settled. Contrary to this, the midwinter minimum represents the culmination of the Atlantic storm intensity and influence.

During the summer months (July–August), the mean atmospheric pressure was fairly constant at 1006 hPa, with only slight variation from year to year. In autumn and early winter, the atmospheric pressure increased slightly as the snow-cover builds up. A minimum atmospheric pressure of 953 hPa

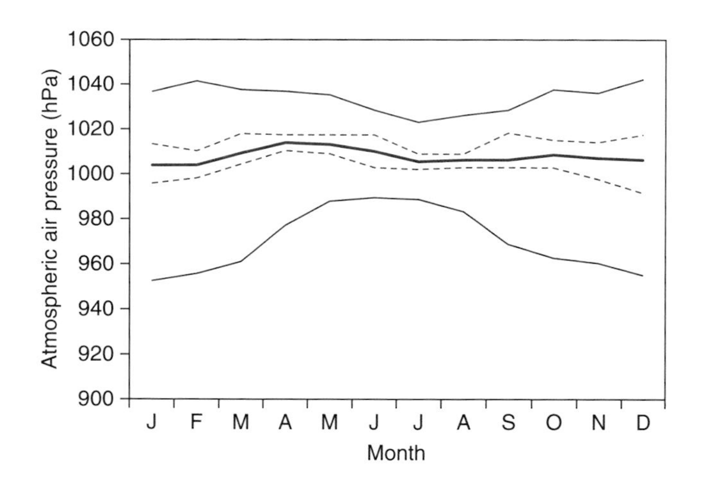

Figure 9 Mean monthly air pressure (bold line), maximum mean monthly air pressure (upper dashed line), minimum mean monthly air pressure (lower dashed line), absolute maximum air pressure (upper thin line) and absolute minimum air pressure (lower thin line) at Zackenberg 1996–2005.

was measured on January 25, 1997. A maximum atmospheric pressure of 1042.5 hPa was measured on December 16, 2001.

V. WIND DIRECTION AND SPEED

Winds are partly the result of large-scale atmospheric circulations. In addition, local factors such as geography, orography and topography (altitude and relief) may significantly influence direction and speed (Przybylak, 2003). To some extent, Zackenberg lies sheltered from the strong winds at the outer coast as well as the strong winds from the ice sheet. The average annual wind speed at Zackenberg is 2.8 m/s compared to 4.5 m/s at Daneborg, but the seasonal variations of both wind speed and wind directions are considerable (Figure 10).

In winter, sea and land surfaces are much the same, since they are both covered more or less continuously in ice and snow. Small values of absorbed solar radiation result from low sun angles, and high-surface albedo over a homogeneous cover of snow will ensure that differences in energy partitioning between marine and terrestrial surfaces are small. After terrestrial snowmelt, the land surface warms, inducing large temperature differences between land and the still frozen sea, and this temperature gradient induces a cold and humid daytime sea breeze with an average speed of 2.7 m/s (Figure 11). As the sea ice melts, the temperature differences between sea and land get smaller, but because of the continuous input of solar radiation, the sea breeze

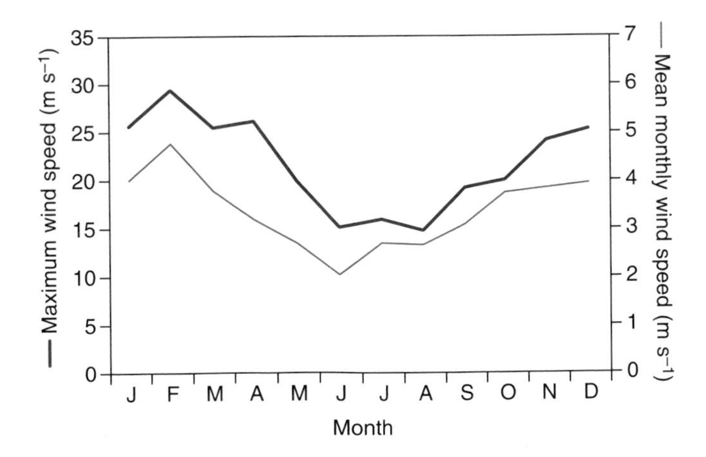

Figure 10 Mean monthly wind speeds (blue line) and absolute maximum wind speeds (red line) at Zackenberg 1996–2005.

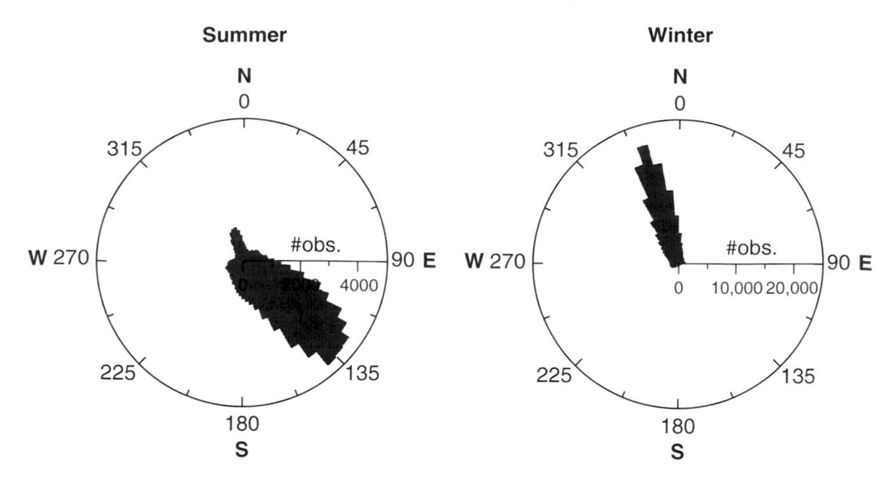

Figure 11 The frequency of wind directions in July (left) and December (right) for the time period 1996–2005. Note different scales in the two diagrams.

will still exist, only in a weaker version with average wind speed close to 2.0 m/s. Such sea breezes have been found all over the Arctic—at Svalbard (Wojcik and Kejna, 1991), in the Ammassalik area, Southeast Greenland (Mernild *et al.*, 2005), on Disko Island, West Greenland (Hansen *et al.*, 2005) and in the Hudson Bay area in Canada (Weick and Rouse, 1991).

During autumn, when the sun for periods disappears behind the highest mountains because of the decreasing sun angle, frost can occur on cloud-free nights already from August. This causes an alteration in the temperature gradient between the land and sea and a change in the wind direction as a land breeze arises. The land breeze is smaller in extent and weaker in intensity (1.5–2.5 m/s) due to the greater stability of the nocturnal atmosphere, and because the land and sea temperature differences are much smaller at night. In fall and early winter, the temperature gradients reverse with cold and snow-covered land and a relatively warm and unfrozen sea. The continuous land breeze gradually transforms to a katabatic (downward going) wind from the ice sheet, which dominates the rest of the winter (Figure 11).

Although holding some of the windiest places of the northern hemisphere, Greenland as a whole has many days with calm (less than 1.5 m/s). At Zackenberg, the calm conditions prevail (30% of the time on an annual basis. In the cold autumn/winter periods, calm prevails only 20% of the time, partly because of the outflow from the ice sheet and partly because the strong topography around Zackenberg generates small local katabatic winds. During the melting season in spring, the isothermal surface causes a pronounced increase of up to 51% of the time with a drop to about 30% in the snow-free summer season.

Local wind systems may be strongly affected by an increased pressure gradient, connected with the passage of weather systems on a synoptic scale. Strong winds connected with such systems have their own patterns, which are very dependent on the topography and on the wind direction in relation to the coast. In Northeast Greenland, the pressure distribution favours northerly winds most of the year except for the short summer period with no snow-cover. Because of friction, northerly winds appear north-north-westerly in the fjords, and the air will more or less be part of the shallow katabatic outflow from the ice sheet, adiabatic warmed (the air mass will be compressed and thereby heated 1 °C for each 100 m of altitude descent) and therefore rather dry. Typically, there will be no low clouds within this flow, but high clouds may occur.

With a pressure gradient initially directed along the coast with the lower pressure to the south, a so-called barrier wind will develop. Typically, this happens in front of a cyclone moving north along the coast, with easterly winds blowing towards the coast where they partly will be lifted up and cause precipitation and partly will be deflected along the coast acting as a barrier, in the direction of the low pressure. In this process, a ridge of high pressure will form over the coast, and the wind will accelerate to a strong northerly low-level jet in a narrow zone along the coast. The associated weather is typically dull with heavy precipitation. In the fjords, the north wind at low levels will appear as a generally more moderate north-north-westerly wind because of friction.

In the Zackenberg area, the wind speed during these situations has been only 8–18 m/s in the summer period, and they lasted only a few hours to 1 day. In winter, the wind may be stronger and the duration of the strong wind may be essentially longer. The maximum wind speed often exceeds 20 m/s, and a maximum of 29.5 m/s was measured on February 14, 1998, when 31.5 m/s was measured at Daneborg closer to the outer coast. Since 1958, the highest maximum wind speed at Daneborg was 34.0 m/s measured on March 8, 1993 (Cappelen *et al.*, 2001).

Strong winds from west or northwest may also be of local origin, that is, a part of the shallow katabatic wind pattern of the ice sheet. Their velocity depends on the steepness and topography of the ice slopes towards the coast, so they occur only at exposed places.

Foehn winds occur all over Greenland. In Northeast Greenland, strong foehn events usually take place somewhere once or twice in a winter. Typically, the synoptic pattern will be a stationary anticyclone over or south of Iceland directing warm air from the Atlantic Ocean northward toward West Greenland and from there further on to Northeast Greenland, causing the temperature to rise well above the freezing point. The strongest foehn at Zackenberg occurred on January 25–27, 2005, the wind speed reached 17.8 m/s, and the air temperature rose from below −16 °C to more than +10 °C. When the wind speed was at its peek, the temperature increase was more than 13 °C/hour, and for more than 14 hours the air temperature was above the freezing point. No change in the snow depth was recorded during this incident, but during the snow density measurements the following spring, a hard crust of ice in the snow pack indicated that melting actually took place.

Mainly in summer, southerly winds (appearing southeasterly in the fjords) on a synoptic scale occur now and then with cyclonic activity over the ice sheet and/or high pressure to the east. This causes an increase in the inflow of cool and foggy surface air from the coastal waters, and the fog may spread far into the fjords (Trans, 1955).

VI. SPATIO-TEMPORAL VARIATIONS IN PRECIPITATION

The amount of precipitation over any area depends on the moisture content of the air, the pattern of synoptic scale weather systems affecting the area and the topography, altitude and character of the underlying surface. At Zackenberg, the relative humidity shows a dependence on atmospheric transmissivity and cloud cover (Figure 3), with low values of 55–70% in the winter period and 75–88% in the summer period (Figure 12). Although the relative humidity is comparatively high due to the adjacent Greenland Sea, the combination with low air temperatures results in limited amounts of precipitation. At Zackenberg, annual precipitation amounts to 261 mm, which is in accordance with the fact that mean annual totals of precipitation over the

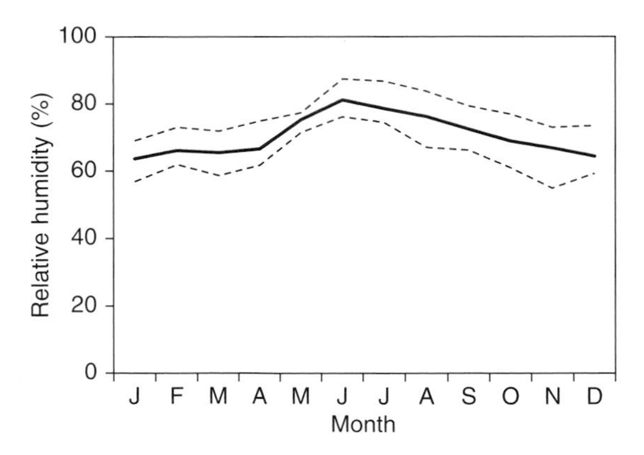

Figure 12 Maximum, mean and minimum monthly relative humidity at Zackenberg 1996–2005.

majority of the Arctic do not exceed 400 mm (Przybylak, 2003; Mernild, 2006; Mernild *et al.*, 2007).

The number of meteorological stations measuring precipitation in Greenland is very limited, and stations with long records are limited to the coastal regions. At Zackenberg, automatic registrations are carried out using a Belfort gauge with Nipher shield to minimize the wind-induced turbulence over the gauge orifice. The lack of daily inspection and cleaning of accumulated snow in the funnel of the gauge unfortunately causes an uncontrollable delay in the registration with a few inexpedient large values in warm periods or at the end of the winter, which makes standard correction of the measurements impossible.

Measuring of snow precipitation is uncertain, especially if it takes place when it is windy and cold (Yang *et al.*, 1999; Mernild *et al.*, 2006). Snowfall in the Arctic is most often connected with strong winds and typically takes the form of fine snow flakes (Sturm *et al.*, 1995). As a result, wind easily lifts and redistributes the snowflakes according to exposure and local topography. Arctic snow usually begins to drift at wind speeds above 5 m/s (Mernild *et al.*, 2006), and during the polar night, it is sometimes difficult to distinguish between a period of snowfall and a period of drifting snow.

At Zackenberg, the fraction of liquid (rain) precipitation and solid (snow) precipitation at different air temperatures was estimated on the basis of observations from different locations on Svalbard (Førland and Hanssen-Bauer, 2003). For air temperatures below −1 °C precipitation is solid in 100% of the events and for temperatures above 3 °C precipitation is liquid in 100% of the events. In between, the fraction of snow and rain is calculated by linear interpolation. Solid (snow) precipitation was calculated from snow depth sounder observations and is assumed to have an accuracy of within ±10–15%. The snow depth at the time when the ground temperature reaches 0 °C in spring is given as "solid precipitation at start of snowmelt" (Table 1) (Rasch and Caning, 2005). Measurements of the density for this compact snow give 400 ± 50 kg/m^3, which is in accordance with studies in other parts of the arctic region (Oelke *et al.*, 2003). By multiplying snow depth in spring with density, approximate values for the winter precipitation can be calculated. These values are in reasonable accordance (0–40% differences) with the precipitation measured with the Belfort gauge during winter in 4 out of 8 years (Table 1). For the two years with the largest amount of snow during winter, the values based on "solid precipitation at start of snowmelt" are more than twice the measured precipitation. The last two years indicate that over-catches can also occur, especially if ice crust is formed early in the winter during warm Foehn situations followed by periods with strong wind during snowstorms. Hence, snow depths measured at the sonic range snow depth sensor are assumed to be representative for the winter precipitation, as drift

Table 1 Annual precipitation (October 1 to September 30 the following year) in millimetre water equivalent at Zackenberg measured and estimated by using NCEP/NCAR reanalysis data

	1997/1998	1998/1999	1999/2000	2000/2001	2001/2002	2002/2003	2003/2004	2004/2005	Average
Measured solid precipitation	182	176	149	177	130	175	238	232	182
Solid precipitation at start of snowmelt	232	364	152	200	344	112	192	144	218
Mixed precipitation	8	27	21	38	1	6	11	21	17
Liquid precipitation	83	25	1	29	27	10	14	29	27
Total precipitation	**323**	**416**	**174**	**267**	**372**	**128**	**217**	**194**	**261**
NCEP solid	187	283	226	201	340	147	195	168	218 ***1.00***
NCEP mixed	53	124	39	47	85	29	33	22	54 ***0.31***
NCEP liquid	88	47	13	45	80	37	29	36	47 ***0.58***
NCEP total	**328**	**454**	**278**	**293**	**505**	**213**	**257**	**226**	**319**

Notes: Solid precipitation is falling as snow, when air temperature is below −1 °C; liquid precipitation is falling as rain, when air temperature is above 3 °C; and mixed precipitation is falling as sleet, when the air temperature is between −1 °C and 3 °C. Solid precipitation at the start of snowmelt is a measurement for the solid precipitation during the winter period and is found by multiplying the snow depth at the time when the entire snow pack is melting and is saturated with water with the density of 400 kg/m^3—see text. Total precipitation is the sum of spring snow depth, mixed and liquid precipitation. The figures in italic in the average column indicate the correction factors, which are used to correct the NCEP/NCAR analysis data to Zackenberg level in Figure 14.

of snow is assumed in total not to accumulate or remove snow from the measuring spot (Rasch and Caning, 2005).

In the Arctic, rainfalls in the summertime are characterized by a very low intensity (Przybylak, 2003), although special but rare synoptic events delivering an essential part of the annual precipitation may occur. At Zackenberg, the highest liquid precipitation rate recorded was as high as 4.8 mm/hour (August 16, 1998, DOY 228). The average liquid precipitation rate was 0.47 mm/hour, but nearly half the liquid precipitation rates (3215 hourly events) in the period 1996–2005 were below 0.2 mm/hour, and only 12% were above 1 mm/hour.

To compensate for the missing daily measurements of precipitation during the winter period, it is necessary to use data from numerical weather models. One of the most widely used sources of numerical weather models in climate research is the dataset from the NCEP/NCAR reanalysis (Kalnay *et al.*, 1996; Oelke *et al.*, 2003; Serreze and Barry, 2005). Most NCEP/NCAR data are provided on a 2.5 × 2.5 degree grid and represent more than 50 years (continually updated) of global atmospheric analyses and surface fields. By downscaling NCEP/NCAR outputs of precipitation and air temperature, it is possible to produce monthly estimates of the solid, mixed and liquid precipitation during the period 1996–2005 and to produce yearly estimates back to 1958. It is well known that the NCEP/NCAR precipitation outputs perform very well during winter (Serreze and Hurst, 1999), but the most significant problem with the NCEP model is a severe oversimulation of summer precipitation over land areas in the Atlantic region, excessive convective precipitation. For the period 1996–2005, the NCEP/NCAR model estimates the average solid precipitation extremely well ($r = 0.857$, $p = 0.002$ and $df = 7$) compared to the "solid precipitation at start of snowmelt" estimates (Table 1), while the mixed and liquid precipitation has to be reduced with 31% ($r = 0.647$, $p = 0.024$ and $df = 7$) and 58% ($r = 0.789$, $p = 0.005$ and $df = 7$), respectively, in order to match the measured average values. For the hydrological year (October 1 to September 30), the average annual precipitation was 261 mm w.eq. (water equivalent) for the years 1996–2005, with a minimum of 128 mm w.eq. in 2002/2003 and a maximum of 416 mm w.eq. in 1998/1999 (Table 1).

The monthly totals of precipitation at Zackenberg are characterized by the highest precipitation occurring mainly in the autumn and winter months—in the autumn when the air temperature is still relatively high (particularly in September–October) and in the winter months when the cyclonic activity is at its greatest (November–February) (Figure 13). The monthly precipitation is lowest in April, when the anticyclonic activity is greatest, and in the summer shortly after snowmelt, when air temperatures increase more rapidly than the humidity. Such annual cycles of precipitation (Yang *et al.*, 1999) occur in arctic areas, where atmospheric circulation is strongest, while in the parts of

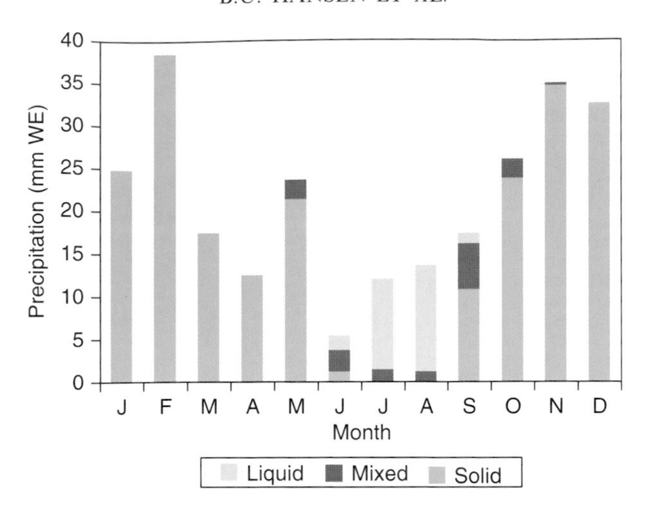

Figure 13 Mean monthly precipitation in millimetre water equivalent (w.eq.) at Zackenberg for the period 1996–2005. Solid precipitation is falling as snow, when air temperature is below −1 °C; liquid precipitation is falling as rain, when air temperature is above 3 °C; and mixed precipitation is falling as sleet, when the air temperature is between −1 °C and 3 °C. The annual precipitation is 261 mm as an average for the period.

the Arctic with the most continental climate (such as in North Greenland), the annual cycles of precipitation strongly depend on air temperature, and here maximum precipitation is in summer and the minimum in winter. On average, mixed precipitation is 17 mm (7% of total) and can occur from May until November depending on air temperature, while liquid precipitation is 27 mm (10% of total) and occurs only from June to September.

For the period 1958–2005, the NCEP/NCAR reanalysis of the annual precipitation shows a significant increase of 1.9 mm w.eq./year ($r = 0.40$, $p = 0.020$ and $\mathrm{df} = 46$) (Figure 14). During this period, the average was 178, 15 and 22 mm w.eq. for solid, mixed and liquid precipitation, respectively. Four dry years (1965/1966, 1967/1968, 1968/1969 and 1977/1978) with annual precipitation of less than 100 mm are evident, while five wet years (1959/1960, 1986/1987, 1992/1993, 1998/1999 and 2001/2002) with annual precipitation higher than 300 mm are clearly distinguished from the rest of the population.

The strongest precipitation gradient in Greenland is the latitudinal gradient along the east coast, which is documented in several studies (Ohmura and Reeh, 1991; Chen *et al.*, 1997; Bromwich *et al.*, 2001; Cappelen *et al.*, 2001; Rysgaard *et al.*, 2003). Nearly 2400 mm of precipitation occur per year at the southern part of the gradient (Figure 15), rapidly decreasing to 500 mm at

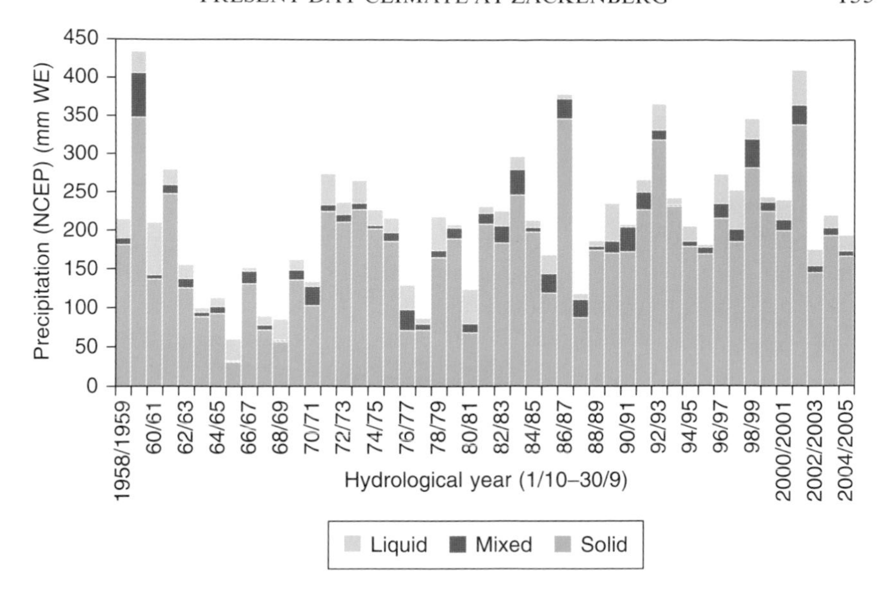

Figure 14 Annual precipitation at Zackenberg based on NCEP/NCAR reanalysis data for the period 1958–2005. The hydrological year is defined as the period October 1–September 30 the following year. Solid precipitation is falling as snow, when air temperature is below −1 °C; liquid precipitation is falling as rain, when air temperature is above +3 °C; and mixed precipitation is falling as sleet, when the air temperature is between −1 °C and +3 °C.

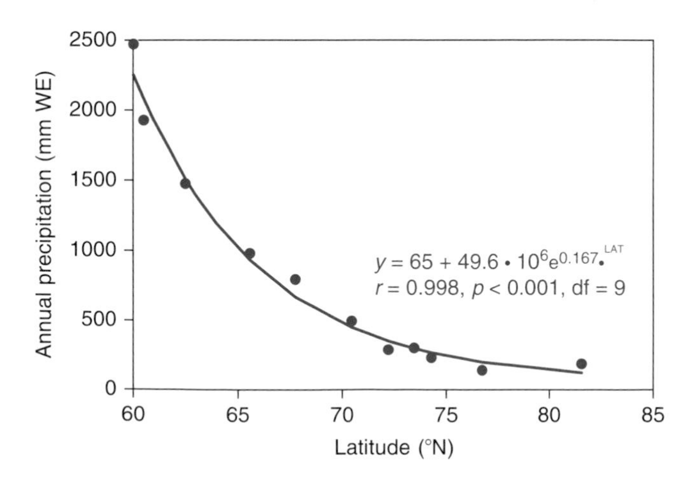

Figure 15 Mean annual precipitation in East Greenland versus latitude (data from Cappelen *et al.*, 2001).

70°N with a more gentle decrease to less than 100 mm at the northern parts of the gradient, all caused by a strong temperature and humidity gradient along the coast.

A further coast-inland gradient is found along the coast perpendicular to the latitudinal gradient (Ohmura and Reeh, 1991). The coast-inland gradient is about 100–200 km wide, and it reaches 1500 mm/year in the south, drops to about 40 mm/year at the latitude of Zackenberg and fades out to less than 10 mm in the north. Very few attempts have been made in Greenland to record altitudinal precipitation gradients, and none in the high-arctic region (Hasholt *et al.*, 2003; Mernild, 2006). In neighbouring high-arctic regions, such as Svalbard (Humlum, 2002), the altitudinal precipitation gradient is 15–20%/100 m in the coastal regions, while it is somewhat smaller, 5–10%, at the inland regions. The precipitation in the region is directly impacted by the onshore flow from the Arctic Ocean. This flow would have relatively high humidity and produces precipitation over the coastal slopes by orographic lifting, but the amount of precipitation caused by this lifting is smaller further inland, as the humidity of the air mass decreases during the precipitation event.

VII. SPATIO-TEMPORAL VARIATIONS IN AIR TEMPERATURE

Air temperature is the most important, and therefore also the most often studied, climatological parameter. The instrumental records of arctic temperature are brief and geographically sparse. Only five time series, all located in Greenland, can be extended back to the second half of the nineteenth century, but spatial distribution and reliable estimates of air temperature characteristics in the Arctic are possible only for the last 50–60 years (Vinther *et al.*, 2006; Cappelen, 2007; Cappelen *et al.*, 2007a, b). However, for the past 20–30 years, satellites have constituted new and extremely powerful sources of information about the spatial distribution of temperature in the Arctic.

At Zackenberg, like most other high-arctic stations, end of July is the warmest period of the year and July is the warmest month of the year with a mean monthly air temperature (MMAT) of 5.8 °C for the period 1996–2005. The coldest summer in the period was 1997, with an MMAT of only 3.7 °C for July (Figure 16), while 2003 became the warmest summer, with an MMAT of 7.6 °C in July. For both flora and fauna, the length of the frost-free period can be quite important. On average, the frost-free period (no minimum temperatures below 0 °C) at Zackenberg was 35 days, lasting from July 13 to August 18, but it was as small as 15 days (July 23 to August 7) in 2001 after a winter with heavy and long-lasting snow-cover and as long as

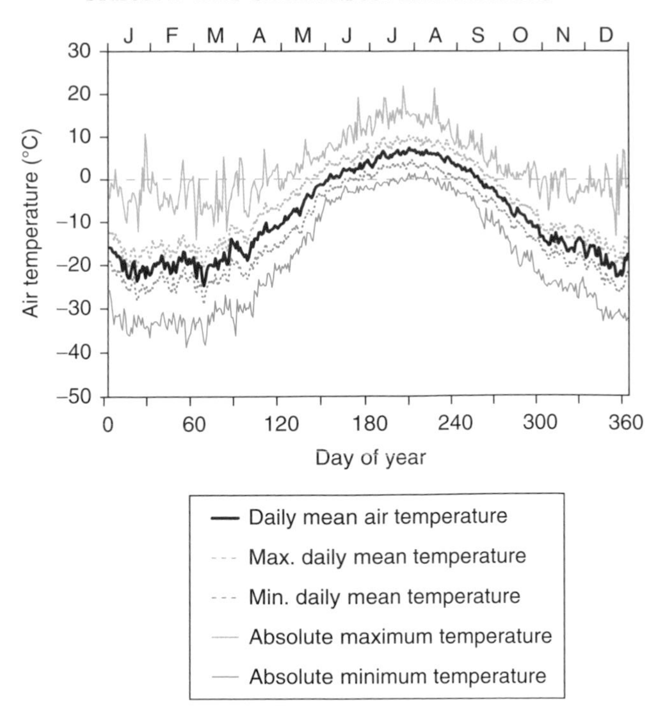

Figure 16 Daily mean air temperature (bold red line), maximum daily mean air temperature (upper dashed green line), minimum daily mean air temperature (lower dashed blue line), absolute maximum air temperature (upper thin green line) and absolute minimum air temperature (lower thin blue line) at Zackenberg 1996–2005.

60 days (June 27 to August 26) in 2003 after a winter with moderate snow-cover and an early snowmelt. An absolute maximum temperature of 21.8 °C was recorded at 1800 hrs on July 21, 2005, during a small Foehn following 2 days with cloud-free sunny weather and no temperatures below 10 °C.

The coldest period at Zackenberg is during and shortly after the polar night (Figure 16), when the period December to March had MMAT just below −20 °C and with −22.4 °C in February as the coldest mean annual air temperature (MAAT) for the period 1996–2005 (Figure 17). In this period, air temperatures can drop well below −30 °C, and an absolute minimum temperature of −38.9 °C was measured at 1900 hrs on February 23, 1998, after several cloud-free days with a small wind speed from northwest. Figure 16 shows that positive air temperatures can occur in short periods during all the winter months, and Figure 17 indicates pronounced warm anomalies in the last two winters, with MMAT between −15 °C and −20 °C. Especially during these two winters, the number of "warm hours" (air

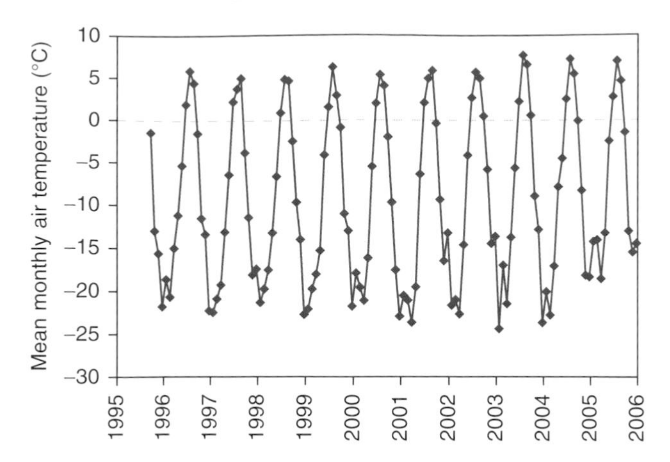

Figure 17 Mean monthly air temperatures at Zackenberg from September 1995 to December 2005.

temperature >0 °C) increased, reaching a maximum in January 2005, with an average air temperature of 4.8 °C during a 13 hours long Foehn situation with air temperatures as high as +10.7 °C at 2300 hours on January 26.

The annual range of temperature (i.e., difference between highest and lowest MMATs) has in previous studies (Przybylak, 2003) been used to distinguish between three well-defined types in the annual cycle of temperatures in the Arctic: (1) maritime, (2) coastal and (3) continental. The maritime type has an annual temperature range of 10–15 °C and the coastal type has a range of 15–25 °C, while the continental and most common type (80% of the Arctic) has an annual temperature range of more than 25 °C. Figure 17 shows that Zackenberg most of the years belong to the continental type, but in the more recent years Zackenberg has moved towards to the coastal type.

The arctic region (Rouse, 2000), and especially Greenland (Cappelen *et al.*, 2001; Rysgaard *et al.*, 2003), is known for several pronounced temperature gradients. The most well-documented gradient is the latitudinal gradient (Figure 18), which exhibits a linear decrease of −0.89 °C/°lat. along the east coast of Greenland. The long period of midnight sun in North Greenland and the presence of sea ice along the coast are the reasons for the more moderate decrease (−0.12 °C/°lat.) in the warmest summer month (July), while in the coldest winter month (February) the absence of sun or low sun elevation, the extension of the sea ice and the prevailing cold northern winds increase the latitudinal gradient to more than −1.4 °C/°lat.

More important is the difference between the outer coasts, where drifting ice and cold water in the summer makes the air cold and humid, and the ice-free inland where the weather is warmer and often sunny. Daneborg and Zackenberg (Figure 19) are part of this gradient. The aerial distance between

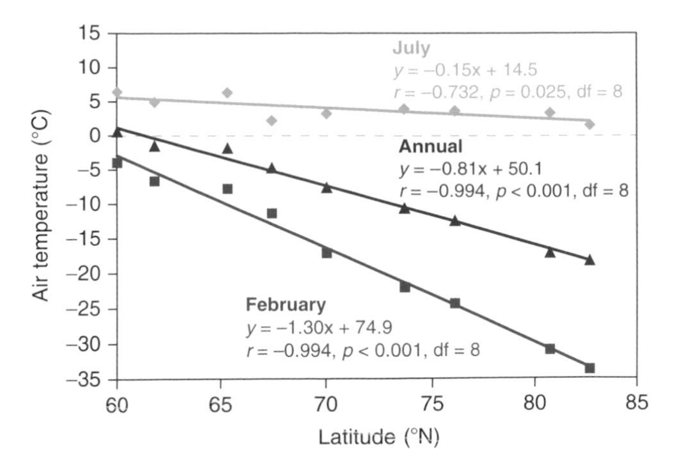

Figure 18 Mean annual, July (warmest month) and February (coldest month) air temperatures versus latitude along the east coast of Greenland.

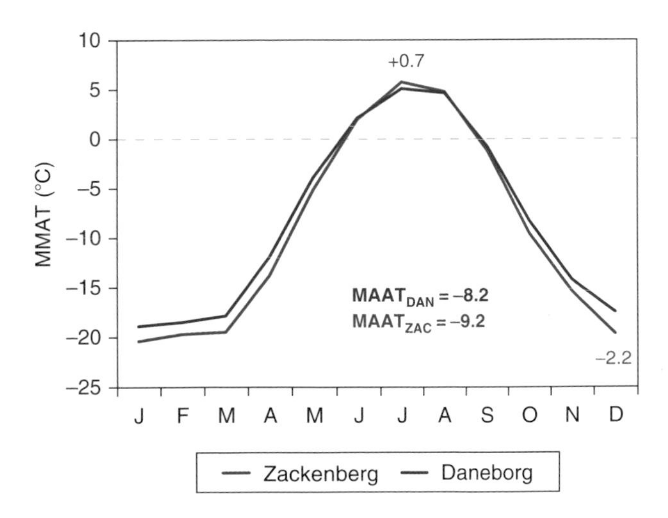

Figure 19 Mean monthly air temperatures (MMATs) and mean annual air temperature (MAAT) at Daneborg (outer coast) and at Zackenberg (inland) 1996–2005.

the two stations is only 23 km, but in the summer, the temperature difference is +0.7 °C, while in the winter the gradient is three times larger with opposite sign due to the presence of partly open water at the mouth of Young Sund. A comparison (Bay, 1992) between the only other inland station in North Greenland, Kap Moltke, and the nearest outer coast station, Station Nord, indicates gradients twice the size both in summer and in winter at a distance eight times greater.

In the snow-free summer season, where onshore winds prevail, a strong temperature gradient is found perpendicular to the fjord coast. If the fjord is ice covered, the temperature just above the ice surface is very close to 0 °C. During the onshore transport, the temperature of the air mass will increase rapidly the first two kilometres inland, and it reaches daily averages of 3–7 °C at the Zackenberg meteorological station, while it seems to reach a maximum of 7.5 °C in Store Sødal, 10 km from the fjord (Table 2). When the fjord is ice free, the surface-layer temperature is 2–7 °C and the temperature gradient the first two kilometres from the coast is much smaller but still increasing. Similar gradients have been found at the coast of Hudson Bay, Canada (Rouse, 1991).

A prominent feature in the arctic environment is the presence of strong low-level temperature inversions (situations in which temperature increases rather than decreases with altitude). Table 3 shows that increasing lapse rates of 0.83–1.24 °C/100 m are dominating in 55–79% of the wintertime because of the low-sun or no-sun, which causes radiation cooling of the snow surface with the consequence of a negative net radiative balance at the surface. The temperature difference between the inversion base and top averages is 10–12 °C and the average thickness of the inversion is about 800–900 m, but maximum thickness very often exceeds 1200 m and more rarely 1600 m (Przybylak, 2003). At Zackenberg, a maximum temperature gradient of 3.25 °C/100 m was measured on March 6, 2004. In the winter period, onset of strong winds results in a dramatic and almost instant temperature increase followed by a more moderate drop in temperature, if the wind calms down again. These episodes are often associated with precipitation and a normal decreasing lapse rate of −0.4 to −0.6 °C/100 m. In spring and early summer, the cooling from the melting snow is the controlling factor in the presence of a surface-based inversion layer, but the thickness of the layer and the temperature gradients within the layer is much smaller (Serreze and Barry, 2005). In the snow-free part of the summer, a prevailing positive net radiation balance causes an intense mixing of the lower atmosphere and results in an increase of low clouds or fog on the slopes of the surrounding mountains. This causes an advective lifting of the inversion base from 300 to 600 m, but surface-based inversions (Table 3) can occur during summertime, if the sun is covered by clouds or hidden by the surrounding mountains.

Table 2 clearly demonstrates the inversion's influence on annual range of MMATs in Zackenbergdalen. In February–March, the MMATs on Dombjerg (1278 m a.s.l.) are the highest, because the rest of the stations are still influenced by shade from the surrounding mountains. During April to July, the MMATs on Aucellabjerg (micrometeorological station M3, 420 m a.s.l.) are the highest due to snowmelt and radiative cooling at the valley floor and due to normal decreasing temperature gradients above M3. In August–September, the inversion layer is almost absent, and decreasing temperature gradients dominate the entire valley system, with the highest MMATs at the main meteorological station (38 m a.s.l.). From October to January, as

Table 2 Mean monthly air temperatures (MMATs) at the four climate stations in the Zackenberg area 2003–2005

Climate station	Altitude (m a.s.l.)	Distance from nearest coast (km)	Distance from outer coast (km)	January	February	March	April	May	June	July	August	September	October	November	December	Annual
Zackenberg	38	2	27	−14.3	−14.1	−18.6	−13.2	−2.5	2.7	**6.9**	**3.5**	**−0.1**	−8.3	−18.2	−18.4	−7.8
Microclimate station M3	420	8	27	**−12.7**	−13.7	−14.5	**−12.0**	**−0.9**	**4.0**	6.8	2.4	−1.6	**−7.9**	**−16.0**	**−17.1**	**−6.9**
Dome station	1278	10	32	−16.8	**−13.3**	**−13.4**	−13.5	−6.8	2.0	5.0	−0.3	−5.3	−11.0	−18.7	−20.5	−9.4
St. Sødal station	78	10	32	−14.8	−14.1	−19.3	−13.8	−3.2	3.1	7.4	3.8	−0.6	−9.1	−18.7	−19.1	−8.2

Notes: The warmest MMAT among the first three stations are in bold. The altitudinal variation of the warmest month is caused by the temperature inversion.

Table 3 Monthly air temperature gradients between the Zackenberg climate station (38 m a.s.l.) and the microclimate station M3 (420 m a.s.l.)

	Increasing lapse rate with altitude		Decreasing lapse rate with altitude		Mean lapse rate
Month	°C/100 m	%	°C/100 m	%	°C/100 m
January	0.89	47.7	−0.59	52.3	+0.11
February	0.91	56.9	−0.59	43.1	+0.26
March	1.24	54.9	−0.57	45.1	+0.43
April	0.66	49.0	−0.60	51.0	+0.02
May	1.05	31.3	−0.66	68.7	−0.12
June	0.67	40.8	−0.54	59.2	−0.05
July	0.34	46.5	−0.47	53.5	−0.02
August	0.32	32.6	−0.54	67.4	−0.19
September	0.47	10.0	−0.72	90.0	−0.60
October	0.79	41.9	−0.57	58.1	+0.02
November	0.83	66.7	−0.46	33.3	+0.40
December	1.16	79.3	−0.51	20.7	+0.82

the surface-based inversion gets more dominating in the valley, the MMATs at micrometeorological station M3 are highest again.

Unfortunately, regular temperature measurements at the nearby Daneborg meteorological station first started in 1958 as the NCEP/NCAR reanalysis does, but the Climate Research Unit (CRU) at Tyndall Centre for Climate Change Research has developed a dataset, CRU TS 2.1 (Mitchell and Jones, 2005), which is much like the NCEP/NCAR reanalysis dataset, only with a spatial resolution of 0.5 × 0.5 degrees and with a monthly time step for the period 1901–2002. Seasonal and annual time series of air temperature data downscaled to Zackenberg level are presented in Figure 20. Air temperature trends in various periods are shown in Table 4, and the latest 15-year period (1991–2005) was characterized by a statistically significant annual warming of 2.25 °C closely followed by a warming of 2.22 °C for the period 1916–1930. No other 15-year period shows statistically significant annual trends. For the winter season (December–February), the period 1916–1930 showed the highest significant warming of 3.40 °C closely followed by 3.37 °C in the period 1976–1990. In spring (March–May), only the period 1916–1930 showed a significant warming of 3.01 °C, and no significant trends were found in the 15-year period for the autumn (September–November). In summer (June–August), the latest period 1991–2005 showed the greatest significant warming of 1.47 °C followed again by 1.16 °C in the period 1916–1930. The period 1946–1960 showed the only significant cooling of 1.00 °C. These trends are similar to trends from other studies along the east coast of Greenland (Przybylak, 2003; Box, 2002).

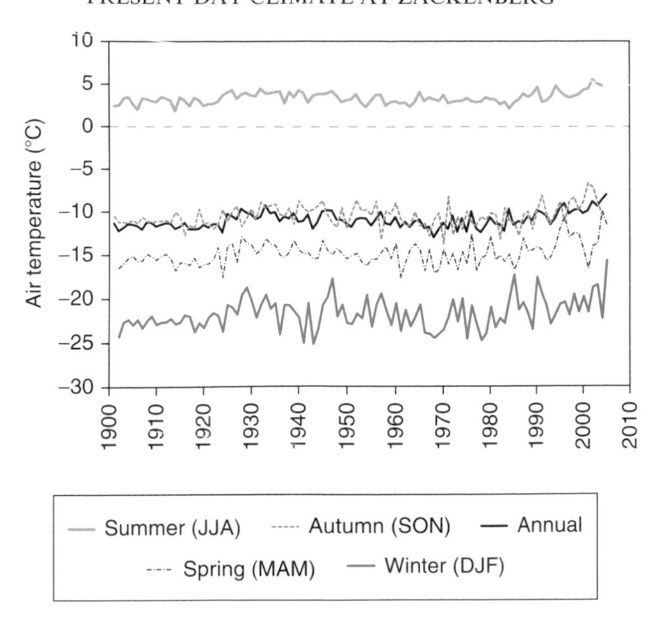

Figure 20 Seasonal air temperatures at Zackenberg. Data from 1996–2005 as measured together with downscaled data from the NCEP/NCAR reanalysis from 1901 to 1995.

The greatest variability occurred in the wintertime series, with a minimum of −25.0 °C in 1941 and a maximum of −15.6 °C in 2005 (Figure 20), while the smallest variability occurred in the summertime series, with a minimum of +1.8 °C in 1915 and a maximum of +5.5 °C in 2003. On an annual basis, 1968 showed a minimum of −12.9 °C, and 2005 a maximum of −7.7, but a ranking of the five warmest annual averages [2005 (−7.7), 2004 (−8.5), 2002 (−8.7), 1996 (−8.9) and 2003 (−9.2)] shows that they all occurred within the last 10 years. The cause of the temporal temperature variability itself is unclear, as extensive temperature studies in Greenland have shown little resemblance to annual NAO, sunspot data or volcanism, but sea-ice extent and concentration increases are found in association with cold periods (Box, 2002) (see Section IX below).

VIII. NAO RELATIONS

Over the period of instrumental records, the arctic climate has exhibited pronounced variability on several temporal scales from interannual, decadal to multi-decadal scales. An important but by no means exclusive source of this variability is the NAO. In its simplest definition, the NAO describes

Table 4 Seasonal air temperature trends at Zackenberg over various periods

Time period	Winter (DJF)		Spring (MAM)		Summer (JJA)		Autumn (SON)		Annual	
1901–1915	+0.45	0.37	+0.01	0.98	−0.10	0.84	−0.22	0.50	+0.10	0.73
1916–1930	***+3.40***	***<0.01***	***+3.01***	***<0.01***	**+1.16**	**0.02**	+1.23	0.18	***+2.22***	***<0.01***
1931–1945	−2.27	0.16	−0.58	0.45	−0.38	0.40	+0.05	0.95	−0.78	0.21
1946–1960	−1.10	0.47	+0.23	0.76	**−1.00**	**0.05**	−0.20	0.87	−0.58	0.30
1961–1975	+0.49	0.74	+0.07	0.96	+0.23	0.55	−1.07	0.43	−0.16	0.81
1976–1990	**+3.37**	**0.09**	+0.36	0.77	+0.42	0.31	−0.62	0.54	+0.91	0.24
1991–2005	+2.53	0.13	+2.21	0.20	**+1.47**	**0.03**	+1.83	0.13	***+2.25***	***<0.01***
1901–2005	***+1.77***	***<0.01***	***+1.81***	***<0.01***	***+0.62***	***<0.01***	***+1.25***	***<0.01***	***+1.39***	***<0.01***
1931–2005	**+1.43**	**0.06**	**+1.36**	**0.02**	+0.23	0.39	+0.42	0.46	**+0.90**	**0.02**

Notes: The left column in each season shows the air temperature trend (°C) during the entire time period, while the right column shows the *P*-value. Trends with statistical significance at or below 10% are in bold, while italic bold values indicate significance at or below 1%. Data from Figure 20 were used in the calculations.

co-variability in sea-level pressure variations between the Icelandic Low and Azores High, the two centres of action in the atmospheric circulation of the North Atlantic. When both are strong, the NAO is taken to be in its positive mode, and when both are weak, the NAO is in its negative phase (see also Stendel *et al.*, 2008, this volume).

During the last decade, a lot of research has been focused on how NAO relates to variability and trends in prominent climate parameters such as temperature and precipitation in the North Atlantic region (Bromwich *et al.*, 1999; Hanna and Cappelen, 2006). An association between NAO and temperature can be found in writings throughout the nineteenth century and is also known as the "seesaw" in surface air temperature over Europe and northeastern North America. Under the positive NAO mode, cold high-latitude air will tend to advect over Greenland and eastern Canada, consistent with the tendency for negative temperature trends in these areas, while an opposing strong advection of warm lower-latitude air over northern Europe and Scandinavia is consistent with positive temperature trends in this area. Under the negative NAO mode, winds over Greenland and eastern Canada are weaker with positive temperature trends, while still southerly, but much weaker winds over northern Europe and Scandinavia bring negative temperature trends.

Since the Zackenberg area and central East Greenland are placed in the middle of the "seesaw," no significant correlation between NAO and surface air temperatures is found along east Greenland. Bromwich *et al.* (1999) found a strong positive association ($r = 0.57$) between the autumn (SON) NAO index and the precipitation amount for central East Greenland for the period 1985–1995, while no significant correlation was found between NAO and the precipitation amount in the other three seasonal periods. The period 1985–1995 was dominated by positive NAO values and high amounts of precipitation, while in periods with negative NAO values and small amounts of precipitation no significant correlation can be found for the central East Greenland area around Zackenberg (Hinkler, 2005; Serreze and Barry, 2005).

IX. CLIMATE–SEA ICE RELATIONS

The sea ice in the Arctic Ocean and adjacent seas consists primarily of two types of ice—the seasonal (or first year) ice and perennial (or multi-year) ice. The seasonal or first year ice ranges from a few tenths of a meter near the southern margin of the arctic marine cryosphere to 2.5 m in the High Arctic at the end of winter, while multi-year ice is about 3–6 m thick, harder and almost salt free after some years. The area of sea ice decreases roughly from 15 million km^2 in March to 7 million km^2 in September, since much of the first-year ice melts during the summer. An average of 10% of arctic multi-year ice exits the Arctic Ocean primarily through the Fram Strait each year. For sometime, it has been recognized that the annual variability in the export of sea ice and low-salinity water through the Fram Strait and in the Greenland Sea has a tremendous influence on the weather along the Greenland east coast and further south. A number of efforts have been made to estimate the ice area/extent in the Greenland Sea and the volume flux through Fram Strait on the basis of mass balance requirements, measurements of ice drift and thickness, and models or a combination of models and observations. Through NASA Goddard Space Flight Center (GSFC), time series of estimated ice area and extent are available on a daily and monthly basis from satellite data since the late 1970s, while mid-month values of sea-ice concentration on a $1 \times 1°$ spatial resolution through the Walsh and Chapman Northern Hemisphere Sea Ice Data Set 1870–1998 from NCAR. Although the older parts of the later dataset consist of infrequent land/sea/air observations, where temporal and spatial gaps are filled with climatological or other statistically derived data, the Walsh and Chapman data are considered as the most reliable dataset of sea-ice variability for the Northern Hemisphere, but variability and trends in the first part of the time series have to be used with care.

A combination of the two datasets is shown in Figure 21. For the Greenland Sea (1.2 million km^2), a considerable seasonal and annual decrease is seen during the past 100 years. For the winter period (December–March), the decrease is about 200,000 km^2/century, and in the summer period (July–August), the decrease is nearly 270,000 km^2/century, while the annual decrease is about 250,000 km^2/century. A comparison between sea ice (Figure 21) and air temperature (Figure 20) indicates that summers with a sea-ice extent above the trend line have a tendency of being cold, while summers with a sea-ice extent below the trend line tend to be warmer and a bit wetter. A comparison between sea ice (Figure 21) and precipitation (Figure 14) indicates that in winter periods (1963–1969), with a large sea-ice extent, the winters at Zackenberg have a shallow snow-cover, while in winters with a smaller sea-ice extent (1989–1995) than normal, the snow-cover may be deep depending on the number of cyclones that passes Zackenberg. Both periods correspond to minimum and maximum phases of the NAO index, respectively (Figure 4), which indicates that NAO also has pronounced influences on the sea ice cover (Cavalieri, 2002; Rogers *et al.*, 2005; Hinkler *et al.*, 2008, this volume).

Studies by Rogers *et al.* (2005) show that cyclones are more frequent along the Northeast Greenland coast and over the adjacent Fram Strait in relatively mild winters with a low ice export through the strait. Anomalously cold winters with a high ice export along the east coast of Greenland move the cyclone activity away from the Northeast Greenland coast and into the Barents Sea and Eurasia side of the Arctic Ocean. Rogers *et al.* (2004)

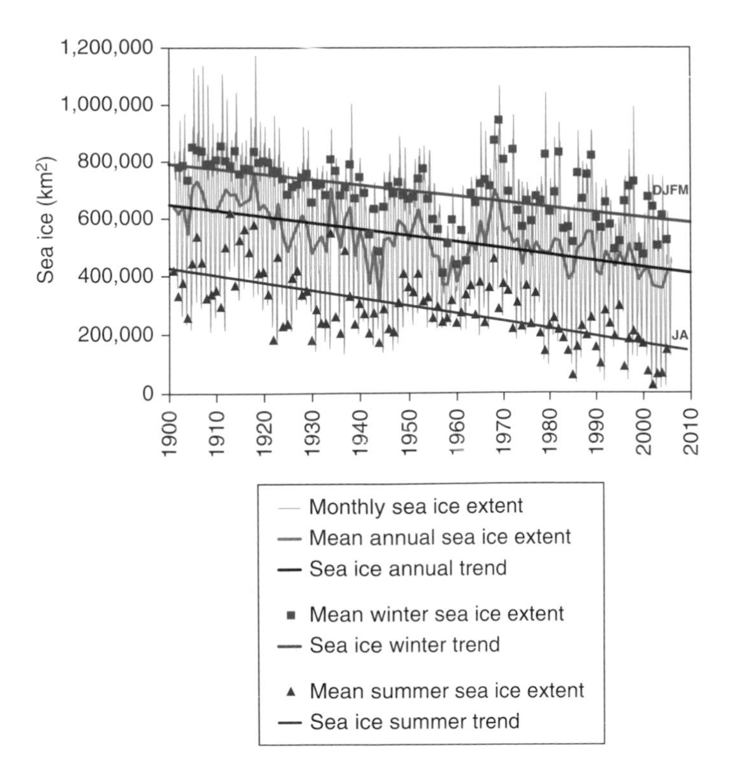

Figure 21 Monthly sea-ice extent (thin grey line) in the Greenland Sea, mean annual sea-ice extent (black bold line) and annual trend (black bold dashed line, -2086 km^2/year, $r = 0.674$, $p < 0.001$ and df $= 104$) for the period 1900–2006. Mean winter (December–March) (blue squares) sea-ice extent and winter trend (blue bold line, -1854 km^2/year, $r = 0.511$, $p < 0.001$ and df $= 104$), mean summer (July–August) (red triangles) sea-ice extent and summer trend (red bold line, -2568 km^2/year, $r = 0.651$, $p < 0.001$ and df $= 104$) (data for 1900–1998 from http://arctic.atmos.uiuc.edu/SEAICE and 1978–2006 from http://polynya.gsfc.nasa.gov).

found that the number of negative NAO months dominated in the cold winters (December–March), while the number of positive NAO months dominated in the mild winters. In cold winters with three or more negative NAO months, the average amount of solid precipitation was ~100 mm at Zackenberg, while it was more than doubled as much in winters with three or more positive NAO months.

Several investigators have documented correlations between winter NAO and sea-ice flux through Fram Strait or sea-ice extent in the Greenland Sea. Dickson *et al.* (2000) found a correlation of 0.77 between NAO and the Fram Strait ice flux over the period 1976–1996, while Vinje (2001) obtained a correlation of only 0.1 for a 50-year time series (1950–2000). The NAO correlation to Fram Strait ice export was also relatively high around 1840–1860 and during 1930–1950 (Schmith and Hansen, 2003), while in other decades it was near zero, and Vinje (2001) even found a negative correlation of −0.32 for the period 1962–1978. Vinje (2001) sums up the situation by stating that recent observational and modelling studies provide evidence of an unstable link between NAO and ice export through the Fram Strait and along the East Greenland coast. Other external factors such as sea-ice compactness and thickness, cloud cover, cyclone frequency, pollution and volcanism have some influence, but NAO and sea ice are still regarded as the two most dominating external factors on the climate in the coastal region of eastern Greenland and around Zackenberg (see further in Hinkler *et al.*, 2008, this volume).

X. CONCLUSIONS AND FUTURE PERSPECTIVES

The presentation of the individual climatological parameters demonstrates large spatiotemporal variations. Surprisingly, the greatest variations occur in winter, when the differentiated influence of the solar energy is low or equal to zero (polar night), but this is connected to the fact that in the cold winter period, the cyclonic activity is more intensive and frequent than in the warmer summer period. In addition, the temperature contrast between the arctic air and the advected air from the mid-latitudes is highest during this period. In turn, the effect of the underlying surface is not large, since snow and sea ice cover almost the entire arctic area.

In the warm summer period, the solar radiation is the most important climatological element and causes the greatest heterogeneity of the meteorological elements in all spatial scales: micro-, macro- and topo-climatic. The albedo of the underlying surface, which is significantly differentiated (snow/ice, bare soil, vegetation and water), increases the influence of solar ration in the radiation balance. However, because of the attenuated influence of the atmospheric and oceanic circulation and the large areas of the Arctic Ocean and adjacent seas not covered by sea ice (open water), the climatic spatiotemporal differences are lesser in summer than in winter.

The expected reduction in the amount of sea ice suggests that the physical conditions in Zackenbergdalen will become more similar to present-day conditions further south, for example, at Illoqqoortoormiut/Scoresbysund. Thus, the area extending from Zackenberg and a few hundred kilometres south represents climate gradients probably reflecting the recent century's climate change. This indicates that the north–south transects in this region may be highly valuable in evaluating adaptations in ecological processes to different physical settings in relation to future climate change in the Arctic. To obtain an improved knowledge about the climate variability in the region and the influence on the ecosystem requires that the future research come to grips with the intimate couplings between its atmosphere, land and ocean, and the future research has to be expanded to more gradient studies over a wider area using a better synthesis of observations, new sensors and by capitalizing on the growing capabilities of models and satellite remote sensing.

ACKNOWLEDGMENTS

Monitoring data for this chapter were provided by the ClimateBasis monitoring programme, run by ASIAQ, Greenland Survey and the GeoBasis programme run by Geographical Institute, University of Copenhagen, and the National Environmental Research Institute, Denmark. The programmes are financed by the Greenland Home Rule and the Danish Environmental Protection Agency, Danish Ministry of the Environment, respectively.

REFERENCES

ACIA (2005) *Arctic Climate Impact Assessment,* 1042 pp. Cambridge University Press, New York.

Bailey, W.G., Weick, E.J. and Bowers, J.D. (1989) *Arctic Alpine Res.* **21**, 126–134.

Bay, C. (1992) *Meddr. Grønland, Bioscience* **36**, 1–102.

Bliss, L.C. (1997) In: *Ecosystems of the World* (Ed. by F.E. Wielgolaski), Vol. 3, pp. 551–684. Elsevier, New York.

Box, J.E. (2002) *Int. J. Climat.* **22**, 1829–1847.

Bromwich, D.H., Chen, Q.-S., Li, Y.F. and Callather, R.I. (1999) *J. Geophys. Res.* **104**(D18), 22103–22115.

Bromwich, D.H., Chen, Q.-S., Bai, L., Cassano, E.N. and Li, Y. (2001) *J. Geophys. Res.* **106**(D24), 33891–33908.

Cappelen, J. (2007) DMI Annual Climate Data Collection 1873–2006, Denmark, The Faroe Islands and Greenland—with graphics and Danish summary. *DMI Technical Report No. 07–05.*

Cappelen, J., Jørgensen, B.V., Laursen, E.V., Stannius, L.S. and Thomsen, R.S. (2001) *Danish Meteorological Institute, Technical Report 00–18,* 152 pp.

Cappelen, J., Laursen, E.V., Jørgensen, P.V. and Kern-Hansen, C. (2007a) DMI Monthly Climate Data Collection 1768–2006, Denmark, The Faroe Islands and Greenland. *DMI Teknisk Rapport No. 07–06.*

Cappelen, J., Laursen, E.V., Jørgensen, P.V. and Kern-Hansen, C. (2007b) DMI Daily Climate Data Collection 1873–2006, Denmark and Greenland. *DMI Teknisk Rapport No. 07–07.*

Cavalieri, D.L. (2002) *Geophys. Res. Lett.* **29**(12), 1–4.

Chen, Q.-S., Bromwich, D.H. and Bai, L. (1997) *J. Clim.* **10**, 839–870.

Dickson, R.R., Osborn, T.J., Hurrell, J.W., Meincke, J., Blindheim, J., Adlandsvik, B., Vinje, T., Alekseev, G. and Maslowski, W. (2000) *J. Clim.* **13**, 2671–2696.

Førland, E.J. and Hanssen-Bauer, I. (2003) *Norsk Met. Inst., Report no. 24/02 Klima.* Norsk Meteorologisk Institut, http://met.no/english/r_and_d_activities/publications/2002/klima-02–24.pdf.
Hanna, E. and Cappelen, J. (2006) *Weather* **57**(9), 320–328.
Hansen, B.U., Elberling, B., Humlum, O. and Nielsen, N. (2005) *Danish J. Geogr.* **106**, 45–56.
Hasholt, B., Liston, G.E. and Knudsen, N.T. (2003) *Nord. Hydrol.* **34**, 1–16.
Hinkler, J. (2005) *From digital cameras to large scale sea-ice dynamics,* PhD thesis, 184 pp. National Environmental Research Institute, Denmark.
Hovmøller, E. (1947) *Meddr. Grønland* **144**, 1, 208 pp.
Humlum, O. (2002) *Norwegian J. Geogr.* **56**, 96–103.
IPCC (2007) *Climate Change 2007: The Physical Science Basis* (Ed. by S. Solomon, D. Qin, M. Manning, Z. Chen, M. Marquis, K.B. Averyt, M. Tignor and H.L. Miller), Cambridge University Press, Cambridge, United Kingdom and New York, NY, USA. 996 pp.
Kalnay, E., Kanamitsu, M., Kistler, R., Collins, W., Deaven, D., Gandin, L., Iredell, M., Saha, S., White, G., Woollen, J., Zhu, Y.Leetmaa, A., *et al.* (1996) *Bull. Am. Meteorol. Soc.* **77**, 437–471.
Lafleur, P., Rouse, W.R. and Hardill, S.G. (1987) *Arctic Alpine Res.* **19**(1), 53–63.
Meltofte, H. and Thing, H. (1996) *Zackenberg Ecological Research Operations, 1st Annual Report 1995.* Danish Polar Center, Ministry of Research and Technology, Copenhagen.
Mernild, S.H. (2006) *Freshwater discharge from the coastal area outside the Greenland Ice Sheet,* East Greenland, PhD dissertation, Institute of Geography, Faculty of Science, University of Copenhagen. 394 pp.
Mernild, S.H., Hasholt, B. and Hansen, B.U. (2005) *Danish J. Geogr.* **105**(2), 49–58.
Mernild, S.H., Liston, G.E., Hasholt, B. and Knudsen, N.T. (2006) *J. Hydrometeorol.* **7**, 808–824.
Mernild, S.H., Sigsgaard, C., Rasch, M., Hasholt, B., Hansen, B.U., Stjernholm, M. and Petersen, D. (2007) *Meddr. Grønland, BioScience* **58**, 24–43.
Mitchell, T.D. and Jones, P.D. (2005) *Int. J. Clim.* **25**, 693–712.
Ohmura, A. and Reeh, N. (1991) *J. Glaciology* **37**, 140–148.
Oelke, C., Zhang, T., Serreze, M.C. and Armstrong, R. (2003) *J. Geophys. Res.* **108**(D10), 1–19.
Przybylak, R. (2003) *The Climate of the Arctic*, Vol. 26, Kluwer Academic Publishers, Atmospheric and Oceanographic Sciences Library. 270 pp.
Putnins, P. (1970) In: *Climates of the Polar Regions, World Survey of Climatology* (Ed. by S. Orvig), Vol. 14, 3–123. Elsevier, Amsterdam-London-New York.
Rasch, M. and Caning, K. (2005) *Zackenberg Ecological Research Operations, 10th Annual Report 2004.* Danish Polar Center, Ministry of Research and Technology, Copenhagen.
Rogers, J.C., Wang, S.-H. and Bromwich, D.H. (2004) *Geophys. Res. Lett.* **31**(L02201), 1–4.
Rogers, J.C., Yand, L. and Li, L. (2005) *Geophys. Res. Lett.* **32**(L06709), 1–4.
Rouse, W.R. (1991) *Arctic Alpine Res.* **23**, 24–30.
Rouse, W.R. (2000) *Glob. Change Biol.* **6**, 59–68.
Rysgaard, S., Vang, T., Stjernholm, M., Rasmussen, B., Windelin, A. and Kiilsholm, S. (2003) *Arct. Antarct. Alp. Res.* **35**, 301–312.
Schmith, T. and Hansen, C. (2003) *J. Clim.* **16**, 2782–2791.
Serreze, M.C. and Barry, R.G. (2005) *The Arctic Climate System,* Cambridge Atmospheric and Space Science Series. 424 pp.

Serreze, M.C. and Hurst, C.M. (1999) *J. Clim.* **13**, 182–201.
Sturm, M., Holmgren, J. and Liston, G.E. (1995) *J. Clim.* **8**, 1261–1283.
Trans, P. (1955) *Meddr. Grønland* **127**, 6, 41 pp.
Vinje, T. (2001) *J. Clim.* **14**, 3508–3517.
Vinther, B.M., Andersen, K.K., Jones, P.D., Briffa, K.R. and Cappelen, J. (2006) *J. Geophys. Res.* **111**(D11), 1–13.
Weick, E.J. and Rouse, W.R. (1991) *Arctic Alpine Res.* **23**, 338–348.
Wendler, G. and Eaton, F. (1990) *Theor. Appl. Climatol.* **41**, 107–115.
Wojcik, G. and Kejna, M. (1991) *Acta Univ. Wratisl* **1213**, 351–363.
Yang, D., Ishida, S., Goodison, B.E. and Gunther, T. (1999) *J. Geophys. Res.* **104**(D6), 6171–6181.

Permafrost and Periglacial Geomorphology at Zackenberg

HANNE H. CHRISTIANSEN, CHARLOTTE SIGSGAARD, OLE HUMLUM, MORTEN RASCH AND BIRGER U. HANSEN

SUMMARY

Monitoring of periglacial landforms and processes together with the thermal state of the active layer and the top permafrost is part of the GeoBasis programme at Zackenberg Research Station. An important periglacial condition that has been monitored with high frequency in the Zackenberg lowland is the seasonal thaw progression of the active layer at the ZEROCALM-1 and ZEROCALM-2 sites. The active layer thickness varied between 45 cm and 80 cm during 1996–2005 at these two sites. The permafrost thickness is modelled to be 200–400 m for the Zackenberg area. Temperatures in the top permafrost at 130 cm depth reach −10 °C to −14 °C in midwinter and maximally −2 °C in late summer.

ADVANCES IN ECOLOGICAL RESEARCH VOL. 40 0065-2504/08 $35.00
© 2008 Elsevier Ltd. All rights reserved
DOI: 10.1016/S0065-2504(07)00007-4

At landscape scale, snow-cover duration in the valley Zackenbergdalen has been mapped since 1999 by continuous daily photography. These data show large interannual variations. The effect of snow on ground-thermal conditions was investigated during a 4-year period. Daily photographs provided the distribution, duration and thickness of snow at the ZEROCALM-2 site, where a seasonal snow-patch existed, and where active layer ground-thermal data from different parts of the snow-patch area were collected. These data show large interannual variations determined mainly by late-winter wind activity.

A nivation form–process–sediment model has been developed on the basis of observations in the Zackenberg area demonstrating large-scale Little Ice Age nivation activity. Periglacial landforms, such as ice-wedges, rock glaciers and solifluction lobes, are all part of the GeoBasis monitoring and demonstrate movement rates and/or climatic conditions associated with these landforms. Slope activity in the form of active layer detachment slides has occurred on the Aucellabjerg mountain during the monitoring period. Slope erosion along the river Zackenbergelven and its delta is largely controlled by late-summer peak discharges. Coastal erosion along Young Sund shows increased activity during summers with longer ice-free periods, as in 2001–2003.

I. INTRODUCTION

The Zackenberg landscape exposes a long and interesting geological and geomorphological history covering at least the last 500 million years. Topographically, the research focuses on the wide Zackenbergdalen. A large fault system formed the Zackenberg lowlands, dissecting Caledonian gneiss and granite bedrock exposed in the Zackenberg mountain to the west from younger Cretaceous, Jurassic and Tertiary sedimentary rocks to the east on the gently sloping Aucellabjerg mountain (Escher and Watt, 1976). Today, the high-arctic Zackenberg area is underlain by continuous permafrost, and landscape development is dominated by periglacial processes. Because of dry conditions (Hansen *et al.*, 2008, this volume), glaciers occur only in the mountains.

Periglacial landforms exist in the Zackenberg landscape and include ice-wedges, sorted patterns, rock glaciers, active layer detachment slides, solifluction lobes and sheets, nivation hollows and associated fans and basins together with avalanche fans. Coastal landforms along Young Sund display changes in sea ice cover. The characteristics and activity of all these landforms are an important part of the GeoBasis programme, providing improved knowledge about the development of modern high-arctic periglacial landscapes.

Permafrost is a climatically sensitive thermal state, the top of which is particularly vulnerable to climatic changes. Therefore, monitoring of the thermal state and geomorphological activity in the active layer and top permafrost is part of the GeoBasis monitoring programme. All permafrost monitoring is carried out in the valley bottom, a short distance from the Zackenberg Research Station (Figure 1). This has enabled the collection of a unique summer-thaw-progression data set in two Circumpolar Active Layer Monitoring (CALM) network sites since 1996.

II. PERMAFROST

Permafrost exists where ground temperatures do not exceed 0 °C for at least 2 years in succession (French, 2007). In Greenland, limited research has been carried out on permafrost, and the detailed distribution of permafrost is unknown in most parts of the island. On the circumpolar permafrost map (Brown *et al.*, 1997), Zackenberg is located in the continuous permafrost zone. The overall permafrost distribution in Greenland, based on air temperatures only (Figure 1 in Christiansen and Humlum, 2000), also shows that Zackenberg is located in the zone of continuous permafrost. The empirical thermal limit to discontinuous permafrost is at a mean annual air temperature (MAAT) higher than −6° to −8 °C (Washburn, 1973; French, 2007). With an MAAT in the Zackenberg lowland around −9 °C to −10 °C

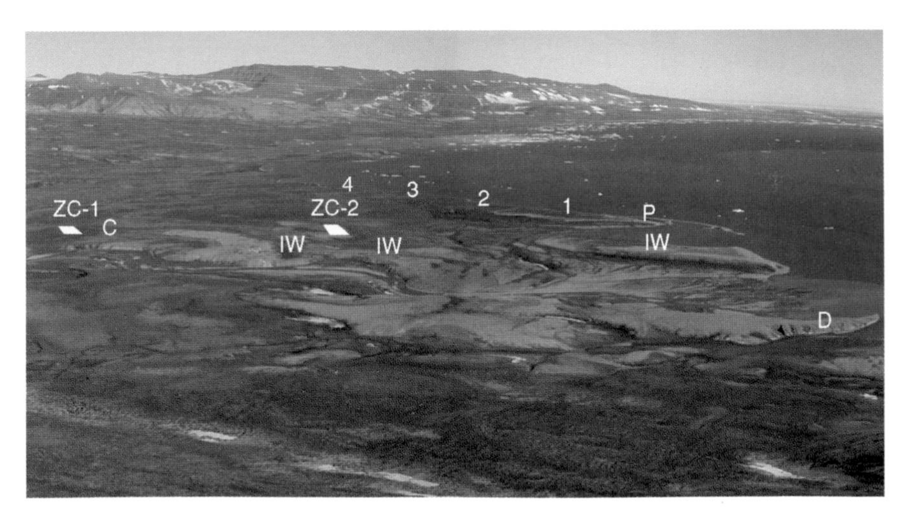

Figure 1 Eastwards view of the lower Zackenberg area, with the raised Zackenberg delta in the centre. Location of most of the periglacial monitoring sites mentioned in this chapter is given. ZC-1 = ZEROCALM-1 site, ZC-2 = ZEROCALM-2 site, 1–4 coastal monitoring sites, P = coastal monitoring profile line, D = Delta monitoring site, IW = ice-wedge monitoring sites, C = meteorological station.

(Hansen *et al.*, 2008, this volume), permafrost is thought to be continuous throughout the landscape. Through the GeoBasis programme, the Zackenberg area has developed to become the best and most intensively studied high-arctic permafrost site in Greenland.

Permafrost thickness is determined by a balance between the internal heat gain with depth below the ground and the heat loss from the terrain surface (French, 2007). The permafrost thickness has never been directly measured in Zackenberg, but, assuming a geothermal gradient of 1 °C/45 m depth, the overall permafrost depth is estimated to be around 400 m. This, however, does not take into consideration any local variations controlled by snow-cover dynamics, sediment, bedrock, rivers or vegetation, all factors affecting the local permafrost thickness.

A. Permafrost and Active Layer Modelling

1. The Virtual World for Windows Model

Permafrost and active layer thicknesses in the Zackenberg area have been calculated by a numerical model "Virtual World for Windows" (VW4W) developed by Ole Humlum. The VW4W model calculates surface and ground temperatures for any complex mountainous landscape, specified by geographical position (longitude and latitude) and spatial orientation, topographic data (digital terrain model), type of landform, type of vegetation, type of sediment, type of bedrock and geothermal heat flow. It is driven by daily values of air temperature, lapse rate, cloud cover, precipitation, wind speed, wind direction and the solar constant. From these input values, the surface energy balance, consisting of short- and long-wave radiation and turbulent fluxes, is calculated in all model points, taking into consideration the effects of topographic shading, cloud cover, exposure to wind and changing position of the sun. Local deviations in wind strength and direction in relation to the overall airflow are modelled from the topography.

The model may be run for any time period with any time step value, but typically the model is run from September 1 to August 31 the following year in the Northern Hemisphere, following the standard glaciological mass balance year, because of the importance of the seasonal snow-cover for many geomorphological phenomena, including ground-thermal conditions.

An empirical database consisting of simultaneous field observations of air temperature, cloud cover, wind strength, incoming and outgoing shortwave radiation, vegetation, landform, surface aspect, surface slope and ground-surface temperature has been established on the basis of observations from Greenland, Faroe Islands, Svalbard and Norway. This database is used to estimate ground-surface temperatures at all points of the model, driven by calculated local values of wind speed, air temperature and radiation balance. Snow is redistributed by wind in the model, and the surface temperature offset derived from snow and vegetation is then calculated using *n*-factors (Smith and Riseborough, 2002). The model also calculates the ground temperature offset, using the Romanovsky–Osterkamp offset (Romanovsky and Osterkamp, 1995). Finally, the seasonal freezing and thawing depth is calculated, using standard values for thermal diffusivity (frozen and thawed state) for bedrock and sediments. Where the calculated freezing depth exceeds thawing, the model estimates the local stable permafrost thickness from the geothermal heat flow. The seasonal thawing depth is used for estimating the local maximum active layer thickness.

Compared to other existing spatial models for seasonal frost and permafrost, VW4W typically calculates a geometrically more complicated distribution for almost any kind of terrain, as, in contrast to many other models, it takes cloud cover, topographic shading, exposure to wind, vegetation, landform and a number of other site-specific factors into account. This makes it possible to select sites in the landscape suitable for testing the model by direct measurements. A simple, yet powerful, test is measuring the ground-surface temperature in selected points in the landscape to compare real measurements with model output. Likewise, the CALM sites at Zackenberg provide good model verification points.

2. *The Modelling Results*

The model run for the period September 1, 1996, to August 31, 1997, in time steps of 3 hour calculates the overall distribution and thickness of permafrost (Figure 2) and active layer (Figure 3). Daily values of air temperature, precipitation, wind direction and speed, as measured by the meteorological station in Zackenbergdalen, were used.

The daily air temperature lapse rate in the study area was estimated by comparing temperature measurements in the main valley at the Zackenberg meteorological station with measurements at the summit of nearby Aucellabjerg (965 m a.s.l.). Unfortunately, there are some gaps in the meteorological data series from the Zackenberg meteorological station. Possible inversion effects in the lower part of the valley appear in subdued form in the value for the average lapse rate calculated. Values on daily cloud cover are also needed, but this type of observation is not carried out in Zackenberg, so simulated cloud-cover data were used in the modelling.

Geothermal heat flow is also not known for this area. As a consequence, a geothermal gradient of 2.5 °C/100 m has been adopted. Observations on wind speed and direction from the meteorological station in Zackenbergdalen are likely to be affected by topography. Measurements from a mountain

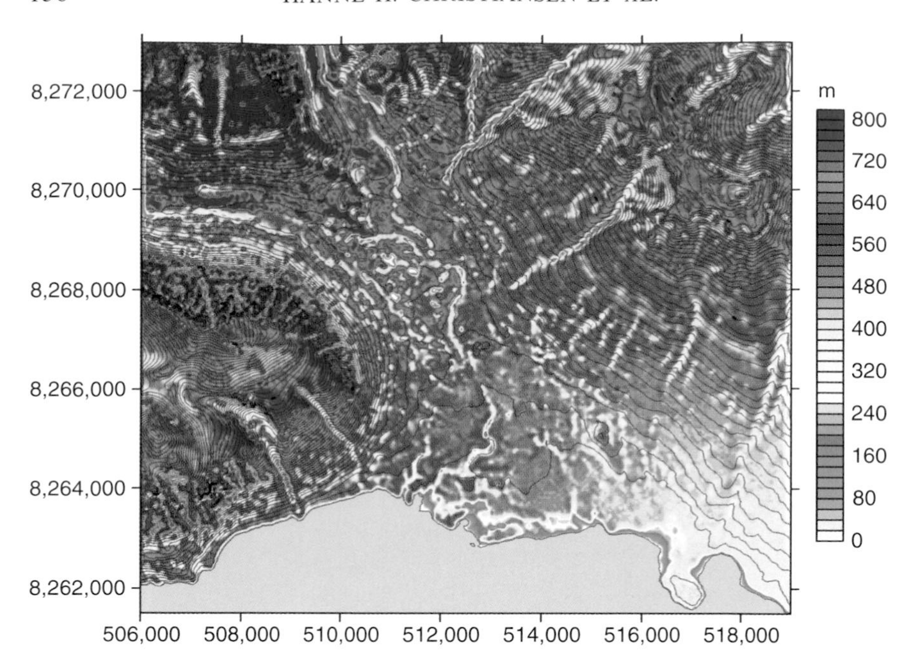

Figure 2 Calculated stable permafrost thickness in meters in the Zackenberg area, using meteorological daily values for the year September 1, 1996, to August 31, 1997. The grid size is 50 m for the topographical model used. Local variations in calculated permafrost thickness are caused by different exposures to wind, topographic shadow, accumulation of snow, heat from running water, etc. The topographic contour interval is 20 m.

station with less topographical influence clearly would have been preferable for the modelling. With these limitations, the model output shows the calculated stable distribution and thickness of permafrost, assuming no change in the meteorological conditions of 1996–1997 for a thousand years. Clearly, only the overall features of the modelling are relevant.

Permafrost is calculated to be 300–500 m thick in the mountains on either side of Zackenbergdalen, and 200–300 m in the main valley area (Figure 2). Along the shoreline, along main river courses and below larger lakes, the permafrost is thinner. Also permafrost is relatively thin in the many small valleys and gullies. This is probably due to the insulating effect of thick snow accumulation at such sites during winter. At wind-exposed sites, relatively thick permafrost is modelled.

The calculated thickness of the active layer (1997) indicates typical values of 40–65 cm in the main valley bottom and on the southwestern slope of Aucellabjerg, and 30–50 cm on the Zackenberg mountain (Figure 3). The lowland values fit well with the CALM data, as demonstrated in Section III. Local effects due to topographic shading, degree of wind exposure, sediment and bedrock type, snow-cover thickness and duration, etc. are the

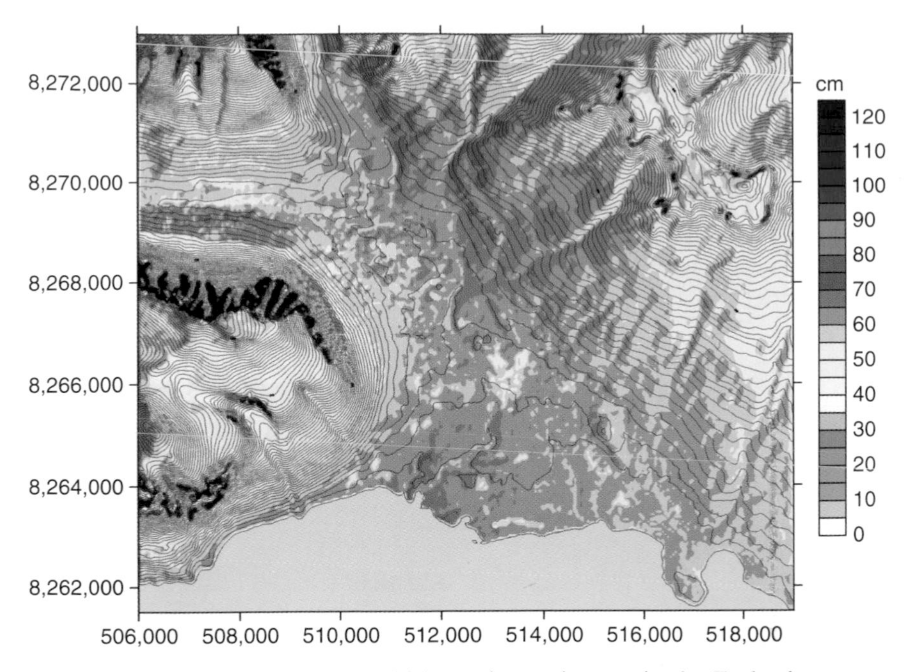

Figure 3 Calculated active layer thickness in centimetres in the Zackenberg area, using meteorological daily values for the year September 1, 1996, to August 31, 1997. The grid size is 50 m for the topographical model used. The topographic contour interval is 20 m. Shallow (green) active layer thickness is generally caused by long-lasting snow, for example, accumulated by wind or by avalanches.

background for especially thin (10–30 cm) and thick (80–100 cm) modelled active layers. Exposed bedrock generates positive active layer thickness anomalies in all cases because of high thermal conductivity.

B. Thermal State of the Top Permafrost

The near-surface permafrost temperature is measured at the meteorological station at 10 different depths (0, 2.5, 5, 10, 20, 40, 60, 80, 100 and 130 cm) from the ground surface down to 130 cm. Thermistors were installed in a tube that was inserted into the top permafrost in the 1995 summer. Monitoring takes place between the two masts of the meteorological station, with hourly recordings for five of the sensors at the east mast and the other five at the west mast. Data have been collected until 2003 (Figure 4).

The active layer thickness, as determined by the 0 °C isotherm, is 90–95 cm in all years. This value is significantly larger than the modelled values (presented earlier) and those obtained at the ZEROCALM monitoring sites (see next paragraph). Ground freezing, seen as the zero-curtain period (French, 2007) with isothermal conditions around 0 °C in the active layer, lasts for 40–60 days in the autumn, ending at the latest around November 1 (Figure 4). Freezing is mainly from the top of the active layer and downwards, and only very little contemporary freezing from the permafrost and upwards. After this period, significant ground cooling quickly occurs. In early winter, when the snow-cover is still thin, large temperature variations are seen in the upper part of the active layer. In mid to late winter, when a thicker snow-cover isolates the ground, smaller temperature variations are seen, and near

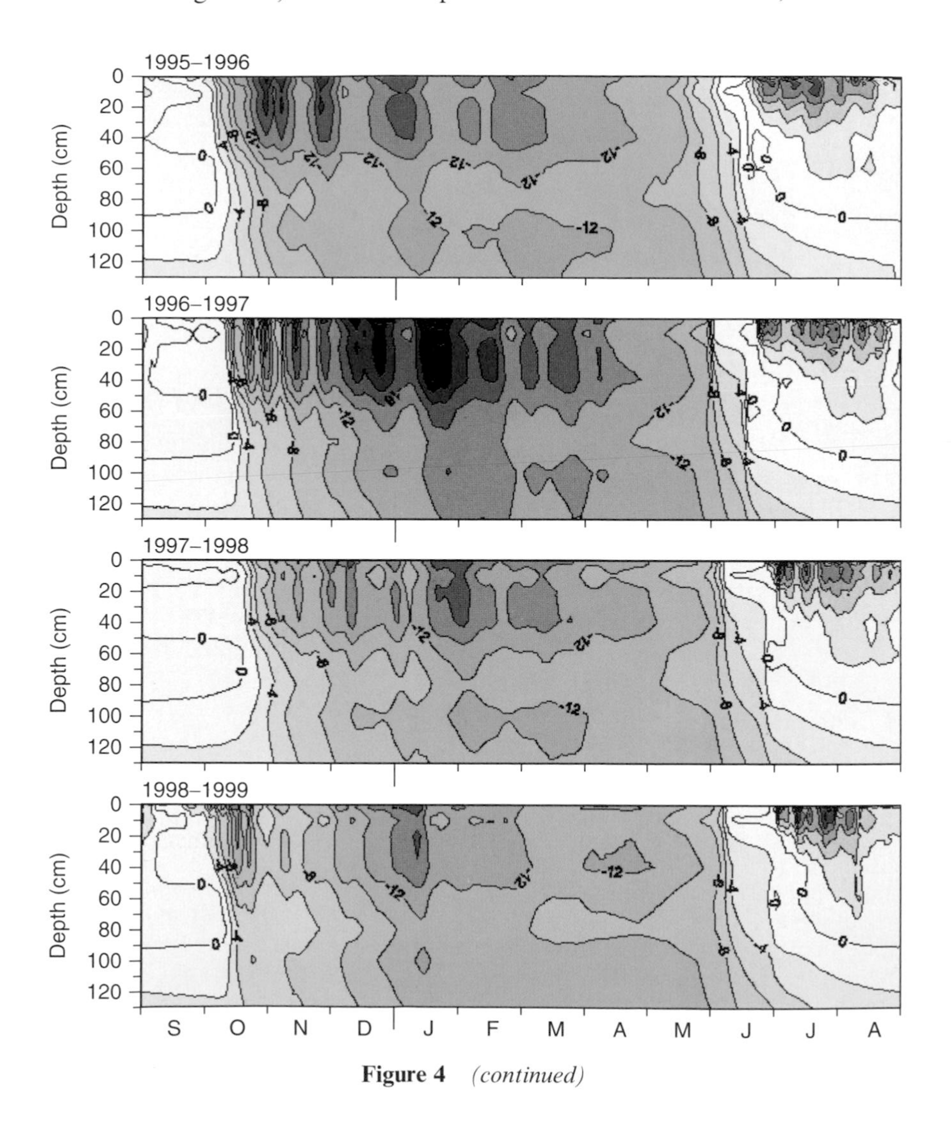

Figure 4 *(continued)*

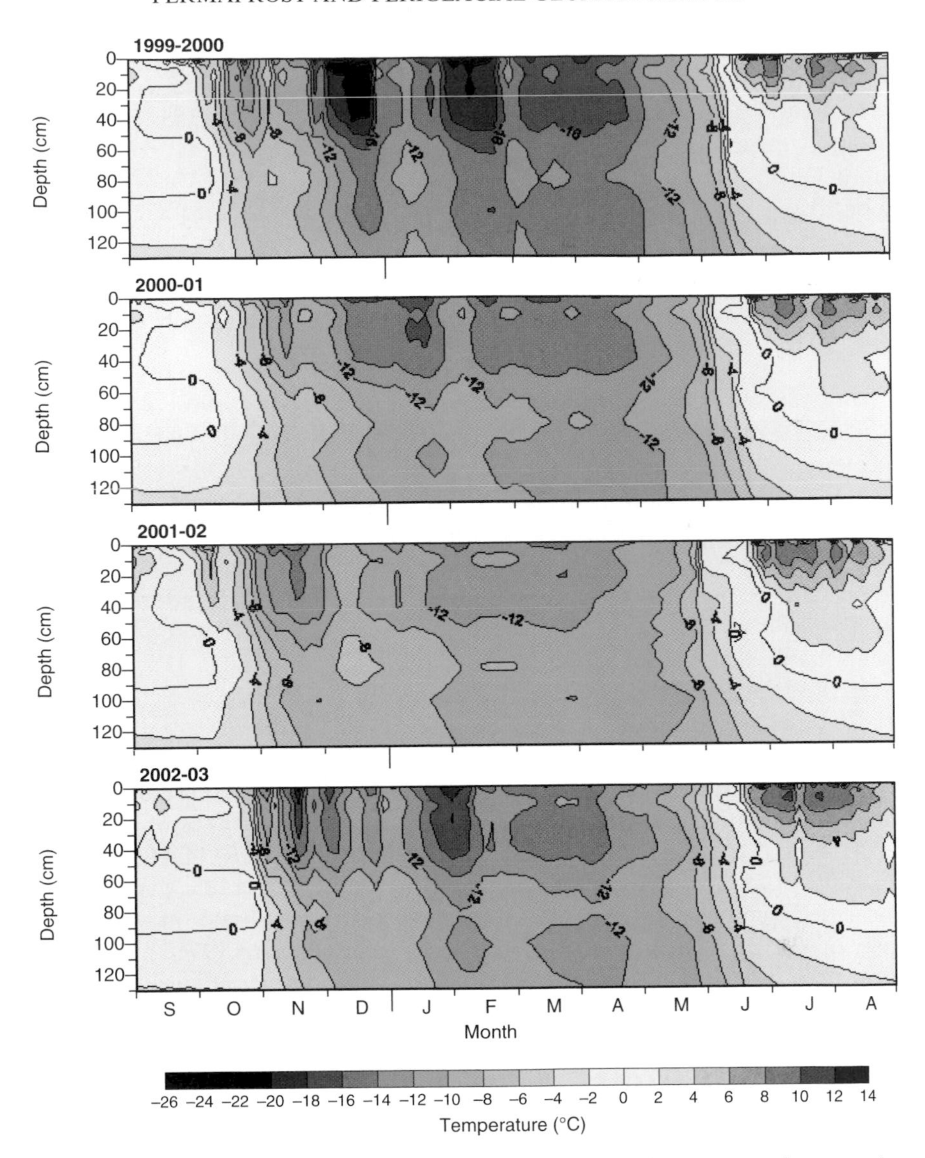

Figure 4 Ground temperatures in the active layer and top permafrost at the meteorological station at Zackenberg from 1995–1996 to 2002–2003. Timescale from September 1 to August 31 the next year.

isothermal conditions around –12 °C to –14 °C occur. Only during the colder winter in 1996–1997 did the ground cool to –14 °C to –18 °C, while in 1999–2000, it was –14 °C to –16 °C. The spring and early summer thawing is initially fast in the upper part of the active layer (Figure 4). This may be caused by meltwater percolating down the thermistor string in the tube, but is also a natural process due to large insolation during the summer.

Ground-surface temperatures in midwinter are around −15° to −20 °C. Monthly ground temperature means are presented in almost all Zackenberg annual reports. The top permafrost temperature at 130 cm varies between −10 °C and −14 °C in midwinter and is stable around −2 °C in late summer, in all years.

III. ACTIVE LAYER THICKNESS—ZEROCALM

Changes in active layer thickness are thought to be an important indicator of climatic change, particularly in high-latitude environments where climatic variations are expected to be amplified (Brown *et al.*, 2000; Stocker *et al.*, 2001). Therefore, the long-term CALM network was initiated in the late 1980s in collaboration with the International Tundra Experiment (ITEX) and was formally established in the mid-1990s (Brown *et al.*, 2000) as an important activity of the International Permafrost Association. It now includes more than 125 sites in the Arctic (Nelson *et al.*, 2004), where the late-summer active layer thickness is recorded.

Here, we present the 10-year record of ZEROCALM data collected as part of the GeoBasis programme. The controls exerted on the CALM data by air temperature, wind, snow-cover and precipitation are discussed. The first Greenlandic CALM sites were established in the central part of Zackenbergdalen in 1996 (Christiansen, 1999, 2004). Today only three CALM sites exist in Greenland, of which two are at Zackenberg, ZEROCALM-1 and ZEROCALM-2, and these two sites are the only Greenlandic CALM sites measured each year. During the summer season, the seasonal thaw progression of the ZEROCALM sites has been measured since 1996, from early June to late August, between 6 and 13 times each summer (Figure 5).

ZEROCALM-1 (100 × 100 m) is located on slightly sloping (0–2°) marine-abraded ground moraine, while ZEROCALM-2 (120 × 150 m) covers a 14° south-facing slope (Figure 2 in Christiansen, 2004). Both have 10 m grid node spacing, with 121 and 208 grid points, respectively, in which thaw depths are determined by mechanical probing. Further details on the two sites are presented by Christiansen (2004), including detailed snow-cover duration at both sites.

The 1996–2002 thaw data from the two ZEROCALM sites show a relatively constant active layer thickness around 60–71 cm in flat parts of the landscape with average snow-cover, despite that the active layer was between 80 cm and 100 cm at the ground-thermal profile (Figure 4) very close to the ZEROCALM-1 site. This difference is most likely caused by meltwater percolation down the thermistor tube. For the 1996–2002 period, however,

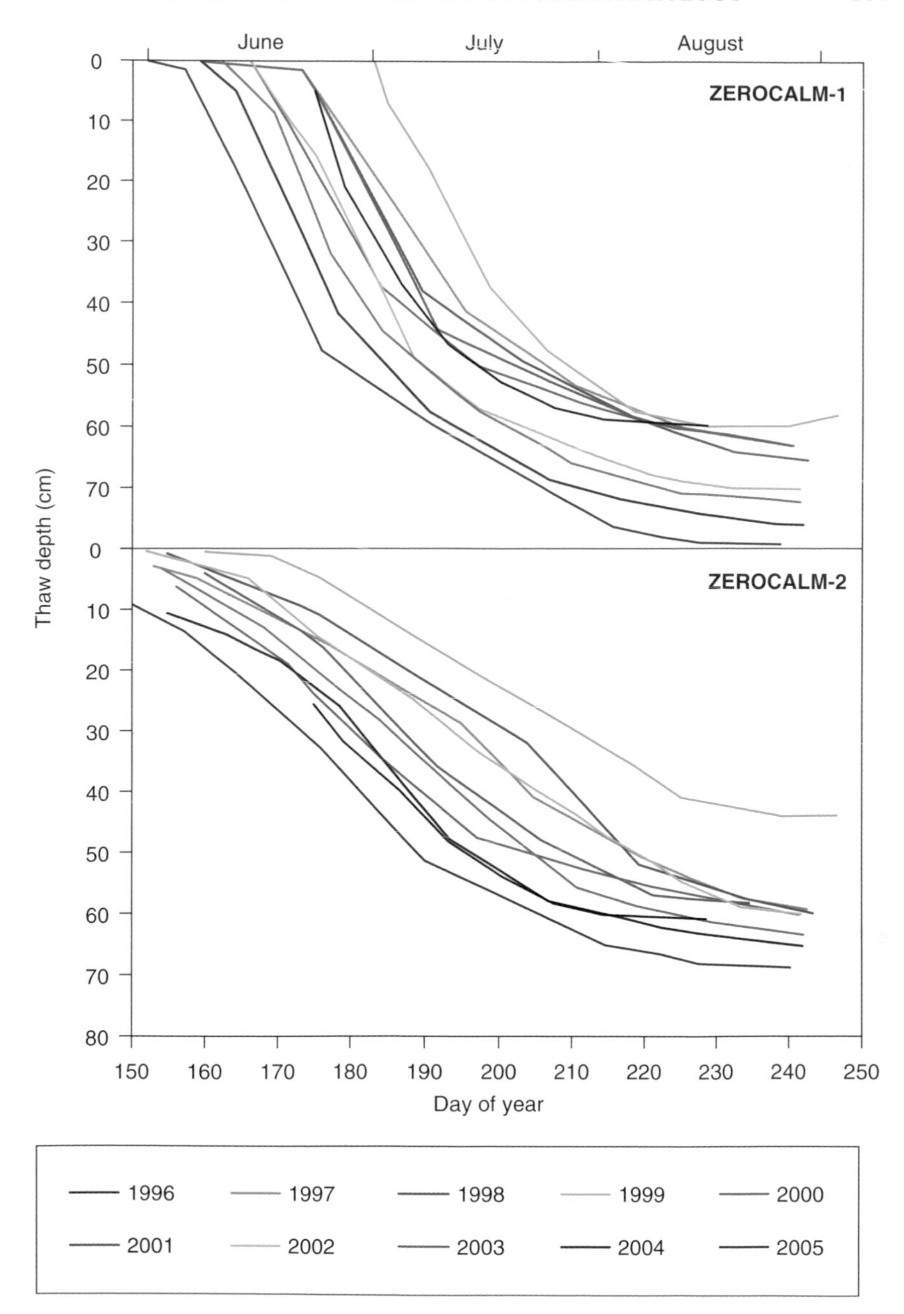

Figure 5 ZEROCALM-1 and ZEROCALM-2 thaw progression from June 1 to September 7. Each data point controlling the shape of the curves is based on averages of the 121 and 208 grid point measurements, respectively.

thawing varied from 44 cm to 61 cm in lee sites with snow-patches such as the ZEROCALM-2 site (Christiansen, 2004). This shows that significant inter-annual variation occurs, and stresses how sloping terrain is more sensitive to changes in winter meteorology, mainly late-winter wind speed and direction, than flat areas, which are more sensitive to changes in summer meteorological conditions, mainly air temperature (Christiansen, 2004).

Data from the entire 10-year period, 1996–2005, show annually increasing active layer thicknesses, especially since 2002, for ZEROCALM-1 and from 2003 for ZEROCALM-2 (Figure 5), reaching maximums of 80 cm and 70 cm, respectively, in 2005. The average active layer thickness increased by 19 cm in ZEROCALM-1 from 1996 to 2005, however, remaining at 60–66 cm until 2002. In ZEROCALM-2, the active layer thickened only by 8 cm from 1996 to 2005, mainly increasing since 2003. However, the large variation (25 cm) between the shallowest and deepest active layer in 1999 and 2005 in ZEROCALM-2 is mainly controlled by the snow-patch duration (Figure 6).

Ground thawing starts only when the ground surface is snow-free. At the ZEROCALM-1 site, this varied from around June 1 to early July, with a tendency for increasingly earlier snowmelt since 2003. The snow-cover at this flat site is homogenous. Normally the snowmelt is almost simultaneous. Thawing of the ground is quickest during the first month in this flat site, and

Figure 6 The ZEROCALM-2 site on August 24 showing the interannual variation in the snow-patch extension in the monitoring site, and on the Zackenberg mountain as seen in the background. The photographs are obtained by an automatic digital camera (Christiansen, 2001).

slower during the rest of the summer. At the ZEROCALM-2 site, the seasonal snow-patch causes a significantly differential snowmelt (Christiansen, 2004). This explains the slower thaw progression lasting the entire summer. The snow-patch accumulates during September or October mainly on the snow-patch back wall, thereby reducing ground cooling. It grows in the downwind direction until its maximum size, which is normally reached around April and May. An automatic camera standing on a 1.5 m high tripod in the ZEROCALM-2 site was completely snow-covered from February to June in both 1999 and 2002. Snowmelt starts in May. Daily, continuous photographs for the period late August 1998 to June 2002 showed that the snow-patch melted away as early as July 23 in 2000, whereas in 1999, it became perennial because of increased late-winter snow drifting (Christiansen, 2004). In 2005, it had melted by July 14, the earliest date recorded.

The presented seasonal thaw progression data show that the active layer is established from mid-August to mid-September at Zackenberg, depending on meteorological conditions during the summer thaw season and the previous winter.

IV. PERIGLACIAL LANDFORMS AND PROCESSES

The geological difference between the sides of Zackenbergdalen controls the distribution of periglacial landforms in the area. Weathering of bedrock leads to coarse-grained debris, while weathering of sedimentary rocks causes fine-grained sediments. This difference explains why rock glaciers exist only in the western part of the Zackenberg area, and ice-wedge polygons occur only in the eastern part. Several periglacial landforms are monitored to study landform activity.

A. Nivation

Snow-patches are widespread all year round in the Zackenberg landscape, stressing that most precipitation falls as snow during winter and is significantly redistributed by the dominating northerly winds. Large seasonal and perennial snow-patches therefore exist mainly on south-facing slopes in the Zackenberg area. A high-arctic geomorphological nivation model has been established for sedimentary environments on the basis of studies at Zackenberg describing the different processes, sediments and landforms occurring at the snow-patches (Christiansen, 1998a). Large parts of the eastern sedimentary Zackenberg landscape are actively being formed by nivation, with snow-patch hollows, pronival meltwater channels and sediment fans or basins (Christiansen, 1998a). Sedimentary, stratigraphical and geomorphological studies show large nivation activity since the Little Ice Age, including avalanches on the southeast corner of the Zackenberg mountain (Christiansen, 1998b), stressing the active periglacial landscape development (Bennike *et al.*, 2008, this volume).

B. Rock Glaciers

Active rock glaciers are frequent in the Zackenberg area (Christiansen and Humlum, 1993, 1996; Humlum, 1998, 1999), glacier-derived as well as talus-derived types. These permafrost-controlled landforms occur in proximity to both glaciers and large ice-cored moraines presumably deposited during the Little Ice Age. Rock glaciers are found in a small valley on the southern side and along the northeast corner of the Zackenberg mountain, along the eastern side of Dombjerg and further north in Lindemansdalen. Most talus-derived rock glaciers develop within a comparatively narrow altitudinal range from 200 m a.s.l. to 400 m a.s.l., while equilibrium lines on glaciers are located within a much broader range of 500–1100 m a.s.l., depending upon exposure to the dominating northerly winter winds. The glacier-derived rock glaciers have initiation line altitudes of 500–600 m a.s.l. Large, angular blocks typically cover the surface of the rock glacier. Most are 0.5–3 m in size, but a few measure as much as 8 m across. The rock glaciers are devoid of natural sections, but clear ice has been observed filling voids at the bottom of the surface layer in one talus-derived rock glacier (Humlum, 1998). Meteorological measurements within the Zackenberg area indicate a typical MAAT of about $-12 \pm 1\ ^{\circ}C$ at the initiation line for talus-derived rock glaciers, and about $-13 \pm 1\ ^{\circ}C$ at the glacier-derived rock glaciers (Humlum, 1999). The mean annual precipitation is presumably about 250 ± 50 mm w.eq. at 200–600 m a.s.l., although the effect of drifting snow may generate local values considerably higher or lower than this.

C. Ice-Wedges

Large-scale ice-wedge polygons can be seen in the eastern part of the Zackenberg lowland. During severely cold winter periods, ground-surface contraction cracking occurs. Later, meltwater runs down into the open cracks and freezes to form vertically foliated wedges of ice in the top permafrost.

To study ice-wedge activity, monitoring sites across three well-developed ~1–1.5 m wide, ice-wedge troughs and associated ramparts were established in summer 1995 (Figure 7). Each site had three 50 cm steel rods installed 40 cm into the active layer in the rampart tops on each site of the ice-wedge troughs, enabling three ice-wedge trough distances of between 2.4 and 3.4 m to be monitored. Some frost heaving of the steel rods had occurred by

Figure 7 Ice-wedge monitoring site with steel rods, 1 m apart, along the rampart tops on both sides of the ice-wedge trough. Open cracks in the rampart indicate a growing ice-wedge. The ruler is illustrating the location of the distance measurements. July 2006. Photo: The GeoBasis monitoring programme.

summer 2006, causing them to be significantly longer than 10 cm above the ground. Between 1995 and 2006, hardly any distance changes occurred across the ice-wedges, maximally +0.1 to −0.1 cm. The fairly open rampart cracks (Figure 7) most likely illustrate active ice-wedge growth, but this does not cause any horizontal displacement of the active layer. Excavation in 1992 of an ice-wedge, close to one of the monitoring sites, revealed a 1.7 m wide ice-wedge below the entire ice-wedge trough. However, to monitor ice-wedge growth directly, steel poles must be installed through the active layer into the top permafrost in order to measure ice-wedge growth.

Ice-wedges may cause large changes to the landscape, if the active layer thickness increases, melting the ice-rich permafrost top, or when local geomorphologically caused erosion exposes ice-wedges (Figure 8).

D. Solifluction

Several large solifluction lobes and sheets exist in the Zackenbergdalen area. Two monitoring sites were established to register the rate of solifluction activity. One was established in 1992 on the western side of the hill Ulvehøj

Figure 8 Left: Backwards erosion actively developing a 2 m deep gulley in the northern part of Zackenbergdalen in the 1999 summer. The erosion is caused by nival meltwater coming from the Aucellabjerg slopes above the lowland. Photo: The GeoBasis monitoring programme. Right: At the upward end of the gulley, an ice-wedge was exposed, suggesting that the gully has most likely developed along an ice-wedge. Photo: Hanne H. Christiansen.

Figure 9 Solifluction lobe ~1.5 m high on the western side of the hill Ulvehøj south of Aucellabjerg (background). The monitoring profile is covering the outermost 5 m of the lobe front. July 2006. Photo: The GeoBasis monitoring programme.

in the eastern Zackenbergdalen area (Figure 9). Here, 10 fixed points were established in a small profile line perpendicular to the solifluction lobe front. The profile fixed points started 40 cm in front of the *c.* 1.5 m high solifluction lobe front, and the profile length was 548.3 cm when established in

August 1992. Measurements of the individual distances between the 10 fixed points in summer 2006, 14 years later, showed forward movement mainly in the lobe front. The total length was, in summer 2006, reduced to 519.1 cm, and the fixed point right in front of the lobe had been overridden. The lower front had moved forward with an average speed of 1.25 cm/year. All the distances between fixed points on the lobe were reduced by between 0.1 and 0.3 cm/year, showing compression in the lobe front. The distance between the upper two fixed points increased by a total of 3.1 cm, indicating an annual 0.22 cm/year extension on the lobe surface.

E. Active Layer Sliding

Small-scale sliding in the active layer has most likely occurred over the slopes of Aucellabjerg since deglaciation and is a natural process, most likely with increased intensity during the Little Ice Age (Christiansen, 1998b). After an early summer >15 cm snowfall on June 15–16, 2001, followed by raising positive air temperatures to 2–3 °C, eight slides occurred in the active layer on the southwest slopes of Aucellabjerg between 400 and 600 m a.s.l. Each slide was from 50 to 500 m long, 3 to 60 m wide and only 20 to 40 cm deep. At the front of the elongated slide scars, depositional lobes up to 1 m high were deposited. The slides took place on 12–16° slopes, with vegetation mainly consisting of scattered old individuals of mountain avens *Dryas* spp. and arctic willow *Salix arctica*, indicating that the surface had been stable for some time.

Before the event, the affected slopes had been snow-free for 1–2 weeks. The snow-cover melted away during 10 days, and meltwater must quickly have saturated the shallow thawed layer, causing a rapid rise in pore-water pressure and a reduction in shear resistance, which consequently led to the active layer failures.

F. Slope Erosion in the River Zackenbergelven and its Delta

The steep slopes along the main river Zackenbergelven, and the cliffs of the raised Zackenberg delta that Zackenbergelven erodes into have experienced significant erosion since 1996, mainly during maximum summer discharge events.

The delta consists of unconsolidated silt and clay-dominated marine bottomsets, together with fluvial foresets and topsets mainly in the sand and gravel fraction deposited largely during the Holocene Climatic Optimum (Christiansen *et al.*, 2002). Delta sediments are perennially frozen below the active layer. Particularly, the *c.* 10 m high and 200 m long western side of the

Figure 10 Block detachment along crevasses parallel to the delta cliff creates clear-cut frozen blocks (left) in the western side of the present delta of Zackenberg. The white appearance on the block surfaces is due to formation of rime, when the frozen soil is exposed to the atmosphere (right). Photo: The GeoBasis monitoring programme.

present delta has experienced erosion because of a combination of fluvio-thermal undercutting of the cliff creating thermo-erosional niches along the flood water level towards the lower cliff part, and thermal erosion of the permafrost through wide open cracks developing in the active layer along the cliff top parallel to the extent of the cliff (Figure 10). Eventually, the undercut frozen sediment collapses along failure planes, and large-scale frozen blocks slump off the cliff. Horizontal recession rates of up to 10 m/year have been observed. On average, a total retreat of 30–40 m has taken place during the last 10 years along this particular part of the delta edge, causing the outer, southernmost part of the raised delta to be only 5–10 m wide at present. Slumped sediment accumulating at the slope base protects the cliff from further fluvial erosion until the material is fluvially eroded and transported further into the fjord.

Along Zackenbergelven, riverbank erosion is largely dependent on flood events. Figure 11 shows erosion along the riverbanks resulting from an extreme flood situation that occurred on July 25, 2005.

G. Coastal Activity

The Zackenberg area has a relatively long coastline affected by periglacial coastal activity. In the western part of Zackenbergdalen in the present delta of Zackenbergelven, the inner part of the shore consists of relatively narrow beaches extending seaward in an intertidal flat with a width of *c.* 100–200 m.

Figure 11 Fluvial erosion along the banks of Zackenbergelven caused by a late summer large flooding on July 25, 2005. Formation of thermo-erosional niches (left) and block slumping (centre and right). Photo: The GeoBasis monitoring programme.

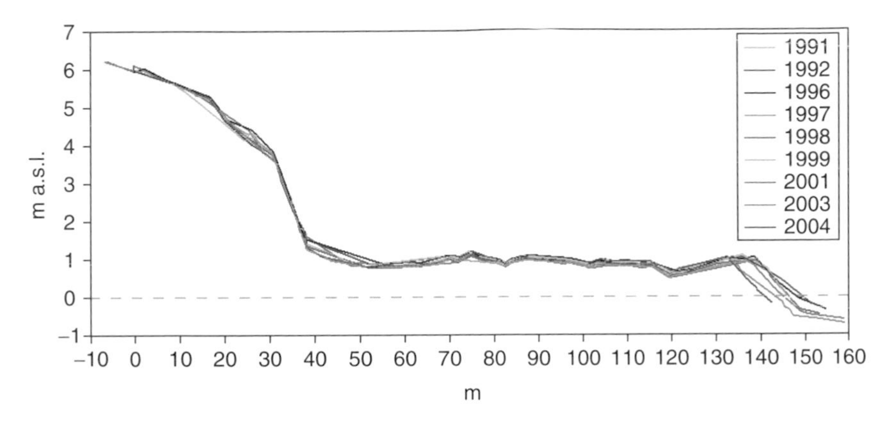

Figure 12 Coastal monitoring profile 2.

Outside the coastal zone, water depth increases steeply to more than 100 m b.s.l. Along the eastern part of Zackenbergdalen, a cliff coast with low (<5 m) bluffs dominates. This bluff terminates in a cuspate foreland consisting of a sandy spit accumulating towards the former active eastern part of the delta of Zackenbergelven (Figure 1).

Changes in the shoreline position have been monitored along the south coast of Zackenbergdalen since 1992, when two topographic profiles, crossing the coastal spit, were established at the cuspate foreland. These profiles were supplemented in 1996 with sites for measurements of coastal cliff retreat east of the cuspate foreland. Data from the coastal monitoring in Zackenberg are reported annually to the international programme, Arctic Coastal Dynamics, coordinated by the International Permafrost Association.

No significant changes in shoreline position have occurred along the cuspate foreland in the entire period of measurements (Figure 12). However, cliff coastal retreat of up to 1.4 m/year has occurred mainly in the period

Table 1 Coastline recession (m) at four monitoring sites (see Figure 1)

	Site 1	Site 2	Site 3	Site 4
1996–1997	0	0	0.3	1
1996–1998	0	0	0.3	1.3
1996–1999	0	0	0.3	1.3
1996–2000	0	0	0.5	1.4
1996–2001	0	0	0.5	1.4
1996–2002	0	0	0.7	2.8
1996–2003	0	0.4	1.6	3.2
1996–2004	0	0.5	1.7	3.2
1996–2005	0	0.7	1.7	3.2
1996–2006	0	0.9	1.8	3.2

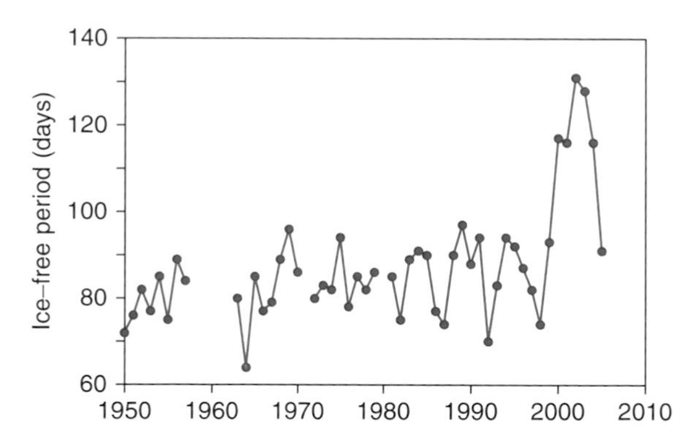

Figure 13 Length of the ice-free period as observed in Daneborg by the Sirius Dog Sledge Patrol in the period 1951–2006 (S. Rysgaard, unpublished data).

2001–2003 (Table 1), so that from 1996 to 2006 a total retreat of 3.2 m was observed especially along the eastern part of the cliff. This retreat is probably due to an increased ice-free period in the fjord, Young Sund, and a reduced concentration of old pack ice from the Arctic Ocean ("Storisen," normally drifting southwards along the east coast of Greenland) in the Greenland Sea, outside the fjord. When sea ice decreases, less attenuation of swell will occur and larger waves with more erosive power will occur along the shores. The Sirius Dog Sledge Patrol in Daneborg has registered the ice-free period since 1951. This shows a significant increase in the length of the ice-free period from 2000 to around 2004, concurrent with the observed maximum coastal retreat rates (Figure 13).

V. ZACKENBERG AS AN IMPORTANT FUTURE GREENLANDIC PERMAFROST OBSERVATORY

Some of the first CALM data were collected in Svalbard and Sweden in the North Atlantic Region (Christiansen *et al.*, 2003). A comparison of the CALM data in this region until 2002 showed that the active layer is thinnest and has the smallest interannual variation at Zackenberg (ZEROCALM-1 only) for mineral soils, and also no clear correlation with the amount of thawing-degree days (TDD) (Christiansen *et al.*, 2003). This stresses that control by local factors is as important as summer air temperature. The North Atlantic CALM data comparison also clearly showed that the relatively shallow active layer in Northeast Greenland is largely controlled by the overall cooling effect of the East Greenland Current causing colder summers than in the eastern part of the region including Svalbard.

The response of the periglacial landforms at Zackenberg to climatic changes according to the warming scenario of Kiilsholm *et al.* (2003) can be predicted only partly, but some lessons can be learned on the basis of the existing interannual meteorological variations and how these affected the periglacial landform activity during the 10-year monitoring period. With decreasing continentality, the active layer will most likely thicken because of warmer summers mainly in the flat parts of the landscape. This will enable deeper active layer detachment sliding, and more nivation sediment transport. As the active layer will also be affected by late-winter wind activity, as demonstrated by the ZEROCALM-2 event in 1999, and the predicted increase in amounts of precipitation associated with delayed snowmelt in snow accumulation sites, both conditions will lead to shallower active layers, particularly in lee site positions where larger snow-patches will exist.

Results of modelling (Figure 14) show the calculated increase in active layer thickness during the twenty-first century, using the warming scenario of Kiilsholm *et al.* (2003), which is an earlier version than the scenario by Stendel *et al.* (2008, this volume). In the main valley bottom and on the slopes of Aucellabjerg, an increase of 8–12 cm in maximum thaw depth is expected. In the northwestern part of the area and at higher elevations, a greater increase is expected, from 20 cm to 30 cm. These modelled changes especially reflect differences in ground-thermal characteristics, changes in local snow-cover and exposure to wind. Where the terrain consists of fine-grained sediments and the topmost part of the present permafrost is ice-rich, an active layer increase of 10 cm or more may cause increased frequency of active layer detachment slides and more solifluction movement on sloping ground. This especially applies to the southwest-facing slopes of Aucellabjerg.

Future permafrost changes are by nature slow, and changes in air and terrain surface temperatures will be felt only at the bottom of for example

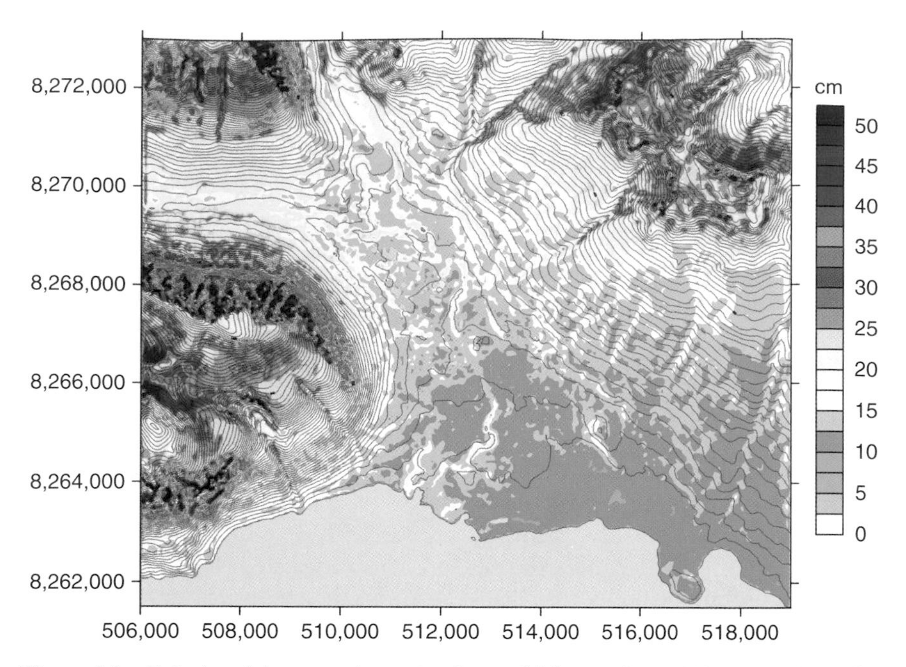

Figure 14 Calculated increase in active layer thickness from 1997 to 2097, using climate change modelling for Greenland as specified by Kiilsholm *et al.* (2003). The grid size is 50 m for the topographical model used. The topographic contour interval is 20 m.

400 m thick permafrost with a delay of several centuries. The permafrost top will react more rapidly, and the above modelled change in active layer thickness (Figure 14) indicates the likely amount of permafrost thaw during the twenty-first century using the Kiilsholm *et al.* (2003) scenario. Future higher wind speeds may complicate this picture, especially where strong northerly winds are enhanced by canalizing effects due to topography. Changes of wind and snow distribution remain very important for the future geomorphological development within the Zackenberg area.

To predict the effect of climatic changes on permafrost, long-term observations are needed. Today there are no ground temperature data below 130 cm at Zackenberg. If the active layer thickens, there will be no data available on permafrost temperatures, since there are no boreholes measuring permafrost temperatures on the entire east coast of Greenland. It would be very useful and natural, given the large number of parameters that are being monitored already at Zackenberg, to establish such permafrost thermal monitoring here, such as suggested in the International Polar Year (IPY) project Permafrost Observatory Project: A Contribution to the Thermal State of Permafrost.

A minimum of one or two 10–30 m deep boreholes with continuous temperature monitoring in different geomorphological settings would be necessary. Together with the existing near-surface thermal monitoring and meteorological recording, this would constitute an important permafrost observatory.

ACKNOWLEDGMENTS

Monitoring data for this chapter were provided by the GeoBasis programme, run by the Geographical Institute, University of Copenhagen, and the National Environmental Research Institute, University of Aarhus, and financed by the Danish Environmental Protection Agency, Danish Ministry of the Environment. Several of us, HHC, OH, MR and CS, have acted as GeoBasis programme managers since 1995, and our collaboration has enabled the presented analyses of much of the GeoBasis monitoring data. HHC acknowledges financial support for periglacial research at Zackenberg from the Danish Natural Research Council 1996–2000. The Danish Polar Center is thanked for providing good logistical facilities at Zackenberg Research Station. The University Centre in Svalbard, UNIS, is thanked for providing logistical support enabling collaboration between the first and second author of this chapter during a visit in March 2006. Søren Rysgaard is thanked for data for Figure 13, and we further want to thank Hugh French and Bo Elberling for constructive review comments.

REFERENCES

Brown, J., Ferrians, O., Heginbottom, J.A. and Melnikov, E.S. (1997) *Circum-arctic map of permafrost and ground-ice conditions,* 1:10,000,000 Map CP-45. Circum-Pacific map series, USGS.

Brown, J., Hinkel, K.M. and Nelson, F.E. (2000) *Polar Geogr.* **24**, 165–258.

Christiansen, H.H. (1998a) *Earth Surf. Processes* **23**, 751–760.

Christiansen, H.H. (1998b) *Holocene* **8**, 757–766.

Christiansen, H.H. (1999) *Geogr. Tidsskr.* **99**, 117–121.

Christiansen, H.H. (2001) *Ann. Glacial.* **32**, 102–108.

Christiansen, H.H. (2004) *Permafrost Periglac.* **15**, 155–169.

Christiansen, H.H., Bennike, O., Böcher, J., Elberling, B., Humlum, O. and Jakobsen, B.H. (2002) *J. Quat. Sci.* **17**, 145–160.

Christiansen, H.H. and Humlum, O. (1993) *Geogr. Tidsskr.* **93**, 19–29.

Christiansen, H.H. and Humlum, O. (1996) In: *Zackenberg Ecological Research Operations, 1st Annual Report, 1995* (Ed. by H. Meltofte and H. Thing), pp. 18–32. Danish Polar Center, Ministry of Research and Technology, Copenhagen.

Christiansen, H.H. and Humlum, O. (2000). In: *Topografisk Atlas Grønland* (Ed. by B.H. Jakobsen, J. Böcher, N. Nielsen, R. Guttesen, O. Humlum and E. Jensen), pp. 32–35. Det Kongelige Geografiske Selskab og Kort and Matrikelstyrelsen.

Christiansen, H.H., Åkerman, J. and Repelewska-Pekalova, J. (2003) In: *Extended Abstracts Reporting Current Research and New Information, 8th International Conference on Permafrost* (Ed. by W. Haeberli and D. Brandova), pp. 19–20. Zurich, Switzerland, 20–25 July 2003.

Escher, A. and Watt, W.S. (1976) *The Geology of Greenland.* The Geological Survey of Greenland.

French, H.M. (2007) *The Periglacial Environment.* John Wiley & Sons, England.

Humlum, O. (1998) *Permafrost Periglac.* **8**, 383–408.

Humlum, O. (1999) *Permafrost Periglac.* **9**, 375–395.

Kiilsholm, S., Christensen, J.H., Dethloff, K. and Rinke, A. (2003) *Geophys. Res. Lett.* **30,** 10.1029/2002GL015742.

Nelson, F.E., Shiklomanov, N.I., Christiansen, H.H. and Hinkel, K.M. (2004) *Permafrost Periglac.* **15**, 99–101.

Romanovsky, V.E. and Osterkamp, T.E. (1995) *Permafrost Periglac.* **6**, 313–335.

Smith, M.W. and Riseborough, D.W. (2002) *Permafrost Periglac.* **13**, 1–15.

Stocker, T.F., Clarke, G.K.C., Le Treut, H., Lindzen, R.S., Meleshko, V.P., Mugara, R.K., Palmer, T.N., Pierrehumbert, R.T., Sellers, P.J., Trenberth, K.E. and Willebrand, J. (2001) In: *Climate Change 2001: The Scientific Basis* (Ed. by J.T. Houghton, Y. Ding, D.J. Griggs, M. Noguer, P.J. van der Linden, X. Dai, K. Maskell and C.A. Johnson), pp. 418–470. Cambridge University Press, Cambridge, United Kingdom and New York, NY, USA, 996.

Washburn, A.L. (1973) *Periglacial Processes and Environments.* A.L. Washburn, Edward Arnold Ltd., London.

Snow and Snow-Cover in Central Northeast Greenland

JØRGEN HINKLER, BIRGER U. HANSEN,
MIKKEL P. TAMSTORF, CHARLOTTE SIGSGAARD AND
DORTHE PETERSEN

SUMMARY

Snow-cover is of significance not only to climate but also to hydrological and ecological systems through its control of the insulating, reflective and water storage properties at the surface of the Earth. In summer, solar radiation is the most important factor in the arctic energy budget. However, the amount of energy available for driving the soil surface systems depends on the albedo of the surface. The albedo of snow is 60–90%, whereas the albedo in snow-free areas is of the order of 10–20%. The energy budget for a given region is therefore strongly coupled to the fraction of snow-covered ground.

By use of different kinds of remote-sensing data (Landsat TM/ETM+, SPOT HRV and Digital Camera Images), the inter- and intra-annual snow-cover distribution in Zackenbergdalen has been monitored. Based on the digital camera data, a model of snow-cover depletion was developed to extend the time series of snow coverage (during melt-off) back in time. The model uses an interpolated data set, which includes information on melt energy and end-of-winter snow amounts, to calculate spatial snow-cover extent during the melting season. The results showed that the date with

ADVANCES IN ECOLOGICAL RESEARCH VOL. 40 0065-2504/08 $35.00
© 2008 Elsevier Ltd. All rights reserved
DOI: 10.1016/S0065-2504(07)00008-6

50% snow-cover in Zackenbergdalen fluctuated within a margin of approximately ±2 weeks around the mean, which was June 21.

During the cold season at high latitudes, short-wave radiation is of minor importance in the energy budget, due to the low sun. Results show that during the winter season, heat advection caused by atmospheric circulation, such as the North Atlantic Oscillation (NAO), may be important in some periods, and that the sea ice conditions in the south-eastern part of the Greenland Sea might be crucial for the winter precipitation budget in Northeast Greenland; that is, significant correlations ($R^2 = 0.5$–0.6) between snow accumulation at Zackenberg and sea ice cover to the east and southeast of Zackenberg was found, showing that reduced amounts of sea ice in the region may be likely to lead to increased snow precipitation. This effect may counteract the expected prolonged average length of the snow-free (growing) season in high-arctic Northeast Greenland in a future climate with higher temperatures.

I. INTRODUCTION

In terms of spatial extent, seasonal snow-cover is the largest component of the cryosphere (i.e., the frozen part of the Earth's surface). With a mean winter maximum extent of 47 million km^2, it corresponds to almost a third of the Earth's total land surface (NSIDC, 2005). Thus, because of its influence on energy and moisture budgets, snow-cover is an important variable at almost any spatial scale in relation to both climate and ecosystems.

A pure snow-cover reflects as much as 80–90% of the incoming solar radiation, whereas a snow-free surface, such as bare soil or vegetation, may reflect only 10–20% (Hinkler *et al.*, 2003). Besides being a strong short-wave ($\lambda \approx 0.5\ \mu$m) reflector, snow is a strong emitter of long-wave ($\lambda \approx 10\ \mu$m) radiation too. Because of these radiative properties, and because of a relatively high latent heat of fusion (i.e., it takes 160 times as much energy to transform one unit of ice to water as it does to raise the temperature of the same unit of ice 1 °C), snow-cover represents a heat sink during melt periods. Consequently, snow generally acts as an effective cooler. On the other hand, because of its poor heat transfer characteristics, snow is a great insulator. Therefore, soil surface temperature is highly dependent on the presence or absence of snow-cover (Groisman *et al.*, 1994).

Under typical winter conditions, during the polar night in Northeast Greenland, both the emissive and the insulating properties of the snowpack are of particular importance. Because of the high emissivity, temperatures can get very low in the vicinity of the snow surface, especially under calm and cloud-free weather conditions. However, because of the insulation, the soil surface organisms (e.g., the vegetation, invertebrates and small mammals) beneath the snow-pack are protected against extremely low temperatures and the resulting frost damage.

In most high-arctic regions, like Zackenberg in Northeast Greenland, virtually all vegetated areas are snow-covered most of the year (because of the presence of the vegetation, snow drifting is limited in these areas). This leaves only a short time window in which the surface is free of snow, where photosynthetic activity can take place, and where herbivores have easy access to food at the surface. Figure 1 is based on data from Zackenberg and depicts the annual bio-climatic variation in a typical high-arctic ecosystem. It emphasizes the particular importance of the short snow-free summer period for the flora and fauna, for example, vegetative activity (Ellebjerg *et al.*, 2008, this volume), and, for example, breeding conditions for shorebirds (Meltofte *et al.*, 2008, this volume). At Zackenberg, the snow-cover might start to form already in September and usually it will start to melt by the end of May. Normally it melts away (or reaches its minimum) before the end of August.

II. SNOW AT ZACKENBERG

A. Monitoring Snow-Cover

The most comprehensive amount of data available for monitoring snow-cover during melt-off at Zackenberg has been obtained using automated digital cameras. The cameras are installed in weatherproof boxes at the

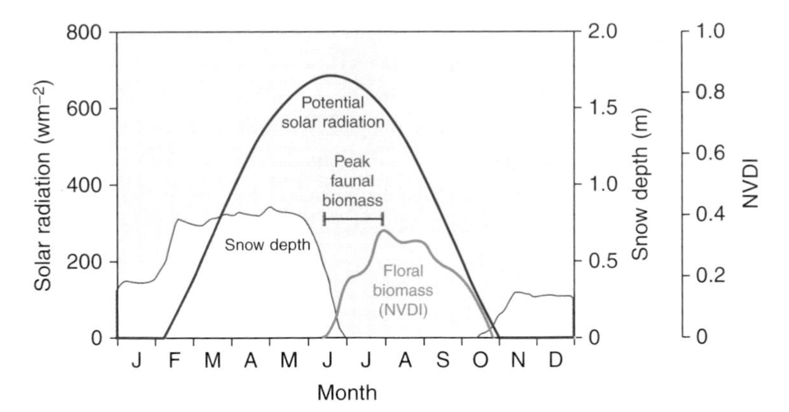

Figure 1 Seasonal characteristics of a high-arctic ecosystem based on climate data (1998 season) from Zackenberg Research Station (74°30′N) in Northeast Greenland. Normalized Difference Vegetation Index (NDVI) was inferred from Advanced Very High Resolution Radiometer (AVHRR) satellite data provided by the National Snow and Ice Data Center, Boulder, Colorado, USA.

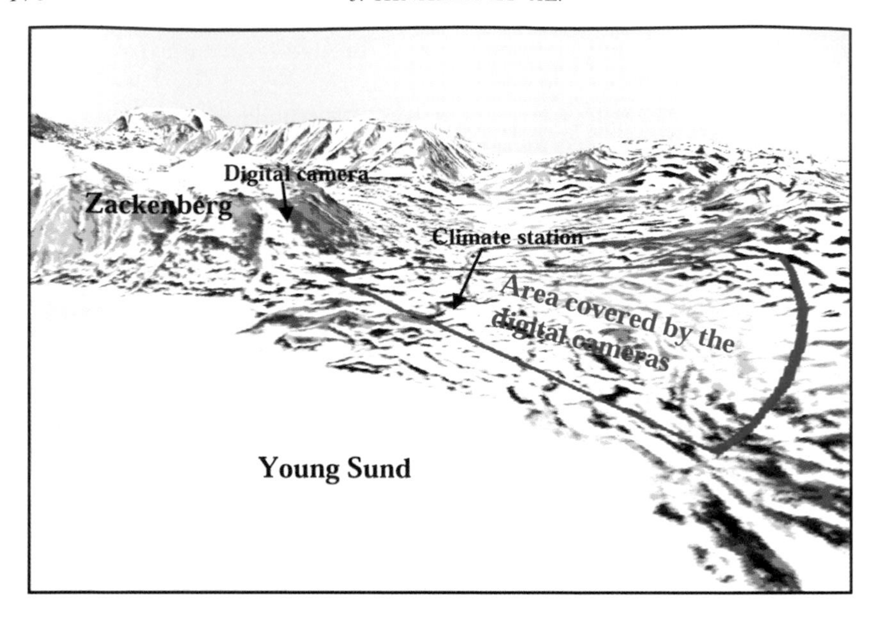

Figure 2 Zackenbergdalen with indication of snow-camera coverage.

eastern slope of the Zackenberg Mountain at ~500 m above sea level. This gives a field of view covering ~17 km^2 of Zackenbergdalen (Figure 2).

The first camera was set up in August 1997, and daily images of the valley have been obtained since then. This gives the opportunity to describe snow-cover during melt-off in Zackenbergdalen in excellent spatial and temporal detail. Through orthographic rectification (projection of oblique images onto a geocoded map) the oblique images taken by the cameras are transformed into digital orthophotos (Hinkler *et al.*, 2002). These images are then used for derivation of binary snow-cover maps, indicating whether pixels are snow-covered or not. The snow classification is done using a *Red Green Blue* Normalized Difference Snow Index (*RGB*NDSI) algorithm, which is described by Hinkler *et al.* (2002).

One of the main advantages of camera-based snow monitoring is that it is relatively insensitive to cloud cover (in contrast to satellite-based techniques). Only low clouds and foggy conditions can make the image data unsuitable for mapping purposes. Thus, for each melting season, snow-cover depletion curves are constructed on the basis of image data obtained at daily frequency. However, in some periods, unfavourable weather conditions (mainly fog coming in from the sea) interrupt the data series. The whole procedure from capturing the images to derivation of snow-cover depletion curves is illustrated in Figure 3.

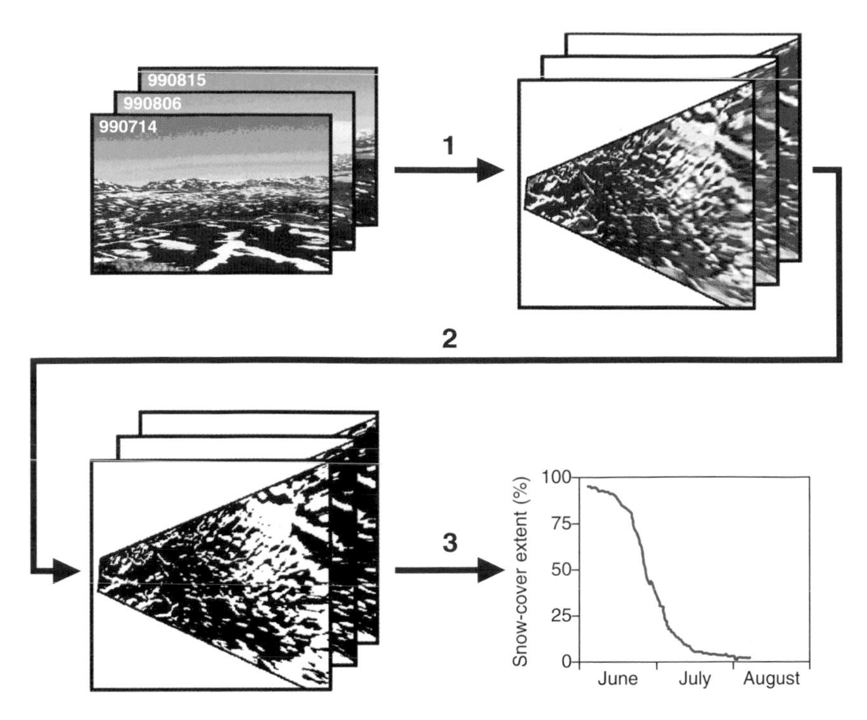

Figure 3 The steps used to extract snow-cover information from digital camera images:1. orthographic rectification, 2. conversion to snow-cover maps using a specially designed algorithm and 3. calculating snow-cover extent for different dates to create a snow-cover depletion curve.

B. Winter: Snow Accumulation and Distribution

Temperatures are generally below 0 °C from September to May and above 0 °C from June to August. Heavy snowfall within the summer period (June–August) has occurred only once (2001 season) since establishment of the research station. This means that virtually all snow-cover depletion takes place during the summer period, whereas accumulation takes place only in winter. In high-arctic areas, this rather sharp division between accumulation and melt periods is relatively common (e.g., Liston, 1999).

In Northeast Greenland, winter precipitation and high wind speeds are usually connected with cyclonic activity over the Greenland Sea. Consequently, at Zackenberg, heavy snowfall is usually connected with both increased air temperatures and increased wind speeds (Figure 4; see also Hansen *et al.*, 2008, this volume).

During winter in Zackenberg, the prevailing wind direction is from north–north-west, and generally the variability around this direction is small

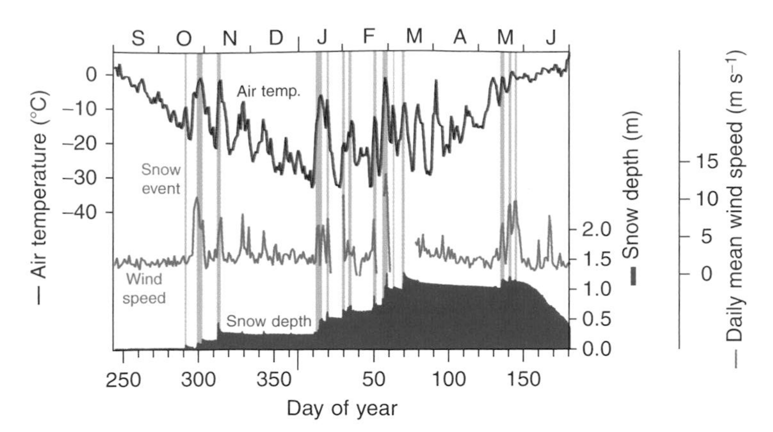

Figure 4 Air temperature, wind speed and snow depth measured at the meteorological station at Zackenberg from September 1, 1998, to June 30, 1999. Major snowfall is usually connected with increasing temperatures and wind speeds.

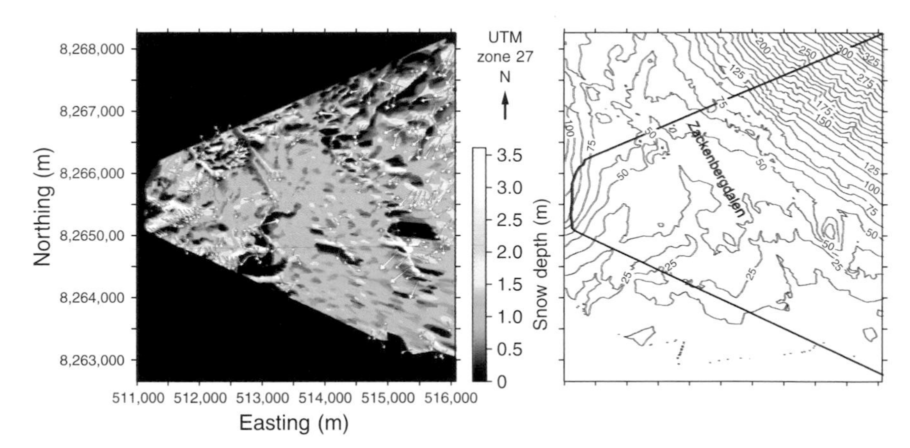

Figure 5 Modelled snow depths as of June 1, 1999, based on binary snow-cover maps and modelled melt rates. White arrows indicate direction and magnitude of the terrain slope at the positions of the larger snowdrifts. The rightmost map shows altitudinal levels of Zackenbergdalen with indication of the modelling area.

(Hansen *et al.*, 2008, this volume). This is important in relation to the distribution of the snow-cover, because it means that the snow-cover distribution patterns in Zackenbergdalen are very similar from year to year, with the larger snow drifts located at the same locations, mainly on southerly facing slopes. The characteristic end-of-winter snow-cover distribution pattern in the valley is shown in Figure 5, which illustrates snow depth distribution patterns for June 1, 1999 (the largest snow accumulation year recorded so far). The mean end-of-winter snow depth within the mapping area was

96 cm and maximum and minimum were more than 3.5 m and less than 20 cm, respectively. It can be observed that the accumulation of the larger snowdrifts on the western valley side is generally orientated to the southeast, whereas on the eastern side it is to the southwest. On the valley floor, where the terrain is more flat, the accumulation is more uniform with only a few large snowdrifts.

The map in Figure 5 was constructed using camera-based snow-cover maps in conjunction with modelled melt rates (Liston and Hall, 1995; Liston *et al.*, 1999). For each day in the melting season, every snow-covered pixel in the corresponding binary snow-cover maps was converted to daily melt rates. The accumulated melt rates (accumulated over the melting season for each snow-covered pixel) were then converted to corresponding water equivalents (w.eq.), which were further converted to snow depths using information on the end-of-winter snow density. Because the research station at Zackenberg is unmanned in winter, snow densities have so far only been measured during summer time, and thus there is no information available on density variation during the accumulation (winter) period. The snow density used in the derivation of the map in Figure 5 is therefore based only on density measurements carried out in the beginning of June, and these measurements are considered the best possible representation of end-of-winter snow densities in the area. In connection with snow depth measurements using ground-penetrating radar, Larsen and Karlsen (2003) found the mean snow density within the study area (June 2–5, 2002) to be 517 kg/m^3, whereas measurements carried out in early June 2004 and 2005 indicated a mean value of 475 kg/m^3. Hence, an average density of 496 kg/m^3 was used in the derivation of the accumulation map shown in Figure 5.

C. Summer: Snowmelt/Snow-Cover Depletion

The depletion of the snow-cover in Zackenbergdalen is typical for mountainous areas where the snow is unevenly distributed because of rugged terrain (Hall and Martinec, 1985). Therefore, snow-cover depletion curves (Figures 6 and 7) generally tend to have a "reversely S-shaped" form. This form is most pronounced in snow-rich seasons. When melting starts, most of the valley is snow-covered, and thus in most years, during the initial part of the melting season, the snow-cover extent stays above 90%. The steepest part of the depletion curves then represents the time where extensive areas with uniform snow deposition (mainly on the valley floor) rapidly become free of snow. In the later part of the summer, only a small fraction of the snow-cover, consisting of the remnants of the largest snowdrifts, is left. In snow-poor years, all of the snow-cover usually melts away. In snow-rich years, on the other hand, parts of the largest snowdrifts still remain by the end of

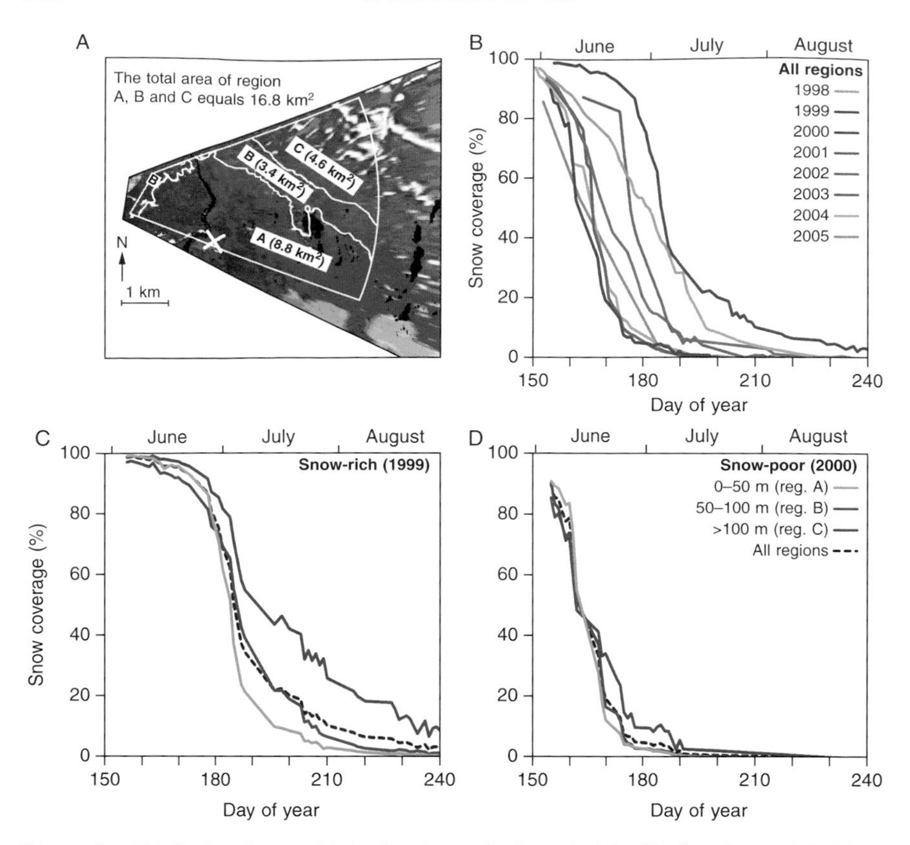

Figure 6 (A) Orthophoto with indications of selected altitudinal regions: A 0–50 m, B 50–100 m, C > 100 m, together with the position of the research station (X). Black areas are hidden when the valley is viewed from the camera position. (B) Snow-cover depletion curves 1998–2005 (day of the year May 30–August 28) for all regions (regions A, B and C viewed as one region). (C) Snow-cover depletion in altitudinal regions A, B, C, and all regions in a snow-rich season (1999). (D) Snow-cover depletion in altitudinal regions A, B, C, and all regions in a snow-poor season (2000).

summer and will therefore stay present during the forthcoming winter and may eventually transform into firn.

Figure 6A shows an orthophoto illustrating three altitudinal levels: 0–50 m (A), 50–100 m (B) and >100 m (C). The snow-cover depletion curves for the entire area are given in Figure 6B, showing that the time at which major snowmelt (steepest part of the depletion curves) occurs has varied considerably by as much as 3 weeks during the period of snow monitoring (1998–2005). This variability has been even larger when a longer time series is considered (Figure 7, more details in next section). The variation in snow accumulation and melt with altitude is shown in Figure 6C–D. Generally,

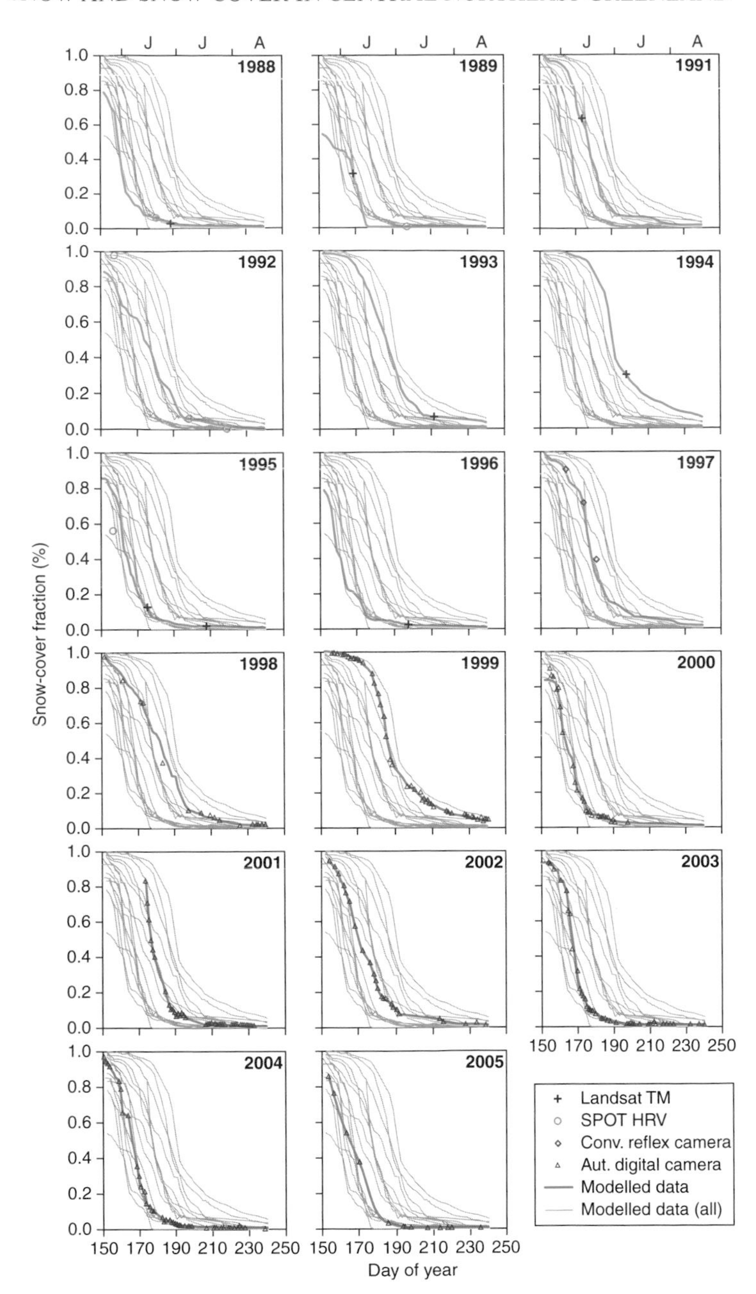

Figure 7 Snow-cover depletion curves for Zackenbergdalen (day of the year May 30–September 7, 1988–2005), measured and modelled data. For each year, the actual graph is shown together with graphs from all the other years given with thin grey lines.

there is a larger accumulation and thereby also a delay in snow-cover depletion at higher altitudes. The delay is much more pronounced in snow-rich (e.g., 1999) than in snow-poor years (e.g., 2000), indicating more intense snowdrift in the snow-rich years. As shown in Figure 5, the larger snowdrifts are located on the valley sides, which explain why snow-cover depletion is particularly delayed at higher altitudes and not on the valley floor.

The depletion curves in Figure 7 are divided into two different sets: 1988–1997 and 1998–2005. The latter one (1998–2005) is based exclusively on digital camera data (binary snow-cover maps derived through orthographic rectification); these curves therefore give the highest precision. Since no digital cameras were installed in the melting seasons prior to 1998, the depletion curves from before 1998 are based on retrospective modelling. The model was constructed using the comprehensive amount of snow-cover information represented by the digital camera data; and it only requires daily temperatures and snow-cover extent from (at least) one day in the melting season as input (see Box 1 for a detailed description of the model). For the

Box 1

Retrospective Snow-Cover Modelling

Modelling snowmelt/snow-cover depletion requires quantification of the energy available for melting. The melt energy balance is formulated as a requirement for conservation of energy:

$$(Q_{sn} + Q_{ln} + Q_h + Q_e + Q_g + Q_p) + Q_m = 0 \quad (1)$$

where

Q_{sn} = net short-wave radiation flux absorbed by the snow
Q_{ln} = net long-wave radiation flux (Q_l incoming + Q_l emitted) at the snow–air interface
Q_h = convective or sensible heat flux from the air at the snow–air interface
Q_e = flux of the latent heat (evaporation, sublimation, condensation) at the snow–air interface
Q_g = flux of heat from the snow–ground interface by conduction
Q_p = flux of heat from rain
Q_m = melt energy

However, in most practical cases, it is not possible to obtain the necessary information to exactly derive all terms of energy input (Q_{sn}, Q_{ln}, Q_h, Q_e, Q_g, Q_p). Therefore, modelling snowmelt normally requires a simplification of the energy balance. This is, in particular, true for

Zackenberg when considering years prior to 1996, where the only way to obtain information on energy available for snowmelt is via air temperatures measured at an automatic weather station in Daneborg about 21 km to the southeast of Zackenberg (see Figure 2). One of the most common simplifications is the degree day model, which uses the sum of temperatures above the melting point (base temperature) as a "melting potential" (Rango and Martinec, 1995).

For the time prior to 1998, detailed information on snow coverage at Zackenberg is available only through a limited amount of satellite data with higher spatial resolution (Landsat TM/ETM + and SPOT HRV). The requirements behind modelling snow-cover depletion curves for the Zackenberg area can thus be stated as follows: Based on digital camera data, it should be possible to construct detailed snow-cover depletion curves, retrospectively, using only daily observations of air temperature in conjunction with a limited number of satellite images (1–3 per melting season).

A snow-cover depletion curve describes the decline in the spatial snow-covered fraction during the melting season, and can be expressed in two different ways–either as a function of time or a function of cumulative melt energy (Menoes and Brubaker, 2001).

The advantage of the melt energy-based snow-cover depletion curve is that the shape of the curve is dependent on only two factors: (1) end-of-winter snow-cover distribution and (2) end-of-winter snow accumulation—provided that there is no snowfall within the melting season.

Because of a more or less fixed wind regime during winter, the end-of-winter snow-cover distribution at Zackenberg is virtually similar from year to year (see characteristic end-of-winter distribution in Figure 5). On the contrary, winter precipitation (and thereby also the end-of-winter snow accumulation) shows great inter-annual variation. This means that for Zackenberg, the shape of a melt energy-based snow-cover depletion curve is dependent on only one factor, namely, the end-of-winter snow accumulation.

In the derivation of the snow-cover depletion model this fact is utilized. The model simply consists of a data set which holds all melt energy-based depletion curves possible at Zackenberg—meaning depletion curves corresponding to any end-of-winter snow accumulation.

The data set (visualized in Box Figure 1) is constructed using the snow-cover observations from the digital camera images, end-of-winter snow depths and air temperatures (as accumulated degree days). The thick dots illustrate the camera observations from the melting seasons 1998–2004 (except 2001, which had snowfall within the melting season), and

(continued)

Box 1 *(continued)*

the data in between these observations were constructed using Kriging interpolation (Stein, 1999).

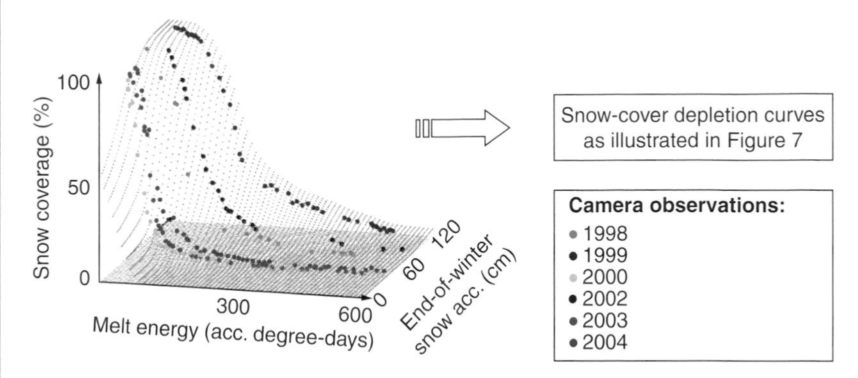

Box Figure 1 Visualization of snow-cover depletion modelling based on digital camera data. The Curvature illustrates the interpolated data set from which any possible snow-cover depletion curve in Zackenbergdalen can extracted.

The concept of the model is, for a given melting season, to select the appropriate melt energy-based depletion curve from the data set using a specially designed computer programme. If the end-of-winter snow accumulation is known, the modelling is an easy task, because there is one and only one melt energy-based depletion curve, which corresponds to a given end-of-winter snow accumulation. However, if the end-of-winter snow accumulation is not known (like in the years prior to 1998), the data set can be used iteratively, provided that there is at least one day in the melting season, at which the snow-cover fraction (from, e.g., a high-resolution satellite image) is known together with the accumulated degree days. The melt energy-based depletion curve that is closest to intersect the given point(s) (acc. deg. days, snow-cover fraction) is then selected from the data set. After the iterative modelling also the end-of-winter snow depth is known, because a certain depletion curve corresponds to one and only one end-of-winter snow accumulation.

period 1998–1997, the snow-cover maps were derived from high-resolution satellite data (Landsat TM and SPOT HRV). The year 1990 is absent because no cloud-free satellite images were available this year.

The large inter-annual variation in snow accumulation in Zackenbergdalen implies large variations in the length of the snow-free season. The variation is clearly expressed when comparing snow-cover depletion curves from different melting seasons (Figure 7).

The date when spring melt has reduced snow-cover to 50% has fluctuated within a margin of approximately ±2 weeks around the mean, which is June 21. Minimum and maximum values were June 5 (1989) and July 9 (1994), respectively.

When the date with 10% snow-cover is considered, the variability increases with a time span of 51 days between maximum and minimum dates (June 21 in 1988 and 1989, and August 11 in 1994). Most likely, this is because in snow-rich years larger snow deposition occurs in the main accumulation areas, resulting in further prolongation of the melting period within these areas. The relation is further emphasized when comparing years with respect to both snow conditions and temperature; for example, 2002 had a large snow accumulation (87 cm on June 1), whereas the year 2003 had one of the smallest accumulations recorded (45 cm on June 1). The day with 50% snow-cover occurred at about the same time in both years (June 19 and 16, respectively). The explanation is that although both years had relatively high temperatures in June, temperatures started to remain above freezing level already in mid May in 2002, whereas snowmelt first started in early June in 2003. However, the date with 10% snow-cover occurred more than 2 weeks later (July 9 vs June 24) in 2002 than in 2003, indicating a more intense formation of snowdrifts in years like 2002 with high snow precipitation.

As indicated above, the inter-annual variability in end-of-winter snow accumulation at Zackenberg is large, and thus highly affects snow-cover and thereby the general ecosystem conditions in Zackenbergdalen during the short summer season (e.g., Ellebjerg *et al.*, 2008, this volume; Høye and Forchhammer, 2008, this volume; Meltofte *et al.*, 2008, this volume). Early spring snow depths, as measured on June 1 (using a sonic depth sensor) at the Zackenberg meteorological station, ranged from 26 cm (2005) to 111 cm (1999), and when modelled values (derived iteratively in connection with snow-cover depletion modelling, see Box 1) are included, the range is even larger: 6 cm in 1989 and 111 cm in 1994 and 1999.

D. Larger Scale Relations and Future Climate Change Perspectives

1. Comparison with NCEP Data

To study how local-scale snow precipitation is related to more regional scale variations, the local end-of-winter snow accumulation in the Zackenberg region (also including Daneborg) was compared to large-scale reanalysis data of w.eq. from the National Centers for Environmental Prediction (NCEP). The comparison is shown in Figure 8.

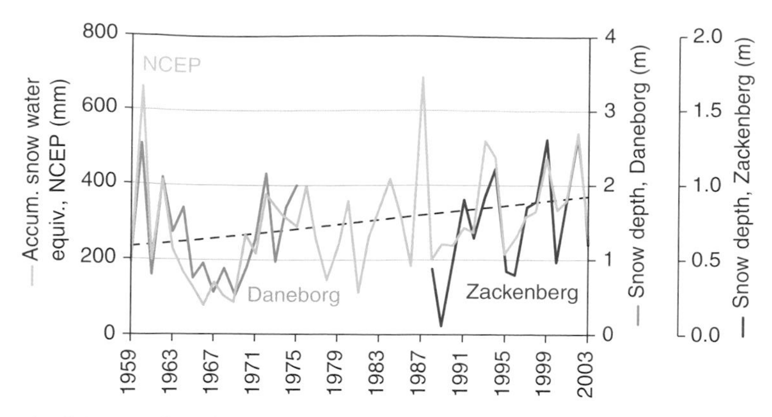

Figure 8 Measured and retrospectively modelled end-of-winter snow depths at Zackenberg and Daneborg shown together with National Centers for Environmental Prediction (NCEP) reanalysis snow water equivalents (w.eq.) (Zackenberg pixel) accumulated over the winter. The dashed line shows the trend in NCEP-based snow precipitation over the whole period.

The NCEP-based end-of-winter w.eq. values covering 1959–2003 are from a 1.904° × 1.875° (latitude × longitude) grid-cell, which includes the Zackenberg region, and are calculated as all precipitation accumulated through October-May, falling when the air temperature is below 0 °C. At the meteorological station in Daneborg, measurements of snow depth ended in 1975, and thus local-scale snow depths from 1976 to 1987 are not available. Generally, the variability in NCEP-based w.eq. corresponds well with ground-based snow depth data from Zackenberg and Daneborg ($R^2 = 0.69$ and 0.60, respectively), and seen through the whole period (1959–2003) there has been a significantly increasing trend ($p < 0.006$) of 35 mm w.eq. per decade. The amount of scatter, which however exists, may be caused by several different factors such as uncertainties in the NCEP analysis/forecast system, snowdrift/melt, complexity in local wind patterns and precipitation distribution (which is likely, because the region is characterized by strong topography), and the fact that a 13,650 km^2 grid-cell may represent a variety of climatic conditions (e.g., Daneborg is exposed almost directly to the Greenland Sea, whereas Zackenberg is situated inside a fjord system). Particularly striking for the NCEP-based end-of-winter w.eq. is the exceptionally high values in the years 1960 and 1987. Unfortunately, no comparison to ground truth was possible for 1987 since no ground-based measurements are available for that year. However, the 1960 observations from Daneborg may suggest that even though this year had one of the highest amounts of snow precipitation ever recorded, the NCEP reanalysis might overestimate precipitation rates in extreme situations (such as in 1960 and 1987). Furthermore, the fact that the number of ground stations is limited in the region might also increase uncertainty in the reanalysis data.

2. *Correlation with NAO*

A change in larger scale snow precipitation rates and distribution patterns naturally requires changes in the atmospheric circulation at synoptic scales and/or in the hydrological cycle. From middle to high latitudes in the northern hemisphere, the North Atlantic Oscillation (NAO) (Hurrell, 1995) is the most prominent weather pattern and explains the major inter-annual variability of the atmospheric circulation over the middle and high latitudes (Stendel *et al.*, 2008, this volume). It is most pronounced during winter and is related to changes in temperature, storminess, sea ice distribution and precipitation patterns (Hall and Martinec, 1985; Hurrell *et al.*, 2003; Johannessen *et al.*, 2004). The correlation between NAO and precipitation over the western and south-western parts of Greenland is thoroughly documented (e.g., Hurrell, 1995; Appenzeller *et al.*, 1998; Bromwich *et al.*, 1999), whereas for high-arctic Northeast Greenland information on the topic is limited.

3. *Correlations with Sea Ice Conditions*

One factor which may significantly impact the hydrological cycle in the Arctic would be reductions in sea ice coverage, which are predicted to occur during the current century (Stendel *et al.*, 2008, this volume). A decreased sea ice coverage would allow greater evaporation from the ocean surface and result in larger moisture transport onto coastal areas and hence, larger amounts of precipitation.

In a recent study by Hinkler (2005), the variability in end-of-winter snow precipitation amounts around Zackenberg was analyzed in relation to the NAO and to sea ice variability within the Greenland Sea. Both of the analyses were made over space and time.

The analysis on NAO versus snow precipitation includes snow precipitation from the entire northern North Atlantic in the form of $1.904 \times 1.875°$ grid-cells focusing, in particular, on the grid-cell containing Zackenberg (Figure 9). In each cell, the running correlation (19-year running windows) between snow precipitation (October–May) and the NAO was calculated. The results indicate that in some periods the NAO might be an important mechanism in controlling snow precipitation in Northeast Greenland. The degree of importance, however, varies significantly over time, possibly in an oscillating manner. During most of the seventies, the correlation between snow precipitation in the Zackenberg grid-cell and the NAO was clearly significant, whereas both before and afterwards it was insignificant. Thus, because the correlation between NAO and snow precipitation is insignificant during two thirds of the time, other (atmospheric as well as oceanic) factors affecting snow precipitation may be more important than the NAO.

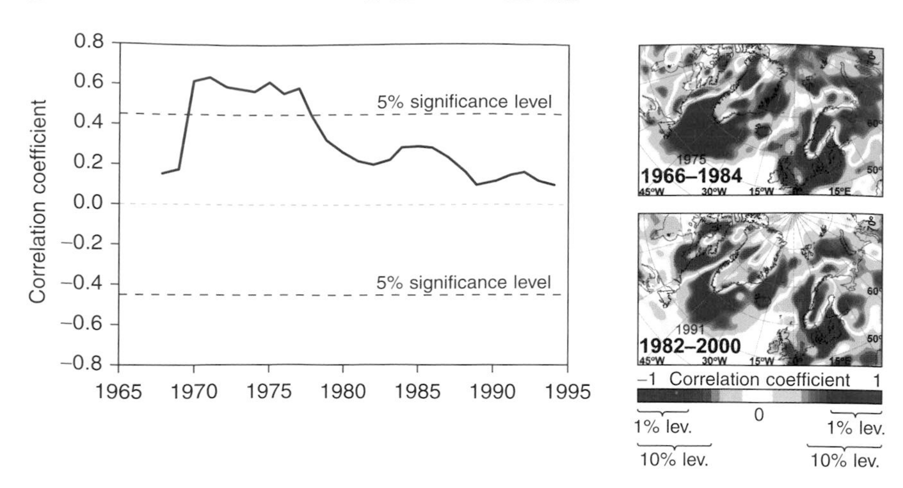

Figure 9 Running (19-year window) correlation between snow precipitation and the North Atlantic Oscillation (NAO). The graph displays the running correlation 1959–2003 (centre years 1968–1994) for the pixel-containing Zackenberg. The maps to the right display it for the entire northern North Atlantic, in two selected windows: 1966–1984 (window with maximum correlation in the Zackenberg grid-cell) and 1982–2000 (the period covered by the sea ice vs snow precipitation analysis), respectively.

The apparent oscillating behaviour of the correlation could indicate that periodically precipitation over Northeast Greenland gets coupled or decoupled to/from the NAO (this is in contrast to the situation in, e.g., West Europe and Southwest Greenland, where the correlations are always significant, see correlation maps in Figure 9). The reason for this is presently not known, and should be subject for further investigation. Hypothetically, it may be related to shifts between phases where more or less continental/coastal climatic regimes predominate.

Interestingly, a significant negative correlation (meaning that reductions in sea ice coverage might lead to increases in snow precipitation) between sea ice duration in the south-easterly part of the Greenland Sea and end-of-winter snow accumulations in the Zackenberg region was found. Sea ice duration is defined as the amount of time (number of days) each year with sea ice coverage above a certain threshold. To depict the spatial variation of the correlation, the Greenland Sea was divided into sectors and sub-sectors based on direction and distance from Zackenberg (Figure 10), and then a correlation coefficient (sea ice duration in each sub-sector vs snow precipitation at Zackenberg) was calculated. The highest correlations were found to the southeast of Zackenberg and reach a maximum in the south-southeast (SSE) region, when ice data up to a distance of 750 km from the research station are included. This may indicate a "centre of action" to be situated within this region around 500 km north of Iceland between Greenland and Jan Mayen. However, because sea ice

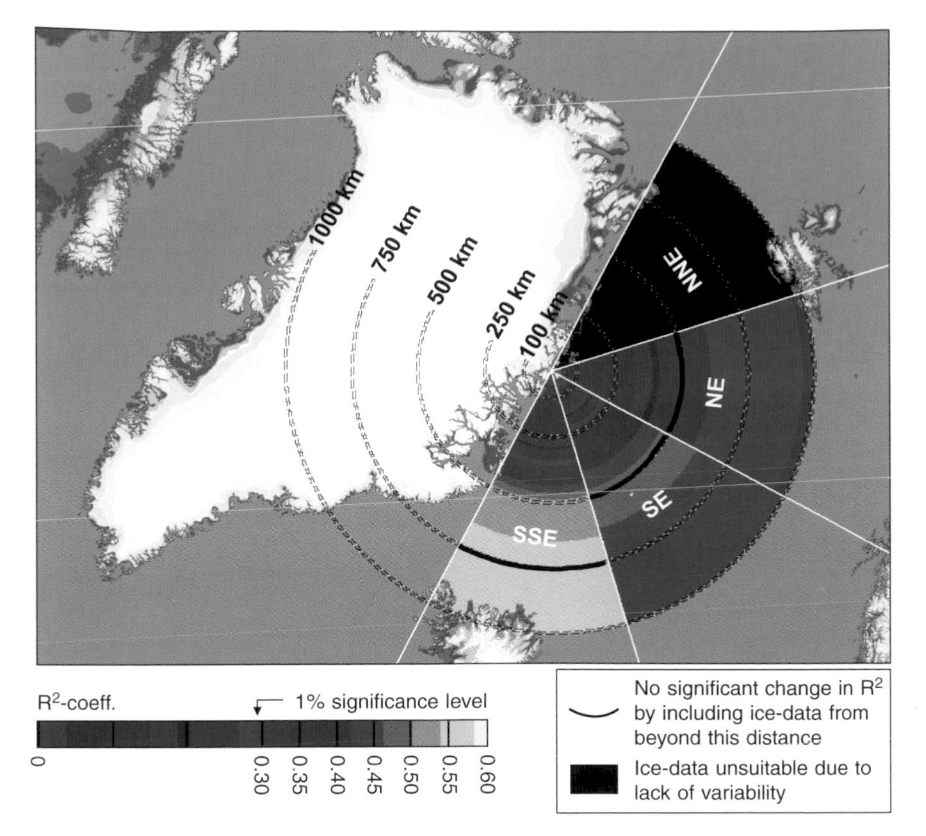

Figure 10 Correlation map showing the correlation (1982–2000) between snow accumulation in the Zackenberg region and sea ice duration in selected sectors of the Greenland Sea (based on direction and distance from Zackenberg).

duration is correlated to surface air temperature in the region, it cannot be disregarded as an option that the air temperature (which may be controlled by some other large-scale atmospheric circulation phenomena than the NAO) may control both sea ice duration and precipitation. However, in view of the fact that more extensive areas of open water should naturally be expected to increase atmospheric moisture content, the results from this study makes it reasonable to suggest that sea ice duration may play a key role for the snow precipitation budget in Northeast Greenland.

4. Modelling the Future Snow-Cover

In light of the above-mentioned discussion and with the prospect of future increases in arctic temperature, precipitation, NAO index and wind speed—especially in winter (Stendel *et al.*, 2008, this volume), future snow-cover conditions in high-arctic regions like Zackenberg still remain uncertain. Based on the IPCC A2 and B2 scenarios of air temperature and snow precipitation, some possible scenarios of the twenty-first-century snow-cover depletion in Zackenbergdalen were calculated (Figure 11), using the snow-cover depletion model from the present study. The modelled temperatures and precipitation (Dethloff *et al.*, 2002; Kiilsholm *et al.*, 2003; Rysgaard *et al.*, 2003) were provided by the Danish Meteorological Institute (DMI).

The results emphasize the importance of the balance between the air temperature during the melting season and the end-of-winter snow accumulation. Compared to present average conditions, the combination of increased temperature/melt rate, particularly in spring, and increased snow precipitation in winter seems to leave average summer snow-cover depletion close to *status quo*, with 2–3 weeks delayed depletion in areas with pronounced snow accumulation. This delay is due to more intense snowdrift, which leads to the creation of larger snowbeds, and thus it is possible that the tendency of delayed melt-off in major accumulation areas will be even more pronounced than predicted by the model, because of the predicted increases in wind speed.

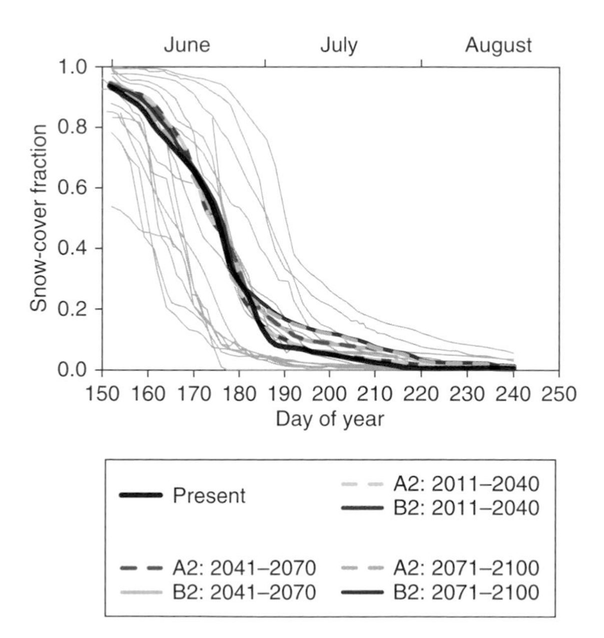

Figure 11 Scenarios (with A2 and B2 scenarios as temperature input in the snow-cover depletion model) of average snow coverage at Zackenberg in 30-year normal periods: day of the year May 30–September 7, 2011–2040, 2041–2070 and 2071–2100 compared to present average conditions (thick black line) and present variability (thin grey lines).

The prospect of increased climatic variability may also increase the frequency of seasons with late as well as early snowmelt in the future. Finally, increased snow precipitation combined with a higher frequency of thaw/freeze events during winter may lead to formation of ice crusts.

III. CONCLUSIONS

As in most arctic regions, snow is one of the most important parameters influencing ecosystem dynamics at Zackenberg. The large inter-annual variability in end-of-winter snow accumulations, which was found to range from a few centimetres to more than a meter, results in large variations in the length of the snow-free (growing) season, according to both measured and modelled results. However, because of a rather fixed wind regime in winter, which is dominated by north-north-westerly winds, the snow distribution pattern in Zackenbergdalen is much the same from year to year. The largest snow accumulation occurs on the valley sides on slopes with a southerly orientation, whereas on the valley floor the accumulation is more uniform and snowdrifts are more stochastically distributed. Snowdrifting is generally more intense in snow-rich than in snow-poor years. This leads to the formation of snowdrifts that are larger in the snow-rich than in the snow-poor years. Thus, when snow amounts increase, the melting season is prolonged more in areas with large snow accumulation.

This phenomenon may be even more distinct in a future climate with higher spring temperatures and more precipitation (snow). The snow-cover depletion model indicates that the combination of increased temperature/melt rate and increased snow precipitation seems to leave average summer snow-cover depletion close to *status quo* during the twenty-first century, with 2–3 weeks delayed depletion in areas with pronounced snow-accumulation, and probably with a much larger inter-annual variation in progress of snowmelt.

When considering climate variability at larger scale, this study indicates increased snow precipitation at Zackenberg with decreasing amounts of sea ice in the south-eastern Greenland Sea, and sometimes with increased NAO index. This hypothesis is, however, complicated by the fact that air temperature (which might be controlled by other large-scale atmospheric circulation phenomena than the NAO) may control both sea ice duration and precipitation. The correlation between the NAO and snow precipitation over Northeast Greenland seems to oscillate over time, being significant in only one third of the time. Hypothetically, this phenomenon could be due to shifts between phases where more or less continental/coastal climatic regimes predominate.

ACKNOWLEDGMENTS

Monitoring data for this chapter were provided by the GeoBasis programme, run by the Geographical Institute, University of Copenhagen, and the National Environmental Research Institute, University of Aarhus, and financed by the Danish Environmental Protection Agency, Danish Ministry of the Environment. We also thank Ulf Pierre Thomas, technician at the Institute of Geography, University of Copenhagen, for indispensable technical support. We thank Kim Have, Technical University of Denmark, for software development. Finally, we thank Niels Tvis Knudsen, University of Aarhus, Denmark, and Jon Børre Ørbæk, Norwegian Research Council, for their very constructive reviews.

REFERENCES

Appenzeller, C., Stocker, T.F. and Anklin, M. (1998) *Science* **282**, 446–449.

Bromwich, D.H., Chen, Q.S., Li, Y.F. and Cullather, R.I. (1999) *J. Geophys. Res. Atmos.* **104**, 22103–22, 115.

Dethloff, K., Schwager, M., Christensen, J.H., Kilsholm, S., Rinke, A., Dorn, W., Jung-Rothenhausler, F., Fischer, H., Kipfstuhl, S. and Miller, H. (2002) *J. Clim.* **15**, 2821–2832.

Groisman, P.Y., Karl, T.R. and Knight, R.W. (1994) *Science* **263**, 198–200.

Hall, D.K. and Martinec, J. (1985) *Remote Sensing of Ice and Snow*. Chapman and Hall, Cambridge, 189 pp.

Hinkler, J. (2005) *From Digital Cameras to Large Scale Sea-Ice Dynamics—A Snow-Ecosystem Perspective*. Ph.D. thesis, National Environmental Research Institute and the University of Copenhagen, Department of Geography, 184 pp.

Hinkler, J., Pedersen, S.B., Rasch, M. and Hansen, B.U. (2002) *Int. J. Remote Sens.* **23**, 4669–4682.

Hinkler, J., Oerbaek, J.B. and Hansen, B.U. (2003) *Phys. Chem. Earth* **28**, 1229–1239.

Hurrell, J.W. (1995) *Science* **269**, 676–679.

Hurrell, J.W., Kushnir, Y., Ottersen, G. and Visbeck, M. (2003) In: *The North Atlantic Oscillation. Climatic Significance and Environmental Impact* (Ed. by J.W. Hurrell, Y. Kushnir, G. Ottersen and M. Visbeck), pp. 1–35. American Geophysical Union, Washington DC.

Johannessen, O.M., Bengtsson, L., Miles, M.W., Kuzmina, S.I., Semenov, V.A., Alekseev, G.V., Nagurnyi, A.P., Zakharov, V.F., Bobylev, L.P., Pettersson, L.H., Hasselmann, K. and Cattle, H.P. (2004) *Tellus A* **56**, 559–560.

Kiilsholm, S., Christensen, J.H., Dethloff, K. and Rinke, A. (2003) *Geophys. Res. Lett.* **30**, DOI: 10.1029/2002GL015742.

Larsen, J.N. and Karlsen, H.G. (2003) In: *Zackenberg Ecological Research Operations, 8th Annual Report, 2002* (Ed. by M. Rasch and K. Caning), pp. 53–54. Danish Polar Center, Ministry of Science, Technology and Innovation, Copenhagen.

Liston, G.E. (1999) *J. Appl. Meteorol.* **38**, 1474–1487.

Liston, G.E. and Hall, D.K. (1995) *J. Glaciol.* **41**, 373–382.

Liston, G.E., Winther, J.G., Bruland, O., Elvehoy, H. and Sand, K. (1999) *J. Glaciol.* **45**, 273–285.

Menoes, M.C. and Brubaker, K.L. (2001) *How Similar Are Snow Depletion Curves from Year to Year? Case Study in the Upper Rio Grande Watershed.* 58th Eastern Snow Conference, Ottawa, Ontario, Canada, Abstract.

NSIDC (2005) *Northern Hemisphere Snow Extent: What Sensors on Satellites are Telling Us about Snow Cover.* The National Snow and Ice Data Center (NSIDC), University of Colorado Boulder, http://nsidc.org/sotc/snow_extent.html.

Rango, A. and Martinec, J. (1995) *Water Resour. Bull.* **31**, 657–669.

Rysgaard, S., Vang, T., Stjernholm, M., Rasmussen, B., Windelin, A. and Kiilsholm, S. (2003) *Arct. Antarct. Alp. Res.* **35**, 301–312.

Stein, M.L. (1999) *Interpolation of Spatial Data Some Theory for Kriging*. Springer Verlag, New York, 280 pp.

Hydrology and Transport of Sediment and Solutes at Zackenberg

BENT HASHOLT, SEBASTIAN H. MERNILD, CHARLOTTE SIGSGAARD, BO ELBERLING, DORTHE PETERSEN, BJARNE H. JAKOBSEN, BIRGER U. HANSEN, JØRGEN HINKLER AND HENRIK SØGAARD

SUMMARY

This chapter focuses on hydrology together with sediment and solute transport in the Zackenberg area in relation to climate variability during 1995–2005. The results indicate decreasing precipitation and increasing evapotranspiration during the study years. Theoretically, this should result in decreasing run-off, but no such trend is found in the run-off from the period. Model calculations justify that the loss of water available for run-off is compensated by an increased contribution of meltwater from glaciers within the catchment. It is found that extreme events can dominate the output of sediment and solutes for the catchment.

ADVANCES IN ECOLOGICAL RESEARCH VOL. 40 0065-2504/08 $35.00
© 2008 Elsevier Ltd. All rights reserved
DOI: 10.1016/S0065-2504(07)00009-8

I. INTRODUCTION

The recent interest in global change has brought the Arctic into focus because of expected large increases in air temperature and precipitation (ACIA, 2005; Stendel *et al.*, 2008, this volume). Consequently, the hydrological cycle will be particularly affected.

With an area of *c.* 2.1 million km^2, Greenland covers a substantial part of the Arctic. However, no long-term systematic investigations of the water balance have been carried out. The first measurements of discharge and its relation to climate in Greenland were carried out in 1958 in Southeast Greenland (Valeur, 1959), and Hasholt (1996, 1997) and Hasholt *et al.* (2005) presented review papers on run-off and sediment transport in Greenland. A few previous studies of the water balance at Zackenberg have been carried out by Søgaard *et al.* (2001) and Mernild (2006).

II. DRAINAGE BASIN

The extent of the drainage basin of the river Zackenbergelven upstream the hydrometric station is 512 km^2. On at least two sites, the drainage divide (Figure 1A) is located on alluvial cones, which might cause shifts of run-off pattern and as a result increase the drainage basin area up to 23 km^2. The total glacier cover in the basin is 101 km^2, that is, *c.* 20% of the catchment area (Figure 1B). The drainage system is dominated by the east-west-oriented valley, Store Sødal. Several river branches originate from the A.P. Olsen Land glacier complex. They are easily distinguished by their ample summer discharge and their milky appearance caused by the fine sediment load. Other streams, originating from the non-glaciated areas upstream Store Sø, have clear water, and the discharge from these streams is decreasing more during summer than in the glacial streams. Streams from the western part of the basin run through the lake Store Sø with a large delta mainly built by sediment from glacier streams. Along both sides of Store Sø, minor streams enter the lake. Especially during the early summer, the streams from the south side dry out later than streams from the north side. Downstream Store Sø, Zackenbergelven runs as a single channel until the confluence with the tributary from Lindemansdalen, that is, Lindemanselven. After Lindemansdalen, two more tributaries, Palnatokeelv and Aucellaelv, draining sedimentary rock areas, enter Zackenbergelven. A number of minor rivers and rivulets drain the western and eastern parts of the valley Zackenbergdalen directly into the fjord. The drainage density is larger in the eastern sedimentary rock area than in the western crystalline rock area (Figure 1B). Lake Store Sø is by far the largest of the lakes in the area, *c.* 4.9 km^2 (*c.* 1% of catchment area),

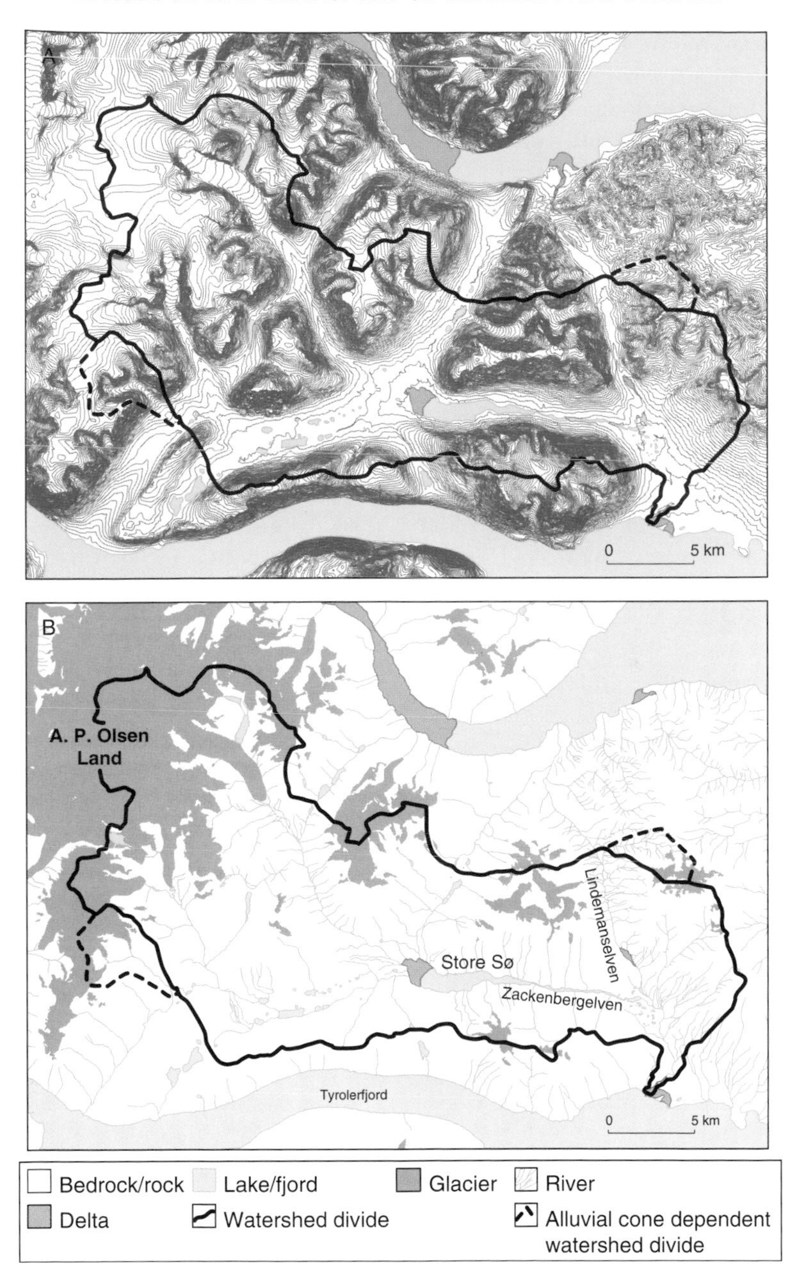

Figure 1 (A) The river Zackenbergelven catchment, topography and drainage divide. (B) Surface types, drainage network.

but other minor lakes are located in the valley. South of the confluence between Zackenbergelven and Lindemanselven, several small lakes are located in a moraine area (Figure 1B). Watercourses passing through lakes are depleted significantly of their sediment load. This is observed from delta deposits at the stream inlets to the lakes and from colour changes of water respectively up- and downstream of the lakes.

Relevant climatological data are given by Hansen *et al.* (2008, this volume).

III. ELEMENTS OF THE WATER BALANCE

In Figure 2, the mean annual values of elements of the water balance within the catchment are presented. Mean values of the elements are given for the period 1995–2005.

SnowModel is an energy balance-driven model that accounts for blowing snow described in Liston and Elder (2006). The Danish NAM model is a lumped conceptual hydrological model operating with a snow and a glacier reservoir.

A. Precipitation and Snowmelt

A thorough description of weather and precipitation is given by Hansen *et al.* (2008, this volume). Here, only the amount and distribution of precipitation relevant to run-off are presented. Maximum monthly precipitation values were 30–40 mm in February and November, and minimum were *c.* 10 mm in April, June and September (Figure 3). The average annual snowfall constitutes 72% of the total precipitation (Mernild *et al.*, 2007a). The annual average amount of corrected point precipitation within the observation period was 219 mm w.eq. y^{-1}, varying from 112 to 438 mm w.eq. y^{-1}. The average correction factor was 1.04. Correction of precipitation is treated by Hansen *et al.* (2008, this volume). Maximum daily rainfall was 29 mm w.eq., whereas it was 25 mm w.eq. for solid precipitation. Corresponding maximum intensities were up to 4.8 mm w.eq. h^{-1}. Total annual precipitation and the end of winter snow water equivalent for 1995–2005 (Figure 4A–C and G) reveals a decreasing trend in corrected total precipitation during the observation period. Significant trends are given in the figures.

Snowmelt and depletion of the snow-cover has been described by Hinkler *et al.* (2003) and Mernild *et al.* (2007b). Normally, the seasonal snow-cover in the main valley disappears (i.e., less than 1% snow left in the study area) between the end of June and the end of July. The difference between north- and south-facing slopes may be up to 21 days. Average summer daily modelled melt values on glaciers ranged from 9 to 17 mm w.eq. d^{-1}, with

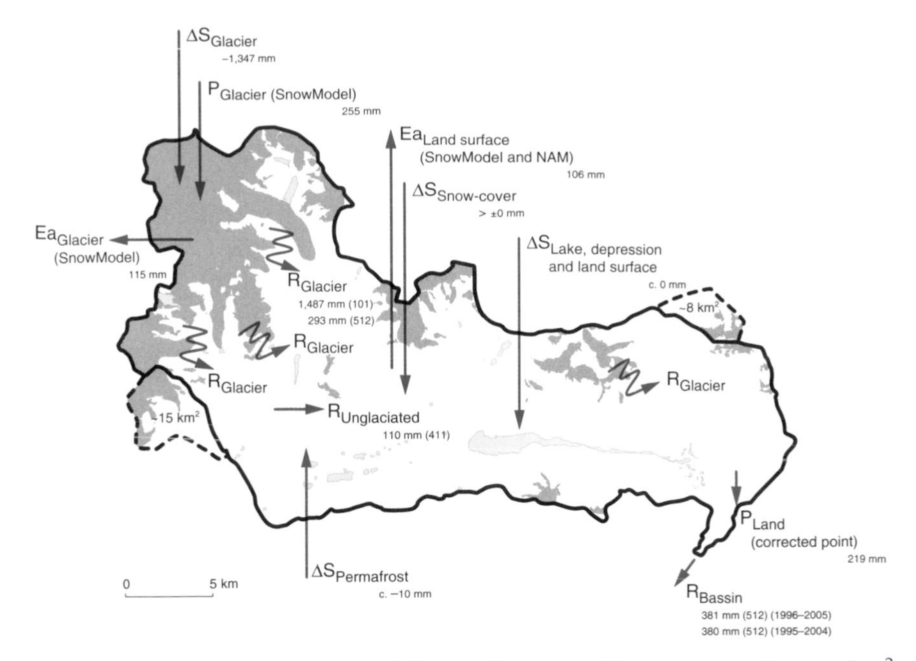

Figure 2 Elements of the water balance given as millimeter per unit area (km^2 calculation area in parentheses). P_{Land} is corrected precipitation at climate station, P_{Glacier} is areal precipitation found in SnowModel, $\text{Ea}_{\text{Land surface}}$ is actual evapotranspiration calculated with the NAM and the SnowModel, $\text{Ea}_{\text{Glacier}}$ is calculated for the glacier area with SnowModel, $\Delta S_{\text{Glacier}}$ is storage change on glacier area, $\Delta S_{\text{Snow-cover}}$ is assumed storage change in snow-cover, $\Delta S_{\text{Lake, depression and land surface}}$ is assumed storage change on surfaces, $\Delta S_{\text{Permafrost}}$ is calculated storage change due to increased active layer depth, R_{Glacier} is calculated run-off from the glacier area alone and distributed over the whole catchment area, $R_{\text{Unglaciated}}$ is run-off from the ice-free areas and R_{Basin} is total run-off from the whole catchment. The two alluvial cone areas that can increase the catchment area are shown (8 and 15 km^2, respectively).

maximum daily values as high as 47 mm w.eq. d^{-1} (Mernild *et al.*, 2007b). It is seen that the model calculated melt of snow from the glacier area decreases, while the amount of glacier melt increased significantly (Figure 4I and J). Annual average modelled snowmelt on glaciers was 207 mm w.eq. y^{-1}, ranging from 372 to 75 mm w.eq. y^{-1}.

B. Evapotranspiration

A reference evapotranspiration is calculated based on Makkink (1957). It is used as an estimate for evaporation from free water surfaces and as a driver in the NAM model. Point measurements of evaporation from water surfaces were carried out using a lysimeter in the summers 1996 and 2002. Data on

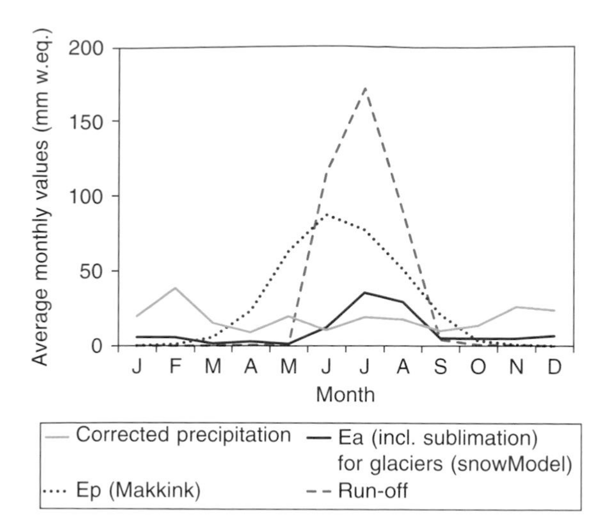

Figure 3 Water balance diagram for the Zackenbergelven catchment giving average monthly values for investigation period in millimetre. Ep is reference evapotranspiration according to Makkink (1957), and Ea is actual evapotranspiration calculated with the NAM and the SnowModel.

evapotranspiration were also available from soil moisture measurements and eddy correlation technique (Søgaard *et al.*, 2001). Sublimation and evaporation from glacier areas were calculated by the SnowModel (Mernild *et al.*, 2007b). Annual evapotranspiration was calculated with the NAM model, with evaporation set to zero when the temperature is below 0 °C.

Particularly in the beginning of the melt season, evaporation from free meltwater surfaces takes place. Measurements in a small pond indicate that daily values of 2.5–6 mm may occur (Søgaard *et al.*, 2001; B. Hasholt, unpublished data), which is in accordance with daily reference evapotranspiration values. Evaporation and sublimation calculated with the SnowModel show a slight increase in evaporation but no significant increase in sublimation (Figure 4E). Sublimation occurs mainly in connection with blowing snow incidents.

Monthly values of reference evapotranspiration calculated according to Makkink (1957) show a bell-shaped curve with a maximum of *c.* 90 mm in June (Figure 3). Average annual glacier area evaporation/sublimation (based on SnowModel) was 115 mm w.eq. (1997–2005) varying from 92 to 135 mm w.eq. between years (Mernild *et al.*, 2007b). Average annual evapotranspiration for the whole drainage basin was 106 mm, with a maximum of 118 mm and a minimum of 87 mm, respectively. Annual model-calculated evapotranspiration components for the period indicate an increase over the period as shown in Figure 4D–F.

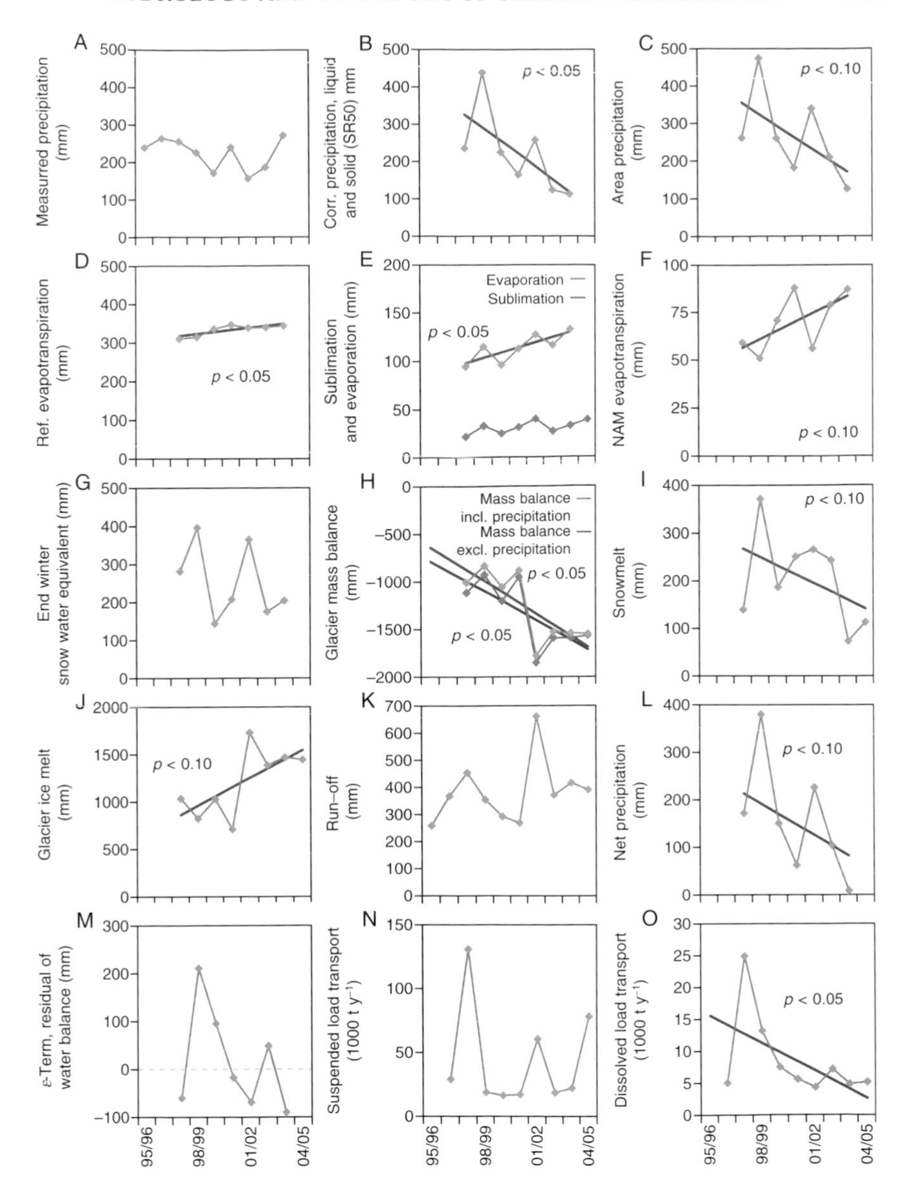

Figure 4 Annual values of variables for the investigation period in mm and tons. When a significant trend is found for a single variable, a trend line is included together with the corresponding p value. (A) Measured precipitation, Belfort, mm; (B) Corrected precipitation, liquid (Belfort) and solid (SR50), mm; (C) Area precipitation, SnowModel, mm; (D) Reference evapotranspiration (Makkink, 1957), mm; (E) Sublimation and evaporation (SnowModel), mm; (F) NAM evapotranspiration, mm; (G) End winter snow water equivalent, mm; (H) Glacier mass balance, mm; (I) Snowmelt, mm; (J) Glacier ice melt, mm; (K) Run-off, mm; (L) Net precipitation ($B - (E$ [winter] $+ F$ [summer]), mm; (M) ε-term, residual of water balance, mm; (N) Suspended load transport, t y^{-1}; (O) Dissolved load transport, t y^{-1}.

C. Storage

1. Surface Water Storage

Surface storage mainly takes place in depressions and permanent lakes. Maximum storage occurs during the melt season and minimum storage occurs during the late summer. Because the lake area is less than 2% and the slopes are generally steep, this quantity is of minor importance for the water balance calculations at Zackenberg.

2. Snow Storage

The snow-cover builds up during the winter period, and snow drifting causes redistribution, as described by Hinkler *et al.* (2008, this volume). Snow storage culminates in mid or late winter, and the end-of-winter storage is available for run-off during snowmelt. Average end-of-winter storage is 241 mm w.eq., with a maximum of 396 mm w.eq. and a minimum of 144 mm w.eq. for the period (Figure 4G), indicating the discharge volume during the thaw period.

3. Glaciers

Glaciers in the basin are not monitored. Change in storage is therefore based on calculations with SnowModel, where the accumulation is found for grid cells from an extrapolation of the corrected precipitation from the meteorological station. The ablation consists of sublimation and run-off loss, both calculated with SnowModel. The annual storage is equal to the net balance of the glaciers, while the glacier contribution to run-off from the whole catchment is calculated as ablation minus sublimation and evaporation. The model was successfully verified for a test area in the valley (Mernild *et al.*, 2007b). Two steady-state assumptions have been applied: (1) All meltwater and liquid precipitation runs off the glacier (i.e., no re-freezing or internal storage), (2) only meltwater leaves the glacier. The actual conditions are probably somewhere in between the two. The average annual SnowModel calculated change in glacier storage was −1,347 mm w.eq. y^{-1} (1997–2005) [equals 266 mm w.eq. y^{-1} for catchment run-off (512 km^2)], while minimum was −952 mm w.eq. y^{-1} (2000/2001) and maximum was −1,844 mm w.eq. y^{-1} (2001/2002), respectively (Mernild *et al.*, 2007b). Results from the individual years during the observation period indicate a negative net mass balance of the glaciers for all years, and there is a significant trend of increasing mass loss from the glaciers in the drainage basin during the study period (Figure 4H).

4. Soil Moisture, Active Layer and Permafrost

The liquid soil moisture has been monitored near the meteorological station by the use of time-domain reflectometry (TDR) techniques with a Techtronics1502 cable tester. Soil moisture contents in six depths are shown in Figure 5. The maximum soil moisture content is reached right after snowmelt when the soil is completely saturated. During the summer, the water content drops because of evaporation and drainage. The most significant decrease is observed in the first weeks after snowmelt. Then moisture content levels out, showing only minor response to precipitation. Soil temperature stays at zero for a certain time before the moisture content increases after the melting and disappearance of the local snow-cover. The storage in the active layer (*c.* 70 cm) can vary from maximum 350 mm (porosity 50%) and 245 mm (porosity 35%) to a minimum of *c.* 70 mm for a dry sandy soil in late summer. The uppermost layer contains more water because of its content of organic matter and finer particles and its higher porosity (Figure 5). Peaks of increase at lower levels indicate the storage of penetrating melt- and rainwater on top of the permafrost.

Changes in soil moisture have also been used for calculating seasonal actual evapotranspiration. Although soil moisture is monitored in single points on the *Cassiope* health, the relative changes through time may apply to simultaneous changes in soil moisture in the basin.

If the thickness of the active layer remains constant, there is no net change in storage from year to year. However, if the active layer increases, water is

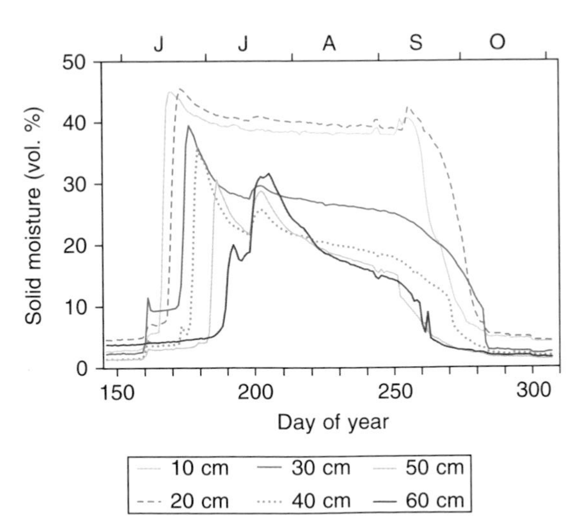

Figure 5 Soil moisture variation at different levels, measured with time domain reflectometry (TDR) at the meteorological station between May 25 and November 6, 1998.

permanently released and will increase the annual run-off. At Zackenberg, the active layer has varied up to *c.* 20 cm during the observation period, with a steady increase during 2002–2005 (Christiansen *et al.*, 2008, this volume). Based on porosity and field capacity of soil samples, the degradation of the permafrost could have released a maximum of *c.* 10 mm $y^{-1}$of free water to the temporary groundwater and streams. Locally, the degradation of permafrost can cause hollows and slides in the landscape, that is, so-called thermokarsts, which is seen along parts of Aucellaelv, Ræveelv, Tørveelv, Kærelv and Grænseelv, and which could have contributed to run-off due to release of frozen water in the degrading permafrost. Yet, at Zackenberg this contribution is considered insignificant.

D. Run-Off

Calculation of run-off in Zackenbergelven is based on a stage/discharge relationship. Discharge is measured with Ott current meters, and stage is recorded on a staff gauge and automatically with a Campbell SR50 sensor and a pressure transducer. Because of snowdrifts in the stream bed and extreme floods, including ice and slush, stage recording has been disrupted temporarily in early spring, and the stage/discharge relationship had to be recalculated because of erosion in the cross section in July 2005. Manual discharge measurements were performed each year on a daily basis until the river cross profile was free of snow and ice (typically a few weeks after break-up), and additional measurements were made throughout the season in order to verify the stage/discharge relationship. The single discharge measurements have an estimated accuracy of ±5–10%, while the stage-discharge calculated values can deviate up to 25% from simultaneous manual measurements. However, the long-term (monthly and annual) discharge is accurate within *c.* 5%. Fifteen-minute values are calculated and presented in the annual reports from the research station (e.g., Sigsgaard *et al.*, 2006) and in Figure 6.

Because of the diurnal variation in melting, streams often have a marked daily variation in discharge. Melting on glaciers in the most remote parts of the drainage basin normally leads to increased water discharge at the hydrometric station around midnight due to travel time for the water through the system. Part of the delay is caused by the lake Store Sø, which delays the water from around 350 km^2 of the basin for about 10 h.

The average monthly run-off for the period of observation has a maximum of *c.* 170 mm w.eq. in July (Figure 3) characterizing the river "regime." According to Pardé (1955), the Zackenbergelven river regime can be classified as glacio-nival, but parts of the basin, close to local glaciers or the A.P. Olsen Land glacier complex, can be classified as purely glacial.

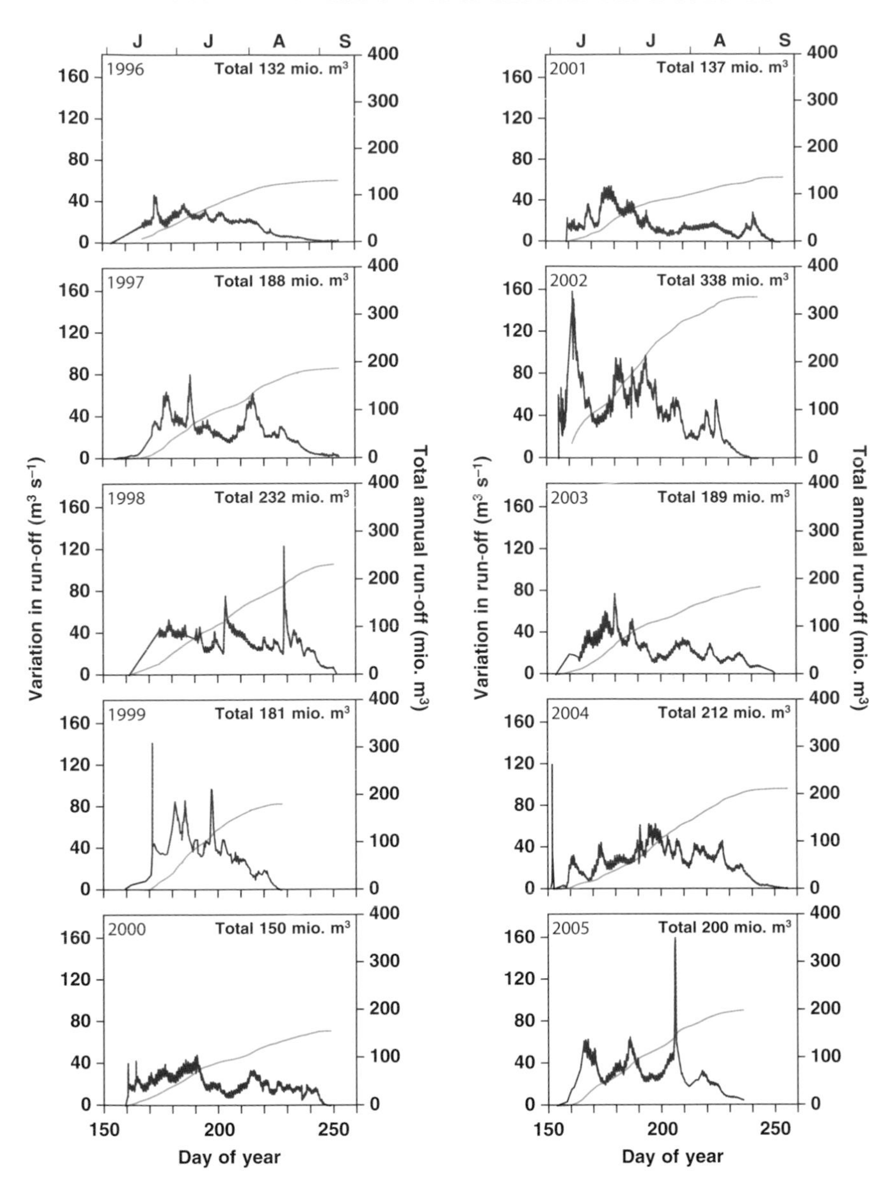

Figure 6 Variation in run-off in Zackenbergelven (15-min values) for the investigation period May 30–September 17 together with total annual run-off. Red lines indicate extrapolated values.

Observed 15-minute values of discharge show diurnal variations of up to 15 $m^3 s^{-1}$ in the early part of the season. Later in the season, when snow in the valley has disappeared, the diurnal variation is less pronounced.

The variation is shown in Figure 6 together with annual sums in million m^3. The duration curves in Figure 7 indicate the percentage of time that the discharge is higher than a certain value. It is seen that the discharge with a duration of 1% vary between 30 and 95 $m^3\ s^{-1}$ from year to year.

Measured (1996–2005) average annual run-off volume divided by catchment area was 381 mm, with a maximum of 658 mm (2002) and a minimum of 257 mm (1998). SnowModel-predicted average annual run-off from catchment glaciers (including rain, snow and ice melt) was *c.*150 million $m^3\ y^{-1}$ (equals 293 mm w.eq. y^{-1}) (1997–2005), with a maximum of *c.* 207 million $m^3\ y^{-1}$ (405 mm w.eq. y^{-1}) (2000) and a minimum of *c.* 104 million $m^3\ y^{-1}$ (203 mm w.eq. y^{-1}) (2001), respectively (Mernild *et al.*, 2007b). The annual run-off values for the observation period show high variability and no significant trend (Figure 4K). The annual daily maximum discharge varied from 47 to 159 $m^3\ s^{-1}$ (92–311 $l\ s^{-1}\ km^{-2}$). An extreme value distribution graph for the observation period is shown in Figure 8. The duration is too short for long-term predictive purposes, but it demonstrates that the three highest values group together separately indicating that different mechanisms are responsible for the maximum events. An analysis of the climatic conditions before and under the extreme events indicates causes for the observed events: (1) a period of very warm weather together with rainfall releasing large amounts of meltwater (August 1998), (2) a glacial meltwater outburst from the A.P. Olsen Land glacier complex (July 2005) and (3) collapse of snow dams initiating large amounts of saturated snow resulting in a flood of slush, ice and meltwater when the river breaks up after spring melt (June 1999 and June 2004).

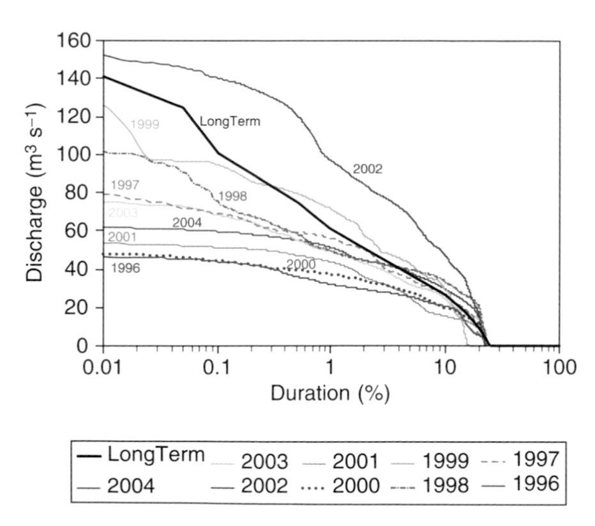

Figure 7 Percentage of time that discharge in Zackenbergelven exceeded values on the *y*-axis. Duration curves for individual study years (thin lines) and for the entire period 1996–2004 (bold line).

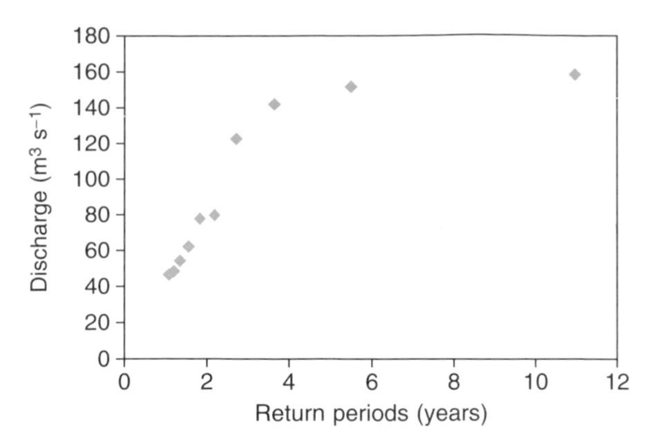

Figure 8 Yearly extreme discharges in Zackenbergelven for the period 1996–2005 shown with return period calculated by use of the Weibull formula.

IV. WATER BALANCE AND MODEL CALCULATIONS

The water balance equation is

$$P - \mathrm{EA} - R + \Delta S + \Delta \mathrm{SS} + \Delta \mathrm{SM} + \Delta \mathrm{PF} + \Delta \mathrm{Gl} = \varepsilon,$$

where P is precipitation, EA is evapotranspiration, R is run-off, ΔS is change in snow storage, ΔSS is change in surface storage, ΔSM is storage change in the active layer, ΔPF is storage change in the permafrost layer and ΔGl is change in glacier mass balance. The term ε should be zero if all the terms in the equation are correct. The monthly values of the main elements of the water balance are given in Figure 3. The catchment precipitation is approximated by the different estimates mentioned by Hansen *et al.* (2008, this volume). The methods for determination of evaporation, sublimation and evapotranspiration are not fully independent, because the reference evapotranspiration is a common input in the calculation models, that is, the value presented is the best estimate. Run-off is the element measured with highest accuracy. The first four storage terms are of minor importance for annual water balances in high-arctic areas. The glacier storage changes, on the other hand, are long-term changes that can explain the observed surplus run-off. To represent the "true" annual catchment precipitation, the SnowModel precipitation was chosen. The catchment annual evapotranspiration was found as the sum of the SnowModel-calculated sublimation and evaporation for the period October 1–June 30 and the NAM evapotranspiration for the period July 1–September 30. The net precipitation (precipitation minus evapotranspiration) should equal the run-off, if no storage changes take place. The net precipitation for the single years in the period indicates a slight but significant negative trend over the period (Figure 4L). This should lead to a falling trend in the run-off over the period, which is not the case. It is also seen that the net precipitation in several years is less than the run-off, indicating a storage loss. The model-calculated mass balance for the glacier indicates a significant negative trend over the period, explaining the storage loss.

The ε term is calculated assuming all other storage changes being zero. The ε term is slightly positive without any significant trend over the period (Figure 4M). It is therefore concluded that a negative mass balance for the glacier areas generally compensate for the negative trend in net precipitation, resulting in a more or less constant run-off. The positive values of the ε term presumably indicate that the applied corrected area precipitation is close to reality or slightly overestimated. The large variation in the ε parameter may be due to the assumption that other storage changes are zero being wrong. The most probable error could be caused by errors in the assumption concerning the snow storage change term, because accumulated snow could remain unmelted in the drainage basin from one hydrological year to the next. The reallocation of snow due to drifting and possible changes in snow water equivalent during winter because of warm spells and precipitation have been calculated using the SnowModel. The melting has been calculated using an energy balance concept. Modelled average annual run-off from the glacier area is 1,487 mm w.eq, with a maximum of 2,051 mm w.eq. and a minimum of 1,031 mm w.eq. (Mernild *et al.*, 2007a). Average annual change in storage on the glacier area, based on water balance calculations, is −1347 mm, with a maximum of −1844 mm and a minimum of −927 mm, corresponding to a loss of storage in the whole basin of −266 mm, −364 mm and −183 mm, respectively.

V. TRANSPORT OF SEDIMENT AND SOLUTES

Sediment and solute transport is strongly related to the surface run-off, because the water erodes bed and banks along the watercourses and act as a transport medium for both sediment and solutes.

A. Sediment Transport and Dissolved Load

Sediment is made available for transport by mechanical erosion by frost action, by glacial erosion and erosion by running water on surfaces and in stream beds and banks. Sediment sinks in the catchment are alluvial cones, inland deltas, and lake basins.

Water samples have been collected from Zackenbergelven every day around 8 a.m. with a depth-integrating sampler (Nilsson, 1969). Experiences from Zackenberg have shown that sediment concentration can have a diurnal variation, so that sampling at 8 am, which has been chosen because of other duties to be taken care of by the operator, will tend to cause an underestimation of the calculated sediment load. Results from measurements of sediment concentration at 2-h intervals during a 24-h period show a diurnal variation in concentration of discharge and sediment load (Figure 9). A standardized correction cannot be applied because diurnal variation is not straightforward. Therefore, the transport values below are presumably 10–15% too low. Figure 10 shows the suspended transport calculated by multiplying the concentration of the daily water sample with diurnal discharge values.

Annual average (9 years) of suspended sediment transport was 42,500 t, with a maximum of 130,700 t and a minimum of 15,000 t, corresponding to 83, 254 and 29 t km^{-2} y^{-1}, respectively. Average annual transport of suspended organic matter was 2,908 t, with a maximum of 8,570 t and a minimum of 1,100 t, corresponding to 5.7, 16.7 and 2.1 t km^{-2} y^{-1}, respectively. Average values for transport of solutes (based on conductivity measurements and a calibration against sum of ions) are 9,900 t (maximum 30,600 t and minimum 3,900 t), which corresponds to 19.4, 59.6 and 7.6 t km^{-2} y^{-1}, respectively.

Total annual transport of suspended sediment shows a large variability (Figures 4N and 10). From Figure 10, it is obvious that short-term extreme events have a strong influence on the distribution over the years. As an example, seven to eight of the peak periods have carried an amount of sediment similar to several of the remaining total annual transports. The most significant event occurred during August 16–22, 1998, and resulted in a transport of 105,000 t, which is almost 8 times as much as in the minimum year. Maximum daily transport during this event was 72,000 t (August 16, 1998), and the peak concentration of suspended sediment was 47 g l^{-1}, indicating sources close to the gauging station. Most of the time concentrations of suspended sediment ranged between 100 and 300 mg l^{-1}. Maximum concentration during the flood July 25, 2005, was 5.8 g l^{-1}. A negative trend is observed in the dissolved annual loads (Figure 4O).

The total load is the sum of suspended and dissolved load. The annual average percentages of suspended load and dissolved load were 80 and 20% of the total, respectively. The load of organic matter was 6% of the total. The annual mean load of bi-carbonate (HCO_3^-) carried to the fjord, Young Sund (1997–2005), was 2,506 t y^{-1}, with a maximum of 7,388 t and a minimum of 1,339 t (Table 1). The bed load has not been monitored. Most bed load is trapped in lakes, so that only a minor amount originating from Lindemanselven and from local bank erosion will pass the gauging station during normal conditions. During the extreme floods, bed load might constitute a significant part of the total sediment transport, because of more severe erosion in bed and banks.

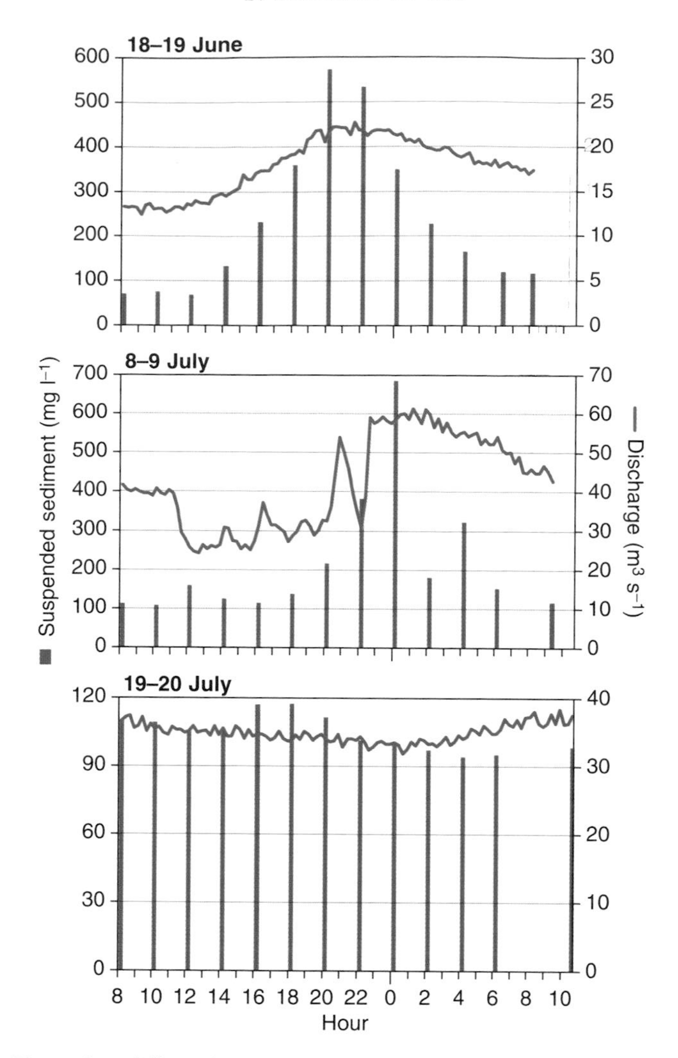

Figure 9 Examples of diurnal variation in concentration of suspended sediment and discharge in Zackenbergelven, 2004.

B. Water Chemistry

1. Sources of Dissolved Elements

Main sources include rainwater, soil erosion and terrestrial weathering products from mineralization of organic matter and weathering of dominant silicate minerals (Rasch *et al.*, 2000). Each of these sources has a characteristic geochemical composition, which further may change with time.

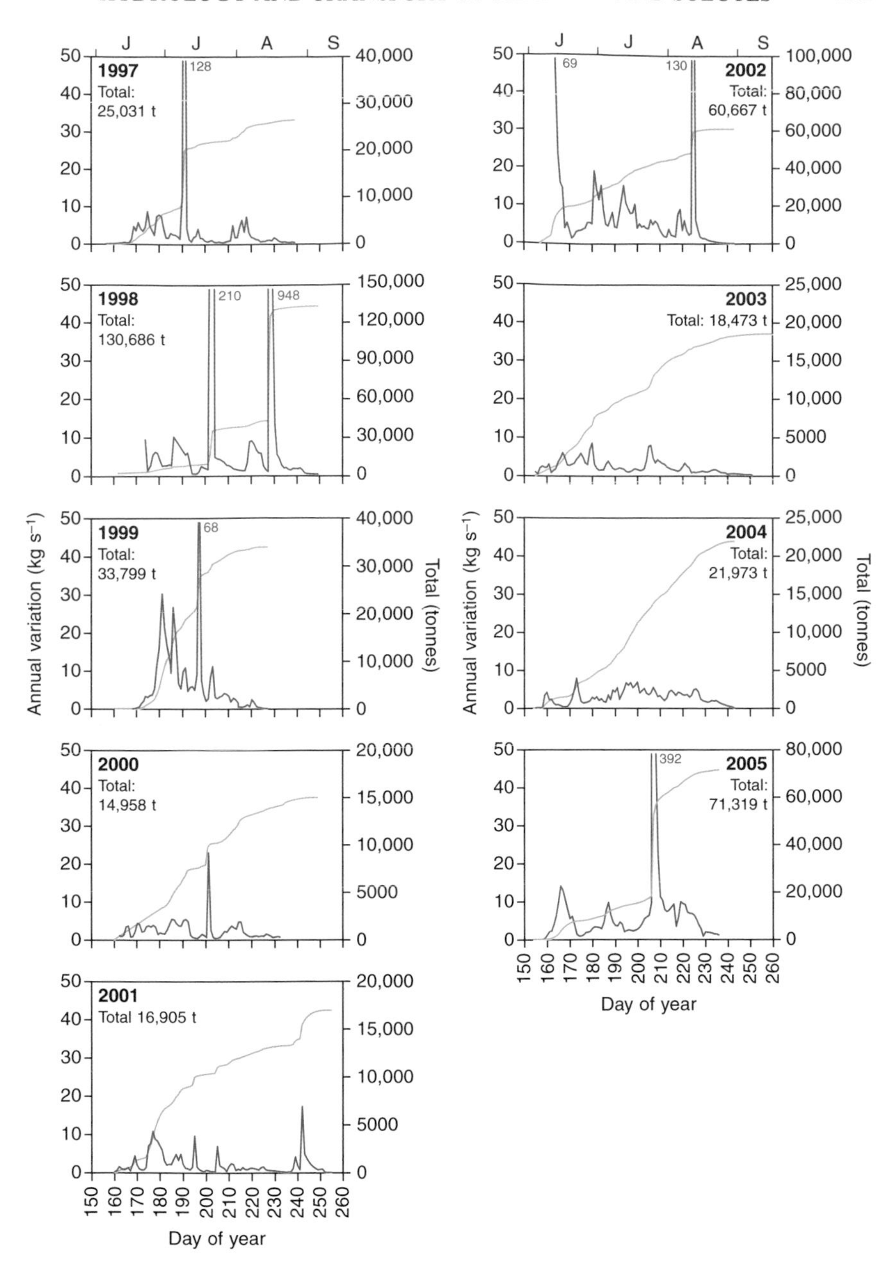

Figure 10 Annual variation of suspended sediment transport in Zackenbergelven during May 30–September 17.

Table 1 Total run-off (millimetre equivalent in water catchment area) and transport of chemical solutes and suspended sediment in the river Zackenbergelven 1996–2005

Year	1996	1997	1998	1999	2000	2001	2002	2003	2004	2005
Run-off (mm)										
June	83	87	99	87	78	103	277	138	90	130
July	131	156	192	232	119	92	291	138	194	195
August	42	118	153	32	94	66	90	83	125	66
September	1	5	8	0	1	6	0	8	4	–
Total run-off	257	366	451	352	292	267	658	368	412	391
Maximum peak event ($m^3 s^{-1}$)	47	80	123	>142	49	54	159	78	>100	>152
Date and time for maximum peak	20/6 21:15	7/7 09:15	16/8 19:15	20/6 13:15	8/7 19:45	27/6 23:00	10/6 19:45	29/6 02:15	1/6 05:00	25/7 05:00
Suspended sediment (ton)[a]		25,031	130,686	33,799	14,958	16,905	60,667	18,473	21,973	71,319
Suspended organic matter (ton)[a]		1,565	8,565	2,385	1,186	1,101	3,299	1,370	1,388	3,475
Cl^- (ton)		513	469	146	79	76	169	56	138	98
NO_3^- (ton)		84	26	31			15	13	27	23
SO_4^{2-} (ton)		1,015	1,987	900	566	1,004	1,128	790	1,000	995
Na^+ (ton)		179	217	193	83	116	199	107	114	131
Mg^{2+} (ton)		132	260	129	74	108	194	116	143	155
K^+ (ton)		147	126	66	53	54	130	84	115	117
Ca^{2+} (ton)		625	922	377	440	436	817	626	504	541
Fe^{2+} (ton)		109	52	59		14	83	35	53	60
DOC (ton)								430	260	1,557
HCO_3^- (ton)		2,178	7,388	1,880	1,339	1,638	2,366	1,794	1,970	1,999

[a]Data are based on a re-evaluation of data and may differ from earlier publications.

Concentrations and total annual loads of elements (Table 1) indicate a dominance of marine sources (NaCl) and weathering products from sulphide oxidation (Fe and SO_4^{2-}) and silicates (base cations). The analysis suggests that significant amounts of base cations and sulphate are due to terrestrial weathering processes. This is consistent with an annual load (1997–2004) of Cl that is only partly correlated to Na ($R^2 = 0.46$). The poor correlation is a result of an additional and variable amount of Na released from silicate weathering processes. It has not been possible to differentiate between dissolved matter as a result of erosion of the river bank and dissolved matter as a result of on-going soil processes releasing soluble weathering products.

2. *Changes in Concentrations During the Run-off Period*

Temporal trends in water chemistry during the growing season result from temporal trends in the mixture of different sources as well as trends within each of the sources. A marked shift is the contribution of elements from winter precipitation (snow accumulation) compared to elements mainly derived from soil weathering processes. In the early summer melting, snow is the primary source of water. During the initial thaw, the content of ions in the snowpack is washed out during a so-called "ion pulse" (Williams and Melack, 1991; Rasch *et al.*, 2000). At Zackenberg, the ion pulse gives rise to the highest concentrations of Cl, SO_4, Na, Ca and Mg (average of 4.8, 41.6, 3.0, 15.1 and 3.4 mg l^{-1}, respectively) during the summer and is observed during the initial snowmelt in early June. A significant variation in the maximum values is due to the relative importance of inflow from the Lindemansdalen sub-catchment dominated by sedimentary rocks and the Store Sødal catchment dominated by crystalline rocks. After the initial peak, the ion concentrations are more constant but sensitive to events such as rainstorms. In most years, an increasing positive deficit in the ion balance has been noted from mid-summer and onwards. This has previously been considered as a result of increasing concentrations of organic anions from surrounding soils as the active layer develops (Rasch *et al.*, 2000). This signals the effect of an increasing contribution of soil water on river water chemistry during the run-off period.

3. *Changes in Annual Loads 1997–2004*

Events and the total run-off are the two main factors controlling variations in element concentrations and annual loads between years (Table 1). Effects of total run-off are evident for alkalinity, which correlates significantly ($R^2 = 0.64$, $p < 0.05$) with discharge over the period (1997–2004). All other dissolved ions correlate with discharge over short periods of time but not between years. The effects of events can be illustrated by high-resolution measurements from 1997 (Rasch *et al.*, 2000). The first peak in June 1997 was during a time where water has not been in contact with the frozen soil. This was in contrast to the second peak in water discharge during July 1997 where the active layer had a thickness of 10–20 cm. The peak discharge was followed by delayed high concentrations of Ca, Mg and SO_4. Subsequent peaks by the end of July and in August 1997 indicated a limited increase in Ca, Mg and SO_4 concentrations, while Cl concentrations dropped significantly. Thus, the time of rainfall during the summer is crucial in order to relate specific rainfall events to flushes of ions. During a 1-week rain event in 1998 (maximum 25 mm in 1 day), an extraordinary pulse of sediments and ions was noted, which was in the order of 2–4 times the average annual loads of most elements. Dominated by this one event, the year 1998 becomes important in order to illustrate and characterize maximum loads of elements, and further emphasizes the need of high-resolution data for additional years in order to provide robust estimates of annual element fluxes and expected impacts of climatic extremes and climate changes. In contrast, the years 1999 and 2000 were rather dry years without any major rain events and therefore show minimum loads of most elements.

VI. DISCUSSION AND CONCLUSIONS

The hydrology of the Zackenberg catchment area has certain distinct characteristics which determine the distribution of water and the timing of run-off. First of all, the hydrological conditions in the area are strongly dependent on the presence of glaciers, which represents a major storage of water. It is also shown that the catchment area may change through time, since the present drainage divide in several cases is located on alluvial cones and glaciers. Depending on the deposition of sediment, the channels on a cone may shift direction abruptly, so that water from the upstream area is directed to another catchment. Such relocation may increase the present catchment area by 23 km^2, or 4% in the future. Another characteristic feature is the dominating east-west valley Store Sødal, which cause "asymmetric" hydrological conditions. North-facing slopes are snow-free 3–14 days later than south-facing slopes at the same altitude and having the same snow depth. As a result, the river receives meltwater later and often for a longer period from north-facing slopes than from the opposite south-facing slopes. The difference in moisture conditions is clearly visible on the two sides of the lake Store Sø.

Point measurements of precipitation at the Zackenberg Research Station in the period 1995–2005 show a slight but significant decrease in the annual amounts. This is not in accordance with an expected increase predicted by global and regional climate change models (Stendel *et al.*, 2008, this volume), and investigations of long-term trends document a long-term increase in precipitation in the area since 1958 (Hansen *et al.*, 2008, this volume). Therefore, it is concluded that the observed trend is a short-term natural variation. However, as precipitation is the principal input to the hydrological cycle, it must be expected that run-off through the period has been reduced because of this negative trend.

The average annual Makkink reference evapotranspiration and the combined evapotranspiration [SnowModel (winter) + NAM (summer)] was 331 and 106 mm, respectively. It is shown that sublimation in connection with blowing snow events accounts for 31 mm (Mernild *et al.*, 2007b).

The Makkink reference evapotranspiration, the SnowModel sublimation/evaporation, and the NAM evapotranspiration indicate a positive trend. The annual evapotranspiration from the glacier area, found by modelling, is 115 mm. The results are in the same order of magnitude as the evapotranspiration measured by eddy correlation techniques (105 mm) for Zackenbergdalen and the corresponding basin value (79 mm) (Søgaard *et al.*, 2001).

The combined effect of the calculated trends is a decreasing net precipitation available for run-off. However, there was no significant trend in measured run-off, and there was no significant correlation between summer temperature and run-off or between precipitation and run-off, implicating that compensating contributions of water occur in the form of storage changes. Measured average annual run-off (1996–2004) is 381 mm. It is seen that the input precipitation (255 mm) is 232 mm less than the output sum of evapotranspiration and run-off (106 + 381 = 487 mm). A significant part of this difference can be explained by calculated storage losses because of a negative mass balance on the glaciers. However, the results could also indicate that our precipitation data from the meteorological station in Zackenbergdalen are not representative for the entire catchment area, and hence that a better basis for calculation of a correction of the measured precipitation could be needed from other parts of the basin in order to explain some of the apparent surplus run-off. The annual run-off regime is dominated by melting of snow and glaciers. Furthermore, the diurnal variation in discharge is caused by diurnal variation in melt rates, which are delayed because of travel distance from melt sources and a further delay of *c.* 10 h by passing the lake Store Sø, causing the maximum to occur around midnight and the minimum late in the morning. Short-term abrupt changes in discharge (Figure 9) at the gauging station can be caused by strong winds either stowing the water or lowering the stage at the outlet from the lake, for example, during Foehn situations. Other short-term changes are related to summer precipitation and glacial outbursts. Extreme discharges are caused by normal snowmelt/rain events and by stochastic breakdowns of snow dams or long-term cyclical outbursts from ice-dammed lakes situated in the A.P. Olsen Land glacier complex.

Change in permafrost could supply water; a rough estimate shows that an increase in the thickness of the active layer by 20 cm through the period could increase the annual run-off from the basin to *c.* 10 mm y^{-1}.

Because of lack of direct measurements from catchment glaciers, a hydrologic glacier model has been applied. Model calculations show a negative mass balance for the glacier area.

Assuming near steady state of water flow through the glacier, the glacier run-off can account for 50–55% of the present discharge at the catchment outlet, and this is the main factor in explaining the excess discharge compared to net precipitation.

Since glaciers covering *c.* 20% of the area deliver *c.* 50% of the present run-off, areas without glaciers should have significantly less run-off, and the discharge from the catchment would decrease abruptly if the glaciers disappeared. The ε term in the water balance is slightly positive, indicating that the applied overall correction is in the correct order of magnitude. The large variation in the ε term through the period indicates that the assumption that all other storage terms than glaciers are zero is not correct. An obvious storage that could explain this is snow remaining in the drainage basin from one year to another.

The calculated average suspended sediment transport is 42,500 t y^{-1} or 83 t km^{-2} y^{-1}. Estimated from short periods of measurements of diurnal variation in sediment concentration, these values could be 10–15% too low, because water samples are collected when the concentration often is close to minimum. Suspended load is *c.* 80% of the total load, while dissolved load is 20%. Bed load is of minor importance except during extreme floods, because of trapping in upstream lakes.

The suspended load exhibits short-term extremes not always related to very high discharges, which could be caused by landslides in the sedimentary part of the basin close to the gauging station. During the initial thaw period, a "ion-pulse" is observed. Extremes are shown to deliver several times the total annual transport during years with normal and low transport. Based on observations of water chemistry since 1997, it is unlikely that decreasing summer precipitation (or shift in summer to winter precipitation) will affect the chemical composition of run-off on a short-term basis. On the other hand, it is highly likely that more summer rain will at least on a short-term basis increase terrestrial weathering processes and annual chemical loads.

This indicates that long-term monitoring is important to evaluate the overall effects of this type of events and calls for new studies tracing these extreme events back in time, for example, by using sediment records and tracer techniques.

The advantage of model calculations to cover periods and areas where measurements are missing is clearly demonstrated. However, an important precondition for using such models is that they can be calibrated against high-quality measurements. Therefore, a calibration of precipitation using snow courses and glacier mass balance for a test period is recommended.

Recent results from 39 high-latitude (50°–80°N) catchments have been analysed (Kane and Yang, 2004), and suggest a decrease in precipitation of *c.* 16 mm year^{-1} degree^{-1} latitude for continental areas, but somewhat blurred by coastal stations. Moving northwards, the ratio of run-off to precipitation increases. This is mostly due to the snowmelt event, with substantial run-off being the dominant hydrological event each year. Evapotranspiration also decreases with latitude, *c.* 12 mm year^{-1} degree^{-1} latitude. This reduction is primarily due to the lack of energy for phase change, because lack of moisture is rarely the case in this environment. In the case of important changes in storage, quantification of this variable is very demanding in water balance computations at time scales of less than a year because of logistic problems. The average annual precipitation for the same latitude as Zackenberg (*c.* 74°N) was *c.* 200 mm (Kane and Yang, 2004). The corresponding average annual evapotranspiration cluster around 70 mm for this latitude, while the ratio of annual average run-off to precipitation was close to 1. The catchment areas included in the review by Kane and Yang (2004) were rather small; only seven were larger than 100 km^2 and the largest was 432 km. The values for the precipitation and evapotranspiration from Zackenberg are slightly higher than the values found in the review by Kane and Yang (2004). However, the run-off ratio from Zackenberg is *c.* 1.4, again stressing the importance of the glacial contribution in this catchment at the moment.

Results from Zackenberg are comparable to results from another catchment in Greenland, the glacier-dominated Mittivakkat basin (Southeast Greenland) having a ratio of *c.* 1.6. This suggests the importance of a variable and generally negative mass balance for the run-off and suggests that the glaciers and perennial snow-covers in the Zackenberg basin also play an important role for the excess run-off. Although melting dominates the annual run-off distribution, summer rainfall determines the secondary peaks. Although the catchment is humid, the evapotranspiration creates local dry ecosystems.

The climate scenarios presented by Stendel *et al.* (2008, this volume) indicate that the future climate would be warmer, moister and windier. This will also influence the hydrological cycle, but the net effect is less clear. If the winter precipitation increases by 40–60%, the crucial question is to what extent the expected warmer spring conditions are able to melt the snow. Because of the more moist conditions, an increase in evapotranspiration is expected to be moderate and not capable of balancing the increase in winter precipitation. If the warmer conditions in spring are able to melt all the snow or more, the run-off will increase accordingly. However, the increase of run-off because of larger snowmelt may be balanced out by a lack of contributions from glacier melt so that the run-off will be at the same order of magnitude as today. It is expected that the extent of snow-free areas will decrease; this implies that snow will be left on the glaciers, and that the contribution to run-off from glacial melt accordingly will diminish or disappear. The glaciers will then have a positive mass balance and the resulting run-off may then be slightly lower than today.

ACKNOWLEDGMENTS

Monitoring data for this chapter were provided by the GeoBasis and Climate Basis programmes, run by the Geographical Institute, University of Copenhagen, ASIAQ (Greenland Survey), and the National Environmental Research Institute, University of Aarhus, Denmark, and financed by the Danish Environmental Protection Agency, Ministry of the Environment, and the Greenland Home Rule. Thanks are due to FNU (Danish Agency for Science, Technology and Innovation) grant 21–00–0065 and to COGCI for supporting part of the project. The Danish Polar Center is thanked for providing logistical support at Zackenberg Research Station.

REFERENCES

ACIA (2005) *Arctic Climate Impact Assessment.* pp. 1042. Cambridge University Press.

Hasholt, B. (1996) *IAHS Publication* **236**, 105–114.

Hasholt, B. (1997) *Proceedings of the Northern Research basins, 11th International Symposium & Workshop.* pp. 71–81. Prudhoe Bay to Fairbanks, Alaska, USA. August 18–22.

Hasholt, B., Bobrovitskaya, N., Bogen, J., McNamara, J., Mernild, S.H., Milburn, D., and Walling, D.E. (2005) In: *15th International Northern Research Basins Symposium and Workshop,* (Ed. by A.S. Davies, L. Bengtsson and G. Westerström), pp. 41–68. Luleå , Sweden, 29 August – 2 September, 2005.

Hinkler, J., Pedersen, S.B., Rasch, M., and Hansen, B.U. (2003) *Int. J. Remote Sens.* **23**, 4669–4682.

Kane, D.L. and Yang, D. (Eds.) (2004) *IAHS Publication* **290**, 271 pp.

Liston, G.E., and Elder, K. (2006) *J. Hydrometeol.* **7**, 1259–1276.

Makkink, G.F. (1957) *Neth. J. Agric. Sci.* **5**, 290–305.

Mernild, S.H. (2006) *Freshwater Discharge From the Coastal Area Outside the Greenland Ice Sheet, East Greenland.* Institute of Geography, University of Copenhagen PhD thesis, 387 pp.

Mernild, S.H., Hasholt, B., and Liston, G.E. (2007a) *Hydrol. Process.* DOI:10.1002/hyp.6777.

Mernild, S.H., Liston, G.E., and Hasholt, B. (2007b) *Hydrol. Process.* **21**, 3249–3263.
Nilsson, B. (1969) *UNGI Rapport 2.* Uppsala Universitet, 74 pp.
Pardé, M. (1955) *Fleuves et Rivières*, 3rd edn. Colin, Paris, 224 pp.
Rasch, M., Elberling, B., Jakobsen, B.H., and Hasholt, B. (2000) *Arct. Antarct. Alp. Res.* **32**, 336–345.
Sigsgaard, C., Petersen, D., Grøndahl, L., Thorsøe, K., Meltofte, H., Tamstorf, M., and Hansen, B.U. (2006) In: *Zackenberg Ecological Research Operations, 11th Annual Report, 2005* (Ed. by A.B. Klitgaard, M. Rasch and K. Caning), pp. 11–35. Danish Polar Center, Ministry of Science, Technology and Innovation, Copenhagen.
Søgaard, H., Hasholt, B., Friborg, T., and Nordstrøm, C. (2001) *Theor. Appl. Climatol.* **70**, 35–51.
Valeur, H. (1959) *Geogr. Tidsskr.* **58**, 54–65.
Williams, M.W., and Melack, J.M. (1991) *Water Resour. Res.* **27**, 1575–1588.

Soil and Plant Community-Characteristics and Dynamics at Zackenberg

BO ELBERLING, MIKKEL P. TAMSTORF, ANDERS MICHELSEN, MARIE F. ARNDAL, CHARLOTTE SIGSGAARD, LOTTE ILLERIS, CHRISTIAN BAY, BIRGER U. HANSEN, TORBEN R. CHRISTENSEN, ERIC STEEN HANSEN, BJARNE H. JAKOBSEN AND LOUIS BEYENS

SUMMARY

Arctic soils hold large amounts of nutrients in the weatherable minerals and the soil organic matter, which slowly decompose. The decomposition processes release nutrients to the plant-available nutrient pool as well as greenhouse gases to the atmosphere. Changes in climatic conditions, for example, changes in the distribution of snow, water balance and the length of the growing season, are likely to affect the complex interactions between plants, abiotic and biotic soil processes as well as the composition of soil micro- and macro-fauna and thereby the overall decomposition rates. These interactions, in turn, will influence soil–plant functioning and vegetation composition in the short as well as in the long term.

ADVANCES IN ECOLOGICAL RESEARCH VOL. 40 0065-2504/08 $35.00
© 2008 Elsevier Ltd. All rights reserved DOI: 10.1016/S0065-2504(07)00010-4

In this chapter, we report on soils and plant communities and their distribution patterns in the valley Zackenbergdalen and focus on the detailed investigations within five dominating plant communities. These five communities are located along an ecological gradient in the landscape and are closely related to differences in water availability. They are therefore indirectly formed as a result of the distribution of landforms, redistribution of snow and drainage conditions. Each of the plant communities is closely related to specific nutrient levels and degree of soil development including soil element accumulation and translocation, for example, organic carbon.

Results presented here show that different parts of the landscape have responded quite differently to the same overall climate changes the last 10 years and thus, most likely in the future too. Fens represent the wettest sites holding large reactive buried carbon stocks. A warmer climate will cause a permafrost degradation, which most likely will result in anoxic decomposition and increasing methane emissions. However, the net gas emissions at fen sites are sensitive to long-term changes in the water table level. Indeed, increasing maximum active layer depth at fen sites has been recorded to gether with a decreasing water level at Zackenberg. This is in line with the first signs of increasing extension of grasslands at the expense of fens.

In contrast, the most exposed and dry areas have less soil carbon, and decomposition processes are periodically water limited. Here, an increase in air temperatures may increase active layer depth more than at fen sites, but water availability will be critical in determining nutrient cycling and plant production. Field manipulation experiments of increasing temperature, water supply and nutrient addition show that soil–plant interactions are sensitive to these variables. However, additional plant-specific investigations are needed before net effects of climate changes on different landscape and plant communities can be integrated in a landscape context and used to assess the net ecosystem effect of future climate scenarios.

I. INTRODUCTION

Arctic soils hold large stocks of elements including organic carbon (Post *et al.*, 1982), and arctic ecosystems have long been considered among the most sensitive among ecosystems in general with respect to climatic changes (Oechel *et al.*, 1993, 1998; Maxwell, 1997). Arctic soil nutrient distributions, nutrient stocks and mineralization rates are closely linked to vegetation and climate and interactions within the soil–plant–atmosphere system are important for understanding biogeochemical element cycling, biological diversity and ecosystem stability (Coyne and Kelley, 1974; Oberbauer *et al.*, 1996). Soil formation, including the decomposition of soil organic matter and the release of weathering products, is sensitive to several site-specific environmental conditions, for example, the abundance, redistribution and quality of substrates (Nadelhoffer *et al.*, 1991; Fahnestock *et al.*, 2000) and climatic factors, in particular soil temperature (Fang and Moncrieff, 2001), water content (Howard and Howard, 1993; Kirschbaum, 1995) and snow (Welker *et al.*, 2000). Availability of oxygen, soil microbial community composition as well as faunal abundance and activity are additional factors influencing the decomposition (Lomander *et al.*, 1998).

The decomposition of soil organic matter releases nutrients essential to plants to the soil and several greenhouse gases, including methane and carbon dioxide to the atmosphere (see Elberling *et al.*, 2008, this volume, for details). Spatial and temporal trends in the decomposition of organic matter result in substantial variation in rates of weathering of inorganic soil components and concurrent availability of nutrients to plants. This causes variations in arctic plant communities over small distances in relation to micro-topography, snow-cover and hydrology (Oberbauer *et al.*, 1996). Consequently, temporal changes in plant community structure, composition and distribution in the landscape may represent an additional feedback mechanism for long-term climate changes (Zimov *et al.*, 1996). This feedback also involves the complex interactions between plants, the soil organic matter stocks and mineralization, which is linked through the availability of nutrients for plants and micro-organisms competing in nutrient-limited arctic ecosystems (Evans *et al.*, 1989; Oberbauer *et al.*, 1991).

Therefore, the assessment of present element cycling and long-term trends within the soil–plant ecosystem in arctic regions requires improved understanding of the interacting factors at a landscape scale including soil communities, nutrient and energy availability, soil water distribution, snow-cover, plant communities and soil organic matter quality. While it is still debatable to what extent the arctic region will act as a source or a sink for atmospheric carbon in the future, relatively little has been reported on plant community-controlled variations in the amount of soil organic carbon and on the potential for future shifts in the distribution of plant communities as a result of climate changes in the High Arctic.

The present study summarizes and elaborates on earlier monitoring and research to evaluate spatial and temporal trends as well as environmental controls of the interactions of soil and characteristic plant communities found within Zackenbergdalen in Northeast Greenland. Finally, this study discusses the effects of climatic changes on the interacting soil–vegetation processes.

II. SITE CHARACTERISTICS

The study sites are situated in Zackenbergdalen near the Zackenberg Research Station in Northeast Greenland (74°30′N, 20°30′W). Sites have been selected to represent major plant communities reflecting the hydrological, pedological and geomorphological features present in the valley. A meteorological station located within the study area was established in 1995, and results are described in detail by Hansen *et al.* (2008, this volume). Based on data from the station, the mean annual temperature is about −9.5 °C and the annual precipitation varies from 150 to 200 mm. Minimum air temperatures during winter are close to −40 °C, and mean temperatures 2.5 cm below snow-free soil surfaces are as low as −18 °C for about 4 months of the year. Winds during winter are typically from the north, while south-easterly winds dominate during the growing season.

A. Topography and Landscape Dynamics

Zackenbergdalen can be separated into a western part dominated by gneissic bedrock and an eastern part with soils originating from sedimentary and basaltic bedrock (Figure 1A). The topography (Figures 1A, 2A), landscape forms and wind direction are main factors controlling both the snow distribution and accumulation patterns (Figure 1B) as well as the drainage pattern of melt-water. These patterns are found on both a large (landscape) and a small scale (100–200 m) and can therefore be illustrated conceptually as a transect across typical landscape forms in the valley from hilltops to depressions (Figure 2). The top of the hills are windblown and exposed throughout the year with little or no accumulation of snow. From the hilltops towards the depressions, there is an increase in soil water content from dry conditions (even arid conditions and salt accumulation at the soil surface) at the hilltops to wet conditions in the bottom of the depression. The dominant wind pattern during winter leaves large snow-patches on the south-facing slopes, ensuring high surface and soil water contents during a large part of the growing season. The landscape forms in Zackenbergdalen are described in more detail by Christiansen *et al.* (2008, this volume).

Given the low precipitation during summer, the water availability during the growing season is mainly controlled by the topography and snow distribution patterns, resulting in the densest vegetation beneath snow-patches and along rivulets and small drainage channels. This is illustrated in Figure 1C, where areas with a high greenness are associated with the snow accumulation patterns. Because of increasing air temperatures in the last decade, the snowmelt takes place earlier in the year than previously, which has resulted in fewer long-lasting snow-patches. Hence, a smaller area is influenced by snow-patches in the late growing season than was the case in earlier years. Hinkler (2005) found that a base-temperature of 2 °C is most valid for summed temperatures in relation to potential snowmelt. With the increasing air temperatures at Zackenberg, this corresponds to a change in

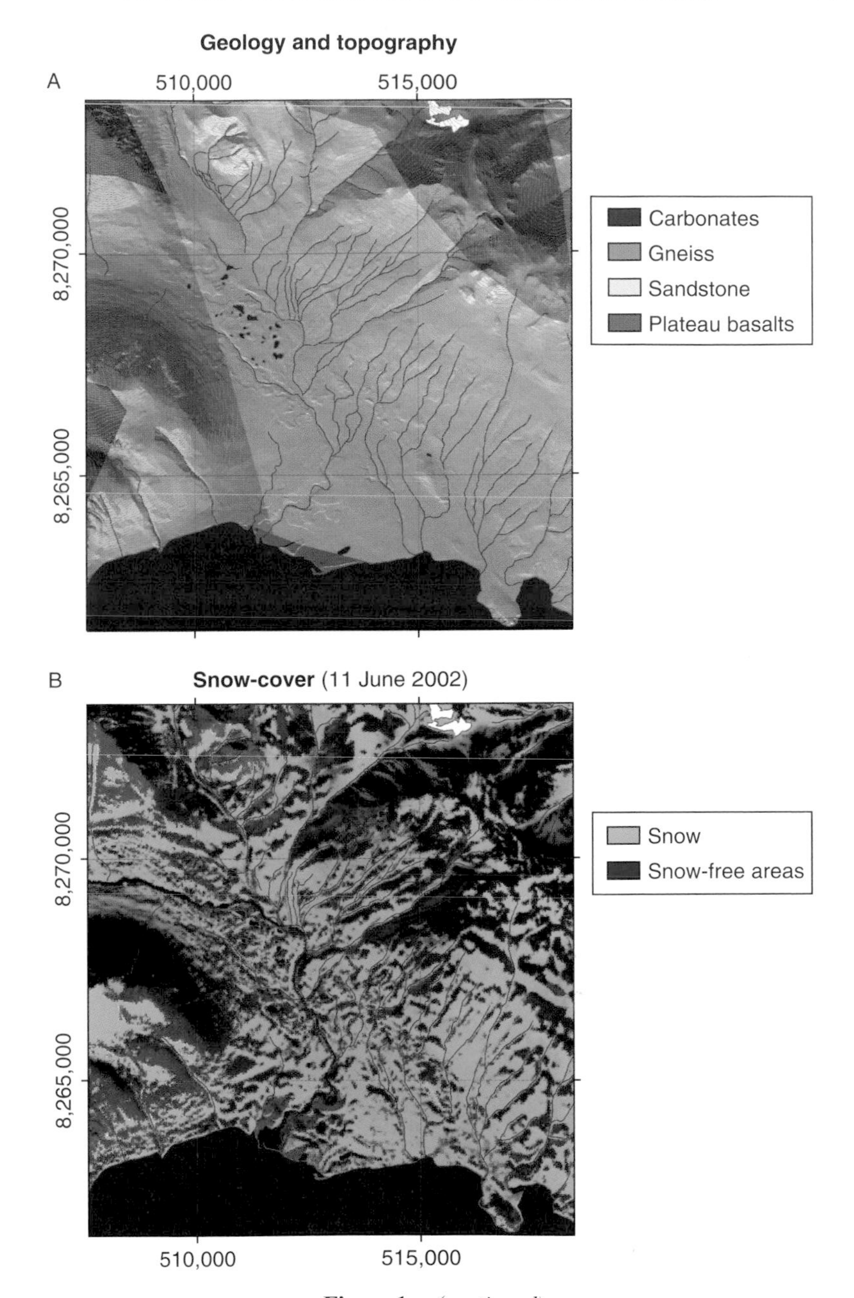

Figure 1 *(continued)*

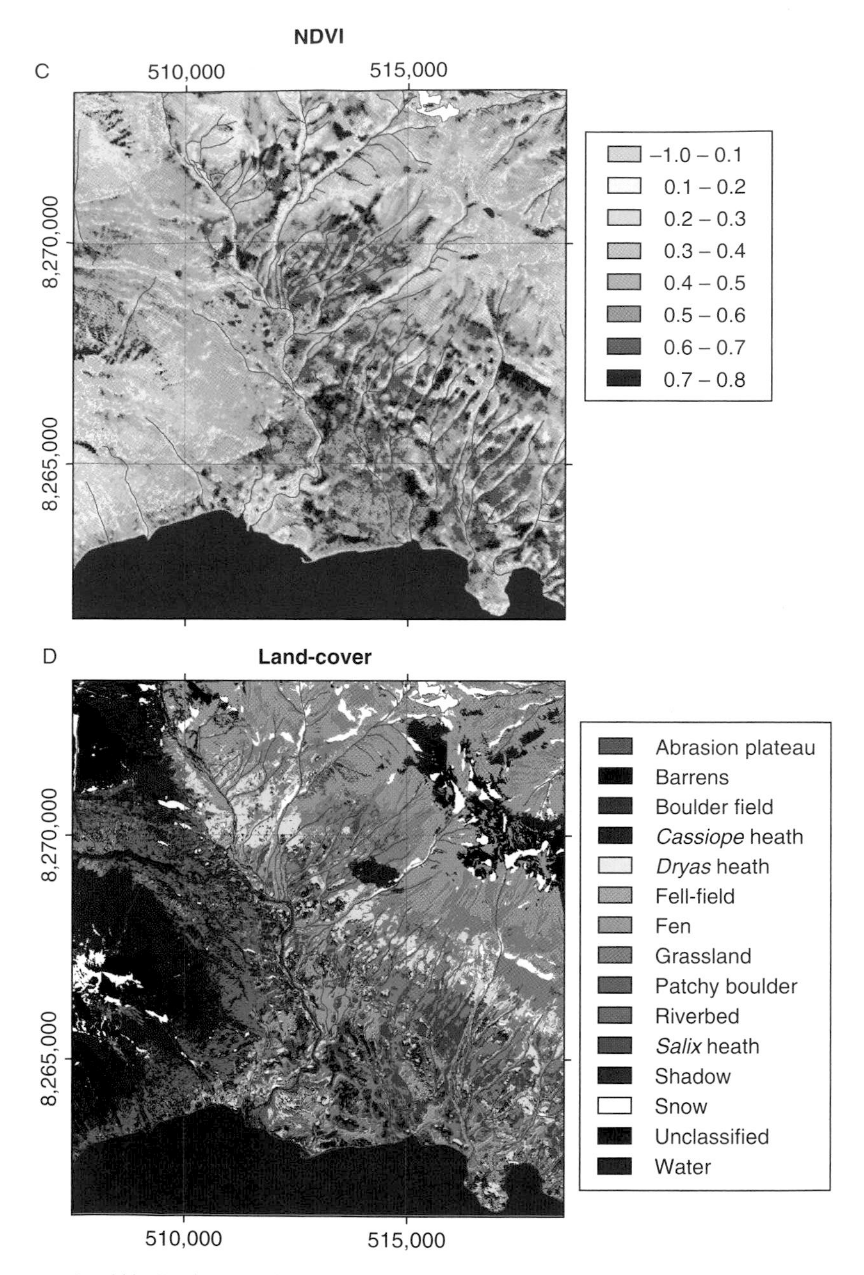

Figure 1 (A) Geology and topography of the study area around the valley Zackenbergdalen, ortophoto August 7, 2000, resolution 20 cm, (B) snow-cover June 11, 2002 (light blue), (C) false colour view of Zackenbergdalen July 30, 1993, with normalised difference vegetation index, NDVI (green), snow and lakes (blue) and bare surfaces (beige) and (D) plant communities.

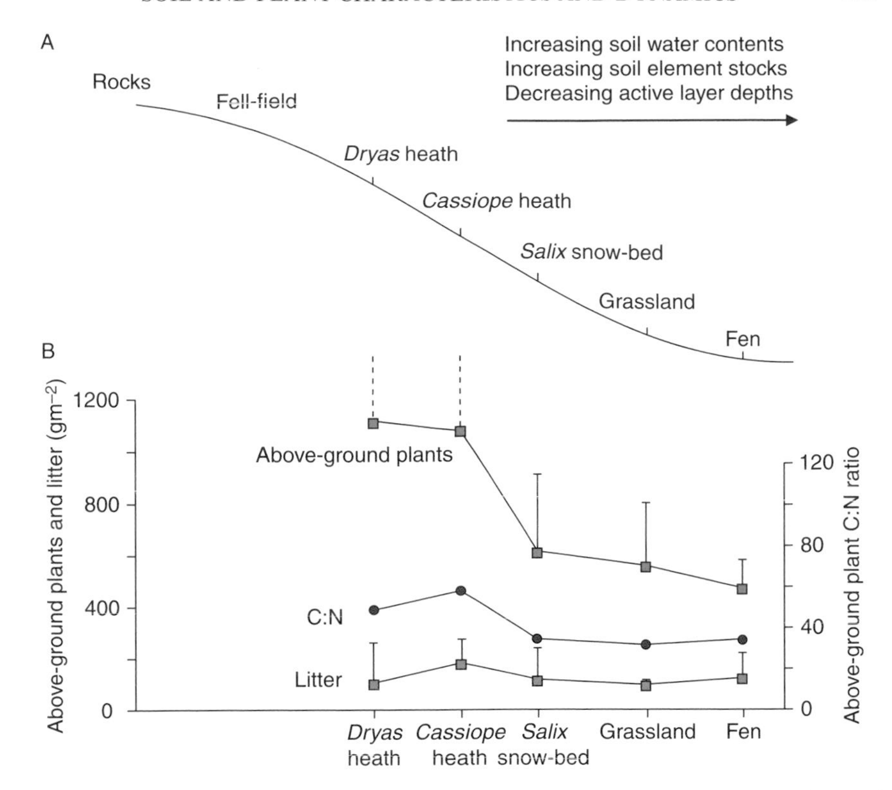

Figure 2 (A) Conceptual gradient from mountain barren ground to wetland found at Zackenberg and (B) spatial trends in above-ground plants and litter amounts and C: N ratio. Additional data on soil physical and biogeochemical properties of major plant communities are given in Table 1. Symbols are mean values and vertical bars represent one standard deviation ($n > 4$); in two cases bars are dashed as they are not shown in full length.

potential snowmelt from just above 4 m in the late 1990s to above 8 m in 2003 and 2004. Consequently, snow-patches that normally supplied melt-water to the vegetation during the entire growing season have disappeared already in late-July, limiting the continuous supply of water. This is in line with a decreasing water table level in fen areas and the first observations of a corresponding shift in plant communities as described later.

B. Plant Communities

Bay (1998) described and classified the plant communities in the central part of Zackenbergdalen and mapped their distribution using aerial photos. The vegetation mapping was extended and adjusted in 2005 using hyper spectral imagery with 5 m spatial resolution, resulting in a land cover map with 14 land cover classes (Figure 1D). Overall, the vegetation forms zones ranging from fens in the depressions and ending with fell-fields and boulder areas towards the hilltops. In addition, there is a variation in vegetation between the two geological features separated by Zackenbergelven, the main river of the valley (Figure 1A). East of Zackenbergelven the lowland is dominated by white arctic bell-heather *Cassiope tetragona* heaths mixed with arctic willow *Salix arctica* snow-beds, grasslands and fens, the latter occurring in the wet, low-lying depressions, often surrounded by grassland. On the transition from the lowland to the slopes of Aucellabjerg (50–100 m a.s.l.), the vegetation is dominated by grassland. Between 150 and 300 m, open heaths of mountain avens, *Dryas* sp., are dominating, and gradually the vegetation becomes more open with increasing altitude towards the fell-fields with a sparse plant cover of *S. arctica* and *Dryas* sp. Grassland, rich in vascular plant species and mosses, occurs along the wet stripes from the snow-patches in the highland (250–600 m). On the west side of the river, arctic blueberry *Vaccinium uliginosum* is more abundant among the many scattered boulders, and the fens have higher species diversity (Figure 1D). Although the bedrock is gneiss, local deposits of marine clay may be the reason for the higher species diversity.

Changes in species composition and distribution of plant communities are monitored regularly along a ZERO-line transect across 128 plant community interfaces from sea level to 1040 m a.s.l. at the summit of Aucellabjerg. The most recent analysis in 2005 showed no distinct signs of floristic and distributional changes since 2000, although the major snow-patches in the area have been diminishing during the last 5 years. Along the ZERO-line, the lichen cover in some cases, however, showed signs of changes, indicative of drying of the environment based on observations of changed community composition of epigaeic lichens (Hansen, 2006).

1. Plant Communities

There are at least five principle plant communities in Zackenbergdalen. Fens occur on level terrain in the lowland, where the soil is saturated by water throughout the growing season (Figure 1D). They are often related to semi-permanent snow-patches, which supply the low-lying areas with water, or to watercourses. Late winter snow depth on fen areas is about 55 cm on average. Common species are *Dupontia psilosantha* and white cottongrass *Eriophorum scheuchzeri*. The peak season cover of vascular plant species and mosses is on average 10% and 100%, respectively, and the vascular plant leaf and the moss biomass is 60 and 270 g m^{-2}, respectively (Arndal, 2006).

Grasslands are dominated by graminoids, that is, plant species belonging to Cyperaceae, Juncaceae and Poaceae. The cover of dwarf shrubs varies, and in some places *S. arctica* is abundant—especially in tussocky areas and on the lower parts of the slopes of Aucellabjerg. Grasslands occur mostly on slightly sloping terrain with adequate supply of water early in the season, but later the soil water regime changes from wet to moist. The late winter snow depth varies from zero to more than 1 m. The main species in the grasslands are wideleaf polargrass *Arctagrostis latifolia*, tall cottongrass *Eriophorum triste* and alpine foxtail *Alopecurus alpinus*. The percentage cover of vascular species is about 25%, while the moss cover is 55%, and the average vascular plant leaf and moss biomass is about 35 and 370 g m^{-2}, respectively.

S. arctica snow-beds (Figure 1D) occur mostly on sloping terrain, often below the *C. tetragona* belt on the slopes, with prolonged snow-cover. The late winter snow depth is on average 95 cm (Hansen *et al.*, 2008, this volume). The most abundant species is *S. arctica*. Other common species are *A. alpinus*, tundra starwort *Stellaria crassipes*, northern woodrush *Luzula confusa*, arctic woodrush *Luzula Arctica* and arctic tundra grass *Poa arctica*. The cover of vascular plants and mosses is about 40% and 50%, respectively. The vascular plant leaf and moss biomass is about 40 and 200 g m^{-2}, respectively. A cover of mosses or organic crust occurs, where the *S. arctica* cover is open.

Locally on south- or west-exposed slopes, *V. uliginosum* is the dominating species in heath areas (Patchy boulder heath in Figure 1D). *Vaccinium* heaths are rare on the east side of Zackenbergelven, but more abundant on the west side. Here they are common in the westernmost part, in which the heath is fragmented by eroded bedrock. On the slopes of Aucellabjerg, *Vaccinium* heaths occur typically on south-exposed slopes just below the *Dryas* heaths and above the *S. arctica* snow-beds. The snow-cover is thinner and the plants are snow-free earlier in the season than in the below-lying snow-beds. The cover of shrubs is about 50%, whereas the moss cover is about 30% (Bay, 1998).

C. tetragona heaths occur only in the lowland on moist, level ground and are abundant below an elevation of 50 m. The *Cassiope* plants are between 5 and 10 cm tall and the vascular cover is about 80%. The most common herbs are *L. arctica* and *L. confusa*. The moss cover is about 30%. The vascular plant leaf and moss biomass is 95 and 235 g m^{-2}, respectively (Arndal, 2006). The *C. tetragona* heath on the west side of Zackenbergelven is more open, and the average cover is low, because the terrain consists of eroded gneiss, and the vegetation is restricted to the areas where fine textured soil is deposited. The late winter snow depth is on average 80 cm.

Dryas heath occurs as two subtypes: (1) together with bellardi bog sedge *Kobresia myosuroides*, curly sedge *Carex rupestris*, and glaucous bluegrass *Poa glauca* on wind-exposed places, where the soil is dry, and (2) with *S. arctica*, moss campion *Silene acaulis* and alpine bistort *Polygonum viviparum* patches on more snow-protected places in the upland above *c*. 150 m, where the soil is moist during the summer season. The late winter snow depth in this community type is on average 61 cm and the cover of vascular plants and mosses in the *Dryas* heath (type 2) is about 90% and 5%, respectively. The vascular plant leaf and moss biomass is both about 60 g m^{-2}.

In the upland above 300 m a.s.l. on moist soils, the landscape is dominated by an open plant community with a vascular plant cover less than 5%. These areas are covered by a variable, but in some places thick, snow-cover. The average late winter snow depth is 140 cm. *Dryas*, *S. arctica* and arctic mouse-ear chickweed *Cerastium arcticum* are the most frequent species, but several other species are also common, for example, Hooker's cinquefoil *Potentilla hookeriana*, arctic poppy *Papaver radicatum*, *L. confusa* and three-flowered lychnis *Melandrium triflorum*.

On the abrasion plateaus in the lowland, the plant community is characterised by a low vascular plant cover, less than 5%, on extremely dry soils. Abrasion plateaus occur in wind-exposed sites, especially on ridges and tops of moraine hills, and are either snow-free or have a very thin snow-cover during winter. Floristically they are characterised by species associated with dry soils and a low snow-cover, for example, *K. myosuroides*, red-stemmed cinquefoil *Potentilla rubricaulis*, short bluegrass *Poa abbreviata*, Hooker's cinquefoil *P. hookeriana* and spike sedge *Carex nardina*.

C. Soils

Zackenbergdalen is a flat valley dominated by non-calcareous sandy fluvial sediments. The soil development is weak and the soil has been classified as Typic Psammoturbels (Elberling and Jakobsen, 2000). Soil characteristics of replicate profiles from the five main plant communities are summarized in Table 1 (modified from Elberling *et al.*, 2004). The soils are slightly acidic to neutral (pH range 5.0–7.2), with pH generally increasing with depth. Most sites are more or less cryoturbated with discontinuous soil horizons. The maximum active layer depth is reached by the end of August (1999–2001) and varies from about 40 cm at fen sites to about 80 cm at *Dryas* sites (for details see Christiansen *et al.*, 2008, this volume). The highest proportion of fine material and less sorted sediments are at *Salix* sites (25% finer than 2 μm; $d_{50} = 15$ μm) and fen sites (20% finer than 2 μm; $d_{50} = 75$ μm). The grain sizes and the degree of sorting increase continuously through *Cassiope* sites (10% finer than 2 μm; $d_{50} = 150$ μm) to *Dryas* sites (5% finer than 2 μm; $d_{50} =$ 360 μm). The grain size distribution, drainage conditions and the distribution of snow-patches coincide roughly with a variation in near-surface water saturation from about 40% at *Dryas* sites to 100% at fen sites in July–August.

The thickness of litter layers varies between 0 and 2 cm, with an average of 0.8 cm at the *Cassiope* and *Salix* snow-bed sites, and litter is almost lacking at

Table 1 Physical and biogeochemical properties of major soil/water/vegetation types at Zackenberg (±1 standard deviation)

Vegetation type	*Dryas* heath	*Cassiope* heath	*Salix* snow-bed	Grassland	Fen
% Water saturation (per vol 0–5 cm)	40–60	60–80	65–90	60–100	90–100
Approximately % of total area[a]	6.1	3.8	7.1	19.8	2.6
Approximately % of vegetated area[a]	9.3	5.7	10.9	30.2	4.0
Total soil C 0–20 cm (kg C m^{-2})	4.1 (±0.4)	6.3 (±3.0)	10.5 (±1.8)	7.6 (±2.2)	5.6 (±3.2)
Total soil C 0–50 cm (kg C m^{-2})	6.1 (±1.1)	8.5 (±2.6)	21.2 (±4.1)	12.3 (±3.4)	11.3 (±3.8)
Soil temperature 5 cm (–°C)[b]	9.1 (±3.8)	5.5 (±1.8)	7.6 (±3.8)	Nd[c]	7.6 (±3.8)
Soil CO_2 efflux (μmol m^{-2} s^{-1})[b]	0.84 (±0.2)	0.52 (±0.1)	1.22 (±0.3)	Nd[c]	2.48 (±0.5)
Maximum thaw in August (cm)	80	65	45	45	40

[a]From Arndal (2006).
[b]Average of hourly readings in July and August 2001.
[c]Not determined.

the *Dryas* and fen sites. However, layers of peat occur at fen sites, although mixed with substantial amounts of mineral material, and are therefore considered to be part of the mineral soil profile.

1. Soil Element Stocks

The soil profiles have maximum concentration of carbon (C) and nitrogen (N) near the surface and decreasing concentrations with depth (Table 1) with the exception of buried organic-rich layers in some profiles. A relict A-horizon (A_b) of 1–5 cm thickness was found in most soil profiles at depths between 15 and 30 cm and is in some places associated with a well-developed Podzol. Dating of the buried horizons have shown that they represent a soil development during the Holocene Climate Optimum starting at least

5000 years ago (Christiansen *et al.*, 2002). The buried layers occur at different depths and have varying thicknesses and carbon contents both among replicate sites at each plant type and among plant communities. The largest differences in thickness and depths of the buried layers are between *Cassiope* and fen sites. At fen sites, the buried organic layer is within the permafrost. Plant remains (birch leaves) were found in all buried organic layers.

Soil bulk densities in the valley increase with depth from about 0.8 g cm^{-3} to about 1.6 g cm^{-3} (Figure 3). Variations in soil bulk densities within each plant communities are of the same order of magnitude as between plant communities, except for the consistently lower densities at the fen sites.

For each soil layer, the total amount of soil organic carbon has been estimated by multiplying the average C concentration by the average soil density and thickness of the layer. The total amount of soil organic C to certain depths has subsequently been estimated by adding the sum of layers above and is referred to as the soil C stock to a certain depth. The mean and one standard deviation of C stocks to 20 and 50 cm depth are given in Table 1. Although large variations in stocks are observed between profiles of each plant communities, variations between different plant communities were larger. Maximum variations within a single plant community are at the *Cassiope* sites, consistent with an irregular distribution of the buried A_b-horizon.

The depth used for calculating stock is critical for both the estimates of absolute stock values and comparisons between vegetation sites. Figure 4 shows that using 50 cm as the depth for stock estimates includes only 55% of the soil organic carbon found in the upper 1 m at fen sites, whereas 92% is

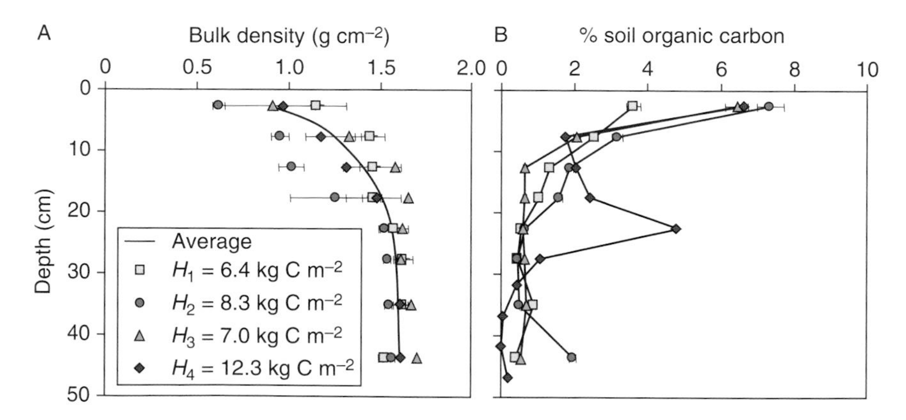

Figure 3 Profiles (replicates) of soil bulk density and soil organic content at four *Cassiope* sites showing the maximum spatial variation observed at Zackenberg within a single plant type. One standard deviation (±) of tree replicates per depth is shown as horizontal lines. The total amount of soil organic carbon is calculated to a depth of 50 cm (modified from Elberling *et al.*, 2004).

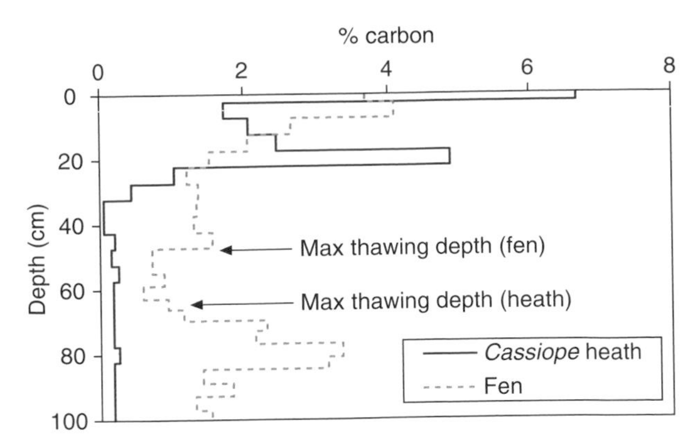

Figure 4 Concentrations and distribution of soil organic carbon in *Cassiope* and fen sites (modified from Elberling *et al.*, 2004).

included at *Cassiope* sites. In contrast, the 50 cm depth stock value represents 100% of the carbon found within the active layer at the fen sites (maximum thaw depth about 40 cm) but 95% at *Cassiope* sites (maximum thaw depth about 65 cm). If buried surface layers are present, the standard deviations become relatively high as at, for example, the *Cassiope* sites (Figure 2B).

The C stocks estimated to depths of 20 and 50 cm (Table 1) indicate that *Dryas* had the lowest C stock within the upper 50 cm with 6.1 kg soil-C m^{-2}, followed by *Cassiope* sites with 8.5 kg soil-C m^{-2}, fen sites with 11.3 kg soil-C m^{-2}, grassland sites with 12.3 kg soil-C m^{-2} and *Salix* sites with 21.2 kg soil-C m^{-2}. The vegetation-specific differences in C stock were only slightly affected by less than 10% when calculated without taking buried A_b-horizons into account, except the *Cassiope* sites where the C stock was reduced by up to 25%. Taking the area distribution of each plant type in the valley into account (see Table 1), the average (±SD) amount of soil organic C to a depth of 50 cm is 8.5 ± 2.0 kg soil-C m^{-2} or 85 t soil-C ha^{-1}. Within the upper 20 cm, the amount of soil organic C is 4.9 kg m^{-2} of which 37% was found at grassland sites, 35% at *Salix* sites, 8% at fen sites, 12% at *Cassiope* sites and 9% at *Dryas* sites. Compared to other fractions of ecosystem carbon pools (vegetation, litter, debris and roots), the soil organic pools represent more than 94% of the total ecosystem organic carbon.

2. *Soil Water Content and Nutrient Availability*

Although concentrations of dissolved ions in soil water collected at three depth intervals at the *Dryas*, *Cassiope*, *Salix* and grassland plant communities in 2003, 2004 and 2005 vary strongly among replicate plots within plant communities and between years (Figure 5 and Table 2), some patterns emerge.

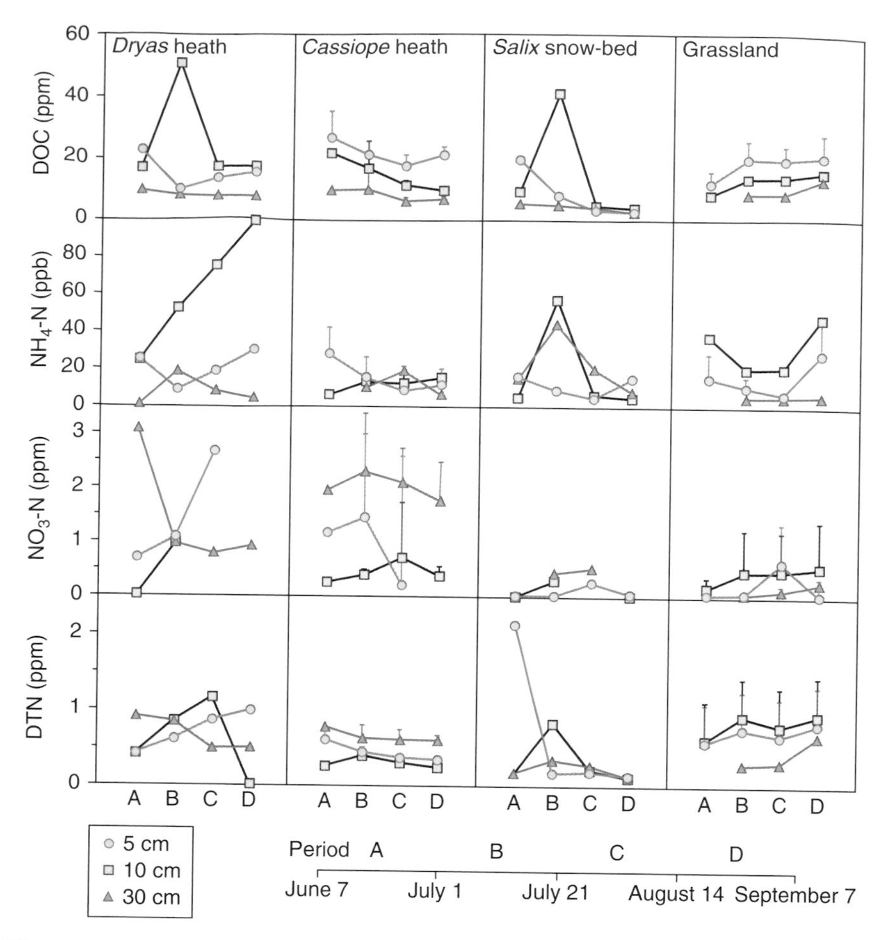

Figure 5 Temporal and spatial trends in soil water chemistry including NH_4-N (ppb), NO_3 (ppm), dissolved organic carbon (DOC, ppm) and dissolved total nitrogen (DTN, ppm). Data are shown for four different periods: (1) from June 7 to July 1, (2) from July 1 to July 21, (3) from July 21 to August 14 and (4) from August 14 to September 7. Values are average values measured during 2003–2005. When values are based on more than two replicate years the standard deviation is shown as vertical lines.

Both NH_4 and dissolved organic carbon generally showed higher concentrations in surface soils in June than later during summer (Figure 5), most likely due to microbial lysis and release of microbial constituents immediately after snowmelt. This is consistent with laboratory studies showing nutrient release from microbial biomass after thaw (Grogan *et al.*, 2004). Across plant communities, the ammonium (NH_4) concentration in the upper 5 cm of soil (with highest fine root biomass) consistently showed a minimum in July to

Table 2 Soil water chemistry in major vegetation types at Zackenberg. Values are annual means of four samples per year from 2003 to 2005

Vegetation type	*Dryas* heath			*Cassiope* heath			*Salix* snow-bed			Grassland		
Soil water chemistry in 3 depths (cm)	5	10	30	5	10	30	5	10	30	5	10	30
NO_3-N (ppm)	1.5	0.5	1.4	0.9	0.4	2.0	0.1	0.1	0.5	0.2	0.4	0.1
NH_4-N (ppb)	20.7	62.2	7.9	15.8	11.7	11.7	11.1	18.2	21.6	15.4	31.5	5.5
Dissolved organic C (ppm)	15.2	25.3	8.1	21.7	14.9	8.2	8.5	14.8	4.4	18.0	12.8	10.1
Dissolved total N (ppm)	0.7	0.8	0.7	0.4	0.3	0.6	0.7	0.3	0.2	0.7	0.8	0.4

mid-August (Figure 5), reflecting high plant N uptake during the period of maximum plant growth. Nitrate (NO_3) concentrations were variable but generally higher in the *Dryas* and *Cassiope* heaths than in *Salix* snow-beds and grasslands (Table 2), most likely because both *Dryas* and *Cassiope* have low capacity to utilize NO_3 (Michelsen *et al.*, 1996). As a consequence, N losses (both as NO_3 and as dissolved total N, DTN) at 30 cm depth, that is, below the rooting zone, are highest in the *Dryas* and *Cassiope* plant communities (Table 2), and these ecosystems may, hence, contribute importantly to N input into the wetter grasslands and fen plant communities down-slope (Figure 2B).

III. SOIL-PLANT INTERACTIONS AND TEMPORAL TRENDS

A. Biomass Element Stocks, Seasonal Trends and Litter Production

The total aboveground biomass varies almost threefold among the five major plant communities (Figure 1), and the total carbon content in the vegetation generally follows the total biomass pattern (Figure 6). On an area basis, significantly more carbon was found in the vegetation of the *Dryas* and *Cassiope* heaths compared to fen, grassland and *Salix* snow-bed (Figure 6).

The leaf biomass C in the grassland, fens and *Salix* snow-beds (Figure 6) increases in mid-June, followed by a period of stable pools during July and a decrease by the end of August. This pattern is explained by a rapid expansion of new leaf tissue production and maturation, followed by a phase of stable biomass and finally leaf dieback. The *Dryas* and *Cassiope* heaths have a more stable seasonal pattern in biomass and leaf biomass C pools as they shed their leaves more gradually and start the growing season with over-wintering leaves, which are able to photosynthesize earlier than expanding leaves in deciduous plants.

In a vegetation study at Zackenberg in 1997, Bay (1998) found a total average aboveground vascular plant biomass of 266 and 214 g m^{-2}, respectively, in the *Cassiope* and *Dryas* heaths. Arndal (2006), however, estimated the stem plus leaf biomasses of 613 and 300 g m^{-2}, respectively, in 2004, that is, considerably higher than in previous estimate of 1997. If this difference is real, it could be because the *Cassiope* heath had suffered from frost injury in early 1997 and has been recovering since then (L. Illeris, personal observation). In the grasslands and fens, Bay (1998) estimated the biomasses to 37 and 134 g m^{-2}, respectively, compared to estimates of 53 and 110 g m^{-2} by Arndal (2006). In *Salix* snow-bed, the above-ground vascular plant biomass was estimated to 76 g m^{-2} by Bay (1998) compared to 136 g m^{-2} estimated

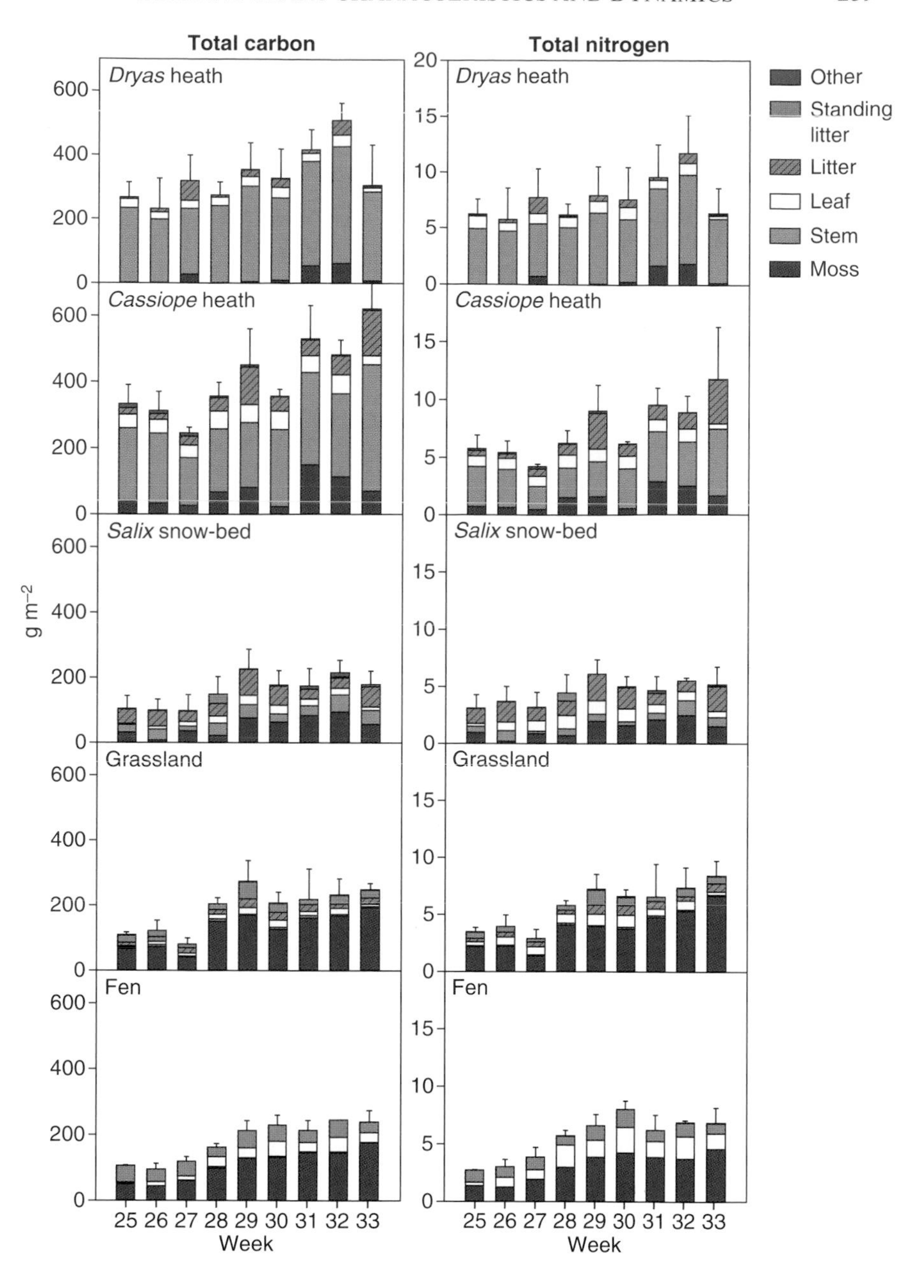

Figure 6 Temporal trends in total carbon and nitrogen pools in different vegetation compartments during the growing season mid-June to mid–August 2004. Data are means ±1 standard error; $n = 3$ per plant community. The stem fraction in *Dryas* includes *c.* 50% mass of attached dead leaves. Data are from Arndal (2006).

by Arndal (2006). The biomasses, particularly in vegetation with little woody tissue, are likely to vary both spatially and from year to year depending on climatic conditions and may explain part of the differences in the estimates in 1997 and 2004. Indeed, the summer of 2004 was among the warmest recorded since the monitoring began in 1996, which may have resulted in higher plant production than the average.

The moss biomass in the five plant communities (Figure 6) varied from a total mean of 60 g m^{-2} in the *Dryas* heaths to 370 g m^{-2} in the grassland, and mosses dominated especially the total biomass in the non-heath plant communities. Consequently, in grasslands, fens and *Salix* snow-beds, the mosses made up the major part of the total biomass and C. The higher proportion of moss biomass is in agreement with reports from Alaska, where the photosynthetic biomass of mosses may exceed that of vascular plants by two- or threefold (Williams and Rastetter, 1999).

The two heath plant communities had similar nitrogen content without significant seasonal variation (Figure 6). Evergreens continue to accumulate nitrogen in leaves and stems throughout the season, and the relatively stable nitrogen content suggests that there is a steady loss and uptake of nitrogen in the heath plant communities. Even though the leaf nitrogen concentration in the grasslands, fens and *Salix* snow-beds decreased over the summer, it was accompanied by an increase in the total standing stock until late July (Figure 6) because of the increasing leaf biomass. Because of the large stem mass with high N content, the highest total nitrogen content was found in the heaths, while mosses contributed strongly to the total nitrogen content in the non-heath vegetation. The fens had significantly higher leaf nitrogen content per unit area than the other plant communities, possibly reflecting higher nutrient availability.

Root biomasses have rarely been reported from high-arctic ecosystems, partly because the patchy distribution of the roots requires intensive sampling strategies. However, measurements from Zackenberg show biomasses of about 250 g m^{-2} in *Cassiope* heaths and 310 g m^{-2} in *Dryas* heaths (Elberling *et al.*, 2004) and a much higher mean of 1500 g m^{-2} in *V. uliginosum*-dominated heaths (Rinnan *et al.*, 2005).

B. Soil Ecosystem Effects of Soil Manipulations

1. Controls of Organic Matter Turnover and Effects of Changes in Temperature, Water and Nutrient Supply

Decomposition of organic matter in arctic ecosystems is closely controlled by temperature and soil moisture, and by the quality of the organic matter (Hobbie, 1996; Robinson *et al.*, 1997). These controls have also been documented quantitatively for Zackenberg (Christensen *et al.*, 1998; Elberling, 2003; Elberling and Brandt, 2003; Elberling *et al.*, 2004). Despite the relatively high number of studies which have focussed on the controls on soil respiration, C emission and C turnover in soils of the High Arctic, including the Zackenberg area (Elberling *et al.*, 2008, this volume), our knowledge of soil C stocks and turnover in different high-arctic plant communities is still incomplete.

In mesic to wet tundra soils, the depth of the water table constitutes a strong control of C dynamics (Christensen *et al.*, 1998). As a consequence, the soil C stocks increase along a gradient from *Dryas* and *Cassiope*-dominated heaths towards snow-beds in the Zackenberg area (Table 1; Elberling *et al.*, 2004). In addition to the soil water content, higher soil surface temperature in drier systems leads to higher C losses, as shown by gradient studies (Christensen *et al.*, 1998) and studies of seasonal variation (Elberling *et al.*, 2004). The temperature-controlled C emission may lead to accelerated losses of C from the soil, if the growing season is prolonged and average summer temperatures increase in the region.

About one third of the ice-free Arctic is covered by polar desert and semi-desert, and in the high-arctic region these ecosystems hold about 40% of the regional terrestrial C stores (Bliss and Matveyeva, 1992). Such dry ecosystems, represented at Zackenberg, for example, by open, patchy *Dryas* heaths with *K. myosuroides* and *S. arctica* surfaces, are likely to represent a high biological activity, indicated by high abundance of soil microorganisms, high numbers of individuals in the mesofauna (Sørensen *et al.*, 2006) and high CO_2 emission compared to mesic heaths (Elberling *et al.*, 2004; Sørensen *et al.*, 2006). Long-term manipulations of such plant communities at Zackenberg have shown that increased precipitation, corresponding to an additional 8 mm weekly during summer, may increase surface soil C emission and microbial biomass by as much as 40% and 24%, respectively (Illeris *et al.*, 2003). This will make a considerable difference to the carbon budget and soil C stocks of these dry ecosystems in a future climate change prospective, provided that plant biomass productions do not respond to the same extent to wetter high-arctic summers. Preliminary experimental results from Zackenberg suggest that vascular plants in dry ecosystems respond slowly to water addition (A. Michelsen, unpublished data), in line with results from other high-arctic sites (Wookey *et al.*, 1995).

The C:N ratio in arctic vegetation is generally found to be around 40 (Rastetter *et al.*, 1997), and studies at Zackenberg showed corresponding vegetational C:N ratios (Figure 7; Elberling *et al.*, 2004).

The deciduous plant communities contain more N relative to C than the wintergreen and evergreen plant communities (Figures 2 and 7). This plant-specific difference is mirrored in the soil, as soil from the deciduous fen, grassland and *Salix* plant communities have lower C:N ratios than soil from

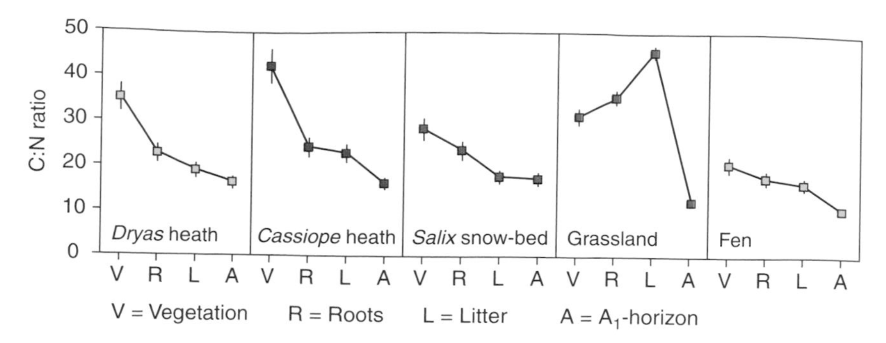

Figure 7 Average C:N ratio for five plant communities including above-ground plant material, roots, litter and near-surface organic A-horizons (modified from Elberling *et al.*, 2004). Vertical bars represent ±1 standard error of mean.

the predominantly wintergreen *Dryas* and the evergreen *Cassiope* heath communities (Elberling *et al.*, 2004). The longevity of leaves from wintergreen and evergreen plant species is longer than the longevity of leaves from deciduous species and graminoids, and hence, they often invest more resources in structures and carbon-based metabolites used as defence against frost plus herbivore and insect attacks (Chapin and Shaver, 1988; Hansen *et al.*, 2006). Also, contrary to the deciduous species, the wintergreen and evergreen species spend carbon resources on structural tissues in form of stems, which also is mirrored in the soil C:N ratios. However, other factors such as decomposability of the existing C in soil may be more important than the C:N ratio for the soil decomposition rate. For example, Elberling *et al.* (2004) found in laboratory experiments that when adding glucose to soil samples from *Cassiope* heath, *Dryas* heath and *Salix* snow-bed, the production of CO_2 increased, indicating that soil microbial activity was limited by the availability of labile carbon. Hence, shifts in vegetation composition and, consequently, altered litter C:N ratios may have major impact on C turnover (Hobbie *et al.*, 2000).

Nutrient inputs exceeding annual plant demand do not seem to affect C pools strongly. Both in dry and wet ecosystems at Zackenberg, nutrient (N and/or P) additions did not lead to significant changes in soil C emission (Christensen *et al.*, 1998; Illeris *et al.*, 2003). However, in laboratory studies of soil from Zackenberg, simultaneous addition of labile carbon plus N and P led to increased C release from the soils (Sørensen *et al.*, 2006). This suggests that microbes and decomposition processes are primarily limited by availability of labile C stemming from root exudates, lysed microbes and soil animals, and easily degradable plant matter. However, if labile C is readily available, which may be the case immediately after snowmelt, as indicated by high soil DOC concentrations (Figure 5), the decomposition processes may periodically be limited by nutrient (N and P) availability.

Information regarding the composition and mass of soil microbe, micro- and mesofauna communities in Greenland is limited. Data on diatom assemblages (Van Kerckvoorde *et al.*, 2000) and testate amoebae communities across five plant communities (Trappeniers *et al.*, 2002) have, however, been reported from Zackenbergdalen. Differences in communities were both plant and soil specific, and at least testate amoebae assemblages seem ecologically distinct across short distances in the high-arctic landscape.

Similar to sub-arctic ecosystems, the total soil microbial biomass C in Zackenberg soil constitutes 1.0–2.5% of the soil organic matter, with lowest mass in dry and highest mass in wet heaths (Illeris *et al.*, 2003; Rinnan *et al.*, 2005). Of the soil total N, 1.5% is found in the microbial biomass in Zackenberg heaths (Michelsen *et al.*, 1998). Hence, the microbial N concentration in the soil is about 50 times higher than the plant-available NH_4^+ concentration in the soil. As a consequence, small fluctuations in microbial biomass, for example, caused by thawing in spring and subsequent lysis of microbes may lead to high release of N, shown by high NH_4^+ in lysimeter water collected in the root zone in June (Figure 5). The high release of NH_4^+ at spring thaw probably is an important source of N to the plants during their rapid growth in mid-summer, shown by the consistent minimum in NH_4^+ in lysimeter water from the upper soil layer of all plant communities in July.

2. Soil Fauna

Soil nematodes and other components of the micro- and mesofauna of the soil food web feed on soil microbes and detritus and are therefore important for organic matter turnover. The abundance of collembola and mites was similar in Zackenberg communities to that of other high-arctic sites, or even higher (Sørensen *et al.*, 2006). For instance, there were up to 45,000 mites m^{-2} in dry heaths (Sørensen *et al.*, 2006). Mites, collemboles and nematodes at high altitude sites in the sub-arctic respond to experimentally increased summer temperature and nutrient availability with changes in community structure (e.g., of collembola and mites; Sjursen *et al.*, 2005) and increased abundance (nematodes; Ruess *et al.*, 1999). Thus, it is likely that the responses of these soil animals to climatic changes may affect decomposition processes and C turnover also in high-arctic plant communities.

IV. CONCLUSIONS AND FUTURE CONSEQUENCES OF CLIMATE CHANGES

Long-term scenarios for climate changes in the Zackenberg region (Stendel *et al.*, 2008, this volume) suggest a future warmer climate, particularly with increasing winter temperatures and moister conditions mainly as increased winter precipitation. Important for plant–soil interactions are also increasing atmospheric CO_2 concentrations and UV radiation as well as several indirect effects of increasing spring temperatures, such as increasing active layer thickness and increasing weathering rates, which are not included in the future scenarios of soil–plant feedbacks to climate changes (Stendel *et al.*, 2008, this volume). It is likely that net ecosystem effects of future climate changes will occur in the Zackenberg area, but also that the impact of climate changes will cause different responses in different plant communities and landscape types with different soil formation. Furthermore, impacts will probably be seen on both a short-term basis (few years) as well as substantial shifts/transitions in plant community structure and area distribution of plant communities on a longer timescale as noted by Oberbauer *et al.* (1996) and Grogan and Chapin (2000).

There are wide ranges of physiological responses of arctic plants as shown in this chapter, which can be directly linked to climate changes and associated with important plant–soil interactions. Among these are, for example, plant responses to increased nutrient supply due to increasing weathering, nitrogen fixation (Sorensen *et al.*, 2006) and decomposition (Robinson *et al.*, 1997) rates as well as increasing active layer depths and length of growing season. Many plants are likely to increase their productivity and alter the system's nutrient cycling, possibly creating positive feedbacks on plant growth. In particular, in nutrient-limited plant communities, the presently scarce plant cover may increase within the present vegetation layer, and new layers of, for example, shrubs may establish. In contrast, minor light attenuation by changes in cloudiness and small changes in summer precipitation may not affect arctic plants profoundly on a short-term basis. As species respond differently to climate changes, the classification into plant functional types at Zackenberg is important. However, because of limited experimental data from Zackenberg so far, it is difficult to use such a classification in Zackenberg to generalize responses and provide predictions of net ecosystem effects. Thus, the following brief discussion focus on contrasting the important plant communities at Zackenberg.

A positive plant response to future climate scenarios for Zackenberg is likely at the dry exposed vegetation sites. Here, plants will benefit from both a longer growing season and in particular more soil water available during the growing season, as plants are already water limited. These exposed sites will most likely show a denser plant cover. Furthermore, intrusions of other plant species, now found at warmer latitudes further south, are likely to occur over time.

The effects of higher spring temperatures and a thicker active layer on element cycling, for example, at *Cassiope* heaths, can be expected to be even stronger than predicted when using only a simple temperature sensitivity relationship. The reason is that increased decomposition of organic matter stored in buried A-horizons (Figure 3) may act as an additional source of nutrients. *Cassiope* may, however, be stressed in years with lots of snow and subsequently wet soil throughout the growing season. Lack of snow in early winter is an additional possible effect, which may affect many plant species that need a protective snow-cover. Frost-damaged *Cassiope* have been noted during the last 10 years, which is consistent with a late date for the first snow the previous autumn.

Salix snow-bed sites hold very large quantities of nutrients stored as organic matter (Table 1). Decomposition rates of the organic matter are sensitive to the water content, as decreasing water contents during the growing season are likely to increase both the availability of oxygen and soil temperatures and thereby increase decomposition rates. However, this may to some extent be counteracted by a thicker snow-cover at the end of the winter and thus a shorter period of plant growth and weathering processes during summer. Because of the current predictions of more winter precipitation and later snowmelt in snow accumulation areas, it is concluded that the snow-bed vegetation will be a more dominating plant community in the area but not necessarily characterised by an increasing plant growth.

With respect to wet sites holding large stocks of organic matter, the fen sites are the most extreme, holding a highly weatherable stock of carbon. Furthermore, several indicators suggest that fen sites may be among the plant communities in Zackenberg which have responded to a warmer climate during recent years: the higher air temperatures mainly during summer have resulted in an increasing maximum thaw depth and a lowering of the water table. Visual inspections suggest that parts of the fen area have changed so that it presently can be classified as grassland. As fen sites are partly water-saturated throughout the growing season, increasing air temperatures will enhance decomposition rates and nutrient availability directly, but may also indirectly (in combination with changes in water balance) influence the depth to the water table and thereby change soil redox conditions. A shift from anaerobic to partly or temporarily aerobic conditions will limit the production of methane but generally increase the CO_2 production rates (as discussed further by Elberling *et al.*, 2008, this volume).

From the discussion above, it appears that water content, temperature and substrate quality are all crucial parameters controlling the presently observed spatial variation in soil–plant element cycling at Zackenberg. Indirect and more slowly responding factors may have a larger impact. Such effects are likely to be the result of changes on a landscape scale, for example, changing wind pattern, snow distribution, draining and formation of thermokarsts (Evans *et al.*, 1989; Hobbie *et al.*, 2000). Future studies should focus on interactive effects of environmental factors, investigate long-term responses to manipulations and incorporate interactions with other trophic levels.

ACKNOWLEDGMENTS

Thanks are extended to the staff at the laboratories at Institute of Biology and Institute of Geography (University of Copenhagen) for making most of the analyses. Monitoring data for this chapter were provided by the GeoBasis programme, run by the Institute of Geography, University of Copenhagen, and the National Environmental Research Institute, University of Aarhus, and financed by the Danish Environmental Protection Agency. Additional funding was obtained from the Danish Natural Sciences Research Council and the Aage V. Jensen Foundation. Thanks are extended to students taking part in the collection of field data, in particular J. Søndergaard, colleagues who participated in soil and plant sampling and discussions as well as two anonymous reviewers providing very helpful comments to an early draft of this chapter.

REFERENCES

Arndal, M.F. (2006) In: Seasonal variation in gross ecosystem production, biomass, and carbon, nitrogen and chlorophyll content in five high arctic vegetation types. MSc thesis, Institute of Biology, University of Copenhagen, 101 pp.

Bay, C. (1998) Vegetation mapping of Zackenberg valley Northeast Greenland. Danish Polar Center and Botanical Museum, University of Copenhagen, 29 pp.

Bliss, L.C. and Matveyeva, N.V. (1992) In: *Arctic Ecosystems in a Changing Climate. An Ecophysiological Perspective* (Ed. by F.S. Chapin, III, R.L. Jefferies, J.F. Reynolds, G.R. Shaver and J. Svoboda), pp. 59–89. Academic Press Inc., San Diego.

Chapin, F.S., III and Shaver, G.R. (1988) *Oecologia* **77**, 506–514.

Christensen, T.R., Jonasson, S., Michelsen, A., Callaghan, T.V. and Havström, M. (1998) *J. Geophys. Res* **103**, 29,015–29,021.

Christiansen, H.H., Bennike, O., Böcher, J., Elberling, B., Humlum, O. and Jakobsen, B.H. (2002) *J. Quat. Sci.* **17**, 145–160.

Coyne, P.I. and Kelley, J.J. (1974) *Nature* **234**, 407–408.

Elberling, B. (2003) *J. Hydrol.* **276**, 159–175.

Elberling, B. and Brandt, K.K. (2003) *Soil Biol. Biochem.* **35**, 263–272.

Elberling, B. and Jakobsen, B.H. (2000) *Can. J. Soil Sci.* **80**, 283–288.

Elberling, B., Jakobsen, B.H., Berg, P., Søndergaard, J. and Sigsgaard, C. (2004) *Arct. Antarct. Alp. Res.* **36**, 509–519.

Evans, B.M., Walker, D.A., Benson, C.S., Nordstrand, E.A. and Petersen, G.W. (1989) *Holarctic Ecol.* **12**, 270–278.

Fahnestock, J.T., Povirk, K.L. and Welker, J.M. (2000) *Ecography* **23**, 623–631.

Fang, C. and Moncrieff, J.B. (2001) *Soil Biol. Biochem.* **33**, 155–165.

Grogan, P. and Chapin, F.S. (2000) *Oecologia* **125**, 512–520.

Grogan, P., Michelsen, A., Ambus, P. and Jonasson, S. (2004) *Soil Biol. Biochem.* **36**, 641–654.

Hansen, A.H., Jonasson, S., Michelsen, A. and Julkunen-Tiitto, R. (2006) *Oecologia* **147**, 1–11.

Hansen, E.S. (2006) In: *Zackenberg Ecological Research Operations, 11th Annual Report, 2005* (Ed. by M. Rasch and K. Caning), pp. 57–68. Danish Polar Center, Ministry of Science, Technology and Innovation, Copenhagen.

Hinkler, J. (2005) From digital cameras to large scale sea-ice dynamics—A snow-ecosystem perspective. MSc thesis, National Environmental Research Institute and the University of Copenhagen, Department of Geography.

Hobbie, S.E. (1996) *Ecol. Monogr.* **66**, 503–522.

Hobbie, S.E., Schimel, J.P., Trumbore, S.E. and Randerson, J.R. (2000) *Global Change Biol.* **6**, 196–210.

Howard, D.M. and Howard, P.J.A. (1993) *Soil Biol. Biochem.* **25**, 1537–1546.

Illeris, L., Michelsen, A. and Jonasson, S. (2003) *Biogeochemistry* **65**, 15–29.

Kirschbaum, M.U.F. (1995) *Soil Biol. Biochem.* **27**, 753–760.

Lomander, A., Kätterer, T. and Andrén, O. (1998) *Soil Biol. Biochem.* **14**, 2017–2022.

Maxwell, B. (1997) In: *Global Change and Arctic Terrestrial Ecosystems* (Ed. by W.C. Oechel, T. Callaghan, T. Gilmanov, J.I. Holten, B. Maxwell, U. Molau and B. Sveinbjörnsson), Ecological studies 124, pp. 21–46. Springer, New York.

Michelsen, A., Schmidt, I.K., Jonasson, S., Quarmby, C. and Sleep, D. (1996) *Oecologia* **105**, 53–63.

Michelsen, A., Quarmby, C., Sleep, D. and Jonasson, S. (1998) *Oecologia* **115**, 406–418.

Nadelhoffer, K.J., Giblin, A.E., Shaver, G.R. and Laundre, J.R. (1991) *Ecology* **72**, 242–253.

Oberbauer, S.F., Tenhunen, J.D. and Reynolds, J.F. (1991) *Arctic Alpine Res.* **23**, 162–169.

Oberbauer, S.F., Gillespie, C.T., Cheng, W., Sala, A., Gebauer, R. and Tenhumen, J.D. (1996) *Arctic Alpine Res.* **28**, 328–338.

Oechel, W.C., Hastings, S.J., Vourlitis, G.L., Jenkins, M., Riechers, G. and Grulke, N. (1993) *Nature* **361**, 520–523.

Oechel, W.C., Vourlitis, G.L., Brooks, S., Crawford, T.L. and Dumas, E. (1998) *J. Geophys. Res.* **103**, 28,993–29,003.

Post, W.M., Emanual, W.R., Zinke, P.J. and Stangenberger, A.G. (1982) *Nature* **298**, 156–159.

Rastetter, E.B., McKane, R.B., Shaver, G.R., Nadelhoffer, K.J. and Giblin, A. (1997) In: *Global Change and Arctic Terrestrial Ecosystems* (Ed. by W.C. Oechel, T. Callaghan, T. Gilmanov, J.I. Holten, B. Maxwell, U. Molou and B. Sveinbjörnsson), Ecological Studies 124, pp. 437–451. Springer, New York.

Rinnan, R., Keinänen, M.M., Kasurinen, A., Asikainen, J., Kekki, T.K., Holopainen, T., Ro-Poulsen, H., Mikkelsen, T.N. and Michelsen, A. (2005) *Global Change Biol.* **11**, 564–574.

Robinson, C.H., Michelsen, A., Lee, J.A., Whitehead, S.J., Callaghan, T.V., Press, M.C. and Jonasson, S. (1997) *Global Change Biol.* **3**, 37–49.

Ruess, L., Michelsen, A., Schmidt, I.K. and Jonasson, S. (1999) *Plant Soil* **212**, 63–73.

Sjursen, H., Michelsen, A. and Jonasson, S. (2005) *Appl. Soil Ecol.* **30**, 148–161.

Sorensen, P.L., Michelsen, A. and Jonasson, S. (2006) *Arct. Antarct. Alp. Res.* **38**, 263–272.

Sørensen, L.I., Holmstrup, M., Maraldo, K., Christensen, S. and Christensen, B. (2006) *Polar Biol.* **29**, 189–195.

Trappeniers, K., Kerckvoorde, A.V., Chardez, D., Nijs, I. and Beyens, L. (2002) *Arct., Antarct. Alp. Res.* **34**, 94–101.

Van Kerckvoorde, A., Trappeniers, K., Nijs, I. and Beyens, L. (2000) *Polar Biol.* **23**, 392–400.

Welker, J.M., Fahnestock, J.T. and Jones, M.H. (2000) *Climate Change* **44**, 139–150.
Williams, M. and Rastetter, E.B. (1999) *J. Ecol.* **87**, 885–898.
Wookey, P.A., Robinson, C.H., Parsons, A.N., Welker, J.M., Press, M.C., Callaghan, T.V. and Lee, J.A. (1995) *Oecologia* **102**, 478–489.
Zimov, S.A., Davidov, S.P., Voropaev, Y.V., Prosiannikov, S.F., Semiletov, I.P., Chapin, M.C. and Chapin, F.S. (1996) *Climatic Change* **33**, 111–120.

Inter-Annual Variability and Controls of Plant Phenology and Productivity at Zackenberg

SUSANNE M. ELLEBJERG, MIKKEL P. TAMSTORF, LOTTE ILLERIS, ANDERS MICHELSEN AND BIRGER U. HANSEN

SUMMARY

Results from monitoring and experimental studies representing various elements of climatic change are presented to evaluate the concurrent dynamics of vegetation types as well as species-specific responses to climatic variations. The studies were carried out in the high-arctic valley of Zackenbergdalen in Northeast Greenland. Vegetation type dynamics and land surface phenology were studied through the use of hand-held sensors of reflection of vegetation, from which the far red normalised difference vegetation index (NDVI-FR) can be inferred. Furthermore, species-specific dynamics were studied through measures of timing and magnitude of flowering through a period of 10 years.

Time of snowmelt and temperature were the major controlling factors for the timing of the phenology in the six vegetation types: fell-field, *Dryas* heath, *Cassiope* heath, *Salix* heath, grassland, and fen. Snowmelt had a linear positive effect on the timing of the maximum of the growing season, with late snowmelt causing a later occurrence of the maximum. While higher summed temperatures during the green-up period (time from snowmelt to

ADVANCES IN ECOLOGICAL RESEARCH VOL. 40 0065-2504/08 $35.00
© 2008 Elsevier Ltd. All rights reserved
DOI: 10.1016/S0065-2504(07)00011-6

maximum) also was shown to be positively related to the timing of the maximum, enhanced flower production seemed to cause lower vegetative biomass production and hence a later maximum.

The seasonal vegetative production, expressed as the seasonal integrated NDVI-FR (SINDVI), had a linear negative relation with the temperatures during the previous summer. This was probably due to higher temperatures causing more flowers the following year, leading to a lower NDVI-FR.

A strong negative trend in maximum NDVI-FR is documented in all six vegetation types during the years from 1999 to 2006 with a decrease of 0.01 NDVI-FR per year. The main reason could be drying of the upper soil layers due to earlier snowmelt and higher evapotranspiration during recent years.

Some general trends in phenological characteristics were recognised, although responses varied among species. The time of snow disappearance was the main determinant for onset of flowering. Shrubs seemed to be superior in taking advantage of an early snowmelt with respect to initiation of flowering. In addition, most species developed flowers and seed capsules faster when temperatures increased. More flowers increase the chance of cross-pollination and hence, offer an increased possibility for sexual reproduction and genetic exchange. Most species increased the number of flowers in years following a warm growing season in the previous year, although some species also depended on climatic conditions in the current year prior to onset of flowering.

Experimental manipulations of growing season length, temperature and incoming radiation in plots with the dominant dwarf shrub, white arctic bell-heather *Cassiope tetragona*, confirmed that this species significantly reduced flowering when subjected to shading or reduced growing season length.

Based on current predictions for climatic changes in the Zackenberg area (Stendel *et al.*, 2008, this volume), it is expected that the plant species currently present will be able to put more effort in sexual reproduction, thus ensuring increased species adaptedness to changing climate through enhanced genetic variance. However, this is provided that seeds are able to germinate, that species are not out-competed by faster growing and canopy-forming invading species, and that species are able to continue flowering even under reduced light levels caused by an increased cloud-cover.

I. INTRODUCTION

Detailed knowledge of the factors controlling vegetation phenology in the High Arctic is necessary to be able to provide adequate input regarding growth dynamics and primary production to global circulation models. Predictions of vegetation dynamics based on specific knowledge about these factors are of particular importance, especially because climate change here is expected to be greater than at lower latitudes (ACIA, 2005). Thus, with this information added to the current model framework of global circulation models, it will be possible to be even more accurate in spatial predictions of future climate changes and their effects.

The timing of flowering within the short growing season is of crucial importance for sexual reproduction in plant species growing in high-arctic ecosystems. First, the growing season in the High Arctic may last only 60–100 days (Jefferies *et al.*, 1992). This corresponds well with an observed duration of the growing season of 64–113 days in the different vegetation types at Zackenberg. Consequently, the time available for completion of a full reproductive cycle is short, and a delayed start may result in lack of time for completion of fruit development. Second, a larger number of flowers occurring at any one time for any given plant species is likely to increase the chance of successful cross-pollination (Totland, 1994), which is necessary for species evolution and thereby species survival in a changing environment.

In general, wind- and self-pollination are more widespread among arctic flowering plants than among flowering plants elsewhere (ACIA, 2005), but it has been shown that, for instance, entired-leafed mountain avens *Dryas integrifolia* and purple saxifrage *Saxifraga oppositifolia* are dependent on insects for maximal seed set (Kevan and Baker, 1983). However, differential responses to climate changes across trophic levels may result in a potential mismatch in the relationship between plants and their pollinators, seed-dispersers, as well as herbivores (Dunne *et al.*, 2003; Berg *et al.*, 2008, this volume; Høye and Forchhammer, 2008, this volume). Thus, timing of flowering during a short growing season is of great importance for reproductive success and hence, genetic exchange between individuals of the same species undergoing adaptation to a changing climate.

Timing of vegetative growth in plants is important both for the plants themselves and for animals, dependent on plant forage. As with flowering, animals in the High Arctic are highly dependent on timing in their life cycle. For instance, the progression of the calving season in two populations of caribou from Alaska and West Greenland, respectively, was shown to be highly synchronised with phenological development in the plant forage (Post *et al.*, 2003). The plants themselves need to time the acquisition of resources so as to obtain enough resources to secure flowering and/or growth in forthcoming years when growing conditions may be less favourable. Also, many plants flower almost immediately following snowmelt (Sørensen, 1941), which means that resources for flowering need to be built up the year prior to flowering. Thus, knowledge about the environmental and biotic factors that control the seasonal timing of vegetative growth allows for predictions of resource allocation in plants and the possible consequences for vegetation development and forage quality.

One way to study the timing of vegetative growth is through analysis of vegetation greenness obtained from hand-held meters that measures reflected red (RED) and near-infrared (NIR) light. Vegetation reflects light differently in the RED and NIR wavelengths resulting in a measure, far red normalized difference vegetation index (NDVI-FR, see Box 1), which is highly correlated with the biophysical variables (Myneni *et al.*, 1995). This allows for monitoring of vegetative development across time on the scale of local vegetation types. Furthermore, if a relationship between hand-held measures of vegetation growth and satellite image data can be established, it enables the use of satellite images for vegetation monitoring in remote or very large areas with limited possibility for human access (Hope *et al.*, 2003). Land surface phenology, defined as the seasonal pattern of variation in vegetated land surfaces, is used in this study of vegetation growth dynamics in Zackenbergdalen.

Individual species may not respond in the same way to alterations in growth conditions. Several studies have been carried out on how flowering and seed set in individual plant species in alpine and sub-, low- and high-arctic areas react to differences in duration of snow-cover, growing season temperature and water availability. Although many of these document some general trends in the responses across species, such as an earlier leaf bud burst and earlier flowering, shorter pre-floration time (the time from snowmelt to onset of flowering) and/or faster capsule development, increased flowering

Box 1

NDVI and Land Surface Phenology

All green vegetation has a characteristic spectral reflectance feature. The chlorophyll and pigments in plants in general absorb light strongly at red wavelengths (RED), while the mesophyll at near-infrared wavelengths (NIR) causes scattering and higher reflectance beyond 800 nm. This characteristic feature varies with different pigment content and plant biomass and is therefore used for long-term monitoring of land surface phenology sequentially in the same plots. One of the most widely used methods for measuring this spectral feature is the normalised difference vegetation index (NDVI). NDVI correlates with biophysical variables like biomass, leaf area index, CO_2 flux, etc. (e.g., Myneni *et al.*, 1995). Box Figure 1 shows an example of the spectral reflectance of two vegetation types with very different biomass and vegetation cover. NDVI is calculated as the difference in reflection between the NIR and the RED spectral bands using the following equation (Rouse, 1973):

$$\frac{\sigma_{\mathrm{NIR}} - \sigma_{\mathrm{RED}}}{\sigma_{\mathrm{NIR}} + \sigma_{\mathrm{RED}}}$$

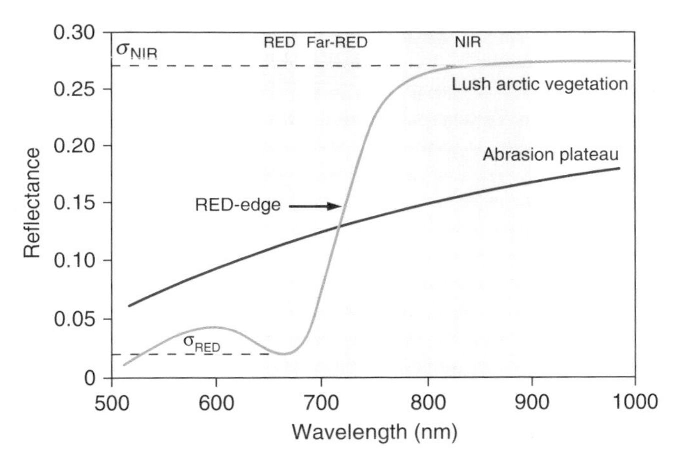

Box Figure 1 Spectral reflectance used for monitoring land surface phenology. Two typical arctic vegetation types with different greenness are shown. Normalized difference vegetation index is calculated from the reflectances at the red and near-infrared wavelengths (see text for formula). RED, red; NIR, near-infrared; NDVI, normalized difference vegetation index.

where σ_{NIR} is the reflection at the NIR wavelength and σ_{RED} is the reflection at the RED wavelength. Depending on the sensors available, NDVI can be calculated from spectral bands with different band centre and width to capture specific properties of the plants. Frequently NIR is exchanged with far-red reflectance measures (centred at 730 nm) directly on the "red-edge," which is the sharp transition between the RED trough and the NIR roof (Box Figure 1). Shifts in the red-edge have been shown to indicate plant "health" status. This altered far-red NDVI is referred to as NDVI-FR.

Data from frequent NDVI measurements can describe the land surface phenology using a quadratic model (Box Figure 2). The quadratic model has previously been used in the High Arctic (Marchand *et al.*, 2004; de Beurs and Henebry, 2005). Arctic vegetation is dependent on snowmelt for the initiation of growth and reproduction. The growing season starts around the end of snowmelt (ESM) with an increase in greenness during the green-up period in the beginning of the summer until maximum greenness is reached. The vegetation reaches maximum in mid-summer after which the greenness fades towards the end of the growing season. Typically, arctic vegetation types have an NDVI value of around 0.1 at the start of the growing season (Stow *et al.*, 2003; Jia *et al.*, 2006). This initial value is used in the calculation of the land surface phenology parameters. To calculate the exact date (DOYmax) and level (MaxNDVI) of maximum NDVI, a quadratic model was fitted to the NDVI measurements (Box Figure 2). This model was also

(*continued*)

Box 1 *(continued)*

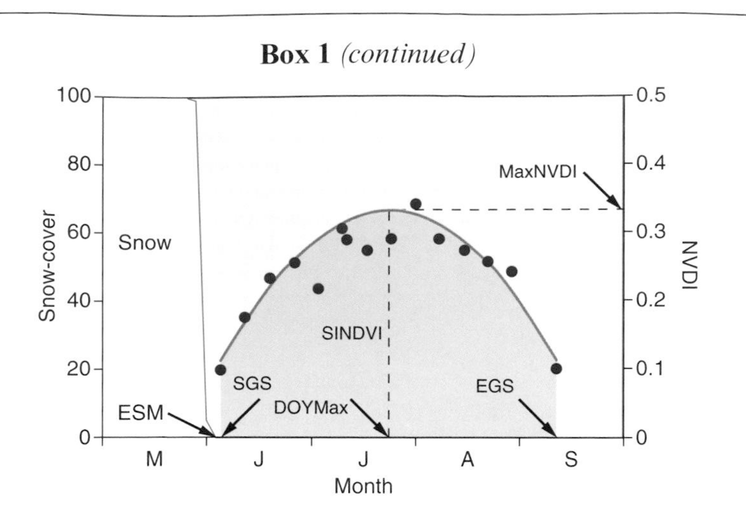

Box Figure 2 The use of NDVI measurements for calculating growing season parameters and monitoring land surface phenology. ESM, end of snowmelt; SGS, start of growing season; DOYmax, day of maximum NDVI; EGS, end of growing season; SINDVI, seasonal integrated NDVI.

used to calculate the seasonal integrated NDVI (SINDVI) from the start of the growing season to the end of the growing season. Stow *et al.* (2003) showed SINDVI to differ characteristically between arctic vegetation types, and it has also been used in primary production studies (de Beurs and Henebry, 2005; Tamstorf *et al.*, 2007).

and increased sexual reproduction with increased temperatures (Wookey *et al.*, 1993, 1995; Woodley and Svoboda, 1994; Johnstone and Henry, 1997; Molau, 1997; Welker *et al.*, 1997; Arft *et al.*, 1999; Sandvik and Totland, 2000; Rayback and Henry, 2005; Høye *et al.*, 2007a), most of these studies have been carried out for one or a few species only, thus limiting generalisation of results. Indeed, in a meta-analysis of tundra plant responses to experimental warming, Arft *et al.* (1999) showed that both vegetative and reproductive responses varied among life-forms. In addition, a species does not necessarily react in the same way to altered growth conditions across its entire geographical distribution. For instance, different factors influenced the vegetative growth in *Cassiope tetragona* subjected to experimental manipulations of growth conditions in three different altitudinal and latitudinal sites (Havström *et al.*, 1993). At a sub-arctic tree-line site, growth was mainly increased by nutrient addition, whereas temperature enhancement caused growth increase at a high altitude fell-field and a high-arctic site.

Apart from the varying responses to climatic changes within species, differential responses among species, in contrast, may cause a shift in the entire composition of the plant communities and hence, in plant community function. In support of this view, a shift towards an increased abundance and cover of shrubs has taken place in northern Alaska (Tape *et al.*, 2006). This emphasises the importance of species-specific knowledge to predict ecosystem changes and thereby elucidate possible differences in ecosystem productivity. In summary, it is important to study and be able to evaluate land surface phenology, vegetation type dynamics and species-specific responses in flowering to the predicted climate changes in the High Arctic. This study summarizes the current knowledge on vegetation type dynamics and species-specific reproduction in the high-arctic valley Zackenbergdalen and evaluates the likely plant responses to predicted climate changes.

II. STUDY SITES AND METHODS

This chapter presents results from monitoring and experimental work carried out at Zackenberg Research Station in Northeast Greenland (74°30′N, 20°30′W). All study sites included are located in the central part of Zackenbergdalen, with a meteorological station placed in the centre of the valley. Here, the landscape is dominated by *C. tetragona* heaths mixed with arctic willow *Salix arctica* snow-beds, grasslands and fens (Bay, 1998). In this chapter, we use *Salix* heath for the combined vegetation types that cover both snow-beds and drier heath areas dominated by *S. arctica*. A detailed description and classification of the vegetation types in Zackenbergdalen was made by Bay (1998), and a description of the common plant communities and updated vegetation map for Zackenbergdalen is available in Elberling *et al.* (2008a, this volume). Plots for monitoring of growth dynamics and plant reproduction were established in 1995 as part of the BioBasis monitoring programme and includes weekly monitoring of seven plant species. See Box 2 for details on species and sampling procedures.

To evaluate responses in *C. tetragona* to different climatic factors, experimental plots with manipulations of duration of early season snow-cover, growing season temperature and incoming light were established in 2004 (Figure 1). Observations were carried out once or twice per week during the growing seasons of 2004, 2005 and 2006 following the same sampling procedures as for the monitoring data. Five plots of 1 m^2 each were established for each treatment. All plots were established in a relatively uniform dry *Cassiope* heath (Table 1).

The land surface phenology and growth dynamics of six major vegetation types were evaluated for Zackenbergdalen for the period 1999–2005 (Tamstorf *et al.*, 2007) and updated to 2006 in this chapter. Measurements from the permanent monitoring plots (see Box 2) were used to evaluate the seasonal changes in plant biomass as expressed by the NDVI-FR (see Box 1).

Box 2

Monitoring Species-Specific Reproductive Dynamics

The BioBasis monitoring programme includes seven plant species chosen to represent the most common species and growth forms present within the plant communities in Zackenbergdalen. In addition, the species are all common and widespread in the Arctic. The species are white arctic bell-heather *Cassiope tetragona* (evergreen dwarf shrub), arctic willow *Salix arctica* (deciduous shrub), arctic poppy *Papaver radicatum* (perennial herb, rosette), purple saxifrage *Saxifraga oppositifolia* (perennial herb, prostrate), moss campion *Silene acaulis* (perennial herb, cushion), entired-leafed mountain avens/mountain avens *Dryas integrifolia/octopetala* (wintergreen) and arctic cotton-grass *Eriophorum scheuchzeri* (sedge). The present chapter focuses on six of the seven species and is using phenological data from Ellebjerg *et al.* (2008), while further details on the remaining species, *Dryas*, can be found in Høye *et al.* (2007a).

The monitoring included weekly observations of time of snowmelt, timing of flowering and flower density (total number of flowers m^{-2} plot) from May 1 to September 1 each year from 1996 to 2005. For all species, except *E. scheuchzeri*, 200 flowers (if present) were counted in each plot once a week, and the counts were divided into fractions of buds, open flowers, senescent flowers and open capsules. Registrations of flower density were made once per season at the time of peak flowering. For *E. scheuchzeri*, only registrations of flower density were made. Plot locations were spread out across the valley to cover the entire ecological amplitude of the species in question. Hence, plant density varied among plots, causing that the species in question could be dominant in some plots and sub-dominant in others. This also resulted in varying plot sizes (Box Table 1). Number of plots monitored varied among species, and in

Box Table 1 Plot sizes and number of plots included in analyses of timing variables and flower density for *Cassiope tetragona*, *Salix arctica*, *Papaver radicatum*, *Saxifraga oppositifolia*, *Silene acaulis* and *Eriophorum scheuchzeri*

Species	*C. tetragona*	*S. arctica*	*P. radicatum*	*S. oppositifolia* and *S. acaulis*	*E. scheuchzeri*
Plot size (m^2)	2, 2, 2, 2½, 3, 3	36, 150, 300	90, 91, 105, 150	6, 7, 10	6, 8, 10, 10
Timing	4 plots	3 plots	4 plots	3 plots	4 plots
Flower density	6 plots	3 plots	4 plots	3 plots	4 plots

some instances, not all variables were monitored in all plots of a given species. Hence, the number of plots included in analyses of timing and flower density were not necessarily the same. An overview of available data is presented below and in Figure 4. For further details on plot locations and sampling procedures, see Meltofte and Berg (2006).

To evaluate the typical vegetation types in the area, each monitoring plot was classified to a vegetation type according to Bay (1998). A quadratic model as opposed to more complex ecosystem models was found to be the best for simple and quick estimation of growth dynamics for the relatively sparse high-arctic vegetation with weekly or fewer NDVI-FR measurements. The analysis was based on data sets from 1218 separate positions scattered in Zackenbergdalen. Each position was measured once a week using hand-held NDVI red/far red reflectance sensors. The growing season length, timing of maximum growth and seasonal integrated normalised difference vegetation index (SINDVI) was calculated for six vegetation types: fell-field, *Dryas*

Figure 1 View of experimental plots in a *Cassiope tetragona* heath in Zackenbergdalen with increased temperature (open-top plastic tents), shading (with sack-cloth) and prolonged or shortened growing season (snow removal or addition, respectively, all snow was melted at the time the photo was taken). Photo: Anders Michelsen.

Table 1 Overview of experimental treatments performed in the same plots throughout the growing seasons of 2004, 2005, and 2006 in a *Cassiope tetragona* heath area in Zackenbergdalen

Experimental treatment	Manipulation method	Abiotic effects
Control (C)	None	None
Shading (S)	Shade-tent made from sack-cloth	Decrease of 50–60% in incoming radiation. Soil temperature decrease of about 2 °C in both 2 and 5 cm depth
Increased growing season temperature (T)	Open top plastic tents	Temperature increase of ~1 °C. Exclusion of predators and larger herbivores
Short growing season (SG)	Manual addition of as much snow as possible, as early in the season as possible	Growing season shorter in 2004
Long growing season (LG)	Manual removal of snow as early in the season as possible	Growing season length significantly increased in 2004, 2005 and 2006. Possible disturbance of vegetation when digging

Each treatment included five plots of 1 m^2 each.

heath, *Cassiope* heath, *Salix* heath, grassland and fen. Generalised additive mixed models (GAMMs) were then used to model each response [e.g., timing of the maximum and for seasonal integrated NDVI-FR (SINDVI)] as a function of the following fixed effects: end of snowmelt, temperature, photosynthetically available radiation (PAR), precipitation, and seasonally integrated NDVI (SINDVI) in the previous year with samples as random effects nested within blocks identified by site id. Further details on the methodology can be found in Tamstorf *et al.* (2007).

III. GROWTH DYNAMICS AND LAND SURFACE PHENOLOGY OF SIX MAJOR VEGETATION TYPES IN ZACKENBERGDALEN

The difference in onset of growing season and in maximum greenness among the six major vegetation types in Zackenbergdalen appears from Figure 2. For each data set, a second-order polynomial GAMM was fitted to the average NDVI-FR of each vegetation type for each week. For each of

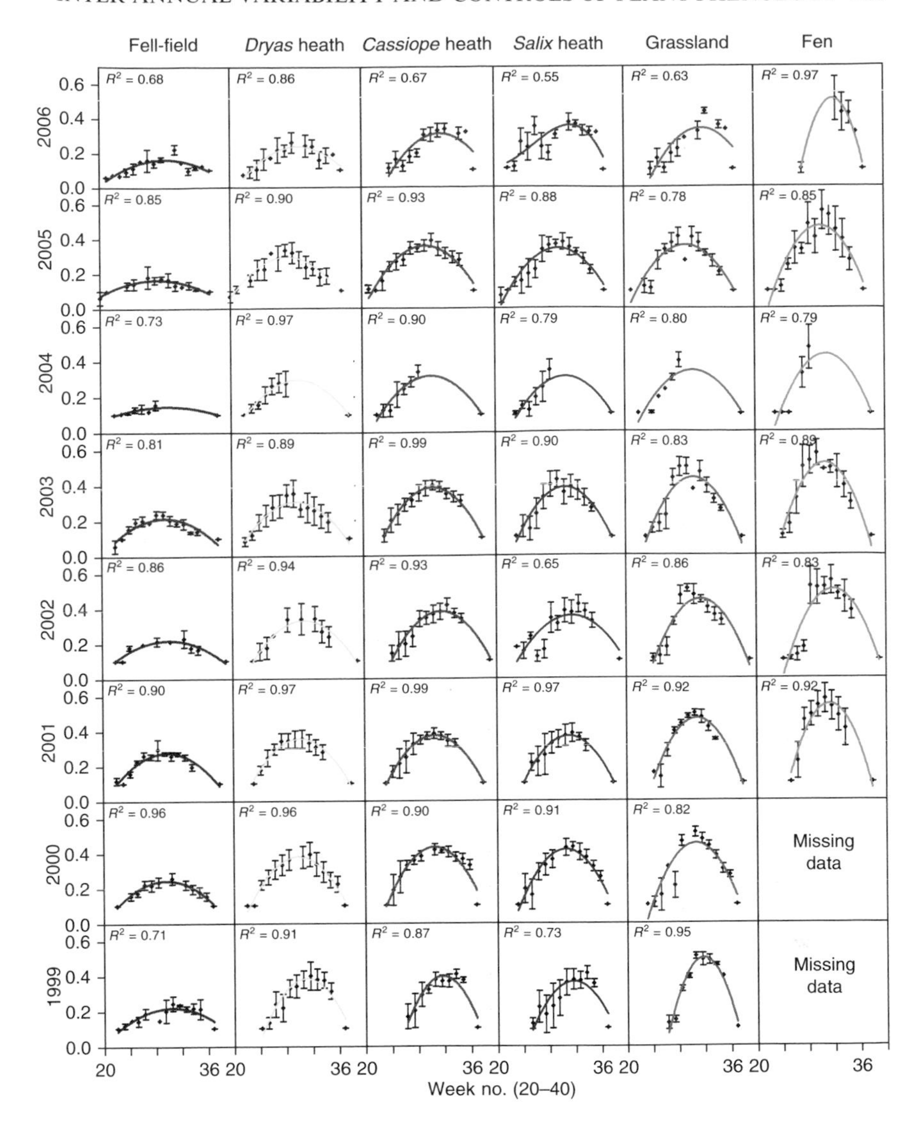

Figure 2 Summary of all far red normalized difference vegetation index measurements for the six vegetation types during May to September 1999–2006. Each curve is an average of all the observations on each given vegetation type in the given year. Error bars indicate one standard deviation (modified from Tamstorf *et al.*, 2007).

the 1218 separate data sets, the polynomial GAMMs explained on average 90% of the variance in NDVI-FR ($0.69 < R^2 < 0.99$).

The vegetation types are exposed to different abiotic conditions resulting in different growth dynamics. Especially the dry vegetation types (fell-field and *Dryas* heath) showed different characteristics than the other types. Fell-field and *Dryas* heath NDVI-FR peak first, followed by *Cassiope* heath, grassland and fen, and finally *Salix* heath. Fell-field and *Dryas* heath are both situated at exposed plateaus, where snow often blows off during the winter months and hence have thinner snow-cover. Here, plant species experience an early snowmelt and hence, an early start of the growing season, and can take advantage of the high levels of PAR around mid-summer. Also, summed temperatures from the climate station (daily mean air temperature at 2 m above the ground summed from January 1) were slightly higher for the dry types than for the other types. On the contrary, *Salix* heaths are often situated around late melting snow-patches in the Zackenberg area (Bay, 1998), and hence, they experience the latest maximum of growth during the season.

The maximum NDVI-FR values (MaxNDVI, Box 1) increase from the relatively dry dwarf shrub heaths (*Dryas* and *Cassiope*) over the moist *Salix* heath to the more wet grasslands and fens (Figure 2). Contrary to the dry types, *Salix* heath has a shorter growing season and hence experience 2–8% lower summed temperatures and 12–20% lower light (PAR). *Salix* heath has a medium MaxNDVI, which varies very little between years.

Fen has the highest maximum values and highest seasonal integrated NDVI (SINDVI, Box 1). For example, in 2006, NDVI-FR peaked around 0.52 in the fens, whereas peak values for *Dryas* heath were around 0.37 (Figure 2). Increasing values of MaxNDVI and SINDVI were found from moss and lichens in fell-fields, over wintergreen and evergreen species in *Dryas* heath and *Cassiope* heath, to deciduous species in *Salix* heath, grassland, and fen. Evergreen and wintergreen species have small, thick and waxy leaves, which can sustain the harsh winter conditions. Hence, they start photosynthesis right after snowmelt, or even before snowmelt (Starr and Oberbauer, 2003). However, these leaves require high amounts of energy to develop and to be sustained (Larcher, 1995) and therefore less green tissue is exposed during the growing season than is the case for the deciduous plant types. This is mirrored in the NDVI-FR values, where *Salix* heath, grassland and fen have the highest occurring values (Figure 2). In contrast, *S. arctica*, fen species and grassland species have a high need of nutrients and water for the development of green tissue during the growing season and hence, grow in moist and wet locations, where sufficient resources are available throughout the growing season.

Tamstorf *et al.* (2007) found that SINDVI and timing of maximum (DOYmax, Box 1) were controlled by different explanatory variables in the six vegetation types, and they found several non-linear responses to the environmental variables. Snowmelt and temperature were both of major importance for the timing of the maximum (DOYmax) within the season as well as for the seasonal growth (SINDVI). In mesic and wet vegetation types, more than 85% of the variance in DOYmax was explained by the models, and a similar result was seen for the SINDVI. All vegetation types, except grassland, showed a positive correlation between time of snowmelt and time of maximal occurring NDVI-FR (DOYmax), that is, the later the snow melted, and hence the later the growing season started, the later the maximum was reached. Other factors than time may, however, influence this period, for example, air temperature. Statistical modeling by Tamstorf *et al.* (2007) showed a positive relationship between temperature and DOY-max in all vegetation types except grassland and fen, meaning that high temperatures in the green-up period (time from end of snowmelt to DOY-max) delayed the maximum. Tamstorf *et al.* (2007) suggested that high temperatures may either inhibit photosynthesis in species growing close to their regional southern limit or increase flowering, which, in turn, may lower the build-up of green plant material.

Temperature in the previous year showed a negative relationship with SINDVI both within and across all vegetation types, meaning that higher temperatures during the growing season caused lower SINDVI the following year (Tamstorf *et al.*, 2007). High temperatures have been shown previously to increase the number of flowers in the following year (Sørensen, 1941; Ellebjerg *et al.*, 2008). This would result in lower NDVI-FR simply because of the brighter colours of the flowers and lower absorption by chlorophyll but also in less greenness since the photosynthates are used for setting flowers and not for biomass growth. Further, the dominant species in some of these vegetation types have a northern distribution pattern in Greenland (Bay, 1992). Hence, they grow close to a regional southern distribution limit, and high temperature increases may reduce plant growth as respiration is increased more than photosynthesis (Havström *et al.*, 1993; Wookey *et al.*, 1994; Graglia *et al.*, 1997), which will decrease SINDVI. Another reason could be the fast melting snow-patches that disappear earlier in the season, causing drying of the soil, which may limit growth and build-up of resources for the following year (Illeris *et al.*, 2004). SINDVI showed a positive relation with SINDVI the following year for all types except grassland (Tamstorf *et al.*, 2007). Most high-arctic plant species are perennials and store resources in storage organs (leaves, stems, roots, rhizomes) from year to year (Berendse and Jonasson, 1992; Shaver and Kummerow, 1992; Brooker *et al.*, 1999). Assuming that SINDVI is a good growth indicator, the year to year relation indicates that resources for growth, acquired during the previous year, are indeed stored and used the following year.

Across years, in 1999 snowmelt was exceptionally late, and 2006 came close to being similar, with deep snow-cover and low spring temperatures causing the growing season to start very late. In contrast, 2005 had low snow-cover and high temperatures both through winter (several thaw events) and during spring, causing an early start of the growing season. Despite the late snowmelt in 2006, there was still a significantly earlier snowmelt of 1.2 days per year during the 8 years ($p = 0.016$). With the increasing temperatures during the period (Figure 3, see Hansen *et al.*, 2008, this volume), one should expect increasing maximal NDVI-FR values and SINDVI values. However, contrary to regional studies of greening in the Arctic (e.g., Stow *et al.*, 2003), there was a clear decrease in the maximum NDVI-FR in all of the vegetation types in Zackenbergdalen (Figure 3). A generalised linear model was used for all six types during 1999 to 2006 for this analysis, and a decrease of 0.01 per year ($p < 0.001$, $r = 0.97$, df = 6) in NDVImax was found. There was no significant difference between the vegetation types (no significant interaction between type and year). The earlier end of snowmelt may lead to reduced availability of water from melting snow at the end of the growing green-up season. The increasing temperatures further enhance the evapotranspiration and snowmelt, leading to a drying of the upper soil layers and, hence, potential water stress. Therefore, although increased greening might be widespread in the Arctic (Zhou *et al.*, 2003; ACIA, 2005), there are areas with opposite trends, depending on the local biotic and abiotic factors.

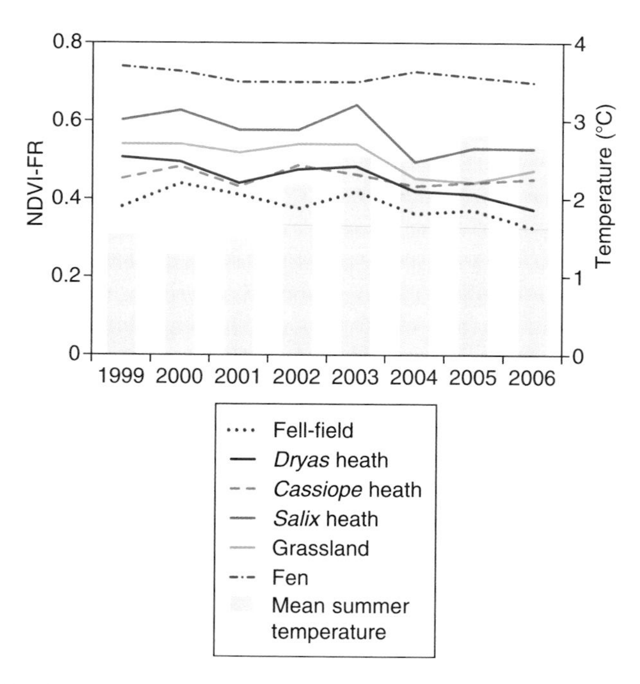

Figure 3 Maximum far red normalized difference vegetation index (NDVI-FR) values from the six vegetation types and mean summer temperature during 1999–2006. Summer is defined as the months in which melting and growth occurs (May 1 to August 31). The slope of −0.01 NDVI-FR per year is significant at $p < 0.001$ (modified from Tamstorf *et al.*, 2007).

IV. SPECIES-SPECIFIC REPRODUCTIVE DYNAMICS

A. Monitoring Studies

It is commonly argued that because most arctic plant species are long lived and slow growing, they are less likely to be able to respond in time to rapidly occurring climate changes. This is further supported by a widespread vegetative reproduction in many arctic species, which does not offer any possibility to evolve with a changing climate (Jonasson *et al.*, 2001). In contrast, if the species are able to increase flower density and production of viable seeds, and these seeds are able to germinate and establish as seedlings, the plant species will be able to evolve concurrently with the changing climate, which will increase the chances of survival for the species. Two recent studies of flowering phenology and flower density based on 10 years of monitoring data from Zackenbergdalen have shown that some dominant plant species may indeed have greater chances of sexual reproduction with the climate changes predicted for the Zackenberg area (Høye *et al.*, 2007a; Ellebjerg *et al.*, 2008), which include increased variability in duration of snow-cover and increased growing season temperature (Stendel *et al.*, 2008, this volume).

Although the growing season in Zackenbergdalen is short, about 2–3 months, there is a very strong variation in the timing of onset of flowering and also in the timing of open seed capsules, of about 1 month between years for most species (Figure 4) concurrent with the variation in timing of MaxNDVI (Figure 2) between years. The relationship between time of snowmelt and onset of flowering differed among species in Zackenbergdalen (Figure 5), indicating that responses to changes in snowmelt pattern may differ. Ellebjerg *et al.* (2008) demonstrated a linear relationship in *C. tetragona* and *S. arctica*, and a second-order relationship in *P. radicatum* (Figure 5), while Høye *et al.* (2007a) demonstrated a second-order relationship in *Dryas* sp.

The importance of timing of snowmelt and onset of flowering has been demonstrated in a large number of other studies, mostly in sub-arctic and alpine settings (Sørensen, 1941; Woodley and Svoboda, 1994; Price and Waser, 1998; Inouye *et al.*, 2002, 2003; Dunne *et al.*, 2003; Molau *et al.*, 2005; Kudo and Akira, 2006). In particular, Ellebjerg *et al.* (2008) found that the relationship between snowmelt and the onset of flowering differed between the two shrubs *C. tetragona* and *S. arctica* on the one hand, and the herb *P. radicatum* and the wintergreen *Dryas* sp. on the other hand. The impact of snowmelt on onset of flowering in *P. radicatum* and *Dryas* sp. was weaker when snowmelt was very early and stronger with later snowmelt, whereas the relationship between snowmelt and onset of flowering in *C. tetragona* and *S. arctica* was linear. This indicates that the two shrubs

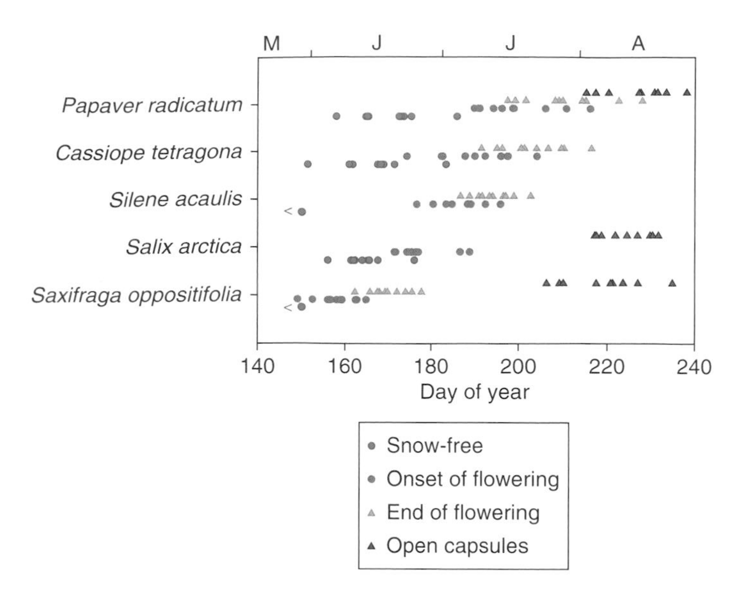

Figure 4 Mean date (day of the year May 20 to August 28) for each of the 10 years of monitoring of 50% snow-free, 50% flowers, 50% senescent flowers (all 1996–2005) and 50% open capsules (1996–2004) in *Papaver radicatum*, *Cassiope tetragona*, *Silene acaulis*, *Salix arctica* and *Saxifraga oppositifolia*. *S. acaulis* and *S. oppositifolia* plots were snow-free prior to June 1 (day 152) in 93% of all observations (modified from Ellebjerg *et al.*, 2008). No timing data were available for *Eriophorum scheuchzeri*. For *Dryas*, see Høye *et al.* (2007a).

are more likely to be able to take advantage of an early snowmelt, whereas *P. radicatum* and *Dryas* sp. seem to require some time for building up of resources before flowering can be initiated. Hence, the two shrubs may benefit more with respect to initiation of flowering than *P. radicatum* and *Dryas* sp. in years with early snowmelt.

If plants are able to vary their length of the pre-floration time (time from snowmelt to onset of flowering) in response to abiotic parameters, they may be able to compensate for adverse climatic conditions such as, for instance, a late snowmelt by developing flowers faster, which will increase their chance of completing sexual reproduction within a growing season (Mølgaard *et al.*, 2002). *C. tetragona* and *S. arctica* had relatively short pre-floration times of 24 and 14 days, respectively (mean duration across 10 years), while *P. radicatum* required more than 29 days to reach flowering (Figure 4). However, it was not the time of snowmelt but rather the amount of incoming radiation (PAR) which was the most important determinant of the length of the pre-floration time in *C. tetragona*, *S. arctica* and *P. radicatum*, with a higher

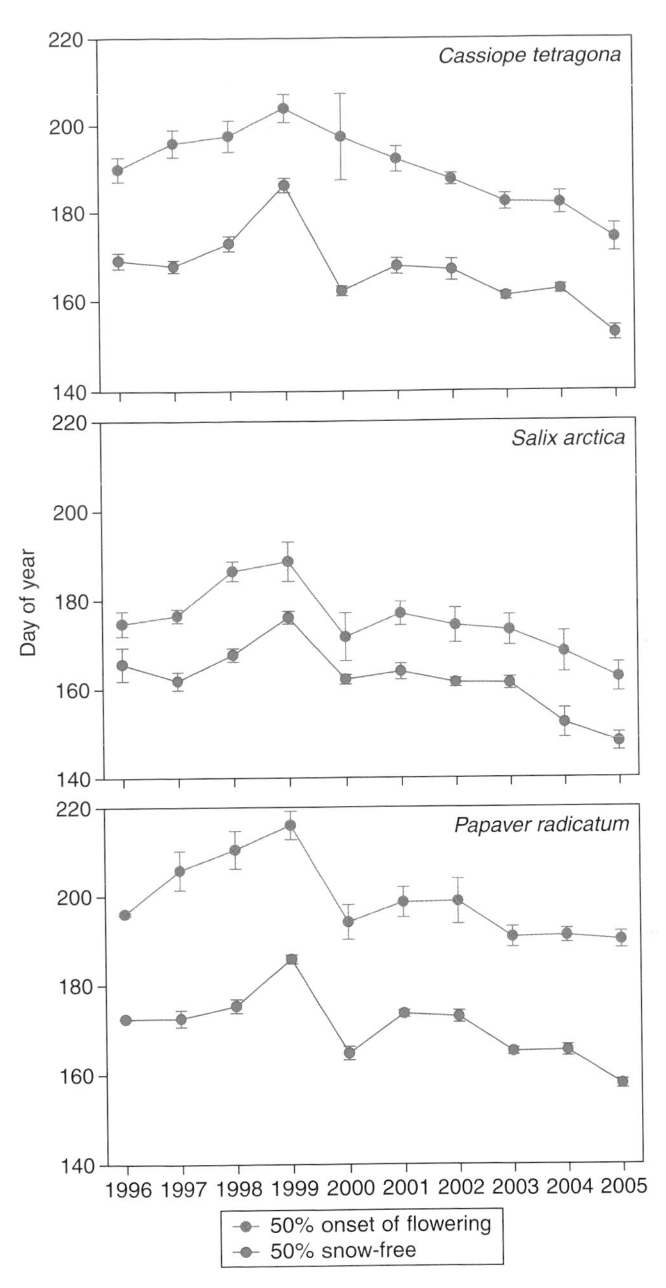

Figure 5 Annual variation in day of year May 20 to August 8 of 50% snow-free and 50% open flowers (mean ± SE) in *Cassiope tetragona* (4 plots), *Salix arctica* (3 plots) and *Papaver radicatum* (4 plots) (Ellebjerg *et al.*, 2008). Data on snowmelt were not available for *Saxifraga oppositifolia* and *Silene acaulis*, and no data on onset of flowering were available for *Eriophorum scheuchzeri*.

total PAR-sum causing a shorter pre-floration time (Ellebjerg *et al.*, 2008). For *Dryas* sp., a higher temperature during the pre-floration time led to shorter pre-floration times in both early and late flowering plots (plots with early and late snowmelt, respectively), with the effect being strongest in early flowering plots (Høye *et al.*, 2007a). Hence, this species may indeed be able to compensate for a shorter growing season by developing flowers faster, if the shorter season is accompanied by a temperature increase. At similar temperatures, the pre-floration time was shorter in early plots than in late. This could be because the early plots with *Dryas* sp. have more light available during the pre-floration time and thus respond in the same way as *C. tetragona* and *S. arctica*. This means that in years when the snow disappears early and thereby enables an onset of flowering closer to summer solstice, these species will be able to take advantage of the longer growing season by initiating flowers faster and thus potentially have a longer time available for completion of the reproductive cycle. However, for *Dryas* sp., this may only be the case for plants growing in areas of relatively early snowmelt.

A faster flower and seed development is beneficial in an unpredictable environment, such as the High Arctic, since it provides a greater chance for completion of seed development. A warmer growing season led to a reduction in the duration of flowering in *P. radicatum* and *S. acaulis*, but did not influence the duration of flowering in *C. tetragona* and *S. arctica*. In addition, a warmer growing season led to a faster development of open capsules and thereby possibly of ripe seeds, in *S. arctica*, *P. radicatum*, and *S. oppositifolia*. The fact that increased temperature did not seem to result in a shorter duration of flowering in *C. tetragona* and *S. arctica* may be because of stronger responses to the date of onset of flowering and PAR. Later onset of flowering in these species led to shorter duration of flowering. Both species responded to PAR in such a way that both high and low PAR-values led to a shorter duration of flowering, although the variance explained by PAR was low. With seasonal progression, PAR-values and duration of flowering decrease, the latter possibly because of increasing drought stress as the season progressed. Soil moisture was observed to decrease in *C. tetragona* heath as the season progressed in 2004 and 2005. The *S. arctica* plots included in the present study are all situated in early snow-free *Salix* heaths and thereby prone to drying out with seasonal progression. In support of this, Woodley and Svoboda (1994) observed prolonged flowering of *P. radicatum* and *C. tetragona* in late melting sites with a more favourable moisture regime in the lowland of Ellesmere Island, Canada. Thus, for *P. radicatum* and *S. acaulis*, a warmer summer will likely lead to more favorable conditions for completion of the reproductive cycle within a given season. However, if the main reason for decreased duration of flowering in *C. tetragona* and decreased duration of flowering and time for seed development in *S. arctica* is drought stress, it is unlikely that this will lead to an overall faster development of ripe seeds.

Most high-arctic plant species initiate flower buds one or more years prior to flowering (Sørensen, 1941). Hence, better growing conditions in the previous year generally will allow the plant to build up more resources, and lead to more flowers in the current year. To investigate this temporal dependence in high-arctic ecosystems, Ellebjerg *et al.* (2008) examined the flower density (total number of flowers m^{-2} plot) of six species common in Zackenbergdalen. The sum of mean diurnal temperatures above 0 °C was used as a measure of the growing conditions in the previous year. For four out of six species examined, a higher total heat sum in the previous year led to higher densities of flowers in the current year with high explained variance for *C. tetragona* (66%), *S. arctica* (44%) and *P. radicatum* (32%) but not for *S. acaulis* (4%) (Figure 6). The latter species seemed to be more influenced by the heat sum in the current year prior to onset of flowering, which rendered *S. acaulis* intermediate between mainly being dependent on last year's heat sum and on current year's heat sum. Likewise, *Dryas* sp., which was examined by Høye *et al.* (2007a), showed a response relating to conditions of both the current and the previous year. In early melting plots of *Dryas* sp., the number of frost hours between snowmelt and flowering was the main predictor of flower abundance (number of flowers m^{-2} covered by *Dryas* sp. in the plot), whereas in late melting plots, it was the number of hours with positive temperature and PAR-values of the previous growing season which determined the flower abundance. The only species being solely dependent ($p = 0.0011$, $R^2 = 0.40$, df = 3) on the time of snowmelt and temperature sum in the current year prior to onset of flowering was *E. scheuchzeri*, which grows in a very wet, and thereby also a cooler, environment than the other species (Ellebjerg *et al.*, 2008). In addition, in this species a higher temperature sum with a threshold of 5 °C as opposed to a threshold of 1 °C in the remaining species was found as an important predictor of flower density. That most of the species respond to the temperature sum of the previous growing season by flowering more vigorously in the current year means that if years of earlier snowmelt becomes more frequent in the future, there will also be more years of vigorous flowering and thereby a greater potential for successful pollination. Thus, more seeds may be developed, which through sexual reproduction will increase chances of species adaptedness to future climate changes.

Morphological or genetic differentiation could be responsible for the observed differences between early and late flowering plots of *Dryas* sp. This was examined by Høye *et al.* (2007a). Based on morphological analysis, the *Dryas* plants in Zackenbergdalen were assigned to two morphotypes, one similar to *D. octopetala* and one intermediate between *D. octopetala* and

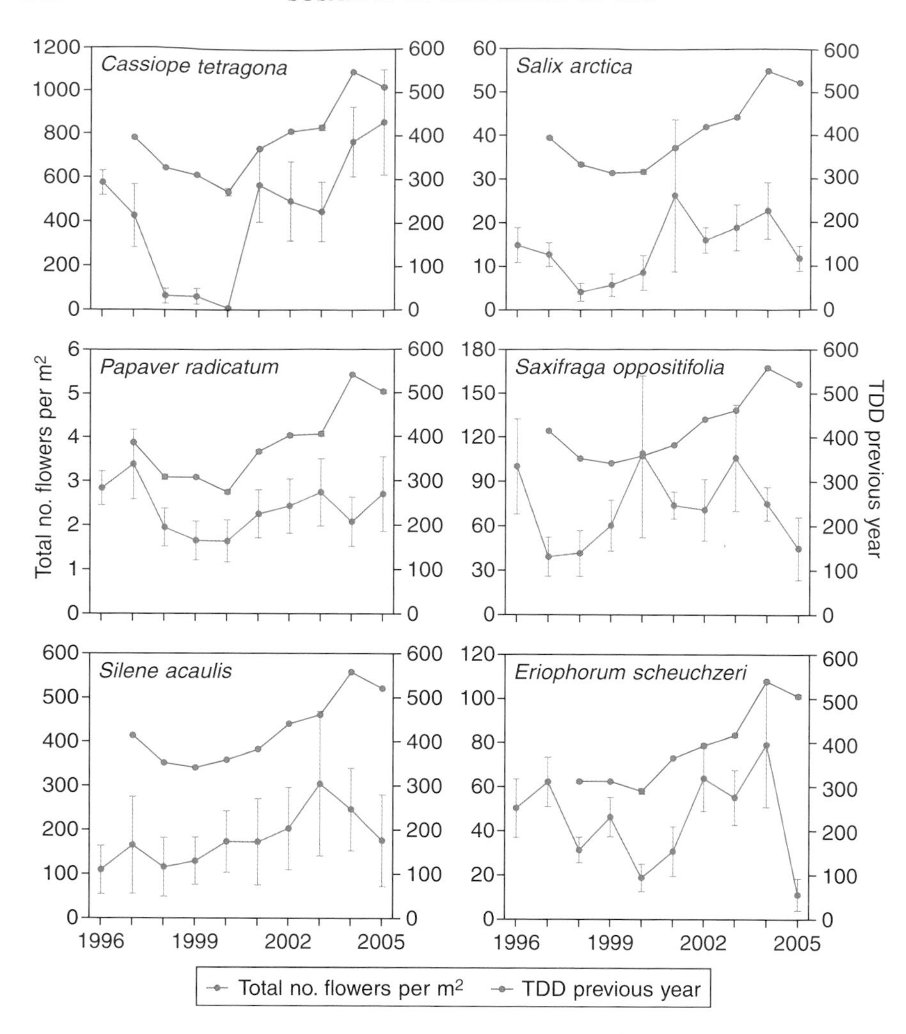

Figure 6 Annual variation in total number of flowers m^{-2} (1996–2005) and thawing degree days (TDD, the sum of positive diurnal temperatures) across the previous growing season from day of 50% snow-free to first day with ambient temperatures below −2 °C in *Cassiope tetragona* (6 plots), *Salix arctica* (3 plots), *Papaver radicatum* (4 plots), *Saxifraga oppositifolia* (3 plots), *Silene acaulis* (3 plots) (1997–2005) and *Eriophorum scheuchzeri* (4 plots) (1998–2005) from Zackenberg. Data for snow-free was lacking in *S. oppositifolia* and *Silene acaulis*. TDD was therefore calculated as the sum of mean diurnal temperature above 0 °C from June 1 until the first day with ambient temperatures below −2 °C (Ellebjerg *et al.*, 2008).

D. integrifolia. The two morphotypes had an almost even distribution across plots, and the allele differences among plots were mainly a result of geographical distance and not of snowmelt patterns. Thus, neither morphological nor genetic differentiation could be related to the apparent phenological responses to differences in snowmelt (Høye *et al.*, 2007a).

B. Experimental Manipulations

To further elucidate species-specific responses to specific climatic changes in *C. tetragona*, experimental manipulations were carried out in a *C. tetragona* heath area at Zackenberg through the 2004, 2005 and 2006 growing seasons (Table 1, Figure 1). Effects of experimental manipulations on growing season length and on total number of flowers m^{-2} were analysed with one-way ANOVA using general linear models (SAS Institute Inc, 2003). For the analysis of total number of flowers, the number of flowers m^{-2} in 2004 was used as covariate to compensate for potential differences in initial *C. tetragona* shoot and flower density.

Flower density across years differed between experimental and monitoring plots. In the experimental plots, there was a general decrease in flower density in all plots from 2004 through 2006 (Figure 7), whereas in the monitoring plots there was no change from 2004 to 2005 (Figure 6). This difference possibly reflects site-specific differences in *C. tetragona* flowering between years.

Manipulations of growing season length only had a limited effect on flower density. Plots where snow was removed manually early in the season [long growing season plots (LG plots)] had a significantly longer growing season of

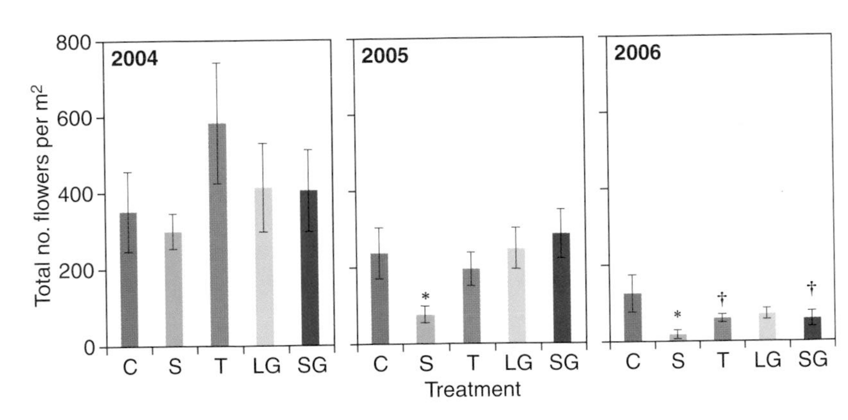

Figure 7 Total number of flowers m^{-2} in *Cassiope tetragona* subjected to five experimental treatments simulating climate change in the summers of 2004, 2005, and 2006. Treatments were C, control; S, shading; T, increased temperature; LG, prolonged growing season; and SG, shortened growing season. The plots are permanent and monitored every year. † denotes a tendency for a treatment effect (Dunnett's test, $p < 0.12$); * denotes a significant treatment effect (Dunnett's test, $p < 0.05$).

1–8 days in all 3 years (Dunnett's test, $p < 0.05$). However, this difference in growing season length was not reflected in a difference in flower density in 2005 and 2006 (Figure 7). This may be because the longer growing season was not sufficient to provide the plants with additional resources to the extent that more flower buds could be formed for the following growing season, as has been shown for species from the monitoring plots. However, there was a tendency for short growing season plots (SG plots) to have lower flower density than controls in 2006 (Figure 7, Dunnett's test, $p < 0.12$), possibly reflecting the tendency towards a slightly shorter growing season in these plots, significant in 2004. Hence, increasing growing season length without a corresponding increase in temperature does not seem to increase flower density in *C. tetragona*, whereas only a small decrease in season length may limit flower development.

The shading treatment simulated effects of a severely increased cloud-cover or decreased light as a consequence of shading from taller canopies. Shading significantly decreased flower density in 2005 and 2006 (Figure 7, Dunnett's test, $p < 0.05$). In 2004, flowering was initiated prior to our establishment of experimental treatments. In 2005 and 2006, the mean decrease in flower density in shaded plots was 77% and 84%, respectively, compared to the control plots. A decreased flowering in response to shading was also observed by Havström *et al.* (1995) in sub-arctic Sweden. At Zackenberg, decreased flowering in *C. tetragona* was pronounced 1 year after start of treatment, and the effect of shading was enhanced through consecutive years. The shading treatment decreased available light by 50–60%, which created a light regime with incoming average PAR during the growing season ranging from ~300 μmol m^{-2} s^{-1} in mid-June to 100 μmol m^{-2} s^{-1} in the end of August (calculation based on PAR data from the ClimateBasis programme, www.zackenberg.dk/data.htm). Hence, the light intensity will often be far below the light compensation point for photosynthesis in *C. tetragona*. It is therefore not surprising that shaded *C. tetragona* does not have enough resources to form flowers, since it is likely to often suffer from reduced light, limiting photosynthesis and affecting plant carbon balance. Hence, our experiments clearly suggest that *C. tetragona* will not be able to produce many flowers if cloud-cover increases strongly and/or these dwarf shrubs are overgrown by a canopy of potentially taller shrubs such as dwarf birch *Betula nana*.

The treatment effects of temperature were less pronounced than those of shading. However, in 2006 there was a tendency for temperature enhanced plots (T plots) to have a lower flower density than the controls (Figure 7, Dunnett's test, $p < 0.12$). This result is somewhat surprising, since a higher total heat sum during the previous growing season was shown to increase the total number of flowers m^{-2} in current year in *C. tetragona* during 10 years of monitoring (Ellebjerg *et al.*, 2008). One explanation for the tendency towards reduced flowering in temperature-enhanced plots could be that the 1 °C warming was too small to have a direct positive effect on the flowering frequency, but that warming led to drying of the soil surface by increased evaporation combined with screening of precipitation.

V. CONCLUSION AND FUTURE PERSPECTIVES

Most of the studied species developed more flowers in years following a warmer growing season, and the results also indicate that the plants generally develop flowers and seeds faster in a warmer environment. With more varying spring and summer climate in the future, the frequency of both years with later and earlier snowmelt than currently will increase. Late snowmelt is likely to have negative effects on the possibilities for sexual reproduction. However, the increased frequency of years with earlier snowmelt means that increased sexual reproduction compared to present may occur. This will cause greater genetic variance, and hence, the overall evolutionary potential of the plant species will increase and thereby increased adaptedness to climatic changes. Also, increased sexual reproduction at least in part will make up for the slow growth of these arctic species. However, it requires that seeds are able to germinate, and that flowering is not impeded by shading, as seems to be the case for *C. tetragona*, for instance, by taller invading species or from an increased cloud-cover.

The analyses from Zackenberg indicate that shrubs are more likely to take advantage of the predicted climatic changes with cloudier summers and increased variability in snowmelt with respect to initiation of flowering than other plant types. That is, they are more likely to complete development of seeds within a season when snowmelt is early. Across the past decade, date of snowmelt has advanced by 14.6 days in the permanent sampling plots in Zackenberg (Høye *et al.*, 2007b). Together, this leads to an expectation of an increased dominance of shrubs within the ecosystem. Such an expansion of shrub cover over the past 50 years has been reported for northern Alaska as well as Canada, Scandinavia and parts of Russia (Tape *et al.*, 2006). However, the strong negative trend in greenness found in all six vegetation types during the last 8 years indicate that other factors, such as soil moisture content, are influencing the phenology and growth of the vegetation in the area. Even earlier snowmelt will further limit the moisture availability, especially during the late-summer season (July–August), unless increased summer precipitation counterbalance this.

Since earlier snowmelt will be more frequent, flowering in many years is likely to occur closer to summer solstice at which time incoming radiation is abundant, and the effect is therefore expected to be of minor importance.

ACKNOWLEDGMENTS

Monitoring data for this chapter were provided by the BioBasis programme, run by the National Environmental Research Institute, University of Aarhus, Denmark, and financed by the Danish Environmental Protection Agency, Danish Ministry of the Environment. We further thank the Danish Polar Center for logistic support during field work at Zackenberg Research Station. In addition, we thank Hans Meltofte for help with data access and use, Marie Arndal, Kristian Albert and Niels Martin Schmidt for field assistance and the Danish Ministry of Environment for financial support. Finally, we appreciate constructive suggestions and comments provided by an anonymous reviewer.

REFERENCES

ACIA (2005), *Arctic Climate Impact Assessment.* Cambridge University Press, New York, 1042 pp.
Arft, A.M., Walker, M.D., Gurevitch, J., Alatalo, J.M., Bret-Harte, M.S., Dale, M., Diemer, M., Gugerli, F., Henry, G.H.R., Jones, M.H., Hollister, R.D. Jónsdóttir, I.S., *et al.* (1999) *Ecol. Monogr.* **69**, 491–511.
Bay, C. (1992) *Meddr. Grønland, Biosci.* **36**, 102 pp.
Bay, C. (1998) *Vegetation mapping of Zackenberg Valley, Northeast Greenland.* Danish Polar Center & Botanical Museum, University of Copenhagen, Denmark.
Berendse, F. and Jonasson, S. (1992) In: *Arctic Ecosystems in a Changing Climate* (Ed. by F.S.I. Chapin, R.L. Jefferies, J.F. Reynolds, G.R. Shaver and J. Svoboda), pp. 337–356. Academic Press, Inc., San Diego, California.
Brooker, R.W., Callaghan, T.V. and Jonasson, S. (1999) *New Phytol.* **142**, 35–48.
de Beurs, K.M. and Henebry, G.M. (2005) *Int. J. Remote Sens.* **26**, 1551–1573.
Dunne, J.A., Harte, J. and Taylor, K.J. (2003) *Ecol. Monogr.* **73**, 69–86.
Ellebjerg, S., Illeris, L., Tamstorf, M.P. and Michelsen, A. (2008) *Submitted.*
Graglia, E., Jonasson, S., Michelsen, A. and Schmidt, I.K. (1997) *Ecoscience* **4**, 191–198.
Havström, M., Callaghan, T.V. and Jonasson, S. (1993) *Oikos* **66**, 389–402.
Havström, M., Callaghan, T.V. and Jonasson, S. (1995), *Effects of simulated climate change on the sexual reproductive effort of Cassiope tetragona.* Global change and arctic terrestrial ecosystems, Ecosystem Report, Report num. 10, EUR 15519 EN, 109–114. European Commission, Luxemburg.
Hope, A.S., Boynton, L., Stow, D.A. and Douglas, D.C. (2003) *Int. J. Remote Sens.* **24**, 3413–3425.
Høye, T.T., König, S.M. and Philipp, M. (2007a) *Arct. Antarct. Alp. Res.* **37**, 412–421.
Høye, T.T., Post, E., Meltofte, H., Schmidt, N.M. and Forchhammer, M.C. (2007b) *Curr. Biol.* **17**, R449–R451.
Illeris, L., Christensen, T.R. and Mastepanov, M. (2004) *Biogeochemistry* **70**, 315–330.
Inouye, D.W., Morales, M.A. and Dodge, G.J. (2002) *Oecologia* **130**, 543–550.
Inouye, D.W., Saavedra, F. and Lee-Yang, W. (2003) *Am. J. Bot.* **90**, 905–910.
Jefferies, R.L., Svoboda, J., Henry, G.H.R., Raillard, M. and Ruess, R. (1992) In: *Arctic Ecosystems in a Changing Climate* (Ed. by F.S.I. Chapin, R.L. Jefferies, J.F. Reynolds, G.R. Shaver and J. Svoboda), pp. 391–412. Academic Press Inc., San Diego, California.

Jia, G.J., Epstein, H.E. and Walker, D.A. (2006) *Global Change Biol.* **12**, 42–55.
Johnstone, J.F. and Henry, G.H.R. (1997) *Arct. Alp. Res.* **29**, 459–469.
Jonasson, S., Callaghan, T.V., Shaver, G.R. and Nielsen, L.A. (2001) In: *The Arctic, Environment, People, Policy* (Ed. by M. Nuttall and T.V. Callaghan), pp. 275–313. Harwood academic publishers, Amsterdam.
Kevan, P.G. and Baker, H.G. (1983) *Annu. Rev. Entomol.* **28**, 407–453.
Kudo, G. and Akira, H.S. (2006) *Popul. Ecol.* **48**, 49–58.
Larcher, W. (1995) *Physiological Plant Ecology*. Springer-Verlag, Berlin Heidelberg.
Marchand, F.L., Nijs, I., Heuer, M., Mertens, S., Kockelbergh, F., Pontailler, J.-Y., Impens, I. and Beyens, L. (2004) *Arct. Antarct. Alp. Res.* **36**, 390–394.
Meltofte, H. and Berg, T.B.G. (2006) *BioBasis: Conceptual design and sampling procedures of the biological programme of Zackenberg Basic,* 9th ed. National Environmental Research Institute, Department of Arctic Environment, 77 pp.
Molau, U. (1997) *Global Change Biol.* **3**, 97–107.
Molau, U., Nordenhall, U. and Eriksen, B. (2005) *Am. J. Bot.* **92**, 422–431.
Mølgaard, P., Forchhammer, M.C., Grøndahl, L. and Meltofte, H. (2002) In: *Sne, is og 35 graders kulde. Hvad er effekterne af klimaændringer i Nordøstgrønland* (Ed. by H. Meltofte), pp. 43–46. National Environmental Research Institute, Denmark.
Myneni, R.B., Hall, F.G., Sellers, P.J. and Marshak, A.L. (1995) *IEEE T. Geosci. Remote* **33**, 481–486.
Post, E., Boving, P.S., Pedersen, C. and MacArthur, M.A. (2003) *Can. J. Zool.* **81**, 1709–1714.
Price, M.V. and Waser, N.M. (1998) *Ecology* **79**, 1261–1271.
Rayback, S.A. and Henry, G.H.R. (2005) *Arct. Antarct. Alp. Res.* **38**, 228–238.
Rouse, J.W. (1973) *Monitoring the Vernal Advancement and Retrogradation of Natural Vegetation.* NASA/Goddard Space Flight Center, Greenbelt, MD.
Sandvik, S.M. and Totland, O. (2000) *Ecoscience* **7**, 201–213.
SAS Institute Inc (2003) *SAS 9.1.3*. Cary, NC, USA.
Shaver, G.R. and Kummerow, J. (1992) In: *Arctic Ecosystems in a Changing Climate* (Ed. by F.S.I. Chapin, R.L. Jefferies, J.F. Reynolds, G.R. Shaver and J. Svoboda), pp. 193–211. Academic Press, San Diego.
Sørensen, T. (1941) *Meddr. Grønland* **125**, 9, 305 pp.
Starr, G. and Oberbauer, S.F. (2003) *Ecology* **84**, 1415–1420.
Stow, D., Daeschner, S., Hope, A., Douglas, D., Petersen, A., Myneni, R., Zhou, L. and Oechel, W.C. (2003) *Int. J. Remote Sens.* **24**, 1111–1117.
Tamstorf, M.P., Illeris, L., Hansen, B.U. and Wisz, M. (2007) *BMC Ecol.* **7**, 9.
Tape, K., Sturm, M. and Racine, C. (2006) *Global Change Biol.* **12**, 686–702.
Totland, O. (1994) *Arct. Alp. Res.* **26**, 66–71.
Welker, J.M., Molau, U., Parsons, A.N., Robinson, C.H. and Wookey, P.A. (1997) *Global Change Biol.* **3**, 61–73.
Woodley, E.J. and Svoboda, J. (1994) In: *Ecology of a Polar Oasis. Alexandra Fiord, Ellesmere Island, Canada* (Ed. by J. Svoboda and B. Freedman), pp. 157–175. Captus University Publications, Toronto.
Wookey, P.A., Parsons, A.N., Welker, J.M., Potter, J.A., Callaghan, T.V., Lee, J.A. and Press, M.C. (1993) *Oikos* **67**, 490–502.
Wookey, P.A., Welker, J.M., Parsons, A.N., Press, M.C., Callaghan, T.V. and Lee, J. A. (1994) *Oikos* **70**, 131–139.
Wookey, P.A., Robinson, C.H., Parsons, A.N., Welker, J.M., Press, M.C., Callaghan, T.V. and Lee, J.A. (1995) *Oecologia* **102**, 478–489.
Zhou, L., Kaufmann, R.K., Tian, Y., Myneni, R.B. and Tucker, C.J. (2003) *J. Geophys. Res. Atmos.* **108,** Art. no. 4004.

High-Arctic Plant–Herbivore Interactions under Climate Influence

THOMAS B. BERG, NIELS M. SCHMIDT, TOKE T. HØYE,
PETER J. AASTRUP, DITTE K. HENDRICHSEN,
MADS C. FORCHHAMMER AND DAVID R. KLEIN

SUMMARY

This chapter focuses on a 10-year data series from Zackenberg on the trophic interactions between two characteristic arctic plant species, arctic willow *Salix arctica* and mountain avens *Dryas octopetala*, and three herbivore species covering the very scale of size present at Zackenberg, namely, the moth *Sympistis zetterstedtii*, the collared lemming *Dicrostonyx groenlandicus* and the musk ox *Ovibos moschatus*.

Data from Zackenberg show that timing of snowmelt, the length of the growing season and summer temperature are the basic variables that determine the phenology of flowering and primary production upon which the herbivores depend, and snow may be the most important climatic factor affecting the different trophic levels and the interactions between them. Hence, the spatio-temporal distribution of snow, as well as thawing events during winter, may have considerable effects on the herbivores by influencing their access to forage in winter. During winter, musk oxen prefer areas with a thin snow-cover, where food is most easily accessible. In contrast, lemmings seek areas with thick snow-cover, which provide protection from the cold

ADVANCES IN ECOLOGICAL RESEARCH VOL. 40 0065-2504/08 $35.00
© 2008 Elsevier Ltd. All rights reserved
DOI: 10.1016/S0065-2504(07)00012-8

and some predators. Therefore, lemmings may be affected directly by both the timing of onset and the duration of winter snow-cover.

Musk oxen significantly reduced the productivity of arctic willow, while high densities of collared lemmings during winter reduced the production of mountain avens flowers in the following summer. Under a deep snow-layer scenario, climate and the previous year's density of musk oxen had a negative effect on the present year's production of arctic willow. Previous year's primary production of arctic willow, in turn, significantly affected the present year's density of musk oxen positively. Climatic factors that affect primary production of plants indirectly, influenced the spatial distribution of herbivores. Additionally, snow distribution directly affected the distribution of herbivores, and hence, in turn, affected the plant community by selective feeding and locally reducing the standing biomass of forage plants.

Although only few moth larvae were observed at Zackenberg, these had in some cases important local effects owing to their foraging on up to 60% of the flower stands on individual mountain avens.

UV-B radiation induces plants to produce secondary plant metabolites, which protects tissues against UV-B damage. This results in lower production of anti-herbivore defenses and improves the nutritional quality of the food plants. Zackenberg data on the relationship between variation in density of collared lemmings in winter and UV-B radiation indirectly supports this mechanism, which was originally proposed on the basis of a positive relationship between UV-B radiation and reproduction in two sub-arctic species of hares (*Lepus timidus* and *Lepus americanus*).

I. INTRODUCTION

In snow-covered ecosystems, such as those in the High Arctic, the distribution of vegetation types is largely governed by clinal variation in snow-cover (Babb and Whitfield, 1977; Walker *et al.*, 2001). In addition to this, the distribution of snow affects the spatio-temporal pattern of flowering and primary production (Høye *et al.*, 2007a; Ellebjerg *et al.*, 2008, this volume). As a result, herbivores in this region are confronted with a spatially and temporally variable food resource, and they adjust their foraging behaviour accordingly. The manner and scale of these adjustments, however, may differ considerably between species. The patterns of dispersion of herbivores are also affected by the risk of predation (Lima and Dill, 1990) together with social interactions especially during the mating season.

Plant–herbivore interactions are reciprocal (Klein *et al.*, 2008, this volume). Herbivores not only depend on plants but also affect their growth and survival directly through grazing and by altering the physical environment by trampling and digging. Despite the relatively low densities of mammalian herbivores in Northeast Greenland, herbivory by small rodents has been shown in other arctic areas to be able to significantly alter the relative abundance of many common plant species in the tundra ecosystem (Olofsson *et al.*, 2002, 2004). Heavy exploitation of the vegetation may, in some cases, result in a decrease in density of herbivores (e.g., Stenseth and Oksanen, 1987; Selås, 1997), and both herbivory and trampling may locally destroy the insulating moss layer, leading to an increase in soil temperature (van der Wal *et al.*, 2004) and hence, induce changes in the plant communities. Defecation by herbivores recycles important nutrients to the vegetation (e.g., van der Wal *et al.*, 2004), and ultimately carcasses of herbivores constitute a large localized source of nutrients for plant growth (Danell *et al.*, 2002).

This chapter concentrates on two mammalian herbivores, collared lemming *Dicrostonyx groenlandicus* and musk ox *Ovibos moschatus*, and their shared food resources, arctic willow *Salix arctica* and mountain avens *Dryas octopetala*, and on one invertebrate herbivore, the lepidopteran larvae of *Sympistis zetterstedtii*, which also feeds on mountain avens. The effects of climate and climate variation on these plant–herbivore systems as well as the interaction between plants and herbivores are here illustrated using data obtained from the long-term biological monitoring programme BioBasis running at Zackenberg Research Station (74°30′N, 20°30′W; Meltofte and Berg, 2006).

II. CLIMATIC PATHWAYS IN THE PLANT–HERBIVORE SYSTEM

There are basically three ways in which climate can influence the interactions between plants and herbivores (Box 1; see also Forchhammer *et al.*, 2008, this volume). All species may be affected directly by climatic variables though not necessarily in the same way or to the same extent. Alternatively, interacting species may be affected indirectly by climatic perturbations mediated through a climatic effect on adjacent or more distant trophic levels. Indirect climatic effects like these may act either bottom–up (i.e., soil–plant–herbivore) and/or top–down (i.e., predators–herbivore–plant). In addition, climate may affect the strength of the intra- and inter-specific interactions. For instance, a decrease in snow-cover will increase the synchrony of growth of arctic willow, which again, to a certain point, will increase the size of musk ox herds (Forchhammer *et al.*, 2005), thereby intensifying the plant–herbivore interaction on the site level.

Although arctic species are adapted to survive under extreme climatic conditions, both plants and herbivores may be directly affected by weather conditions. The annual phenology of flowering of the plants in the

Box 1

Climatic Influence in the Plant–Herbivore System

The influence of climate in a trophic system of interacting species may be either direct (full red arrows in Box Figure 1) or indirect, which is mediated via other adjacent trophic levels (broken red arrows in Box Figure 1). Additionally, climate may not only influence the way the individual trophic levels interact (yellow arrows in Box Figure 1) but also the way individuals of the same species interact (black arrows in Box Figure 1).

Box Figure 1 Conceptual model for the climatic pathways in the simple plant–herbivore system in the Arctic. Black arrows indicate intra-specific interactions, while yellow arrows indicate inter-specific interactions. The direct climatic influence on the various trophic levels is indicated by the full, red arrows, while the indirect climatic influences indicated by the broken, red arrows. See also Forchhammer (2001), Forchhammer and Post (2004) and Forchhammer *et al.* (2008, this volume). Photos: Niels Martin Schmidt.

The indirect climatic effects may be (a) bottom–up processes, where the climatic influence on lower trophic levels is mediated onto the higher trophic levels (Wilson and Jefferies, 1996; Callaghan *et al.*, 2005), for instance, via climate-induced changes in production of secondary plant metabolites (Plesner-Jensen and Doncaster, 1999), (b) top–down processes, where the climatic effects onto the higher trophic levels are mediated onto the lower trophic levels, or (c) via a combination of bottom–up and top–down processes (Turchin *et al.*, 2000).

valley Zackenbergdalen varies widely between years, dependent largely on the date of snowmelt (Høye *et al.*, 2007b; Ellebjerg *et al.*, 2008, this volume). Likewise, the flower production shows large inter-annual variation (Høye *et al.*, 2007b; Ellebjerg *et al.*, 2008, this volume), indicating that climate directly affects not only the timing of and investment in reproduction but also the development of plant tissue.

Thawing events during winter may create ice crust directly hampering the access to food for the larger herbivores such as caribou and musk oxen (Vibe, 1967; Forchhammer and Boertmann, 1993), resulting in increased mortality and reduced fecundity. In contrast, predators may indirectly benefit from the ice crust effects on their prey.

Collared lemmings, by contrast, spend most of their time during winter beneath the snow (Stenseth and Ims, 1993). At Zackenberg, the number of lemming nests is positively correlated with the length of the winter season (Figure 1), indicating that the snow-cover functions as a protective layer against predators and extreme climate variability (Chernov, 1985).

Temperature and precipitation during the autumn movements between summer and winter habitats may influence collared lemming abundance. An early onset of snow accumulation has a positive effect on the winter density of lemming, while late onset of snow accumulation affects the winter population negatively (Figure 1) and expose the lemmings to both increased predation and low temperature (Scott, 1993). Predators also benefit from the increased exposure of lemmings during their spring movements from winter to summer habitats.

Late snowmelt in spring delays the emergence of caterpillars (Morewood and Ring, 1998) and may thus decrease their survival by reducing the time available for resource accumulation before returning to winter dormancy.

In the Arctic, climate influences the decomposition and mineralization of soils and thus the availability of nutrients (Flanagan and Bunnell, 1980; Webber *et al.*, 1980; Elberling *et al.*, 2008, this volume), and these climatic effects may be carried on to the herbivores. For instance, solar radiation influences soil temperature and the decomposition, mineralization and nutrient content of plants, which, in turn, has been found to influence the foraging behaviour of snow geese *Chen caerulescens* in Canada (Wilson and Jefferies, 1996).

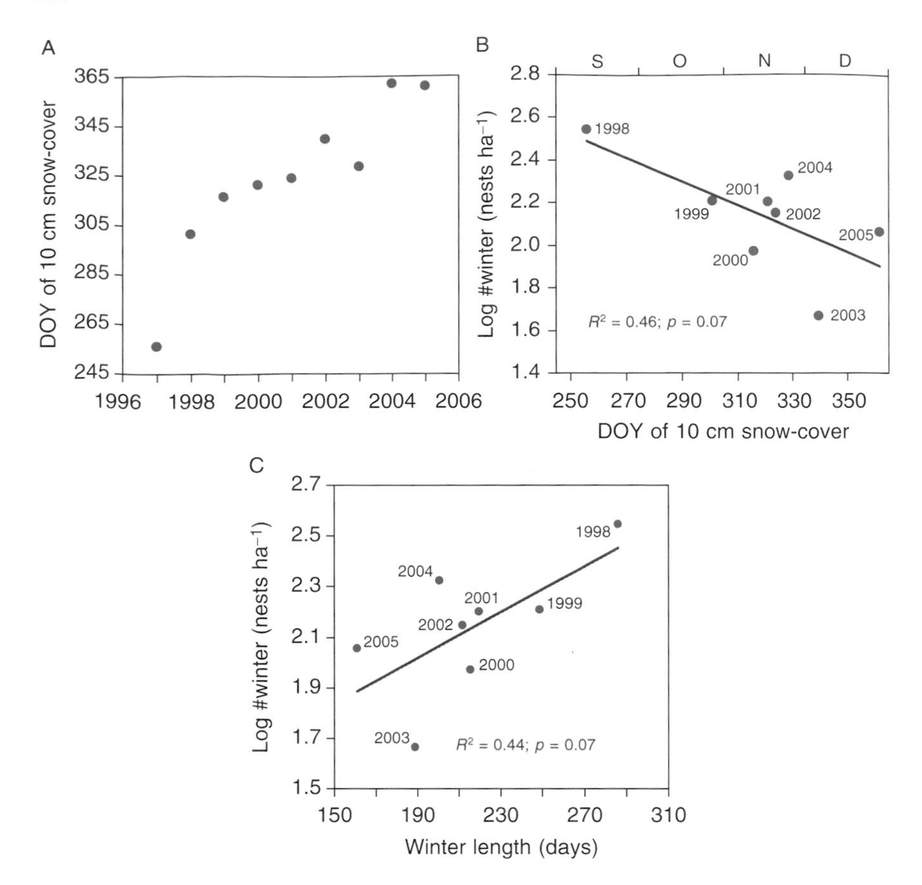

Figure 1 (A) Year to year variation in the onset of winter (defined as first day with >10 cm of snow at the Zackenberg meteorological station) during 1997–2005. (B) The relationship between the number of lemming winter nests and the onset of winter ($R^2 = 0.46$; $p = 0.07$). (C) The length of the winter season (from onset of winter until first day with <10 cm of snow) ($R^2 = 0.44$; $p = 0.07$).

High UV-B radiation may force the plant to allocate resources into UV-B protective secondary plant metabolites (SPM; see Box 2) at the expense of the anti-herbivory targeted SPM (Gwynn-Jones, 1999; Selås, 2006). One consequence of this is increased nutritional value (better assimilation of crude protein) to herbivores.

The climatic pathway in the arctic plant–herbivore system is complex, as different types of tundra respond differently to changes in the abiotic variables. The onset of the winter snow-cover and its duration are probably the most influential on the trophic interactions covered in this chapter both directly and indirectly.

III. PLANT–HERBIVORE INTERACTIONS AT ZACKENBERG

The herbivore system in Zackenbergdalen is relatively simple. The resident vertebrate community consists of one species of rodent (collared lemming), one species of lagomorph (arctic hare *Lepus arcticus*), one species of ungulate (musk ox) and one species of gallinaceus bird (rock ptarmigan *Lagopus mutus*). During summer, two species of geese (pink-footed goose *Anser brachyrhynchus* and barnacle goose *Branta leucopsis*) breed at Zackenberg. In contrast to vertebrates, the exact species composition of the invertebrate herbivore community in Zackenbergdalen is not known, but includes species of Lepidoptera and Hemiptera (Høye and Forchhammer, 2008, this volume).

The BioBasis monitoring programme at Zackenberg has concentrated on collared lemming, musk oxen and larvae of *S. zetterstedtii* and Tenthredinidae sp., but studies of plant–herbivore interactions have focused mainly on the interplay between collared lemmings and musk oxen and their forage.

A. Willow–Musk Ox Interactions

The spatiotemporal utilisation of plant forage by large-bodied and long-lived herbivores, such as the musk ox, is influenced by a range of factors such as variations in quantity and quality of plant forage, social structure, presence of predators and weather conditions (Clutton-Brock *et al.*, 1982; Clutton-Brock and Pemberton, 2004). The relationship between the musk ox and its food resources has been studied intensively in several populations in Greenland (Thing *et al.*, 1987; Klein and Bay, 1991; Forchhammer, 1995; Forchhammer and Boomsma, 1995). Musk oxen move over large ranges (Aastrup, 2004), and fluctuations in the number observed in Zackenbergdalen do not necessarily indicate fluctuations in population size. The actual size of the Zackenberg subpopulation is unknown, but air surveys in 1982–1989 and 2001 estimated the number of musk oxen within Wollaston Foreland and A.P. Olsen Land to be in the order of 500–800 (Boertmann and Forchhammer, 1992; Berg, 2003a). In all study years, musk ox herds at Zackenberg showed an apparent preference for grassland, which constitutes on average 35% of the area used by musk oxen but only 18% of the total available habitat in the valley (Figure 2).

In high-arctic Greenland, the temporal and spatial summer dynamics of musk oxen and one of its main summer food sources, the arctic willow, are linked (Forchhammer, 2003). The production of arctic willow in one year ($t-1$) is positively associated with the abundance of musk oxen the following year (t) (Figure 3A). In contrast, increased numbers of musk oxen in year $t-1$ reduce next year's (t) growth of arctic willow (Figure 3B). This high degree of

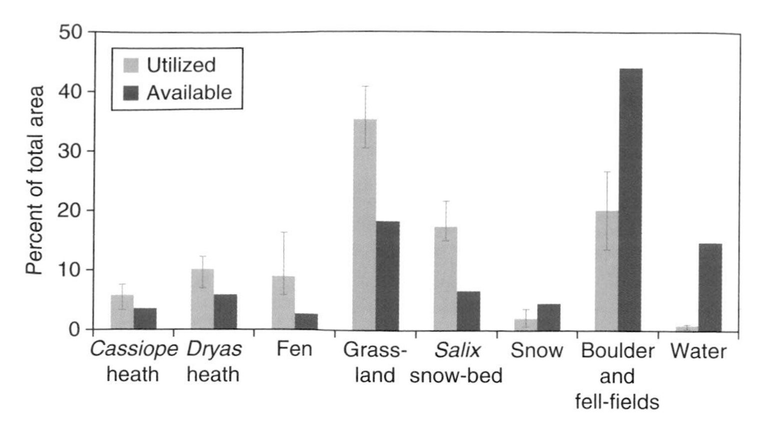

Figure 2 Summer habitat selection by musk oxen in Zackenbergdalen, showing the availability of different vegetation types in the area together with the average percentage musk ox utilisation of each vegetation type (1996–2005). Utilisation of vegetation types is calculated as the fraction of each habitat type within 100-m buffer zones of the herds (D.K. Hendrichsen, unpublished). By using buffer zones, water becomes part of their habitat use, which explains their utilisation of water. Bars indicate the minimum and maximum percentage over the same period of time.

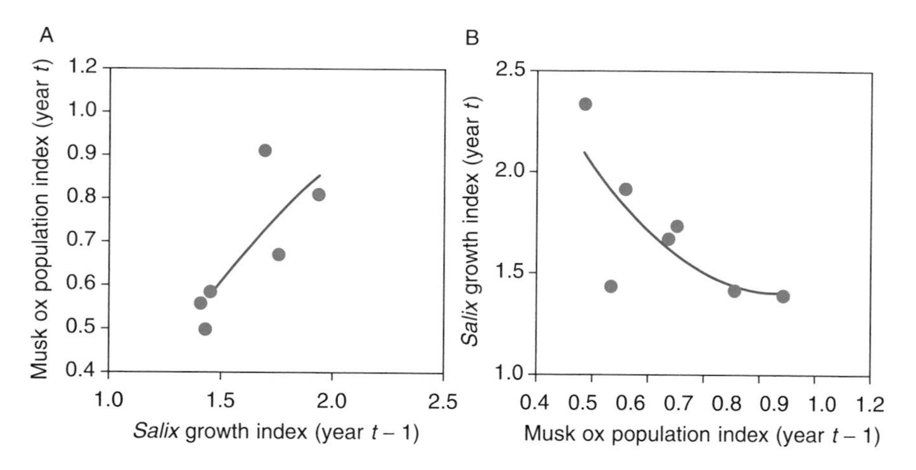

Figure 3 Temporal relationship between (A) current year (*t*) musk ox abundance and last year (*t* – 1) growth of *S. arctica* together with (B) current year growth (*t*) of *Salix arctica* and last year (*t* – 1) musk ox abundance at Zackenberg (modified from Forchhammer, 2003).

connectedness between trophic levels suggests that any direct climate-mediated change in the growth of arctic willow may in turn affect the herbivores, resulting in a significant indirect effect of climate on their abundance. Similarly, any direct climatic effect on the abundance of musk oxen is likely to influence the production of arctic willow (Forchhammer, 2001).

Full evaluations of the plant–herbivore interactions and, hence, of the relative role of climate require analyses performed in an ecosystem or in a community context that integrates all relevant effects simultaneously (Forchhammer, 2001). Hence, analysis of the interaction between musk ox and arctic willow in Zackenbergdalen needs to integrate possible interactive effects of the arctic wolf *Canis lupus arctos*. The dynamics of such a tri-trophic system may be analyzed using a three-dimensional autoregressive model (Post and Forchhammer, 2001), which integrates climatic effects with all intra- and inter-trophic interactions. Schmidt (2006) adopted this approach for the wolf–musk ox–willow system at Zackenberg/Wollaston Forland,

$$\begin{bmatrix} P_t \\ M_t \\ S_t \end{bmatrix} = \begin{bmatrix} a_0 \\ b_0 \\ c_0 \end{bmatrix} + \begin{bmatrix} a_1 & a_2 & 0 \\ b_1 & b_2 & b_3 \\ 0 & c_2 & c_3 \end{bmatrix} \cdot \begin{bmatrix} P_{t-1} \\ M_{t-1} \\ S_{t-1} \end{bmatrix} + \begin{bmatrix} a_4 \\ b_4 \\ c_4 \end{bmatrix} \cdot C_t \qquad (1)$$

where P_t and M_t is growth in the wolf abundance and musk ox abundance, respectively, from year $t-1$ to year t; S_t is the annual growth of arctic willow, whereas C_t is the percent snow-cover on June 10 year t. The regression coefficients at the three trophic levels (a, b and c) express the effects of climate, density dependence and trophic interactions (Schmidt, 2006). For example, the annual change in musk ox abundance (M_t) is a result of the summed influence of changes in last year's wolf abundance (P_{t-1}), density dependence in the musk oxen (M_{t-1}), last year's growth in arctic willow (S_{t-1}) and current year percent spring snow-cover (C_t), that is, $M_t = b_0 + b_1P_{t-1} + b_2M_{t-1} + b_3S_{t-1} + b_4C_t$.

Dividing the data into two time periods (prior to and after the re-invasion of wolves), Schmidt (2006) found that wolves, at the present population density, seemingly did not influence the willow–musk ox system. However, the period prior to wolf reinvasion was also characterised by generally low snow-cover. In this period, musk ox dynamics were mainly governed by density dependence, whereas neither changes in musk ox abundance nor snow-cover influenced the growth of arctic willow (Figure 4A, B). In contrast, during the period after wolf reinvasion, the snow-cover was generally higher and increased growth of willow had a positive influence on the musk ox population (Figure 4C), whereas the arctic willow growth was negatively influence by both increased musk ox population and increased percent snow-cover (Figure 4D). These analyses indicate that changes in climate expressed by variation in snow-cover affect arctic willow directly, whereas the musk ox population at Zackenberg also is affected by climate indirectly through the climatic effects on one of its main food resource, arctic willow. In corroboration, large-scale studies of the dynamics of musk ox populations in North and Northeast Greenland revealed a 3-year delayed influence of the NAO, which is consistent with a climatic influence on fecundity (Forchhammer *et al.*, 2002, 2008, this volume).

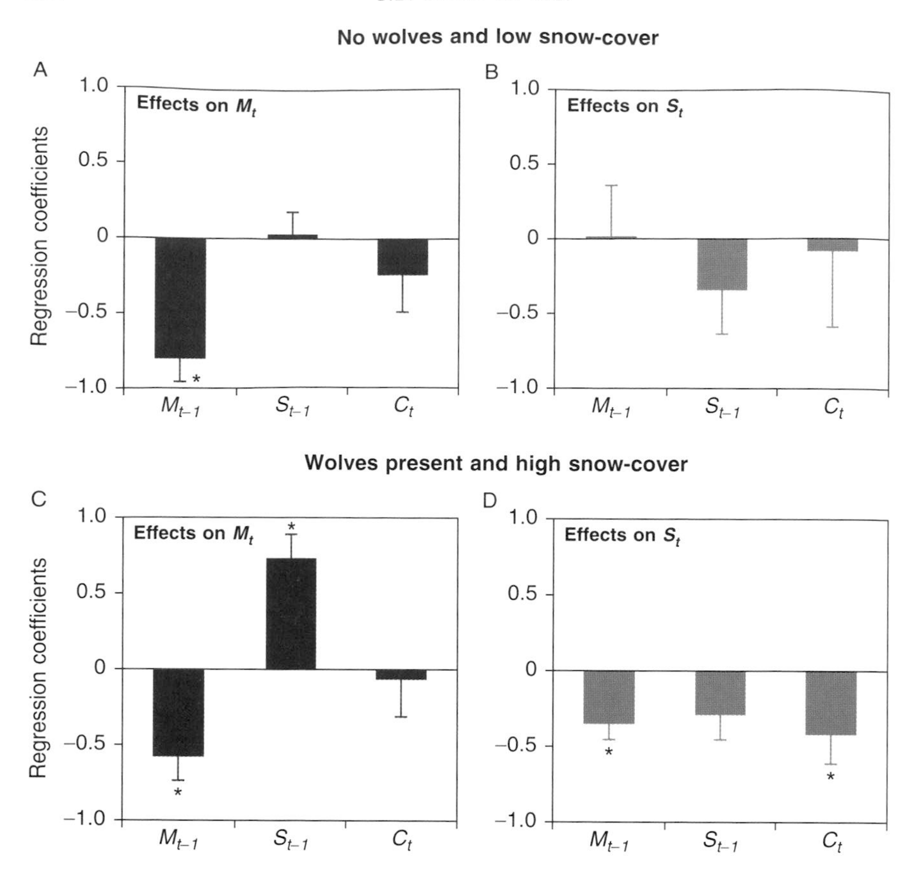

Figure 4 Results from the three-dimensional model (Eq. 1) for musk oxen at Zackenberg. Effects (i.e., model regression coefficients) of last year's musk ox population (M_{t-1}), last year's growth in *S. arctica* (S_{t-1}) and current year percent snow-cover (C_t) on current year musk ox population (M_t) and current year growth of *S. arctica* (S_t) during climatic regimes characterised by low snow-cover (A, B) and high snow-cover (C, D), respectively. Asterisks indicate significant parameter estimates (data from Schmidt, 2006).

B. Plant–Lemming Interactions

The dynamics of the lemming cycle in Northeast Greenland are modulated chiefly by predation (Gilg *et al.*, 2003; Schmidt *et al.*, 2008, this volume), although probably also with a climatic component (Schmidt *et al.*, 2008, this volume) acting via changes in food quantity/quality (Berg, 2003b; Callaghan *et al.*, 2005). The collared lemming eat primarily shrubs and forbs (Batzli, 1993), and mountain avens and arctic willow are by far the most utilised food resource of collared lemmings at Zackenberg throughout the year

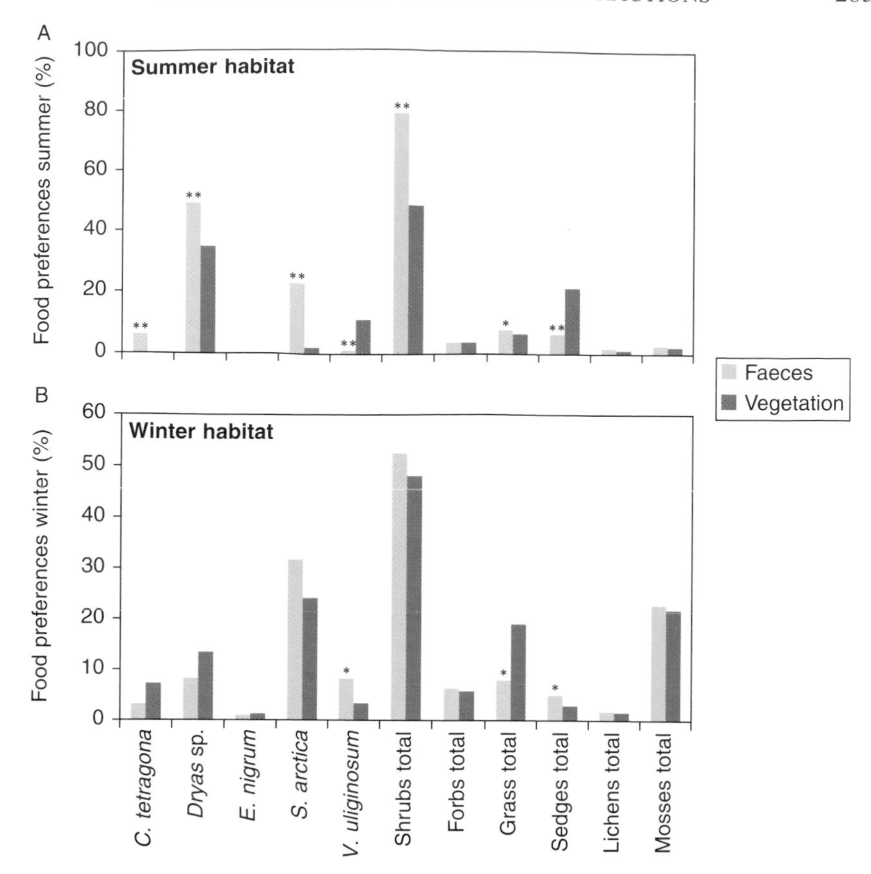

Figure 5 Collared lemming food preference at Zackenberg based on faeces analyses on droppings from nests and burrows compared with vegetation analysis around the respective nest (10 m radius) and burrow (1 m radius). (A) Food plant preferences around summer burrows. (B) Food plant preferences around winter nests. Asterisks indicate significant differences ($p < 0.05$ and $p < 0.01$) (modified from Berg, 2003b).

(Berg, 2003b). These plants are actively selected for during summer (Figure 5A). In winter, their diet reflects the availability of plants and includes a large amount of willow, which is common in their winter habitat (Figure 5B).

One central question is to what extent variation in the abundance of lemmings is influenced by variation in the abundance and quality of their forage. Quality in this perspective is defined as biomass rich in crude protein and modest in concentration of SPM targeted towards herbivory. Grasses are low in SPM (Rhoades and Cates, 1976; Berg, 2003c), and studies on grass–vole relations have not been able to document that the quality of grass is influencing vole abundance (Klemola *et al.*, 2000). Collared lemmings forage, in contrast, contains higher levels of SPM than grasses (Rhoades and Cates, 1976) and, therefore, may negatively affect the herbivores that depend on them.

As in the case of the plant–musk ox interaction, high densities of small herbivores like collared lemmings can negatively affect their preferred food plants. Within the lemming winter habitat, there is an inverse relationship between the density of lemmings and the number of mountain avens flowers the following summer (Figure 6), illustrating that winter herbivory by lemmings may reduce the number of flowers of mountain avens. No such relationship was found for arctic willow. This may reflect the fact that at Zackenberg arctic willow are more evenly dispersed in the winter habitats and so are exposed to lower intensity of grazing per unit area, compared to the more patchily dispersed mountain avens. Despite the few study years, data on plant–lemming interaction in summer suggest that there is a positive relation between the rate of flowering of arctic willow and mountain avens and the density of lemmings in summer habitat. The lack of such positive relation in the winter habitat may relate to differences in the intensity of herbivory in winter and summer habitats. The estimated total area of foraging in winter habitats is on average 18 times greater than the corresponding area in summer (Berg, 2003b). Additionally, the inter-annual overlap in summer forage areas is 25.2% ± 0.02 SE, and within five consecutive years 86.2% of the summer forage area has been reused (T.B. Berg, unpublished analyses). The corresponding figure of inter-annual overlap in winter forage area is expected to be an order of magnitude lower, but no exact field data exists (Berg, 2003b). The intensive foraging during summer is linked to the high predation risk forcing lemmings to forage within 1 m of their summer burrows (Berg, 2003b; Kyhn, 2004; Schmidt *et al.*, 2008, this volume). Consequently, we hypothesise that the concentrations of herbivore-induced SPM is likely to be higher around a

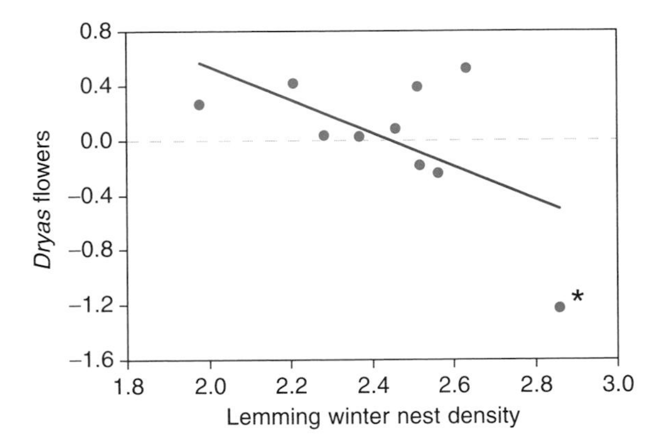

Figure 6 Numbers of *Dryas* flowers at Zackenberg (corrected for the previous year's number of flowers, given as residuals) versus present winter log-transformed lemming density. The outlier (marked with an asterisk) represents the only distinct population peak within the 10-year data ($R^2 = 0.347$, $F_9 = 4.25$, $p = 0.07$).

given summer burrow than around a given winter nest. High concentrations of SPM decrease the digestibility of the forage and the protein uptake by the lemmings and hence, increase their need for compensatory food intake (Seldal *et al.*, 1993), which increases their vulnerability to predators.

Not all SPM are targeted towards herbivores (see Box 2). Some of them are aimed for protection against UV-B. Production of this type of SPM takes up resources for production of SPM aimed for protection against herbivory. Indeed, Selås (2006) demonstrated increased fecundity of both mountain hare *Lepus timidus* and snowshoe hare *Lepus americanus* in response to increased sun screening SPM during high UV-B 2 years earlier. In Zackenbergdalen, Albert *et al.* (2008, this volume) showed an increase in the UV-B-absorbing flavonoids following increased UV-B exposure. Like the hare-UV-B correlation, we found a positive correlation between winter lemming densities at Zackenberg and the UV-B radiation during July and August 2 years earlier (Figure 7), suggesting that a similar relation may exist between UV-B radiation and the reproductive success of lemmings, but more knowledge on the role of SPM on the reproductive success of lemmings is needed, as well as knowledge on how long the effect of these UV-B screening SPM lasts.

Scott (1993) found that low lemming abundance was associated with low temperatures and high levels of precipitation (rain) during freeze-up in October followed by low temperatures and precipitation during November and December. This combination of weather during early winter will lead to high risk of icing events and poor snow-cover and hence, reduced survival of lemmings (extended period of vulnerability to predation by arctic fox and avian predators). At Zackenberg, we found that a late buildup of a deep winter snow layer was coupled to a low winter population of lemmings, and

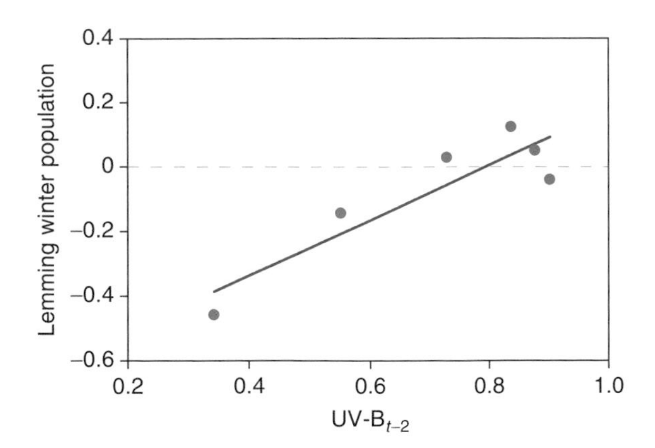

Figure 7 Lemming winter density at Zackenberg adjusted for last year's density (residuals of log-transformed winter nest densities) as a function of the UV-B radiation (during July–August) 2 years earlier.

that a long lasting snow-cover (late onset of snowmelt) was positively related with lemming density (Figure 1). Thawing events during winter, which create icy horizons in the snow pack, prevent the access by herbivores to their forage. On areas with thin snow-cover, severe thawing events destroy the vegetation cover. These effects may eventually lead to a crash of the herbivore population (Callaghan *et al.*, 2005) and disrupt the cyclic dynamics of small rodent populations (Aars and Ims, 2002). The number of days during winter with temperature above 0°C decreases the survival rate of tundra voles (Aars and Ims, 2002), and data from Zackenberg suggest that a similar relationship exists for the collared lemming in Northeast Greenland (Figure 8).

In conclusion, besides the effects of predation (see Schmidt *et al.*, 2008, this volume), data from Zackenberg show that the timing and duration of the snow-cover and positive winter temperatures affect the plant–lemming interaction both directly and indirectly and that these effects act through both bottom–up and top–down pathways. Additionally, UV-B radiation seems to affect the lemming density indirectly with a 2-year delay through food quality as has been found in a hare study by Selås (2006), but more studies are needed.

C. Insect Herbivory

Population dynamics of herbivorous invertebrates in the tundra biome is poorly known due to the lack of long-term data series (Callaghan *et al.*, 2005), but cyclic outbreaks by geometrid moths are know to occur with 10-year intervals in the subarctic hemisphere (Callaghan *et al.*, 2005).

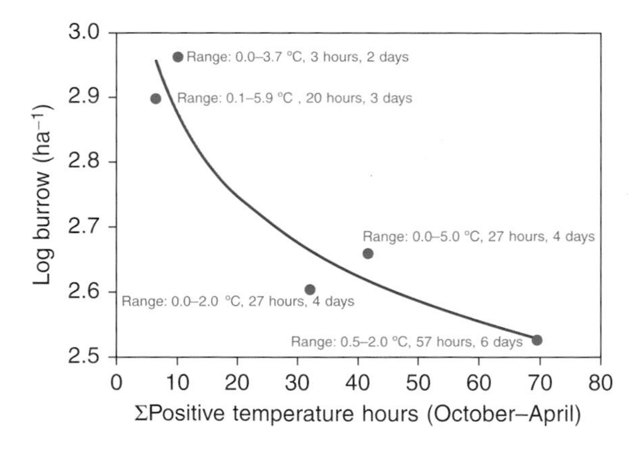

Figure 8 Effect of the sum of positive temperatures measured per hour during winter (October–April) at Zackenberg on the number of active burrows the following summer, $y = 3.3467x^{-0.0659}$, $R^2 = 0.89$. The numbers at each dot show the corresponding temperature range and the number of thawing hours and days. The number of active summer burrows is taken as an index of lemming summer density, which is linked to the winter survival of the lemming population.

No outbreaks of the lepidopteran moth *S. zetterstedtii* occurred during the 10-year time series from Zackenberg. Indeed, only seven larvae of *S. zetterstedtii* were found in all the mountain avens plots. The emergence of *S. zetterstedtii* larvae from winter dormancy takes place the same day their hibernacula become snow-free. Larvae become active by the time mountain avens have begun to flower and therefore can be exploited. The intensity of ovary predation by larvae of *S. zetterstedtii* was recorded as the highest proportion of depredated flowers among open and senescent flowers recorded at weekly checks. Flower ovary predation occurred in those plots that became snow-free in a period of 3 weeks between late-May and mid-June, with an optimum around June 10. Plants that became snow-free after July 1 showed no ovary predation.

From the maximum predation rate and total flower counts, we estimate that a total of 1761 flowers were depredated by *S. zetterstedtii* within the study plots through the study period. Although this is only 4.6% of the total number of flowers produced in the plots during the same period, the distribution is not uniform, and plants in some plots were strongly predated in some years. For instance, flower predation rate exceeded 60% in at least one section of an early snow-free study plot in both 1998 and 1999, meaning that the influence of *S. zetterstedtii* on reproductive success can be substantial on a local scale. It is known from West Greenland that in some years, ovary predation by *S. zetterstedtii* can reach rates of 60–70%, but not what factors may be controlling outbreaks (Philipp *et al.*, 1990).

At Zackenberg, two mountain avens study plots are situated 100 m a.s.l. and 2 km further inland from the other study plots at an average altitude of 35 m a.s.l. The two inland plots had the highest average rates of herbivory, but with considerable variation across years (Figure 9). On the contrary, two late snow-free plots situated in the lowland snow-beds had the lowest average predation rates (<5% in all years), and no more than three flowers were depredated by *S. zetterstedtii* in any year and plot. Dividing observations into a set of early snowmelt seasons and a set of late snowmelt seasons, both of equal range in terms of the date of onset of flowering, the predation rate was largest early in the season ($\text{mean}_{\text{early}} = 4.67$; $\text{SE}_{\text{early}} = 0.93$; $\text{mean}_{\text{late}} = 0.39$; $\text{SE}_{\text{late}} = 0.23$). In this calculation, we excluded the two uphill plots, since data on onset of flowering is available only for 2 years in these plots.

In conclusion, flower herbivory by *S. zetterstedtii* varies considerably from year to year and takes place mainly in areas of early flowering. We did not find any relation between predation rates and flower abundance nor between predation rates and timing of 50% flowering (GLM, $F_{22} = 1.51$, $R^2 = 0.06$, $p = 0.23$). The substantial spatial variability in herbivory rates seems to indicate that the risk of flower herbivory is small and is governed more by local conditions like suitable over-wintering habitats for *S. zetterstedtii* than by climatic effects on a bigger scale as recorded at the metrological station.

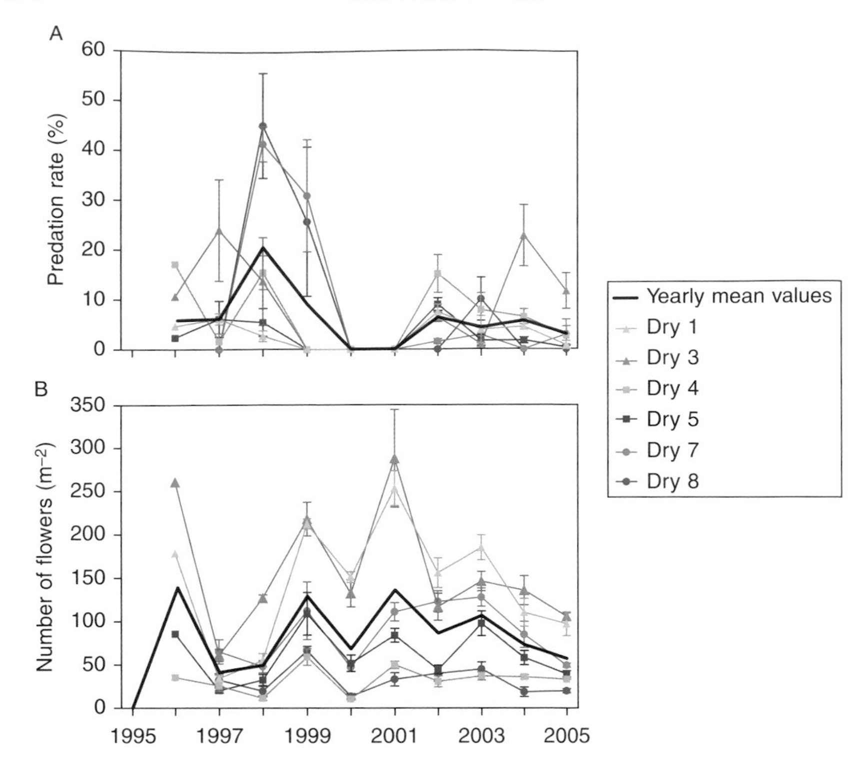

Figure 9 Year to year variation in predation rates on flowers of mountain avens and the corresponding abundance of flowers of mountain avens. (A) Average predation rates of insect flower herbivores in different plots. (B) Density of flowers (flowers m^{-2} in different plots) for mountain avens at Zackenberg. Error bars are one standard error. Data from plots that never exceeded 5% average predation are omitted (Dry 2 and Dry 6). Black lines indicate yearly mean values. The two inland plots (Dry 7 and Dry 8) had the highest average rates of herbivory, but with considerable variation across years.

IV. THE ROLE OF SECONDARY PLANT METABOLITES

Many plants produce various compounds, often referred to as SPM, which may act as anti-herbivore substances or in other ways help protect the plant (Box 2).

In a study conducted at Zackenberg, concentrations of anti-herbivore SPM in arctic willow and mountain avens have been shown to affect the consumption rate (amount eaten/hour) by collared lemming, with more plant material eaten from plants with less concentration of anti-herbivore SPM (Berg, 2003c). Such selective feeding by herbivores may likely affect the plant

Box 2

Secondary Plant Metabolites and Their Role in Plant–Herbivore Interaction

Plants exposed to various stress factors like nutrient shortage, drought, UV-B radiation, climatic extremes, herbivory and parasitism respond by changing their content of secondary plant metabolites (SPM). Some SPM reduces the degree of herbivory, as previously reported for lagomorphs (Bryant, 1987; Robbins *et al.*, 1987; Selås, 2006), small rodents (Batzli, 1983; Seldal *et al.*, 1993; Berg, 2003c; Laitinen *et al.*, 2004) and large cervids (Robbins *et al.*, 1987).

The SPM concentration changes during the growing season in relation to not only the amount of nutrients and energy available but also the reproductive investments of the plant (Batzli, 1983; Lindroth and Batzli, 1986; Laine and Henttonen, 1987; see Fig. 10A). The production of SPM is energetically costly (Soloman and Crane, 1970; Koricheva *et al.*, 1998), and reduction in incoming solar radiation presumably decreases the energy available for production of SPM (see Figure 10B). UV-B is harmful to the plant and therefore plants respond to increased UV-B radiation by allocating amino acids into sun screening on the expenses of herbivore-protective phenolics, resulting in increased levels of leaf nitrogen content (Gwynn-Jones, 1999; Selås, 2006). Hence, the plants increase their nutritional value to herbivores. This is indicative of a climatic influence on the quality of plant tissues as food for herbivores.

Sexual differences may occur in those species having separate sexual individuals like arctic willow (Klein *et al.*, 1998; see Figure 10A). In species of willow, poplar and birch, non-reproducing individuals contained more SPM than reproducing ones, and hence, were less preferred as food items by voles (Danell *et al.*, 1987). The reason is that non-reproducing individuals are able to invest more in SPM than reproducing individuals (Danell *et al.*, 1987). Among reproductive arctic willow, female plants tend to have higher concentration of SPM than male plants, since seed production is more costly than pollen production (Cornelissen and Stilling, 2005), which also has been observed in Zackenbergdalen (see Figure 10A). The reason may be that female plants need more photosynthetically active tissue for seed production than males do for pollen production and hence, females need to protect their tissues more than males.

Food plants with low digestibility increase the time needed for digestion, which escalate the nutritional stress, as stomach size becomes a limiting factor. Hence, food quality may play an important role in

(continued)

Box 2 *(continued)*

lemming ecology as well as in ruminant ecology, where the time for rumination is the limiting factor of food passage. For microtine rodents, consumption of food plants with high levels of proteinase-inhibiting SPM decreases the protein intake significantly, thereby increasing the time needed for foraging and ultimately affecting their condition. In extreme cases, lemmings have been found to die of malnutrition when fed with food plants high in proteinase-inhibiting SPM (Seldal *et al.*, 1993). In general, the quality of a food plant is a trade-off between the SPM content, energy content and the quantity of the given plant (Robbins *et al.*, 1987).

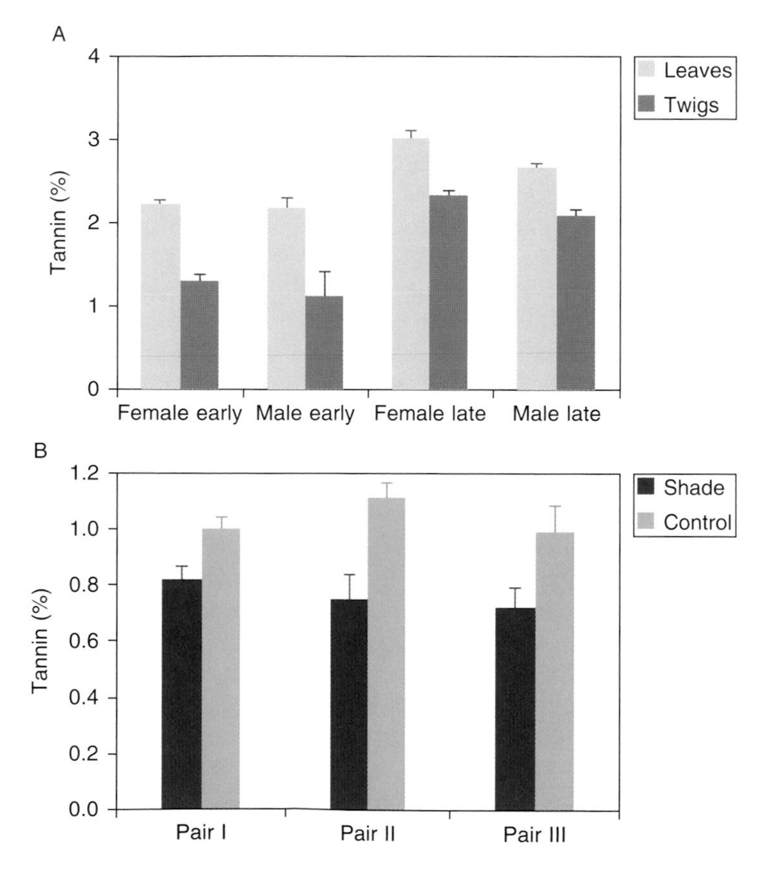

Figure 10 Tannin content in Arctic willow *Salix arctica*. (A) Sexual differences in tannin content and changes in tannin content during the growing season (D. Klein, unpublished data). (B) Three replicate pairs of shaded and unshaded arctic willow plots (1 month of 50% reduction in incoming solar radiation during the growth season) (modified from Klein *et al.*, 1998).

composition in favouring those individuals within a species that have the highest concentration of SPM.

In arctic willow, female-biased sex ratios are found in Zackenbergdalen (Klein *et al.*, 1998) and in arctic Canada (Crawford and Balfour, 1990). Seeds from arctic willow collected in the valley germinate and grow to flowering with a sex ratio close to unity. Nonetheless, the sex ratio of reproducing arctic willow plants in Zackenbergdalen is 60% female biased (Klein *et al.*, 1998). In the valley, female arctic willow plants contain higher concentrations of tannins (an anti-herbivore SPM; Swain, 1979) than male arctic willow plants, especially in late summer (Figure 10A, Berg, 2003c). Similar inter-sexual differences in SPM content have been observed elsewhere in the Arctic (Dawson and Bliss, 1989). Females of arctic willow may thus be better defended against tissue loss to herbivores than male plants, especially during the summer reproductive period (Klein *et al.*, 1998). Arctic willow makes up a large fraction of the diet of musk oxen and lemmings in summer and even in the winter diet of the collared lemming (Figure 11; Klein and Bay, 1991, 1994; Klein, 1995; Berg, 2003c). At Zackenberg, female arctic willow was eaten at significantly lower rates (amount eaten by lemmings per hour) than males (Berg, 2003c). This difference was positively related to the generally higher tannin concentration (mg/g) found in female leaves over the summer (39.28 ± 5.01SE) compared to males (25.87 ± 2.96) (Berg 2003c). Hence, inter-sexual differences in tannin levels may make female willow plants less prone to herbivory than male plants. Cantina experiments conducted at Zackenberg have documented that consumption rates of arctic willow and

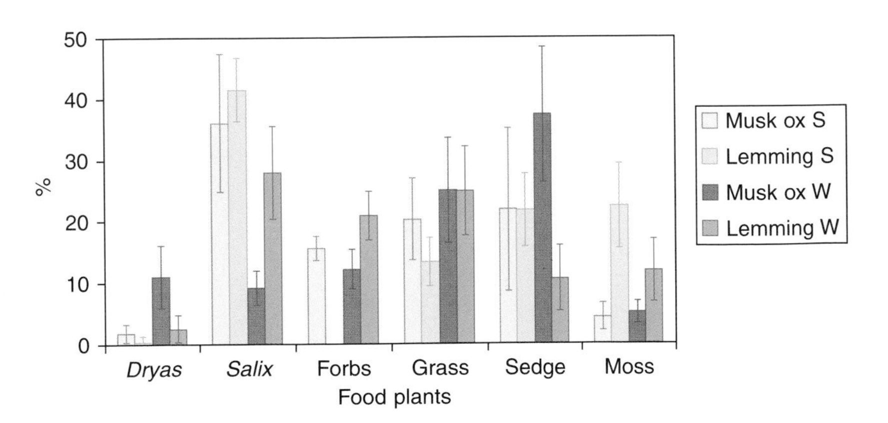

Figure 11 Comparison of fragments of different genera of plants in summer (S) and winter (W) faeces of lemmings and musk oxen in Nansen Land (North Greenland) (modified from Klein and Bay, 1994).

mountain avens by collared lemmings are negatively related to their content of anti-herbivore SPM (Berg, 2003c).

Several small rodent species have been shown to prefer plants with low SPM concentration over plants with high SPM concentrations (Dearing, 1997; Berg, 2003c; Laitinen *et al.*, 2004). Selective feeding on arctic willow genders, that is, sex-biased herbivory, may thus be an important factor in shaping the biased sex ratios observed in arctic willow (Elmquist *et al.*, 1988; Boecklen *et al.*, 1990; Hjältén, 1992; Ueno and Seiwa, 2003). The dietary intake of herbivores is, however, balanced between several elements, and in Zackenbergdalen, the preference of collared lemmings for arctic willow and mountain avens seems to reflect a trade-off between the crude protein content and the content of SPM (Berg, 2003b).

V. CONCLUSIONS AND FUTURE PERSPECTIVES

Beside the impact of predation, climate in terms of timing and duration of the snow-cover is one of the most important variables affecting the plant–herbivore system. Positive winter temperatures that create thawing events with ice horizons in the snow and icing on bare ground can have catastrophic effects on both vegetation and herbivores. Additionally, UV-B radiation seems to be yet another candidate that indirectly affects the herbivore population that depends on plants that produce UV-B-protective SPM, but more studies are needed. At least within the plant–musk ox and the plant–lemming systems, interactions can follow both bottom–up and top–down pathways. Herbivores respond to changes in primary production and plants respond to changes in the intensity of herbivory, with both pathways being affected by climate. Data on plant–insect interaction were able to illustrate only a top–down effect on a small local scale.

The expected future climatic changes in Northeast Greenland are outlined in detail by Stendel *et al.* (2008, this volume). As discussed by Ellebjerg *et al.* (2008, this volume), increased length of the growing season and more precipitation during summer will lead to increased plant biomass. Not all habitat types will be equally affected by the predicted climate change scenario, and climate may affect herbivore-specific winter and summer habitats differently. Wet tundra will be less affected by increased precipitation compared to dry tundra, which most likely will experience a relatively greater increase in biomass compared to the wet tundra (Heal *et al.*, 1998). Dry tundra is intensively used by lemmings during summer and may to great extent be regarded as a bottleneck in terms of survival and population growth. Hence, increased standing plant biomass will have a positive effect on lemmings, not only as an increased amount of forage but also as an increased shelter against avian predators. Increased snow-cover will in general benefit the lemmings, if the snow-covered period stays unchanged. Contrarily, the positive effect of increased snow-cover will be hampered if the buildup of the snow pack is delayed. Musk oxen, in contrast, will benefit from a delayed buildup of the snow-cover, through the prolonged season with relative easy access to forage, if no icing occurs during the freeze-up period. Severe thaw events during winter will destroy the plant cover by ice crust formation on exposed vegetation or prevent the herbivores, like musk oxen and arctic hares, from reaching the vegetation through ice layers in the snow.

Another strong abiotic parameter affecting living species is UV-B radiation. UV-B radiation induces the plants to allocate more resources into UV protection, which does not at the same time protect against herbivores. Hence, higher UV-B radiation will be an advantage for all herbivores as will increased cloud cover, which in general reduces SPM production in plants (Figure 10B).

The well-known lemming cycle will most likely be affected during the expected climate change scenarios (Callaghan *et al.*, 2005; Schmidt *et al.*, 2008, this volume). During the present 10-year data period from Zackenberg, the lemming population has only experienced one well-defined peak (1998). Since this peak, the lemming population peaks have not reached equally high densities. The same holds for the population on Traill Ø, 220 km south of Zackenberg (see Figure 1 in Schmidt *et al.*, 2008, this volume). These two lemming scenarios coincide with the fact that since 1997 the buildup of the snow-cover in Zackenbergdalen and likely also on Traill Ø has been delayed by about 80 days, and seemingly the cyclic fluctuation in lemming numbers has been disturbed. A breakdown of the lemming cycles will probably lead to chaotic fluctuation pattern known from rodent populations under more southerly climates.

ACKNOWLEDGMENTS

The monitoring data used in this chapter were provided by the BioBasis programme, run by the National Environmental Research Institute, University of Aarhus, and financed by the Danish Environmental Protection Agency, Danish Ministry of the Environment. We extend our sincere thanks to Christian Bay, Fiona Danks, Louise Grøndahl, Line A. Kyhn, Per Mølgaard, Eric Post, Charlotte Sigsgaard and Mikkel P. Tamstorf for various inputs in the field and during the writing of this synthesis. We also thank Nick Tyler for valuable comments on an earlier draft.

REFERENCES

Aars, J. and Ims, R.A. (2002) *Ecology* **83**, 3449–3456.

Aastrup, P. (2004) *Polar Biol.* **27**, 50–55.

Babb, T.A. and Whitfield, D.W.A. (1977) In: *Truelove Lowland, Devon Island, Canada: A High Arctic Ecosystem* (Ed. by L.C. Bliss), pp. 589–606. University of Alberta Press, Edmonton.

Batzli, G.O. (1983) *Oikos* **40**, 396–406.

Batzli, G.O. (1993) In: *The Biology of Lemmings* (Ed. by N.C. Stenseth and R.A. Ims), pp. 281–309. Academic Press, London.

Berg, T.B. (2003a) In: *Zackenberg Ecological Research Operation,. 8th Annual Report 2002* (Ed. by M. Rasch and K. Caning). pp. 50–56. Danish Polar Center, Ministry of Science, Technology and Innovation, Copenhagen.

Berg, T.B. (2003b) The collared lemming (Dicrostonyx groenlandicus) in Greenland: population dynamics and habitat selection in relation to food quality. Ph.D. Thesis, National Environmental Research Institute, Denmark.

Berg, T.B. (2003c) *Oecologia* **135**, 242–249.

Boecklen, W.J., Price, P.W. and Mopper, S. (1990) *Ecology* **71**, 581–588.

Boertmann, D. and Forchhammer, M. (1992) *Greenland Environmental Research Institute Report Series, no.* **4**, 1–36.

Bryant, J.P. (1987) *Ecology* **68**, 1319–1327.

Callaghan, T.V., Björn, L.O., Chapin, F.S., III, Chernov, Y., Christensen, T.R., Huntley, B., Ims, R., Johansson, M., Riedlinger, D.J., Jonasson, S., Matveyeva, N., Oechel, W., *et al.* (2005) In: *Arctic Climate Impact assessment, ACIA, 2005,* (Ed. by C. Symon, L. Arris and B. Heal), pp. 243–352. Cambridge University Press, Cambridge.

Chernov, Y.I. (1985) *The Living Tundra*. Cambridge University Press, Cambridge.

Clutton-Brock, T.H. and Pemberton, J.M. (Eds.) (2004) *Soay Sheep. Dynamics and Selection in an Island Population*. Cambridge University Press, Cambridge.

Clutton-Brock, T.H., Guinness, F.E. and Albon, S.D. (1982) *Red Deer. Behavior and Ecology of Two Sexes*. The University of Chicago Press, Chicago.

Cornelissen, T. and Stilling, P. (2005) *Oikos* **111**, 488–500.

Crawford, R.M.M. and Balfour, J. (1990) *Flora* **184**, 291–302.

Danell, K., Elmqvist, T., Ericson, L. and Salomonson, A. (1987) *Oikos* **50**, 396–402.

Danell, K., Bertwaux, D. and Bråthen, K.A. (2002) *Arctic* **55**, 389–392.

Dawson, T.E. and Bliss, L.C. (1989) *Oecologia* **79**, 332–343.

Dearing, M.D. (1997) *Ecology* **78**, 774–781.

Elmquist, T., Ericson, L., Danell, K. and Salomonson, A. (1988) *Oikos* **51**, 259–266.

Flanagan, P.W. and Bunnell, F.L. (1980) In: *An Arctic Ecosystem: the Coastal Tundra at Barrow, Alaska* (Ed. by J. Brown, P.C. Miller, L.L. Tieszen and F.L. Bunnell), US/IBP synthesis series: 12. pp. 291–334. Dowden, Hutchinson & Ross Inc., Stroudsburg, PA.

Forchhammer, M.C. (1995) *Can. J. Zool.* **73**, 1344–1361.

Forchhammer, M.C. (2001) In: *Climate Change Research—Danish Contributions* (Ed. by A.M.K. Jørgensen, J. Fenger and K. Halsnæs), pp. 219–236. Gad, Copenhagen.

Forchhammer, M.C. (2003) In: *Zackenberg Ecological Research Operation, 7th Annual Report 2001* (Ed. by M. Rasch and K. Caning), pp. 60–61. Danish Polar Center, Ministry of Science, Technology and Innovation, Copenhagen.

Forchhammer, M.C. and Boertmann, D. (1993) *Ecography* **16**, 299–308.

Forchhammer, M.C. and Boomsma, J. (1995) *Oecologia* **104**, 169–180.

Forchhammer, M.C. and Post, E. (2004) *Popul. Ecol.* **46**, 1–12.
Forchhammer, M.C., Post, E., Stenseth, N.C. and Boertmann, D. (2002) *Popul. Ecol.* **44**, 113–120.
Forchhammer, M.C., Post, E., Berg, T.B., Høye, T.T. and Schmidt, N.M. (2005) *Ecology* **86**, 2644–2651.
Gilg, O., Hanski, I. and Sittler, B. (2003) *Science* **302**, 866–868.
Gwynn-Jones, D. (1999) *Ecol. Bull.* **47**, 77–83.
Heal, O.W., Callaghan, T.V., Cornelissen, H.C., Körner, C. and Lee, S.E. (Eds.) (1998) *Global Change in Europe's Cold Regions. European Commission, Ecosystems Research Report* **27**, 137.
Hjältén, J. (1992) *Oecologia* **89**, 253–256.
Høye, T.T., Post, E., Meltofte, H., Schmidt, N.M. and Forchhammer, M.C. (2007a) *Curr. Biol.* **17**, R449–R451.
Høye, T.T., Ellebjerg, S.M. and Philipp, M. (2007b) *Arct. Antarct. Alp. Res.* **39**, 412–421.
Klein, D.R. (1995) *Ecoscience* **2**, 100–102.
Klein, D.R. and Bay, C. (1991) *Holarctic Ecol.* **14**, 152–155.
Klein, D.R. and Bay, C. (1994) *Oecologia* **97**, 439–450.
Klein, D.R., Bay, C. and Danks, F. (1998) In: *Zackenberg Ecological Research Operation, 3rd Annual Report 1997* (Ed. by H. Meltofte and M. Rasch), pp. 60–61. Danish Polar Center, Ministry of Research and Technology, Copenhagen.
Klemola, T., Norrdahl, K. and Korpimäki, E. (2000) *Oikos* **90**, 509–516.
Koricheva, J., Larsson, S., Haukioja, E. and Keinänen, M. (1998) *Oikos* **83**, 212–226.
Kyhn, L.A. *A study of the behaviour of wild collared lemmings* (Dicrostonyx groenlandicus *Traill, 1823*). MSc. Term Paper Project, *University of Copenhagen.* 27 pp.
Laine, K.M. and Henttonen, H. (1987) *Oikos* **50**, 389–395.
Laitinen, M-L., Julkunen-Tiitto, R., Yamaji, K., Heinonen, J. and Rousi, M. (2004) *Oikos* **104**, 316–326.
Lima, S.L. and Dill, L.M. (1990) *Can. J. Zool.* **68**, 619–640.
Lindroth, R.L. and Batzli, G.O. (1986) *J. Anim. Ecol.* **55**, 431–449.
Meltofte, H. and Berg, T.B. (2006) *BioBasis: Conceptual design and sampling procedures of the biological programme of Zackenberg Basic*, 9th ed. National Environmental Research Institute, Department of Arctic Environment, 77 pp.
Morewood, W.D. and Ring, R.A. (1998) *Can. J. Zool.* **76**, 1371–1381.
Olofsson, J., Moen, J. and Oksanan, L. (2002) *Oikos* **96**, 265–272.
Olofsson, J., Hulme, P.E., Oksanen, L. and Suiminen, O. (2004) *Oikos* **106**, 324–334.
Philipp, M., Böcher, J., Mattsson, O. and Woodell, S.R.J. (1990) *Meddr. Grønland Biosci.* **34**, 1–60.
Plesner-Jensen, S. and Doncaster, C.P. (1999) *J. Theor. Biol.* **199**, 63–85.
Post, E. and Forchhammer, M.C. (2001) *BMC Ecology* **1**, 5.
Rhoades, D.F. and Cates, R.G. (1976) *Recent Adv. Phytochem.* **10**, 168–213.
Robbins, C.T., Hanley, T.A., Hagerman, A.E., Hjelfjord, O., Baker, D.L., Schwartz, C.C. and Mautz, W.W. (1987) *Ecology* **68**, 98–107.
Schmidt, N.M. (2006) Climate, agriculture and density-dependent dynamics within and across trophic levels in contrasting ecosystems. PhD thesis, Department of Ecology, Royal Veterinary & Agricultural University, Denmark.
Scott, P.A. (1993) *Arctic* **46**, 293–296.
Seldal, T., Andersen, K.-J. and Högsted, G (1993) *Oikos* **70**, 3–11.
Selås, V. (1997) *Oikos* **80**, 257–268.
Selås, V. (2006) *Popul. Ecol.* **48**, 71–77.
Soloman, M. and Crane, F.A. (1970) *J. Pharm. Sci.* **59**, 1670–1672.

Stenseth, N.C. and Ims, R.A. (1993) *The Biology of Lemmings*. Academic Press, London.

Stenseth, N.C. and Oksanen, L. (1987) *Oikos* **50**, 319–326.

Swain, T. (1979) In: *Herbivores—Their Interaction with Secondary Plant Metabolites* (Ed. by G.A. Rosenthal and D.H. Janzen), pp. 657–682. Academic Press, Inc., New York.

Thing, H., Klein, D.R., Jingfors, K. and Holt, S. (1987) *Holarctic Ecol.* **10**, 95–103.

Turchin, P., Oksanen, L., Ekerholm, P., Oksanen, T. and Henttonen, H. (2000) *Nature* **405**, 562–565.

Ueno, N. and Seiwa, K. (2003) *J. Forest Res.* **8**, 9–16.

van der Wal, R., Bardgett, R.D., Harrison, K.A. and Stien, A. (2004) *Ecography* **27**, 242–252.

Vibe, C. (1967) *Meddr. Grønland* **170**, 5, 227 pp.

Walker, D.A., Billings, W.D. and de Molenaar, J.G. (2001) In: *Snow Ecology: An Interdisciplinary Examination of Snow-Covered Ecosystems* (Ed. by H.G. Jones, J.W. Pomeroy, D.A. Walker and R.W. Hoham), pp. 266–324. Cambridge University Press, Cambridge.

Webber, P.J., Miller, P.C., Chapinn, F.S, III and McCown, B.H. (1980) In: *An Arctic Ecosystem: the Coastal Tundra at Barrow, Alaska* (Ed. by J. Brown, P.C. Miller, L.L. Tieszen and F.L. Bunnell), US/IBP synthesis series: 12. pp. 186–218. Dowden, Hutchinson & Ross Inc., Stroudsburg, PA.

Wilson, D.J. and Jefferies, R.L. (1996) *J. Ecol.* **84**, 841–851.

Phenology of High-Arctic Arthropods: Effects of Climate on Spatial, Seasonal, and Inter-Annual Variation

TOKE T. HØYE AND MADS C. FORCHHAMMER

SUMMARY

The short summers of the High Arctic pose a strong time constraint on the annual cycle of all organisms in this region. Although arctic arthropods can complete their development at very low temperatures, the predicted climatic changes may shift their phenology outside its normal range. Hence, arctic arthropods may become exposed to conditions to which they are not adapted. On the basis of long-term data from several plots of pitfall and window traps at Zackenberg in high-arctic Northeast Greenland, we document that the timing of emergence is closely related to date of snowmelt in nine taxa of common surface-active and flying arthropods. Average air temperature seemed to play a lesser role, although the duration from snowmelt to the date when 50% of the individuals in the season were caught (date50) was negatively related to the average daily air temperature during the same time interval in three of the nine taxa. Since short-term weather fluctuations appeared to have a small effect on capture numbers in pitfall and window traps, we suggest that timing of snowmelt is a good predictor of the phenology of most arthropods in high-arctic Greenland. The spatial synchrony of capture numbers between individual traps within plots was high. However, among pairs of plots, the spatial synchrony varied between taxa and habitats and declined with distance between plots for surface-dwelling taxa and with

ADVANCES IN ECOLOGICAL RESEARCH VOL. 40 0065-2504/08 $35.00
© 2008 Elsevier Ltd. All rights reserved
DOI: 10.1016/S0065-2504(07)00013-X

difference in timing of snowmelt for the most abundant families of Diptera (Muscidae and Chironomidae). Detritus feeders (collembolans, mites and most larvae of Diptera) and predators (spiders of the families Linyphiidae and Lycosidae) were abundant throughout the summer season. In contrast, the abundance of more specialized groups, like butterflies (e.g., Nymphalidae) and parasitoid wasps (e.g., Ichneumonidae), was restricted to a narrow seasonal time window in the warmest part of the summer. Because of their narrow phenological range and their host specialization, these taxa may be most vulnerable to trophic mismatch. Furthermore, snowmelt is predicted to become more variable, and this may affect organisms in areas of late snowmelt most severely.

I. INTRODUCTION

Studies of arctic arthropod ecology have focused primarily on the adaptations allowing species to cope with extreme climatic conditions, as well as on the mechanisms responsible for the observed low diversity of arthropods in the Arctic (e.g., Downes, 1964; Strathdee and Bale, 1998). The current attention on climate change has motivated experimental work on the effects of rising temperatures on arthropod assemblages (e.g., Dollery *et al.*, 2006), but studies at the population and community levels are still sparse, especially concerning flying and surface-dwelling arthropods. The consequences of climate change for the life history and population dynamics of many arctic arthropod species are, therefore, still largely unknown (Hodkinson *et al.*, 1998; Bale *et al.*, 2002). Clearly temperature is a central factor in the development and behaviour of poikilotherm organisms like arthropods, and both the timing of emergence and locomotory performance may be constrained by low temperatures (Strathdee and Bale, 1998). Hence, climate is likely to be an important driver of variation in the phenology as well as the abundance of arthropods in the Arctic (Danks and Oliver, 1972).

The environmental conditions of the High Arctic are extremely seasonal, and subzero temperatures prevail during most of the year. As a result, many arthropod species inhabiting this region have a long period of winter dormancy (Downes, 1964; Danks, 2004), and multi-annual life cycles are common in, for example, spiders (Pickavance, 2001; Hammel, 2005), Diptera (Butler, 1982) and Lepidoptera (Morewood and Ring, 1998). Characteristic taxa like Chironomidae and Culicidae have short adult stages serving mainly to complete reproduction and dispersal (Danks and Oliver, 1972; Corbet and Danks, 1973). In insects, the emergence of adults normally coincides with the peak in food resources, availability of mates, or suitable egg-laying habitats (MacLean, 1980), but the actual timing of emergence may vary considerably across taxa (Danks, 2004). Although some species are closely associated with specific host plants (e.g., Hodkinson and Bird, 1998) or host animals (e.g., Kutz *et al.*, 2005), studies of arctic pollinator networks have revealed that generalists are very common (Lundgren and Olesen, 2005). However, in the majority of arctic arthropod species, their detailed behavioural and ecological relationships to adjacent trophic levels remain relatively unknown.

In addition to the direct effect of climate on populations (Forchhammer and Post, 2004), climate change may also induce trophic mismatching (Stenseth and Mysterud, 2002). Trophic mismatch describes the situation where the temporal occurrence of dependent organisms on adjacent trophic levels is not matched, and the consumer–resource linkage is broken. For instance, following a warming climate the peak in insect abundance may already be declining before chicks of insectivorous birds have their peak demand for food (Visser *et al.*, 1998). Such asynchronous shifts in phenology across trophic levels may have even greater effect in the High Arctic because the short summers constrain the phenological flexibility for consumers to changes in timing of food abundance. Thus, to understand the ecological consequences of climatic changes in arctic ecosystems, there is a need to quantify the role of climate in the phenology of arthropods. This group of organisms is central for the functioning of several trophic levels in the Arctic, for example. as food for breeding waders (Meltofte *et al.*, 2007).

Here, we address the question of how climate affects the terrestrial arthropod assemblage in a single ecosystem in Zackenbergdalen in high-arctic Northeast Greenland, based on data from the Zackenberg Basic monitoring programme (Meltofte *et al.*, 2008, this volume). In particular, we focus on the effects of snow and temperature on the phenology of the most abundant taxa of arthropods and on their spatial variation in emergence patterns. First, we provide an overall description of the arthropod fauna and relative differences between sampling plots and years. Because of the multiannual life cycles and the moderate taxonomic resolution of the data we do not provide detailed analyses of inter-annual variation in capture numbers. Instead, we focus our analyses on the phenology of three Diptera families (Chironomidae, Muscidae and Sciaridae), one family of parasitoid wasps (Ichneumonidae), one family of Lepidoptera (Nymphalidae), two families of spiders (Lycosidae and Linyphiidae), collembolans and mites. The importance of temperature and plot-specific dates of snowmelt (defined as date of 50% snow-cover in the plot) for inter-annual variation in phenological events is analysed statistically. In addition, we estimate the spatial synchrony of capture numbers among traps within trapping plots and between pairs of plots. We end with a discussion of how differences in phenological sensitivity to climate across taxa can lead to trophic mismatch within the arthropod food web and its ecological repercussions.

II. THE ARTHROPOD FAUNA AT ZACKENBERG

The entire arthropod fauna in any given locality can be described only by combining results of multiple trap types. Since the monitoring of arthropods at Zackenberg focuses on pitfall and window traps (Box 1), we restrict this description to flying and surface-dwelling species. During the years 1996–2005, a total of 567,644 arthropods were caught using window and pitfall traps (Table 1). Mites and collembolans constituted almost half of these (see Sørensen *et al.*, 2006, for details). Among the remaining specimens, 99.9% were either spiders or belonged to one of four orders of insects: Diptera, Hymenoptera, Hemiptera and Lepidoptera (Table 1). The majority of the mites were caught in plots 2, 5 and 7 (Table 1) where snowmelt occurs early. Collembolans were mainly caught in plots 3, 4 and 5, with much smaller numbers in plot 2 and 7 (Table 1). They are sensitive to changes in moisture (Lensing *et al.*, 2005), but may also respond to anoxia (Hodkinson and Bird, 2004), which is a common condition in areas of late snowmelt. A previous comparison of the density of mite and collembolans in plots 3, 4 and 5 based on soil samples from early August revealed a larger density of collembolans in plot 5, but no difference among the three plots for mites (Sørensen *et al.*, 2006).

In general, Diptera and especially Chironomidae dominate arctic insect communities (MacLean and Pitelka, 1971; Danks, 1981, 1992; Elberling and Olesen, 1999). Indeed, at Zackenberg we found that 34.8% of the total sample of insects and spiders were Chironomidae. The window traps (plot 1) and the pitfall traps in the fen (plot 2) caught the majority of the Chironomidae, suggesting an association to habitats with ponds and soil with high moisture content. The capture numbers of several other Diptera families, for example, Culicidae, Empididae, Phoridae and Scatophagidae, varied considerably between plots, and the plot where the majority of the individuals were caught differed between families. Culicidae (mosquitoes) is a well-known arthropod family in the Arctic, but the number of individuals in the traps at Zackenberg was moderate compared to other families of Diptera, probably related to the trap design. Indeed, a long-term study from arctic Canada used specially designed visual attraction traps, in contrast to the setup at Zackenberg, and demonstrated that differences in timing of emergence were related to pond depth and water temperature (Corbet and Danks, 1973). Muscidae were very common in most of the samples and they were found in all plots in roughly equal numbers. In contrast, the distribution of over-wintering sites of Mycetophilidae has previously been found to be very patchy (N.M. Schmidt and T.T. Høye, unpublished data). This is probably related to their specific habitat requirement for egg-laying and larval development (Böcher, 2001).

Box 1

Arthropod Study Sites and Sampling Design

Arthropods were captured continuously during summer in several different plots with varying average date of snowmelt (Box Table 1).

Box Table 1 Description of the seven different trapping plots

Plot no.	Trap type	Operating period	Description of habitat	Timing of snowmelt
1	Window	1996–2005	Small islet in shallow pond	Early
2	Pitfall	1996–2005	Wet fen dominated by mosses and grasses with *Salix arctica* on the tussocks	Early
3	Pitfall	1996–2005	Mesic heath dominated by lichens with *Cassiope tetragona*, *Dryas* sp. and *Salix arctica*	Late
4	Pitfall	1996–2005	Similar to plot 3, but less *Cassiope tetragona* and more *Dryas* sp. and grasses	Late
5	Pitfall	1996–2005	Arid heath dominated by lichens and *Dryas* sp.	Early
6	Pitfall	1996–1998	Snow-bed dominated by lichens and *Salix arctica*	Late
7	Pitfall	1999–2005	Similar to plot 5, but more exposed and with limited snow-cover during winter. Less *Salix arctica* and more *Dryas* sp. and grasses than plot 5	Early

The trap type and period of operation as well as a short description of the vegetation in the plots is given. The pitfall trap plots each have eight yellow plastic cups with a diameter of 10 cm, while the window trap plot consists of two window traps with windows measuring 20 × 20 cm and positioned perpendicular to each other (Jónsson *et al.*, 1986). All plots are located below 50 m elevation above sea level. Plots were grouped in either early or late snowmelt based on average date of snowmelt. Further information on inter-annual variation in date of snowmelt is given in Box Figure 2.

(continued)

Box 1 *(continued)*

A climate station located within 600 m of the plots provided the weather data (Box Figure 1). Both pitfall and window traps were used (Box Table 1). The sampling in each of the plots started right after snowmelt. Thus, sampling started later in areas and years of late snowmelt than in areas and years of early snowmelt (Box Figure 2). All traps were emptied weekly. Mites and collembolans were only counted but most other arthropods were identified to the family level. Further details regarding sampling procedures are given in Meltofte and Berg (2006), and Bay (1998) provides a more detailed description of vegetation types.

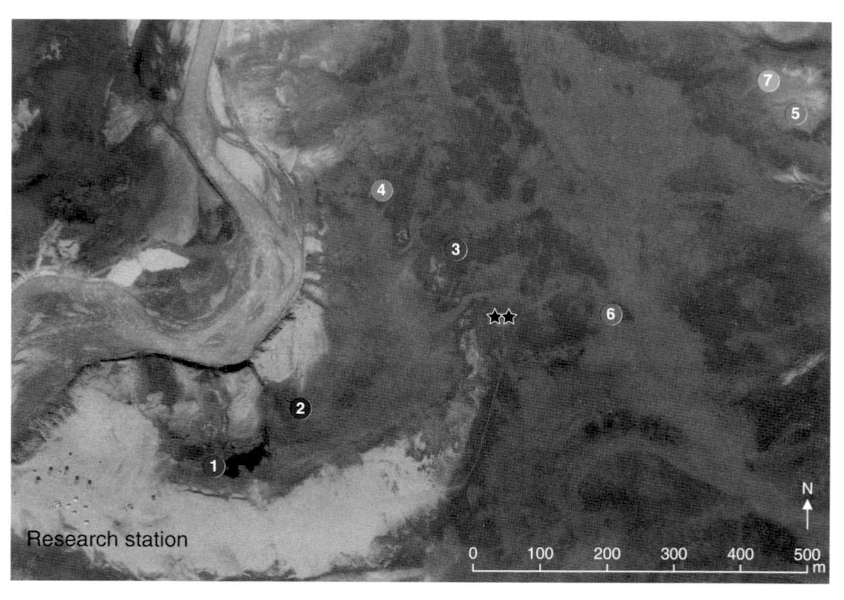

Box Figure 1 Orthophoto of the study area with the sampling plots (plot numbers indicated in dots), climate station (masts indicated by stars) and research station. Abrasion plateaus and snow are light while vegetated areas and ponds are in darker tones.

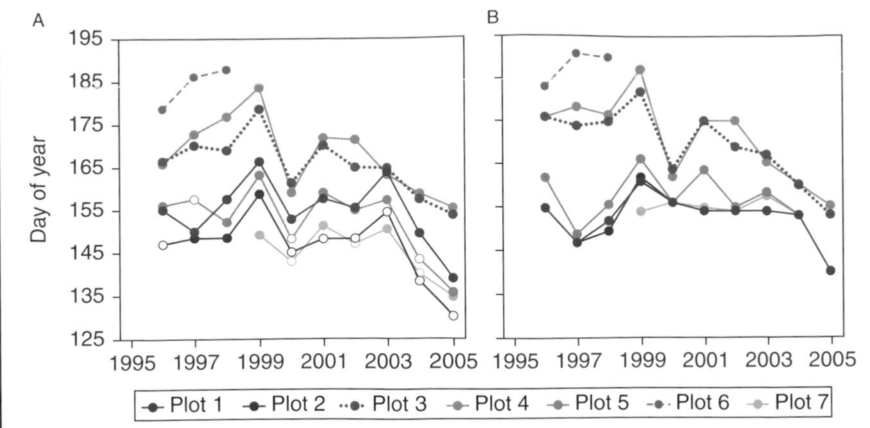

Box Figure 2 (A) Inter-annual variation in the date of ice-melt on the surrounding pond (plot 1) or snowmelt (plots 2–7) and (B) the date of onset of the trapping period (day of the year May 11–July 14). In some years, snowmelt was prior to arrival at Zackenberg and snowmelt was back-estimated from Tiny Tag data loggers (Gemini Data Loggers Ltd., Chichester, UK), snow monitoring images (Hinkler *et al.*, 2002) or estimated from timing of snowmelt in other plots and the relative difference in timing of snowmelt between plots. These observations are indicated by open symbols.

The sampled specimens of Hemiptera were aphids Aphidoidea, scale insects Coccoidea, or the arctic-alpine seed bug *Nysius groenlandicus*. Numbers in all three groups were highly variable among years. Both aphids and scale insects were most numerous in the two plots of late snowmelt (plots 3 and 4), whereas *N. groenlandicus* was caught mainly in the driest plots (plots 5 and 7) during the warmest years. This is in accordance with previous findings for this species (Böcher and Nachman, 2001).

The most common family among Hymenoptera was the Ichneumonidae. Specimens were caught in all plots and all years, but primarily in the fen (plot 2) (Table 1). The second most common group of Hymenoptera was the superfamily Chalcidoidea, and like the arctic-alpine seed bug, this group was caught mainly in 2004 and 2005 (73.4% of the individuals), where the earliest snowmelt of all years was recorded (Box 1, Figure 2).

The majority of the adult Lepidoptera were either arctic fritillary *Clossiana chariclea* or polar fritillary *C. polaris* (Table 1), but arctic clouded yellow *Colias hecla* and species of the family Noctuidae were also found regularly. Although individuals were caught in all plots, only few were caught in the fen area (plots 1 and 2). Arctic clouded yellow was not caught at all in 2001, and since no specimens have been recorded prior to June 24 in other years, most individuals may have been killed by a snowstorm on June 15–16, 2001, as pupae.

Table 1 Numbers of individuals of different taxa caught in seven different plots (see Box 1 for characteristics of the plots)

Order or phylum	Family or species	Plot 1	Plot 2	Plot 3	Plot 4	Plot 5	Plot 6	Plot 7	Sum
Acari		3748	42,731	10,230	12,514	47,564	2545	14,741	134,073
Aranea	Dictynidae	0	18	75	53	94	1	109	350
	Linyphiidae	77	13740	806	1088	1661	689	456	18517
	Lycosidae	17	2329	7363	5504	7506	653	4161	27533
	Thomisidae	0	8	157	220	526	8	390	1309
	Unidentified	0	0	1	2	2	0	0	5
Collembola		876	15,182	24,361	30,566	62,520	1954	12,460	147,919
Diptera	Agromyzidae	146	68	16	22	62	0	66	380
	Anthomyiidae	142	848	548	347	650	106	162	2803
	Calliphoridae	40	77	113	138	95	5	122	590
	Cecidomyiidae	47	7	8	16	21	1	17	117
	Ceratopogonidae	5954	136	54	117	44	–	84	6389
	Chironomidae	61,067	23,238	2572	4289	2688	1413	2036	97,303
	Culicidae	1603	68	97	94	42	9	50	1963
	Cyclorrhapha larvae	0	82	5	9	71	0	107	274
	Empididae	137	6	10	73	11	5	17	259
	Muscidae	10,184	19,602	13,077	14,255	12,889	2973	9858	82,838
	Mycetophiliidae	123	646	66	66	268	353	216	1738
	Nematocera larvae	3	488	37	36	51	3	13	631
	Phoridae	7	9	69	45	704	98	3462	4394
	Scatophagidae	105	543	14	39	14	1	9	725
	Sciaridae	2008	881	2337	1308	1307	4230	1872	13,943
	Syrphidae	76	76	94	91	104	25	73	539
	Tachinidae	34	19	43	54	95	5	73	323
	Other[a]	14	20	4	27	30	8	37	140

Hemiptera	Aphidoidea	14	554	1210	651	93	4	644	3170
	Coccoidea	17	19	2513	1406	272	22	652	4901
	Nysius groenlandicus	21	32	11	9	224	0	323	620
Hymenoptera	Bombus sp.	30	8	27	34	32	1	13	145
	Braconidae	3	30	71	66	153	9	54	386
	Chalcidoidea	21	90	957	392	341	1	236	2038
	Ichneumonidae	421	1838	939	747	948	475	983	6351
	Scelionidae	0	1	180	88	34	0	116	419
	Other[b]	2	11	6	13	20	3	65	120
Lepidoptera	Nymphalidae	24	161	587	721	500	293	273	2559
	Colias hecla (Pieridae)	21	109	79	80	91	21	69	470
	Lepidoptera larvae	0	24	73	59	282	13	214	665
	Noctuidae	13	4	115	106	109	24	79	450
	Other[c]	9	9	4	14	16	0	6	58
Other[d]		33	172	5	14	11	0	3	238
Sum		87,037	123,920	69,187	75,617	142,393	15,954	54,388	567,644

Plots 6 and 7 were in operation only in the years 1996–1998 and 1999–2005, respectively. All other plots were operated in the years 1996–2005. Numbers in parentheses in footnotes indicate number of specimens recorded within taxa. Three pairs of Diptera families were not separated in all years (Anthomyiidae/Muscidae, Chironomidae/Ceratopogonidae and Mycetophilidae/Sciaridae). In these groups, the total number of individuals in each family was estimated from the proportion in each family in years when they were separated. See legend to Figure 2 for details.

[a]Fanniidae (1), Heleomyzidae (25), Piophilidae (3), Tipulidae (60), Trichoceridae (18), Tipulidae larvae (21), Brachycera larvae (11).

[b]Ceraphronoidea (61), Cynipoidea (30), Megaspilidae (5), Tenthredinidae (1), Hymenoptera larvae (21), *Symphyta* sp. larvae (2).

[c]Lycaenidae (37), Geometridae (19), *Plebeius franklinii* (28), Tortricidae (2).

[d]Coleoptera (*Latridius minutus*) (2), Nematoda (9), Ostracoda (152), Siphonaptera (3), Tardigrada (6), Thysanoptera (62).

Zackenbergdalen hosts a total of eight species of spiders: one wolf spider, *Pardosa glacialis*; one Dictynidae, *Emblyna borealis*; five species of Linyphiidae, *Collinsia thulensis*, *Hilaira vexatrix*, *Erigone arctica*, *Erigone psychrophila* and *Mecynargus borealis*; and one Thomisidae, *Xysticus deichmanni* (Larsen and Scharff, 2003). Across years they have been sorted to family level, but in 1999, all spiders were sorted to species level, and a distinction between juveniles and adults was made (Larsen, 2001). The majority of the Linyphiidae were caught in the fen (plot 2) and, based on the data from 1999, most of these belonged to *E. psychrophila*. There was a tendency towards a bimodal distribution of Linyphiidae across the season in all plots, but not equally strong across years. This indicates that adult males die after mating early in the season and juveniles make up a larger proportion of the total capture in the latter part of the season. In corroboration, the bimodal peak was also observed in 1999, where juveniles and adults were separated and the second peak consisted almost entirely of juveniles. *P. glacialis* was mainly caught in plots 3, 4, 5 and 7, and the numbers caught in the fen (plot 2) varied widely between years. Since this plot is much wetter than the other plots and *P. glacialis* generally is less abundant, low soil moisture could be important for the distribution in this species. Indeed, in years of late snowmelt the capture rates of *P. glacialis* in plot 2 were smaller.

III. ENVIRONMENTAL DRIVERS OF ARTHROPOD PHENOLOGY AT ZACKENBERG

Most studies of the phenology of arctic arthropod species have embraced only few years and have been limited to few families of Diptera: Chironomidae (Danks and Oliver, 1972; Danks, 1978; Hodkinson *et al.*, 1996) and Culicidae (Corbet and Danks, 1973), and one species of Hemiptera (Böcher, 1976). A key to this paucity of studies is probably that collecting data over entire seasons and across multiple years in remote areas with limited infrastructure is costly. In addition, there are methodological challenges associated with the tracking of phenological variation in arthropods (Box 2). Hence, little is known about the environmental drivers or abiotic conditions governing phenology of arctic arthropods (but see Høye *et al.*, 2007b).

To characterise the seasonal development of capture numbers and, hence, the phenology of arthropods at Zackenberg, we quantified the date when 25, 50 and 75% of the annual catch was reached (henceforth termed date25, date50 and date75, respectively) for each plot, year and taxon following Corbet and Danks (1973). The timing of capture varied widely across arthropod taxa (Figure 1). For instance, the difference in timing of Anthomyiidae among plots was very pronounced and suggests that different species

Box 2

Phenology Inferred from Pitfall and Window Trap Data

It is well-known that capture numbers from passive traps like pitfall and window traps are a function of both population density and locomotory performance (Southwood and Henderson, 2000). Specifically, in the seasonal arctic environment, density is closely related to the phenological development, and locomotory performance is related to short-term weather fluctuations (Høye and Forchhammer, 2005; Høye and Forchhammer, submitted). The phenological development is typically a bell-shaped function of time through the summer, whereas weather may change on a daily basis. Therefore, we expect capture numbers to vary on both a seasonal and a daily time scale. We have previously used non-linear generalized additive models (GAM; Woods, 2006) to statistically separate the effects of density and short-term weather (Høye and Forchhammer, submitted). The advantage of the GAM modelling approach is that the explanatory power of weather parameters can be assessed statistically by simultaneously taking phenology of the organisms into account without *a priori* assumptions about the exact shape of the phenological development through the summer (Høye and Forchhammer, submitted).

On the basis of both daily and weekly sampling, we constructed four candidate models with each of four weather variables (temperature, solar radiation, wind and precipitation) and a non-linear spline function of capture date. We did this for each taxon in each trapping season in each plot. In this way we quantified the variance explained by a non-linear term estimating the seasonal development and a linear term estimating the effect of short-term weather fluctuations (Høye and Forchhammer, submitted). Lycosidae displayed the most pronounced response to short-term weather fluctuations based on these results. However, the most important weather variable (solar radiation) only explained on average 14.8% of the variation in capture numbers of Lycosidae after accounting for the seasonal development. For Linyphiidae, Muscidae and Chironomidae, the most important weather variable alone explained on average 7.1%, 6.5% and 5.9% of the variation in capture numbers, respectively. At the same time, the variance explained by the non-linear spline function ranged between 71% and 84%. The effect of weather fluctuations was slightly higher in the data set from daily sampling. This indicates that short-term fluctuations in weather do influence locomotory activity, and that this is reflected in variation in capture numbers from pitfall traps. However, the variation arising through changes in locomotory activity is much smaller than the seasonal variation. Therefore, it is expected to have a limited effect on estimates of the timing of the phenological development based on the pitfall and window trap data (Høye and Forchhammer, submitted).

predominate in plots 2, 5 and 7 compared to the plots 3, 4 and 6. In contrast, Nymphalidae, which is represented only by two species, exhibited very little variation in date50 across plots (Figure 1). It is also clear that Lepidoptera and Hymenoptera generally appear later than most families of Diptera (Figure 1),

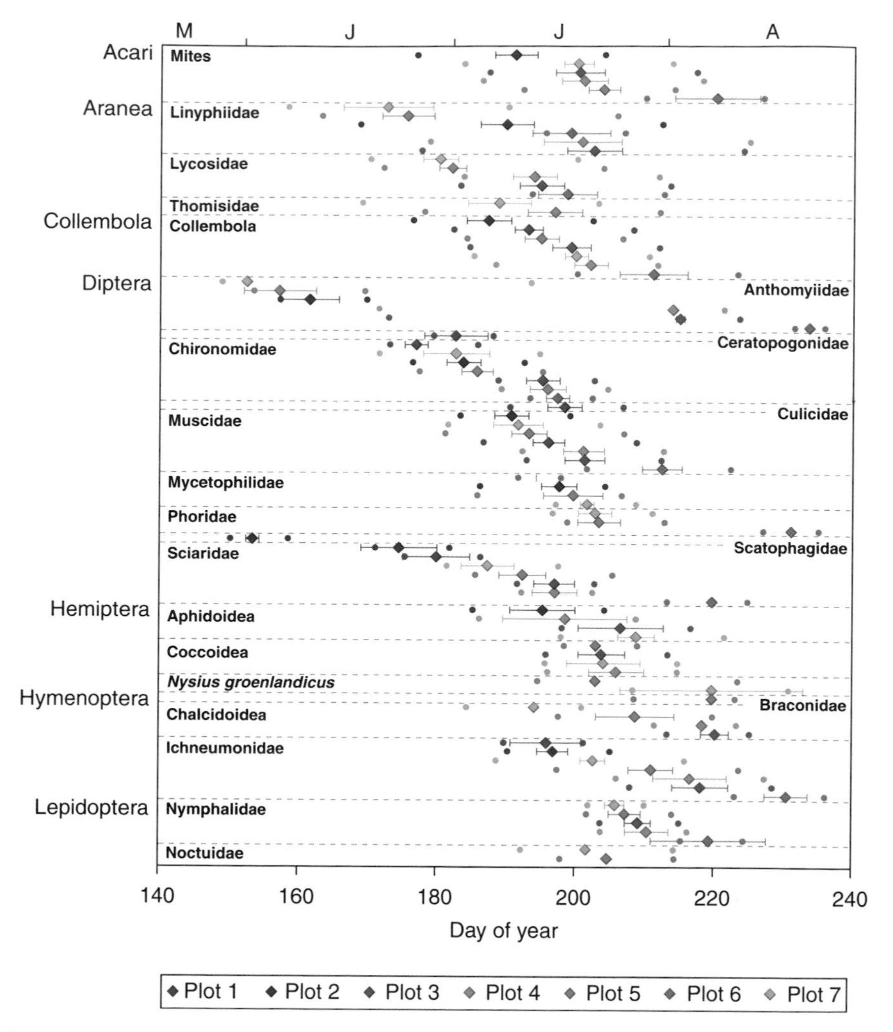

Figure 1 The phenology of all taxa represented by at least 50 individuals per season for each plot (day of year 140–240). Diamonds mark when 50% of the seasonal capture was reached (date50) and dots represent the average date at which 25 (date25) and 75% (date75) of the total capture within one season was reached. Error bars spanning 1 SE of date50 are given when estimates are based on averages across multiple years.

which probably relates to taxa-specific differences in the timing of resource availability and opportunities for reproduction. In some taxa, for example, mites, collembolans and spiders, the life stages are not entirely separated during the annual cycle and therefore individuals can be found throughout the summer season. This is evident in a longer time span between date25 and date75 in these groups (Figure 1).

We identified nine taxa, which were adequately represented in the samples, to allow for detailed statistical analysis of phenological patterns: Chironomidae, Muscidae, Sciaridae, Ichneumonidae, Nymphalidae, Linyphiidae, Lycosidae, collembolans and mites (Figure 2). Clearly, data at the species level would have been ideal, but from an ecosystems perspective information about the phenology and spatial variation of specific taxa is still valuable, especially from the rarely studied High Arctic fauna. We cannot rule out that the species composition within each taxon changes through the season or varies between trapping plots, but given the rarity of this kind of data from the region, our higher-taxon approach provides a necessary starting point. Also, there is a trade-off between taxonomic resolution and the number of individuals caught within each taxonomic unit. Even if the resources had been available, it is likely that sorting the entire data set to the species level would result in many species being represented only by few specimens. A statistical treatment of these species would therefore not have been possible. Hence, the benefit of high taxonomic resolution needs to be weighed against the statistical power of small data sets on each taxonomic unit.

For each of the nine taxa, we analysed the phenological variation in relation to the timing of snowmelt and temperature. We used linear models with date50 as the phenological response contrasted with plot-specific dates of snowmelt and average air temperature in the month proceeding plot-specific snowmelt as predictors. We found that snowmelt was a better predictor of phenological variation than temperature in all taxa (results not shown). Consequently, we proceeded by analysing the effect of timing of snowmelt only and plot number on the date50. This set of linear models demonstrated a significant effect of snowmelt in all taxa except collembolans and Linyphiidae (Table 2). In addition there was a significant effect of plot in mites, collembolans, Chironomidae, Linyphiidae and Lycosidae, indicating that the duration from snowmelt to date50 displayed significant spatial variation in these taxa (Table 2).

Average air temperature at Zackenberg peaks by the end of July, and since the development rate of arthropods is likely to be temperature-dependent, the duration from snowmelt to date50 (i.e., the development period from over-wintering to adult stage) may be faster in late melting plots. To investigate this further, we calculated the number of days from plot-specific date of snowmelt to date50 for each plot and year for all nine taxa. Then we analysed the effect of average temperature and plot for the duration of this development period. In most taxa, the duration of the development differed significantly between plots, and in Nymphalidae, Ichneumonidae and Muscidae warmer development periods were significantly shorter (Table 3).

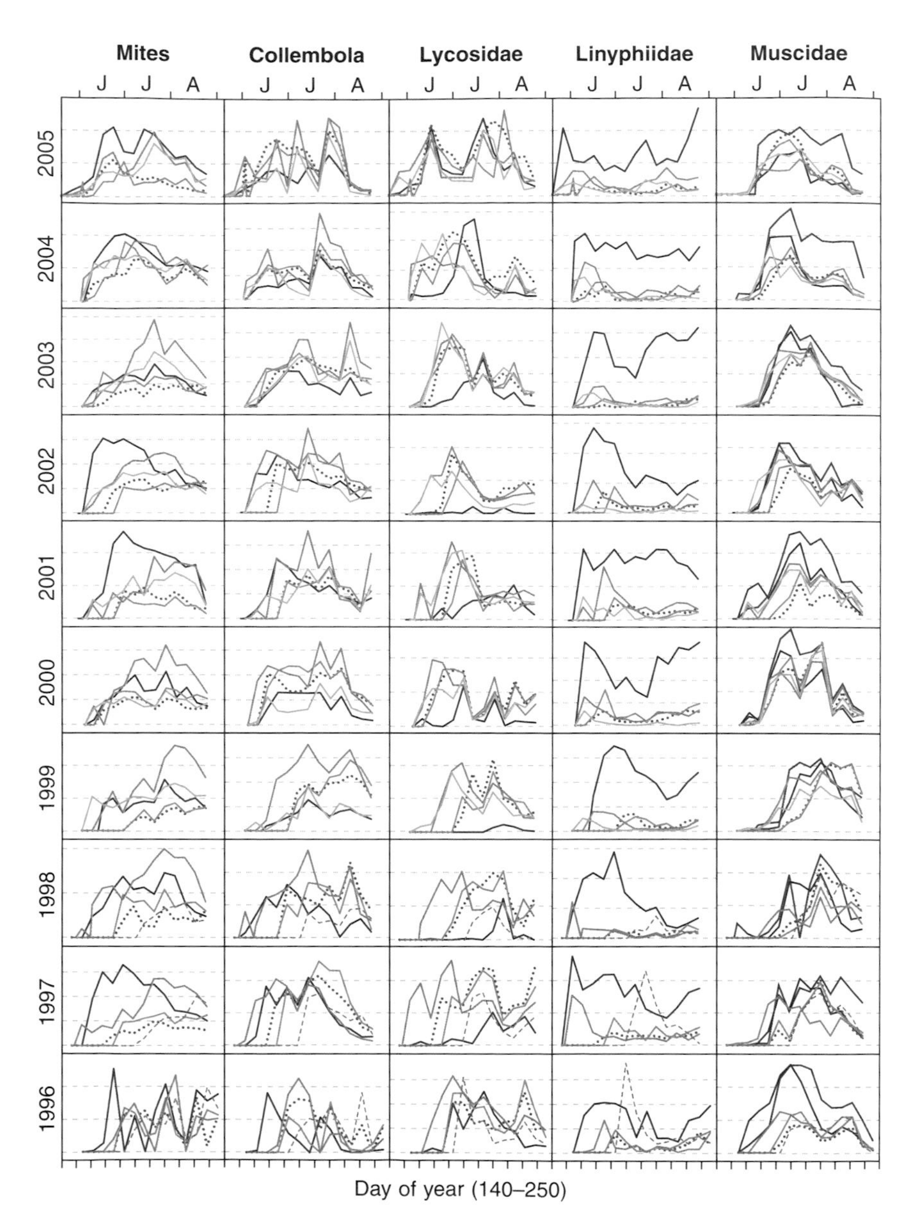

Figure 2 (*continued*)

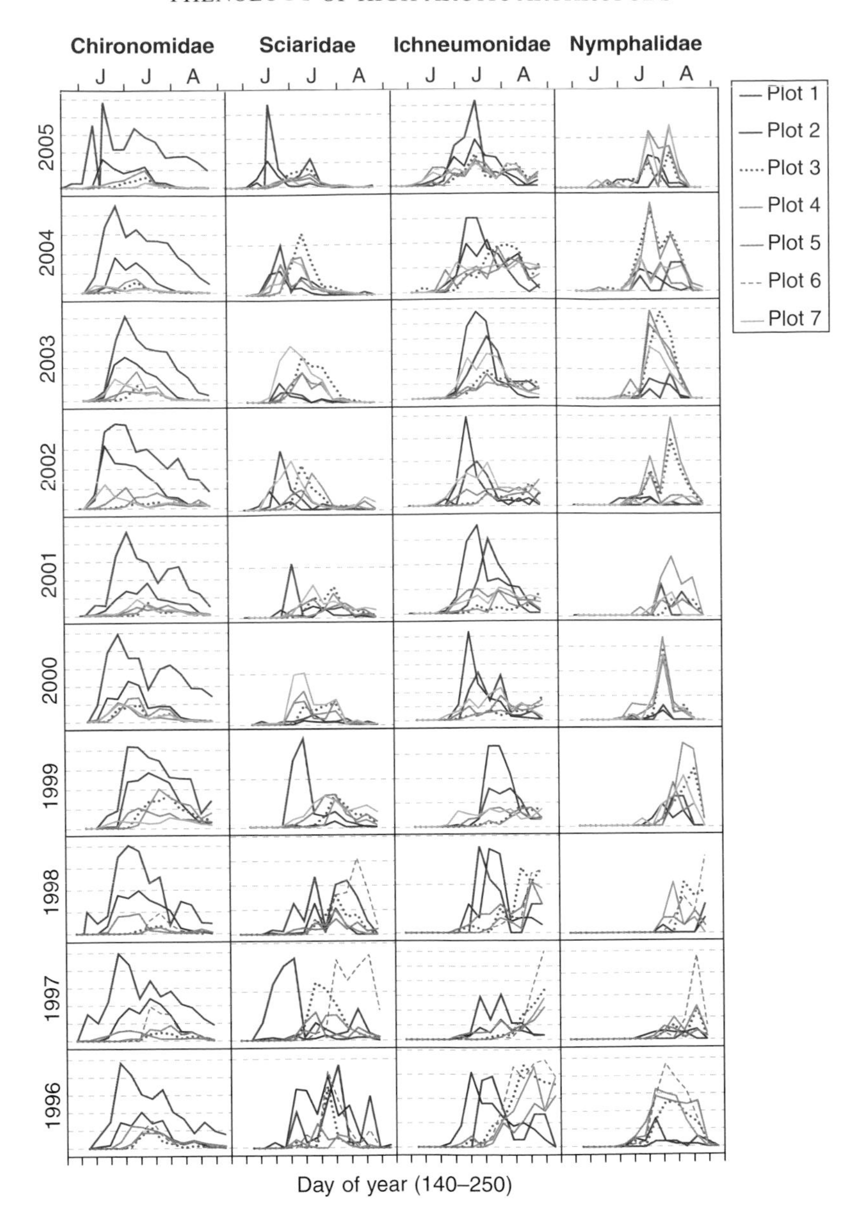

Figure 2 The $\log_{10}$-transformed number of individuals in nine taxa caught through the season (day of year 140–250) in the different trapping plots for the years 1996–2005. Day of year refers to the date when the traps were emptied and the value for any given day represents the catch for the preceding time interval (1 week). Horizontal gridlines are given in increments of 0.5 except in Linyphiidae where they indicate 0.25, and in Nymphalidae and Ichneumonidae where they indicate 0.1. Each sub-panel gives data for all plots in 1 year and each line within a sub-panel represents data from

Although this indicates that air temperature affects the development of arthropods, it is also possible that the relationship is due to individuals moving between plots. However, if organisms emerge mainly from the early melting plots and disperse to late melting plots, the duration between snowmelt and date50 would be shorter in late melting plots. There was an indication of the duration of development to be shorter in late melting plots in Lycosidae, but it was not a general pattern (Table 3). Hence, movement of individuals from early to late melting areas does not seem to be important.

IV. SPATIAL SYNCHRONY IN CAPTURE NUMBERS

Very mobile species of arthropods could easily move between the trapping plots included in this study. This suggests that capture numbers may be closely correlated between plots. Yet, the terrestrial ecosystem at Zackenberg consists of a mosaic of different vegetation types (Bay, 1998), and this could on the contrary reduce the spatial synchrony of captures even between individual traps within a plot. Also, variation in capture numbers is more related to the phenological development than to short-term weather fluctuations (Box 2), and the phenological development is closely coupled to timing of snowmelt, which is spatially variable. Hence, taxonomic differences in the degree of spatial synchrony of capture numbers within and among plots could indicate the responsible processes governing variation in capture numbers in each taxon.

To compare the degree of spatial correlation, we used permutation methods (Quinn and Keough, 2002) to draw sub-samples from the available

---

one trapping plot. In the two spider families Linyphiidae and Lycosidae, juveniles and adults were not separated except for the year 1999, and the ratio of juveniles to adults is not constant across the season (T.T. Høye, unpublished data). Juveniles of Lycosidae cling to the abdomen of the mother during the first week after hatching (Böcher, 2001). Thus, trapping a female wolf spider with juveniles on its abdomen leads to the capture of 50 individuals or more. Likewise, bumblebees can host large numbers of mites on their body and the capture of a mite-infested bumblebee in a trap can result in the simultaneous capture of several hundred mites. After 1998, the capture numbers from individual traps were recorded individually and the spikes in the number of wolf spiders and mites in some traps were easily recognized. For wolf spiders and mites, we changed these spike numbers to the average of the other traps from the same capture period. Ceratopogonidae were not separated from Chironomidae in 1996, 1997, 1998 and 2000, Mycetophilidae were not separated from Sciaridae in 1996 and Anthomyiidae were not separated from Muscidae in 1996 and 2000. Across the years when these families were separated, Ceratopogonidae, Mycetophilidae and Anthomyiidae constituted 6.23, 8.88 and 3.24% of each pair, respectively. Hence, we used pooled numbers of each family pair in years when they were not separated.

Table 2 Summary of final reduced multiple regression models of the date of 50% of annual capture (date50) for nine different arthropod taxa

	SNOW		PLOT									
TAXON	COEF	SE	1	2	3	4	5	6	7	RES DF	R^2	*P*-VALUE
Acari	0.734	0.1293	–	−10.29	−14.81	−16.01	−1.61	−10.87	0.00	38	0.65	<0.0001
Chironomidae	0.706	0.1173	−9.60	2.80	0.87	−0.03	−1.64	−10.23	0.00	48	0.70	<0.0001
Nymphalidae	0.458	0.1088	–	–	–	–	–	–	–	16	0.53	0.0007
Collembola	–	–	–	−12.62	−0.72	2.12	−5.04	11.21	0.00	44	0.39	0.0004
Ichneumonidae	0.819	0.1125	–	–	–	–	–	–	–	42	0.56	<0.0001
Linyphiidae	–	–	–	−4.93	10.58	8.32	−17.83	0.00	–	20	0.47	0.0102
Lycosidae	0.874	0.1133	–	–	−3.46	−6.44	−4.99	−16.15	0.00	34	0.79	<0.0001
Muscidae	0.564	0.0667	–	–	–	–	–	–	–	58	0.55	<0.0001
Sciaridae	0.729	0.0966	–	–	–	–	–	–	–	37	0.61	<0.0001

Observations from years and plots where less than 100 individuals were caught are omitted (less than 50 individuals for Nymphalidae and Ichneumonidae). Full models included date of snowmelt (SNOW), plot (PLOT), and their interaction. Insignificant terms were removed based on *F*-test of type 3 sums of squares ($\alpha = 0.05$). Regression coefficients for date of snowmelt (COEF) with one standard error of mean (SE) and plot (indicated by plot number) if they remained in reduced models are provided. Residual degrees of freedom (RES DF), R^2 and model *p*-values (*P*-VALUE) are given.

Table 3 Summary of final reduced multiple regression models of the number of days between plot-specific date of snowmelt and date of 50% of annual capture (date50) for nine different taxa of arthropods

	TEMPERATURE		PLOT									
TAXON	COEF	SE	1	2	3	4	5	6	7	RES DF	R^2	*P*-VALUE
Acari	–	–	–	–11.10	–20.32	–22.54	–3.54	–22.43	0.00	39	0.66	<0.0001
Chironomidae	–	–	–11.93	2.80	–4.55	–5.93	–3.96	–21.30	0.00	49	0.51	<0.0001
Nymphalidae	–4.38	0.916	–	–	–6.30	–7.69	–2.90	–13.70	0.00	12	0.90	<0.0001
Collembola	–	–	–	–14.54	–21.54	–21.14	–12.74	–28.48	0.00	44	0.39	0.0004
Ichneumonidae	–4.10	1.436	–8.85	–9.18	4.35	3.91	3.85	–0.48	0.00	36	0.35	0.0203
Linyphiidae	–	–	–	–	–	–	–	–	–	–	–	–
Lycosidae	–	–	–	–	–6.05	–9.31	–5.96	–21.11	0.00	35	0.52	<0.0001
Muscidae	–3.46	0.747	–	–	–	–	–	–	–	58	0.27	<0.0001
Sciaridae	–	–	–11.80	–0.64	–10.29	–13.59	–0.73	–9.18	0.00	32	0.41	0.0065

Observations from years and plots where less than 100 individuals were caught are omitted (less than 50 individuals for Nymphalidae and Ichneumonidae). Full models included average temperature in the interval between snowmelt and date50 (TEMPERATURE), plot (PLOT) and their interaction. Insignificant terms were removed based on *F*-test of type 3 sums of squares ($\alpha = 0.05$). Regression coefficients for temperature (COEF) with one standard error of mean (SE) and plot (indicated by plot number) if they remained in reduced models are provided. Residual degrees of freedom (RES DF), R^2 and model *p*-values (*P*-VALUE) are given.

observations of spatial correlation within and among plots. In general, captures of all taxa were highly synchronous within plots (Figure 3A "Intra") with butterflies Nymphalidae and parasitoid wasps Ichneumonidae exhibiting the lowest synchrony. The degree of spatial synchrony among plots varied considerably between taxa (Figure 3A). In addition, the inter-plot correlation differed markedly between pairs of plots. In particular, the correlation was high between the pairwise neighbouring plots 1 and 2, 3 and 4, and 5 and 7 for all taxa (Figure 3A). We did not find any relation to sample size of the data sets from which the sub-samples were drawn, and the result was not sensitive to the exclusion of the largest capture rates. However, we did find that for surface-dwelling arthropods, the spatial synchrony decreased with distance between plots (Figure 3B). For flying insects, synchrony was not related to distance between plots except for Sciaridae (Figure 3B). In this group, different species may be found in different habitats, and they may have different phenological patterns. In the other groups of flying insects, the distances between plots are probably short compared to their range, and differences in spatial synchrony may be more related to timing of snowmelt or the spatial distribution of resources than to dispersal. We found evidence of this for Chironomidae and Muscidae. Capture numbers in these groups were more closely correlated between plots differing by less than 10 days in average date of snowmelt than between plots differing by more than 10 days in average date of snowmelt.

Summarising, the high intra-plot synchrony strongly suggests that average capture rates for each plot adequately describes variation in capture rates of individual traps. Among plots, there was considerable variation in synchrony of capture numbers among pairs of plots. Captures of collembolans, Muscidae and Chironomidae showed the strongest correlation between plots, and the degree of spatial synchrony may to some extent be a function of geographical distance between plots in surface-dwelling arthropods and related to spatial variation in timing of snowmelt in flying arthropods. However, because of the limited taxonomic resolution, differences may to some extent be the result of different species composition among plots and trapping periods.

V. DISCUSSION AND CONCLUSION

The arthropod data set used in this study is probably the most extensive from the entire Arctic, but limitations to the taxonomic resolution prevent a detailed phenological description of the arthropod species at Zackenberg. Nevertheless, compared to studies carried out in other high-arctic locations, such as Taimyr, Siberia (Tulp and Schekkerman, 2007), and Barrow, Alaska (MacLean and Pitelka, 1971; MacLean, 1980), the low capture rate of crane

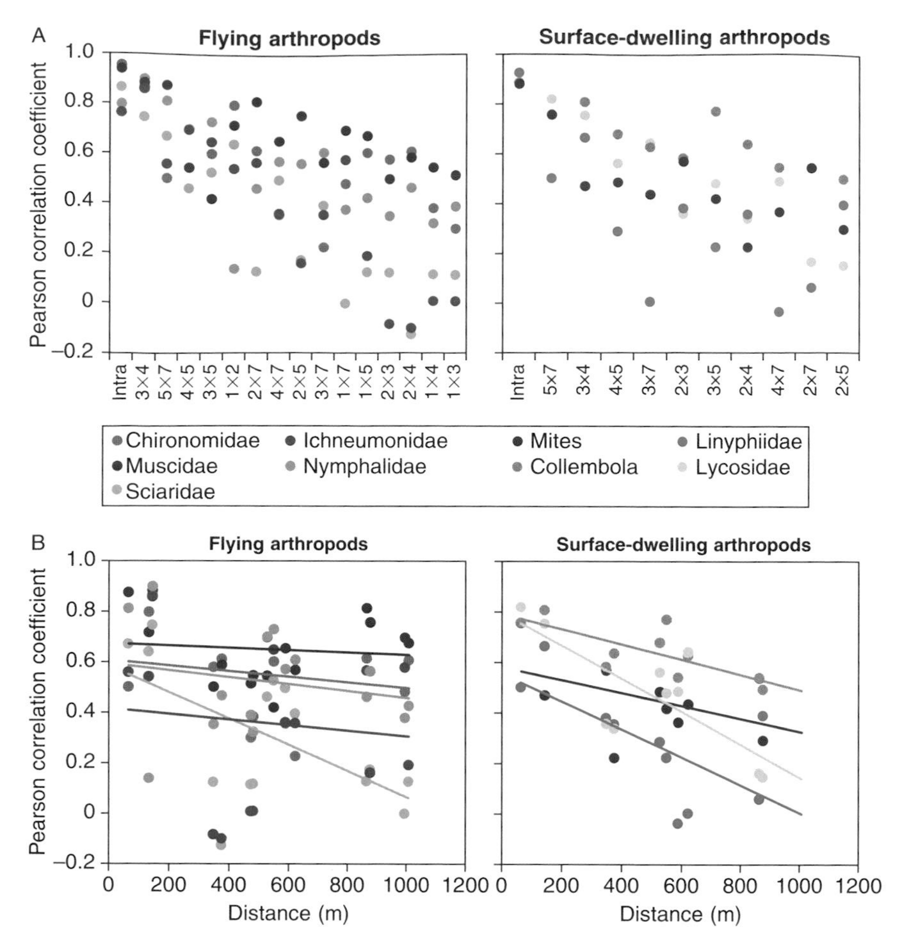

Figure 3 (A) Intra- and inter-plot spatial synchrony of five taxa of flying insects and four taxa of surface-dwelling arthropods. The eight traps in each plot were divided in two groups of four traps, and the weekly capture numbers averaged in each group produced one pair of observations. In this way, all possible pairs of averages of four and four pitfall traps ($n = 35$) were calculated, and pairs where both values were zero were excluded. Each taxon is estimated separately and the values given under "Intra" are the average of 1000 permutated Pearson correlation coefficients. Each coefficient is estimated from a sub-sample of 1000 random observation from the full data set of all possible pairs across years and plots. A similar approach was used to quantify inter-plot spatial synchrony. First, capture rates per trap per day for each trapping period was calculated. Subsequently, a data set including all pairwise combinations between any two plots for any trapping period for any year for any taxon was constructed. All pairs including plot 6 were excluded, since this plot was in operation for 3 years only. All other pairs had at least 50 observations across years and trapping periods, and we calculated Pearson correlation coefficients from 100 random samples of 40 pairwise observations in each taxon. The x-axis gives the plot numbers for each pairwise correlation, for example, "3×4" between plot 3 and plot 4. (B) The relation between inter-plot synchrony and distance between pairs of plots is given separately for flying and surface-dwelling taxa. Regression lines are based on the least squares method.

flies (Tipulidae) at Zackenberg is striking. This family of insects is believed to form an important part of the diet of waders and other insectivorous birds breeding on the tundra in Alaska (Holmes, 1966) and Siberia (Tulp and Schekkerman, 2007). In contrast, the capture rate of Muscidae and spiders appears to be high at Zackenberg compared to Siberia and Alaska. Unfortunately, the sampling procedures differ between studies at these three sites. However, the low occurrence of Tipulidae at Zackenberg has been demonstrated by several different trapping techniques (Meltofte and Thing, 1997; Schmidt and Høye, 2006).

Our results demonstrate that the phenology, that is, the timing of occurrence of terrestrial arthropods at Zackenberg differed considerably among taxa and plots. For instance, the average date when 50% of the annual capture was reached (date50) varied by more than 2 months across taxa (Figure 1). This suggests that different taxa are clearly timed to different parts of the season. At the same time, date50 of the most abundant taxa was significantly related to date of snowmelt (Table 2). In fact, date of snowmelt was a better predictor of date50 than temperature, and only three of the nine taxa included in our analyses (Nymphalidae, Ichneumonidae and Muscidae) had significantly faster development (i.e., the duration from snowmelt to date50) in years where the temperature during this interval was higher (Table 3). We cannot rule out that the species composition within taxa differed between plots, and that this could be responsible for plot differences in duration of development. Alternatively, in some taxa, peak capture rates could reflect increased activity due to mate location behaviour (e.g., Lycosidae), which could be cued to a specific time during the season (i.e., a specific date). In mites and collembolans, the peak in capture numbers during the season was less pronounced than in other groups, but even so, date50 was clearly related to plot-specific timing of snowmelt, which demonstrates that local climatic conditions determine a large part of the inter-annual variation in the phenological development of high-arctic arthropods. Therefore, each taxon may have an optimal period of occurrence, but variation in timing of snowmelt may modify the exact timing of date50 in any given year. In addition, movement of individuals from over-wintering areas to summer habitats could affect estimates of phenology. For example, most Chironomidae over-winter as larvae in ponds, whereas adults can be found in many other habitats (Danks and Oliver, 1972; Danks, 1978). We found that the spatial synchrony of capture numbers is very high at the local scale (within each plot). At the landscape scale (between plots), synchrony varies between taxa. Spatial synchrony at this scale seems to be related to the distance between plots, but may also be influenced by differences in the date of snowmelt between plots. This is most clearly seen in the surface-dwelling species, probably because of their more limited mobility than their flying counterparts (Figure 3). Finally, weather-related locomotory activity as well as changes in population density affects the number of individuals caught in a trap. A statistical separation of the variation in capture rates in one component related to density and another component related to activity demonstrated that although weather does influence activity, its importance is small relative to the phenological development. We conclude that the local phenology is clearly discernable in our data from pitfall and window traps. Specifically, the timing of snowmelt appears to be the most important predictor of phenology of arthropod taxa in high-arctic Greenland. This may be different from other parts of the Arctic, where a more maritime summer climate prevails, for example, on the northern coast of Siberia, where temperature appears to play a much larger role in invertebrate occurrence and activity (Schekkerman *et al.*, 2004).

Although few details are known about the foraging ecology of the arthropod taxa at Zackenberg presented in this study, they belong to several functionally distinct groups. The Diptera are mainly detritus feeders as larvae. As adults, the Chironomidae have limited energetic demands, since the main functions of their adult stage are reproduction and dispersal. Although adult Muscidae are generalists, they may have some importance as pollinators (Philipp *et al.*, 1990; Elberling and Olesen, 1999; Lundgren and Olesen, 2005), similarly to the Nymphalidae. While both collembolans and most mites feed on fungal hyphae, bacteria, or directly on dead organic matter, some mite species are predators or parasites (Böcher, 2001). To some extent spiders are cannibalistic, but judged from the prey selection of Linyphiidae and Lycosidae in general (Nyffeler, 1999), collembolans, mites and Diptera probably form considerable parts of their diet at Zackenberg. This means that there is at least some association between the different taxa in terms of inter-trophic interactions. Mites, collembolans and Diptera families are primary consumers and probably not very dependent on seasonally fluctuating resources. In contrast, Nymphalidae are dependent on timing of flowering, Ichneumonidae are dependent on butterfly larvae among which Nymphalidae species may be important, and spiders are predators on mites, collembolans and small species of Diptera (Böcher, 2001). The largest potential for trophic mismatch in the taxa treated in this study is therefore between pollinators and specific flower species and between parasitoids and lepidopteran larvae. There may, however, also be a risk of mismatch between wader chicks and insect prey (Meltofte *et al.*, 2007).

Over the decade of observations from Zackenberg, the phenology of plants, arthropods and birds has advanced considerably (Høye *et al.*, 2007b). This corresponds with an advancement of the timing of snowmelt in the sampling plots. However, the rates of changes vary greatly within and among groups of organisms. In particular, the trend is generally stronger in areas of late snowmelt than in areas of early snowmelt (Høye *et al.*, 2007b). Timing of snowmelt is predicted to become more variable between years, due to increased variability in winter precipitation and spring temperatures

(Stendel *et al.*, 2008, this volume). Because of snow accumulation this may be particularly pronounced in snow-beds, where snowmelt may occur later than now (Hinkler *et al.*, 2008, this volume). This could have significant effects for arthropods, since late-emerging species seem to be closely inter-trophically timed to the occurrence of their resources on lower trophic levels, for example, butterflies to flowering and parasitoid wasps to presence of host species for egg-laying. Recent studies from Zackenberg indicate that flower abundance in snow-beds will be reduced by later snowmelt (Høye *et al.*, 2007a) with potential negative consequences for pollinators. Overall, the phenology of abundant taxa within Diptera, Hymenoptera, Lepidoptera and spiders are all closely coupled to the timing of snowmelt (Table 2), which suggests that inter-annual variation in date50 of arthropods may increase particularly in areas of late snowmelt.

The consequences for trophic mismatch may be larger in the taxa occurring during a narrow period of the summer season (short phenological range) than in taxa occurring over a larger part of the season. At the same time, if there is no inter-annual variation in timing of occurrence in interacting species, the risk of trophic mismatching is likely to be small. The variation in timing of occurrence is larger in areas of late snowmelt than in areas of early snowmelt at Zackenberg. Furthermore, this difference between areas of

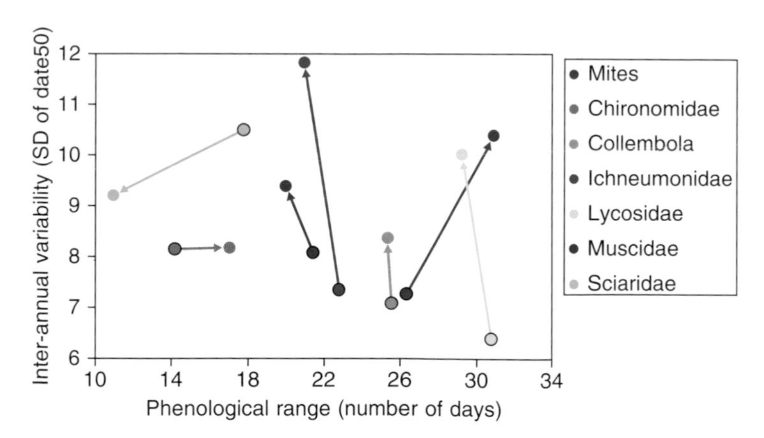

Figure 4 The relation between the phenological range (the average number of days between the date when 25 [date25] and 75% [date75] of the annual capture of a taxon in a given plot is reached) and the inter-annual variation in the date when 50% of the annual capture (date50) is reached. For each taxon, the observations from early (plots 1, 2, 5 and 7) and late (plots 3, 4 and 6) melting plots are separated and the arrow between pairs of points highlights the direction and magnitude of change in the two parameters from early (dots with black circle) to late (dots with no black circle) melting plots. Linypiidae and Nymphalidae were excluded since they were not present in both early and late melting plots.

early and late snowmelt is most evident in taxa with long phenological range (Figure 4). This suggests that although these groups probably are less sensitive to trophic mismatch, because they occur over a large part of the season, they are also the most responsive groups to climatic conditions in late melting areas. A climatic shift towards more variable snowmelt in late melting areas make trophic mismatch most likely in these areas. Climatic changes may in the short term result in a change in the suitability of certain areas as habitat for arthropods, because early and late melting areas will be affected differently. On a longer time scale, altered composition of vegetation types may radically shift patterns of insect biodiversity. Indeed, climate-mediated changes in emergence patterns, abundance and diversity may have strong repercussions through the ecosystem.

ACKNOWLEDGMENTS

Monitoring data for this chapter were provided by the BioBasis programme, run by the National Environmental Research Institute, University of Aarhus, and financed by the Danish Environmental Protection Agency, Danish Ministry of the Environment. The Danish Polar Center provided logistic support at Zackenberg Research Station. We thank Jens Böcher, Hans Meltofte, Niels M. Schmidt, Ingrid Tulp, Hans Schekkerman and one anonymous reviewer for comments on earlier versions of the manuscript and Mikkel P. Tamstorf for providing the map of the study area.

REFERENCES

Bale, J.S., Masters, G.J., Hodkinson, I.D., Awmack, C., Bezemer, T.M., Brown, V.K., Butterfield, J., Buse, A., Coulson, J.C., Farrar, J., Good, J.E.G., Harrington, R., *et al.* (2002) *Globe Change Biol.* **8**, 1–16.

Bay, C. (1998) *Vegetation mapping of Zackenberg valley, Northeast Greenland.* Danish Polar Center & Botanical Museum, University of Copenhagen. 29 pp.

Böcher, J. (1976) *Vidensk. Medd. Dansk Naturh. Foren.* **139**, 61–89.

Böcher, J. (2001) *Insekter og andre smådyr—i Grønlands fjeld og ferskvand.* Atuagkat, Copenhagen.

Böcher, J. and Nachman, G. (2001) *Entomol. Exp. Appl.* **99**, 319–330.

Butler, M.G. (1982) *Can. J. Zool.* **60**, 58–70.

Corbet, P.S. and Danks, H.V. (1973) *Can. Entomol.* **105**, 837–872.

Danks, H.V. (1978) *Can. Entomol.* **110**, 289–300.

Danks, H.V. (1981) *Arctic arthropods. A review of systematics and ecology with particular reference to the North American fauna.* Entomological Society of Canada, Ottawa.

Danks, H.V. (1992) *Arctic* **45**, 159–166.

Danks, H.V. (2004) *Integr. Comp. Biol.* **44**, 85–94.

Danks, H.V. and Oliver, D.R. (1972) *Can. Entomol.* **104**, 661–686.
Dollery, R., Hodkinson, I.D. and Jónsdottir, I.S. (2006) *Ecography* **29**, 111–119.
Downes, J.A. (1964) *Can. Entomol.* **96**, 279–307.
Elberling, H. and Olesen, J.M. (1999) *Ecography* **22**, 314–323.
Forchhammer, M.C. and Post, E. (2004) *Popul. Ecol.* **46**, 1–12.
Hammel, J.U. (2005) *Ökologie und Phylogenie gröenländischer Wolfsspinnen (Lycosidae, Araneae).* MSc thesis, University of Stuttgart, 123 pp.
Hinkler, J., Pedersen, S.B., Rasch, M. and Hansen, B.U. (2002) *Int. J. Remote Sens.* **23**, 4669–4682.
Hodkinson, I.D. and Bird, J. (1998) *Arctic Alpine Res.* **30**, 78–83.
Hodkinson, I.D. and Bird, J.M. (2004) *Ecol. Entomol.* **29**, 506–509.
Hodkinson, I.D., Coulson, S.J., Webb, N.R., Block, W., Strathdee, A.T., Bale, J.S. and Worland, M.R. (1996) *Oikos* **75**, 241–248.
Hodkinson, I.D., Webb, N.R., Bale, J.S., Block, W., Coulson, S.J. and Strathdee, A.T. (1998) *Arctic Alpine Res.* **30**, 306–313.
Holmes, R.T. (1966) *Ecology* **47**, 32–45.
Høye, T.T. and Forchhammer, M.C. (2005) In: *Zackenberg Ecological Research Operations, 10th Annual Report, 2004* (Ed. by M. Rasch and K. Caning), pp. 70–72. Danish Polar Center, Ministry of Science, Technology and Innovation, Copenhagen.
Høye, T.T., Ellebjerg, S.M. and Philipp, M. (2007a) *Arct. Antarct. Alp. Res.* **39**, 412–421.
Høye, T.T., Post, E., Meltofte, H., Schmidt, N.M. and Forchhammer, M.C. (2007b) *Curr. Biol.* **17**, R449–R451.
Jónsson, E., Gardarsson, A. and Gíslason, G. (1986) *Freshwater Biol.* **16**, 711–719.
Kutz, S.J., Hoberg, E.P., Polley, L. and Jenkins, E.J. (2005) *Proc. R. Soc. Lond. B* **272**, 2571–2576.
Larsen, S. (2001) In: *Zackenberg Ecological Research Operations, 6th Annual Report, 2000* (Ed. by K. Caning and M. Rasch), pp. 54–56. Danish Polar Center, Ministry of Research and Information Technology, Copenhagen.
Larsen, S. and Scharff, N. (2003) *Entomol. Medd.* **71**, 53–61.
Lensing, J.R., Todd, S. and Wise, D.H. (2005) *Ecol. Entomol.* **30**, 194–200.
Lundgren, R. and Olesen, J.M. (2005) *Arct. Antarct. Alp. Res.* **37**, 514–520.
MacLean, S.F., Jr (1980) In: *An Arctic ecosystem: The coastal tundra at Barrow, Alaska* (Ed. by J Brown, P.C Miller, L.L Tieszen and F.L Bunnell), pp. 411–457. Dowden, Hutchinson & Ross, Inc., Stroudsburg, PA.
MacLean, S.F., Jr and Pitelka, F.A. (1971) *Arctic* **24**, 19–40.
Meltofte, H. and Berg, T.B. (2006) *BioBasis—Conceptual design and sampling procedures of the biological programmeme of Zackenberg Basic.* National Environmental Research Institute, Denmark. 77 pp.
Meltofte, H. and Thing, H. (1997) *Zackenberg Ecological Research Operations.* Danish Polar Center, Ministry of Research and Information Technology, Copenhagen.
Meltofte, H., Høye, T.T., Schmidt, N.M. and Forchhammer, M.C. (2007) *Polar Biol.* **30**, 601–606.
Morewood, W.D. and Ring, R.A. (1998) *Can. J. Zool.* **76**, 1371–1381.
Nyffeler, M. (1999) *J. Arachnol.* **27**, 317–324.
Philipp, M., Böcher, J., Mattsson, O. and Woodell, S.R.J. (1990) *Meddr. Grønland, Biosci.* **34**, 1–60.
Pickavance, J.R. (2001) *J. Arachnol.* **29**, 367–377.

Quinn, G.P. and Keough, M.J. (2002) *Experimental design and data analysis for biologists*. Cambridge University Press, Cambridge.

Schekkerman, H., Tulp, I., Calf, K.M. and de Leeuw, J.J. (2004) *Studies on breeding shorebirds at Medusa Bay, Taimyr, in summer 2002*. Wageningen, Alterra. 101 pp.

Schmidt, N.M. and Høye, T.T. (2006) In: *Zackenberg Ecological Research Operations, 11th Annual Report, 2005* (Ed. by A.B. Klitgaard, M. Rasch and K. Caning), pp. 94–95. Danish Polar Center, Ministry of Science, Technology and Innovation, Copenhagen.

Southwood, T.R.E. and Henderson, P.A. (2000) *Ecological methods*. Blackwell Science, Oxford.

Stenseth, N.C. and Mysterud, A. (2002) *Proc. Natl. Acad. Sci. USA* **99**, 13379–13381.

Strathdee, A.T. and Bale, J.S. (1998) *Annu. Rev. Entomol.* **43**, 85–106.

Sørensen, L.I., Holmstrup, M., Maraldo, K., Christensen, S. and Christensen, B. (2006) *Polar Biol.* **29**, 189–195.

Tulp, I. and Schekkerman, H. (2007) In: *Environmental forcing on the timing of breeding in long-distance migrant shorebirds*. I. Tulp, PhD thesis, University of Groningen.

Visser, M.E., van Noordwijk, A.J., Tinbergen, J.M. and Lessells, C.M. (1998) *Proc. R. Soc. Lond. B* **265**, 1867–1870.

Woods, S.N. (2006) *Generalized additive models—An introduction with R*. Chapman & Hall/CRC.

Effects of Food Availability, Snow and Predation on Breeding Performance of Waders at Zackenberg

HANS MELTOFTE, TOKE T. HØYE AND NIELS M. SCHMIDT

SUMMARY

The first few weeks after arrival on the tundra in late May and early June appear to be the most critical period in the summer schedule of arctic-breeding waders. Food availability and snow-cover determine population densities and timing of egg-laying, and early egg-laying seems essential, since it increases the chances for re-laying in case of nest failure, optimises timing of hatching of the chicks in relation to the peak period of arthropod food for the young, facilitates early departure of the adults and maximises the time available for the young to grow strong before winter begins in early September.

Conditions for waders in most of high-arctic Greenland seem favourable as compared to several other arctic areas, in that the climate is continental with favourable weather conditions during most summers, and the predation pressure on eggs and chicks is normally moderate. With the projected climate change, the waders of high-arctic Greenland may face more unstable breeding conditions, and in the long term some of the wader species may be hampered by overgrowing of the high-arctic tundra with more lush low-arctic vegetation.

ADVANCES IN ECOLOGICAL RESEARCH VOL. 40 0065-2504/08 $35.00
© 2008 Elsevier Ltd. All rights reserved
DOI: 10.1016/S0065-2504(07)00014-1

I. INTRODUCTION

Arctic tundra is hardly an optimal habitat for adult "arctic" waders. They apparently benefit much more from feeding on rich intertidal coasts or inland wetlands in temperate and tropical areas, where they spend 9–11 months of the year, and where they often find extreme densities of fleshy bivalves and crustaceans (Meltofte, 1996; van de Kam *et al.*, 2004). On the tundra, they have to survive mostly on tiny arthropods, which may abound, but also periodically be very limited, particularly early in the season (Meltofte *et al.*, 2007a).

However, the estimated total of 30 million "arctic" waders (CHASM, 2004) can not breed in temperate and tropical wetlands because of high diversity and density of predators and poor feeding conditions for chicks (Meltofte, 1996). Instead, they migrate to the Arctic to take advantage of the short summer abundance of tundra arthropods and the more limited diversity and density of predators than on southern latitudes.

But the breeding conditions for the waders in the Arctic are constrained by several factors, such as timing of snowmelt and food abundance in spring, poor weather during chick growth in summer and the need for adults and juveniles to leave the Arctic as soon as possible, at least before winter begins in September (Meltofte *et al.*, 2007a). Hence, the adults of most species seem to minimise their stay in the Arctic, particularly so in the High Arctic.

II. THE CRITICAL PRE-NESTING PERIOD AND INITIATION OF EGG-LAYING

Arctic waders depend on rich feeding grounds on their final staging areas in temperate climates to be able to carry out the often thousands of kilometres of non-stop flight to their arctic breeding grounds (van de Kam *et al.*, 2004). However, such optimal areas are most often so far away from the arctic breeding grounds that the weather here has little or no predictive value for progress of spring on the tundra. This means that the waders have to time their arrival in the Arctic according to average acceptable conditions here. In high-arctic Greenland, this is in late May and early June when daily maximum temperatures reach positive values (Meltofte, 1985), and the spring migration is initiated at very much the same date from year to year in the individual birds (e.g., Battley, 2006).

Not only do arctic waders have to build up body stores for their long terminal flights to the Arctic, they also have to secure sufficient body stores for a transformation of organs from "flying mode" to "breeding mode" and for the first critical period on the tundra (Meltofte *et al.*, 2007a).

In preparation for their long flights, waders develop larger flight muscles, while other organs like the digestive system shrink. During their first days on the tundra, waders rebuild their digestive system and "reorganise" other organs in preparation for the breeding season (Piersma *et al.*, 1999). Furthermore, they have to carry sufficient body stores as an insurance against periods of inclement weather like snowfall and hard wind upon arrival. On top of this, female birds have to obtain local nutrients for the production of a clutch, normally of four eggs, which has a total volume of between 50% and 100% of the weight of the female bird herself (Klaassen *et al.*, 2001; Morrison and Hobson, 2004; Box 1). Finally, both mates have to initiate establishment of body stores for the incubation period (Meltofte *et al.*, 2007a).

Box 1

Arctic waders are "income breeders"

Geese and other large waterbirds often carry so rich body stores to the Arctic in the form of fat and proteins that they need few resources from the breeding grounds, before the female can produce a clutch of eggs (Drent and Daan, 1980). They are so-called "capital breeders." This implies that they can lay eggs from a few days after arrival on the often more or less snow-covered breeding grounds.

Analyses of egg and newly hatched chick proteins show that arctic waders are not capable of doing this (Klaassen *et al.*, 2001; Morrison and Hobson, 2004). At Zackenberg, we collected small samples of neck down from 166 freshly hatched chicks from 58 broods besides a few unhatched eggs of common ringed plover, red knot, sanderling, dunlin and ruddy turnstone in 1999 and 2000, which together with similar samples from waders in arctic Canada were analysed for carbon stable-isotope ($^{13}C/^{12}C$) ratios (Klaassen *et al.*, 2001). Carbon isotope ratios in live organisms are markedly different between marine and terrestrial habitats, and since adult waders of the species considered spend virtually all their non-breeding time in staging and wintering areas along marine coasts, the isotope ratios of their eggs and thereby natal down of chicks would reflect this, if the resources for the production of eggs derived from coasts. However, when the waders arrive in the Arctic during late May and early June, the coasts are covered in thick ice, and they feed exclusively on tundra resources, primarily arthropods. In accordance with this, the isotope ratios of natal down were similar to those of juvenile feathers later grown on the tundra and distinctly different from isotope rations of adult feathers grown during the adults' stay on marine coasts (Box Figure 1): they are "income breeders."

(continued)

Box 1 *(continued)*

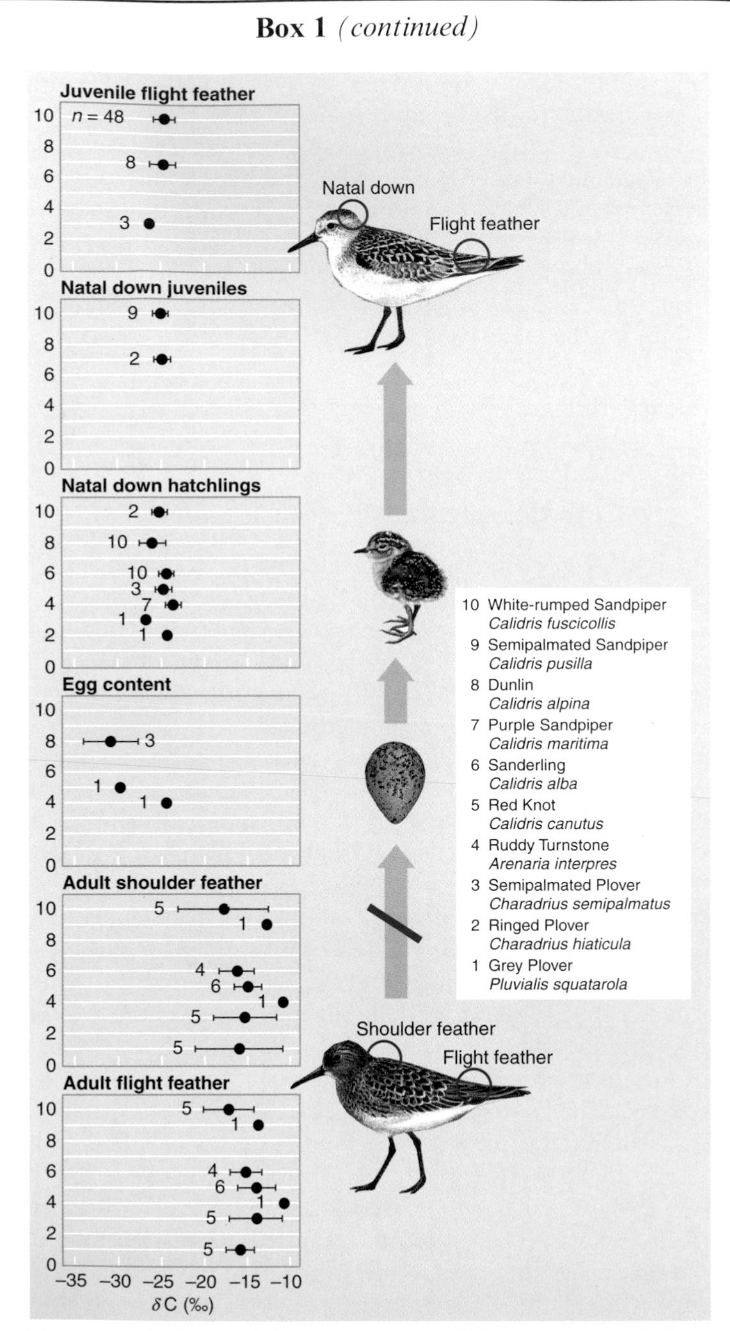

Box Figure 1 Carbon stable isotope ratios (δC) of eggs, natal down and feathers of different species of arctic-breeding waders at different times during the year. All samples were collected in arctic Canada and Northeast

Greenland in 1999 and 2000. Feathers collected from nest-attending adults were grown in winter (adult primary flight feathers) and during spring migration (adult shoulder feathers; data are averages across individuals ±SD). Eggs were collected from deserted nests (egg content), natal down from hatchlings (data are averages across clutches ±SD per species). Data for natal down still attached to the tips of head and neck feathers of freshly hatched chicks, and for secondary flight feathers from independent young are averages across individuals ±SD (reproduced from Klaassen, 2003 with kind permission from Springer Science and Business Media).

The two sampling years at Zackenberg were very much different, in that 1999 had the latest snowmelt recorded during our study years, while 2000 was very early (see Figure 1). Accordingly, egg-laying in waders was much delayed in 1999 as compared to 2000 and other early seasons since then. A minimum of 5–8 days are needed after the arrival of the adult waders on the tundra until they can produce eggs (Roudybush *et al.*, 1979), and this was very much so in 2000, while in 1999, they had to spend 1–2 weeks more on the tundra, before egg-laying could commence (see Figure 1). This means that the waders had been feeding two to three times as long on the tundra in the late breeding season as in the early. But this did not result in any difference in isotope ratios between the years. This means that even in the season with early egg-laying, the resources for the eggs derived from the tundra.

That the pre-breeding period may be critical to arctic waders is further supported by the finding that the four common wader species at Zackenberg used 75–92% of their daytime hours to feed during pre-nesting in an optimal year (Meltofte and Lahrmann, 2006). For comparison, temperate breeding lapwings *Vanellus vanellus* used only between 18% (males) and 32% of their pre-nesting daytime for feeding (Zöllner, 2002), and feeding time used during pre-nesting at Zackenberg was higher or in the high end of what was found in waders on wintering and spring pre-migratory fattening sites (Hötker, 1995, 1999; Kirby, 1997; Ntiamoa-Baidu *et al.*, 1998; Leon and Smith, 1999; Masero and Pérez-Hurtado, 2001; Shepherd, 2001; Scheiffarth *et al.*, 2002).

All this takes place within the first 1–3 weeks on the tundra, so it was no surprise, when our analyses of the first 10 years of data from Zackenberg revealed that initiation of egg-laying in sanderling *Calidris alba*, dunlin and ruddy turnstone was correlated with food availability and snow-cover during the first weeks after the birds' arrival; the more food and the less snow-cover, the earlier egg-laying (Figure 1; Meltofte *et al.*, 2007b). Food availability turned out to explain more of the inter-annual variation in egg-laying than snow-cover in early spring, but snow-cover apparently had a stronger effect in years where the proportion of snow-free land was below 25%.

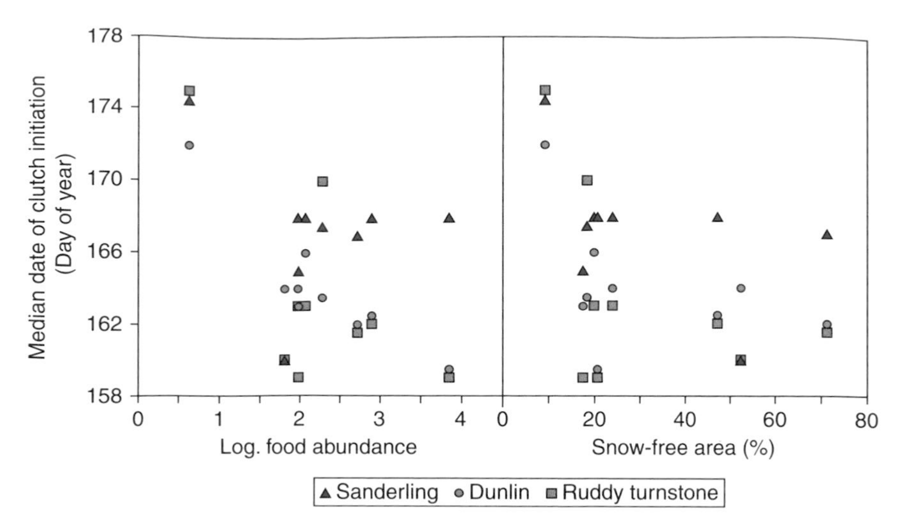

Figure 1 The relationship between median first egg dates (days of the year, June 7–27) of three species of waders and annual food abundance (natural log. transformed number of Diptera and spiders caught between June 3 and 17) and snow-free proportion of land on June 10, respectively, at Zackenberg 1996–2005 (modified from Meltofte *et al.*, 2007b).

III. THE LAYING PERIOD

In sanderling, dunlin and ruddy turnstone, the very first clutches were laid around June 1 and the latest (re-placement clutches) during the first days of July. Median dates varied from around June 8 as the earliest until around June 22 in the latest breeding seasons. Hence, arctic waders seem to breed as early as conditions permit, and a significant impact of spring snow-cover had already been demonstrated by Meltofte (1976, 1985) and Green *et al.* (1977). Snow-cover per se has an impact unrelated to food availability, since many wader territories in the High Arctic are totally snow-covered far into the laying period, preventing the birds from dispersing on the tundra. The waders have to wait not only for patches of land to become snow free but also for the patches to become so extensive that it is not all too easy for the foxes and other predators to scrutinise them for nests (Byrkjedal, 1980). Furthermore, an impact of temperature had been demonstrated by Nol *et al.* (1997), which was interpreted as a temperature effect on food availability.

The very first turnstones and dunlins to initiate their clutches were up to 5 days earlier than the earliest sanderlings (Figure 1). Apparently, a minor part of the sanderling population at Zackenberg performs double-clutching (Meltofte, 2003; Piersma *et al.*, 2006), where the female lay one clutch for a male and then another about a week later for herself. To what extent the later laying dates in this species are the product of such a double-clutching strategy is unknown. Furthermore, the initial laying dates during the last 4 years have been extraordinarily early in that laying dates around June 1 have been recorded only a few times before in high-arctic Greenland (Meltofte, 1985).

Finally, our data confirmed that the total laying period was reduced by up to 2 weeks in late breeding seasons, in that the time around July 1 appears to be the last chance for laying—or re-laying—if the young shall have a chance to fledge and grow strong before winter begins (Meltofte *et al.*, 2007a, Meltofte *et al.*, 2007b). Late egg-laying also resulted in significantly reduced mean clutch sizes in dunlin and ruddy turnstone, but not significantly so in sanderling. This adds to the disadvantage of late breeding, which further includes a more limited time "window" for laying and re-laying in case of failure, increased risk of mismatch between hatching of the chicks and peak occurrence of arthropod food on the tundra, delayed departure of the adults, and reduced time for development of the young before winter begins in September (Meltofte *et al.*, 2007a).

IV. POPULATION DENSITIES AND HABITAT SELECTION

In high-arctic Greenland, wader breeding densities are low, $\leq$2 pairs km^{-2}, in the far north due to the desert-like conditions with often less than 5% vegetation cover and thereby limited food resources (Meltofte, 1976, 1985). Densities are also relatively low, 1–3 pairs km^{-2} but locally up to 10 pairs km^{-2} in the "lush" southernmost parts of Northeast Greenland, where the often long-lasting snow-cover prevents the birds from utilising large expanses of otherwise well-vegetated ground sufficiently early in the season for breeding (Mortensen, 2000). Densities peak with typically 5–10 pairs km^{-2}, but locally up to 16 pairs km^{-2} in central Northeast Greenland, where there is extensive vegetation cover in many places and moderate snow-cover in early spring. Hence, densities are best correlated with the ratio of snow-free, vegetated ground in early June (Meltofte, 1985; Mortensen, 2000). This emphasises the importance of pre-breeding feeding conditions already pointed out above. Pre-breeding feeding conditions determine not only the timing of egg-laying but also population densities. Hence, local population sizes seem to be determined by the amount of land, which allow egg-laying in early/mid June. In such areas, egg-laying may then be delayed until after mid June in year with late snow-cover (Meltofte, 1985; Meltofte *et al.*, 2007b).

Six species of waders breed in the 19.3 km^2 bird census area at Zackenberg with an average of about 260–300 pairs in total (Tables 1), and the density of about 14–15 pairs or territories km^{-2} is among the highest recorded in high-arctic Greenland (Table 2; Mortensen, 2000; Meltofte, 2006a). The individual species have different habitat preferences. This is illustrated by the distribution within different sections of the census area (Figure 2, Table 2). Common ringed plovers are found on poorly vegetated gravel expanses in the lowland and particularly in the areas above 300 m a.s.l., but always close to vegetated areas. The high slopes clear early from snow in spring, but they are

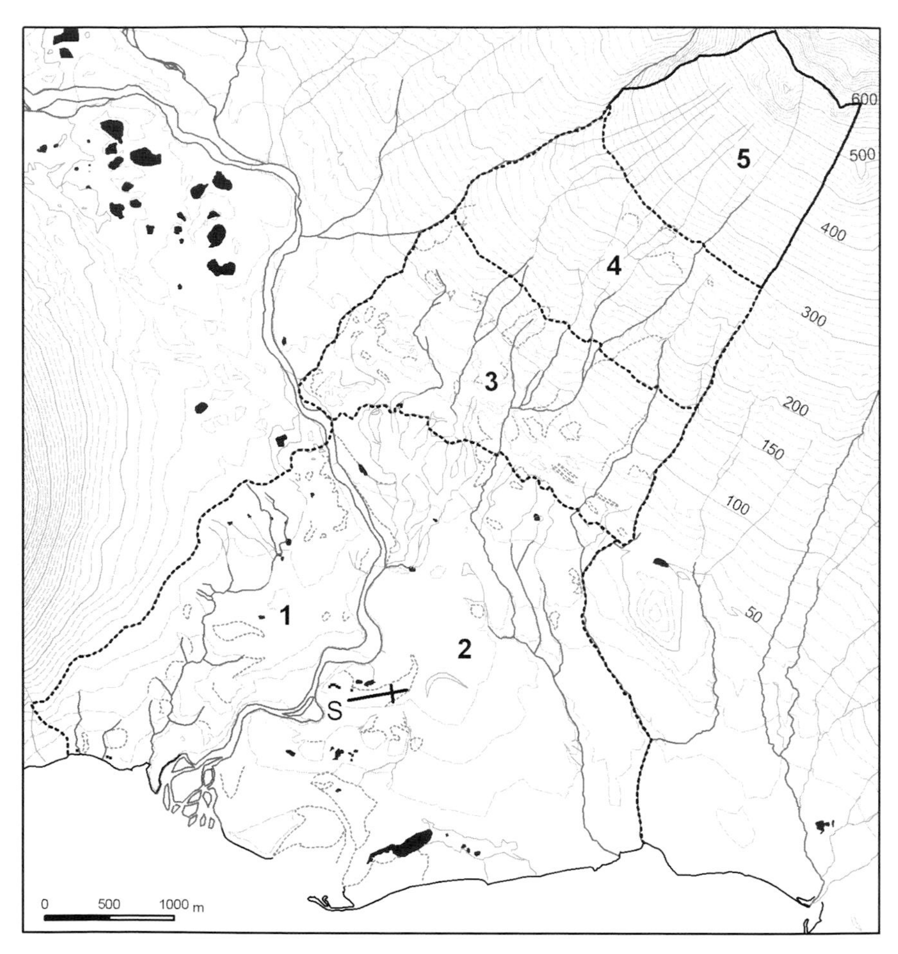

Figure 2 Map of the study area at Zackenberg (74°30′N, 20°30′W) giving demarcation of the individual sections 1–5 together with rivers/streams, lakes/ponds (black) and landscape features (dotted) together with the position of the research station (S) and runway in the lower left centre. Altitudes are given in metres.

often covered in new snow during the summer season, which regularly cover the ground for short spells down to 300–400 m a.s.l. Red knots *Calidris canutus* and sanderlings are mainly found on mesic dwarf shrub heath with low topographic profile—the sanderling apparently on less stony ground than the knot—in the lowlands and particularly on the slopes of Aucella-bjerg. Dunlins breed exclusively in and around wet fens, which primarily are found in the lowlands. Finally, ruddy turnstones breed on gravelly and stony sites, often with hills and ridges. This is why most are found in Oksebakkerne (Musk Ox Hills) between 50 m a.s.l. and 150 m a.s.l., where relatively few sanderlings are found (Meltofte, 2006a). In spite of these differing habitat preferences among the species, the resulting total densities are surprisingly similar below 300 m a.s.l. (Table 2).

According to analyses by Forchhammer *et al.* (2008a, this volume), the inter-annual population variations in common ringed plover, sanderling and dunlin display indications of density dependence, while this is not the case in red knot and ruddy turnstone. Already Meltofte (1985) estimated that the breeding populations of the three former species in high-arctic Greenland were more or less saturated, while the latter two were not. Furthermore, Forchhammer *et al.* (2008a, this volume) analysed that population densities of the ruddy turnstone were negatively affected in years of extensive snow-cover. This again corresponds to previous results for this species in particular (Meltofte, 1985).

V. POPULATION TRENDS

Unlike several other arctic animals, waders at Zackenberg show rather limited year-to-year variation in population sizes (Table 1; Meltofte 2006a). Only common ringed plover and red knot present year-to-year fluctuations of up to a factor 2, while the other species are even more stable. The species involved are relatively long-lived and site tenacious, reappearing in the same territories year after year (Cramp and Simmons, 1983). Even in sanderlings, who show nomadic tendencies in other arctic regions (Tomkovich and Soloviev, 2001), we have recovered adults as well as chicks ringed in our study area in previous years (see also Tomkovich and Soloviev, 1994). Hence, sanderlings were recorded in relatively stable numbers during most study years (Table 1), and the increase seen in dunlins during the first row of study years is probably due to improved census efficiency (Meltofte, 2006a). By contrast, common ringed plovers and possibly even red knots decreased significantly during the study years, while ruddy turnstone showed reduced numbers particularly during 2002–2004. Furthermore, the numbers recorded of red knot and ruddy turnstone showed significant correlation with

Table 1 Annual minimum and maximum estimates of pairs/territories of waders in the 19.3 km^2 bird census area at Zackenberg, 1996–2005

Species	1996	1997	1998	1999	2000	2001	2002	2003	2004	2005
Common ringed plover	54–56	40–48	38–45	51–65	41–43	51–54	37–41	29	35–40	17–20
European golden plover	0	0	0	0	0	1	0	0	0	0
Red knot	33–43	35–44	27–32	25–33	24–27	27–30	24–27	24–25	16–20	30–36
Sanderling	50–63	55–70	62–70	60–67	58–66	58–72	49–55	67–74	61–73	38–49
Dunlin	69–81	75–91	75–94	80–94	98–103	104–111	120–132	105–114	110–122	92–102
Ruddy turnstone	41–51	49–58	56–63	43–49	48–50	45–51	31–37	33–34	45	65–74
Red-necked phalarope	0–1	0–2	1–2	1–2	1–2	1–2	1–2	1–2	1	1
Red phalarope	0	0	0–1	0	0	1	0	0	0	1

Source: Data from Meltofte (2006a).

Table 2 Area size (km^2) and average population densities (means based on annual averages of minimum and maximum estimates of pairs/territories) of waders in five sections of the 19.3 km^2 bird study area at Zackenberg, 1996–2005

Section	Area	*C. hia.*	*C. can.*	*C. alb.*	*C. alp.*	*A. int.*	*All*
5 (300–600 m)	2.24	4.78	0.31	2.28	0.58	0.04	7.99
4 (150–300 m)	2.51	4.04	2.51	5.00	0.88	0.94	13.37
3 (50–150 m)	3.33	1.23	2.66	1.13	4.31	5.29	14.62
2 (0–50 m)	7.77	1.18	1.38	3.67	7.08	2.98	16.29
1 (0–50 m)	3.47	2.19	0.73	3.16	7.41	1.47	14.96
Total	19.32	2.15	1.53	3.16	5.10	2.51	14.45

Note: See map Figure 2 for position of sections.
Source: Data from Meltofte (2006a).

July temperatures 2 years earlier, which suggest an influence of chick survival on population fluctuations 2 years later, when these chicks mature (Meltofte, 2006a; see Section VI.A below).

These conclusions are not without reservations, however, since arctic waders are notoriously difficult to census (Meltofte, 2001a, 2006a), and we need more data to confirm these results. If the decline in common ringed plover is valid and is of more than local bearing, then it corresponds to decreasing numbers found on the wintering grounds of this population in West Africa (Meltofte, 2006a). The same apply to ruddy turnstone, which has been declining on the Northwest European wintering ground during the last decade (Stroud *et al.*, 2004). The most obvious problems having appeared during the study years are three problematic breeding seasons (1999–2001), with extremely extensive and long-lasting snow-cover in 1999, and genuine snowstorms in July and June 2000 and 2001, respectively. In 1999, between one half and two thirds of the ruddy turnstone population apparently did not lay eggs, and the snowstorms in 2000 and 2001 killed many eggs and young (Meltofte, 2000, 2001b, 2003). Taken together, these problematic years could have had an effect, particularly on ruddy turnstone numbers, during the following years.

Few other population change data exist for waders in high-arctic Greenland. In a bird census area at Danmarkshavn Weather Station about 265 km north of Zackenberg, most populations remained much the same between three counts (1969–1975) and four counts (1986–1989) (Boertmann *et al.*, 1991). However, ruddy turnstones halved between the two periods (from 14–17 pairs to 5–9 pairs), and the species did not breed at all in this area in 1907–1908. This is noteworthy, since indications of similar marked population changes have also been found in this species in other parts of high-arctic Greenland (Meltofte, 1985). Turnstones are sensitive to early spring snow-cover, and they are not found in very snow-rich areas. Similar requirements may exist for the red knot, but here we have even less data (Meltofte, 1985).

Other changes in wader populations in high-arctic Greenland involve European golden plover *Pluvialis apricaria* and whimbrel *Numenius phaeopus*, who apparently began to breed in southern Northeast Greenland a few decades ago, possibly as a result of climate amelioration (Meltofte, 1985; Meltofte *et al.*, in press).

VI. BREEDING SUCCESS

Breeding success is hard to measure in arctic waders. Nest success is the most easy parameter to record, but an unknown part of the failures is caused by visits by research workers at the nests, providing olfactory cues for predators like arctic foxes (Tulp *et al.*, 2000). Fledging success is next to impossible to quantify in widely scattered populations, since wader chicks are precocial, leaving the nest within a day after hatching and then wandering widely over the tundra. However, particularly long-tailed skuas *Stercorarius longicaudus* may exert an important predation pressure on wader chicks (Maher, 1970; de Korte and Wattel, 1988).

At Zackenberg, nest success shows no correlation with fox activity as measured by numbers of fox encounters by bird census workers on the tundra during June–July or the number of fox dens with pups in the area (see Schmidt *et al.*, 2008, this volume), and generally waders had more stable reproductive success than the almost "eruptive" breeding success in divers, waterfowl and long-tailed skuas at Zackenberg (Meltofte, 2006b; Meltofte and Høye, 2007). Neither did we find the close negative correlation between alternative prey for the foxes, the lemmings and predation, as has been demonstrated in so many other areas (Meltofte *et al.*, in press). Only in 2004 did we see the classical situation of many lemmings, many foxes and little nest depredation (Thorup and Meltofte, 2005). Nest depredation for all waders pooled has been around 50–60% in most years, which is moderate as compared to many other arctic areas (Meltofte *et al.*, 2007a).

Neither have we seen many clear signs of chick mortality due to inclement weather during chick growth in July, as has been found in other high-arctic areas (Meltofte *et al.*, 2007a). Only in 1997 did we see so unfavourable conditions that we found examples of reduced growth in wader chicks (Meltofte, 1998). The obvious explanation is the more continental climate of Northeast Greenland than, for example, northernmost Siberia, where cold spells with strong wind, sleet and rain are much more common (Tulp and Schekkerman, 2007). Still, a positive correlation was found between temperatures in July, when most chicks grow up, and population size in red knot and ruddy turnstone 2 years later, when these chicks mature (see above). Also in the Siberian Arctic, a significant positive correlation was found between July temperatures and juvenile production (Schekkerman *et al.*, 1998; Soloviev *et al.*, 2006).

Most years we see plenty of juvenile waders on the tundra and on the coasts in August (Hansen and Meltofte, 2006). Only red knots seem to disappear without us being able to observe where they go—or the local population produces very few juveniles (see below).

A. The Red Knot as an Example of Varying Breeding Success

Besides breeding success data from the breeding grounds, the percentage of juveniles in the populations on the wintering grounds serves as an indicator of breeding success (Robinson *et al.*, 2005). Unfortunately, among Greenland waders such data are available only for red knots wintering in Northwest Europe (Boyd and Piersma, 2001). To be able to back-calculate juvenile percentages on the wintering grounds into successful pairs on the tundra, we have used average numbers of fledged juveniles in broods still accompanied by adults. At Zackenberg, we have encountered 70 such wader broods during 1995–2005 with samples of at least 10 broods for sanderling (mean 1.6, $N = 18$), dunlin (mean 1.8, $N = 10$) and ruddy turnstone (mean 2.1, $N = 32$). In red knot, we have six broods averaging 2.5 juveniles, and for all 70 wader broods combined (also including common ringed plovers), the average is 1.93 young produced per successful pair. Here we ignore that some sanderlings may double-clutch, and hence double their annual production.

Little is known about juvenile mortality from fledging to the time of sampling on the wintering grounds. To establish a theoretical calculation of the fraction of successful pairs during different years, we have used estimates of 33% and 50%, respectively (Table 3), and an estimated further mortality of 50% from sampling to their second summer (2 years of age) and thereby supposed arrival on the breeding grounds. Based on Boyd and Piersma (2001), who found that the ratio of juvenile red knots in catches of Nearctic birds wintering in the Wash in southeast England varied from 12.1% in a period of population decline (1969–1977), over 13.8% in a relatively stable period (1985–1995) to 28.8% in a period with population growth (1977–1985), we have used three ratios of juveniles in the wintering population, that is, 10%, 20% and 30%. The adult mortality from summer to midwinter was estimated to be 10%.

These calculations (Table 3) indicate that in periods of low juvenile ratios on the wintering grounds, about 80% of the potentially breeding population

Table 3 Exploration of the parameter space that determines the fraction of red knot pairs arriving back on the tundra breeding areas in early spring that successfully fledge two young based on the known range of juveniles ratios recorded in cannon net catches on the British wintering grounds of Nearctic-breeding red knots (see text for explanation)

Assumptions		Outcome
Post-fledging mortality	Midwinter-juvenile percentage	Percentage of pairs that fledged young
0.33	10	16
0.33	20	39
0.33	30	74
0.5	10	21
0.5	20	51
0.5	30	98

Notes: The fraction of young birds that die between fledging on the tundra and sampling on the European wintering areas 1–9 months later is largely unknown and has here been assessed, for example to be either 0.33 or 0.5. Similarly, we assume that half the average annual adult mortality of 0.2 occurs between breeding and midwinter. We further assume that in their second year, birds remain on the wintering grounds, and that half of the young birds reaching the wintering grounds survive to the next winter.

do not produce any fledged young, be it due to breeding failure or non-breeding in part of the individuals. In stating this, we do not think of the few aberrant years with extremely inclement conditions, for example, in the form of exceptionally late snowmelt or snowstorms in the middle of the breeding season (see, e.g., Meltofte, 1985, 2001b, 2003; Ganter and Boyd, 2000), but of the lengthier periods of poor survival and reproduction described by Boyd and Piersma (2001). Of the examples of 20% and 30% juveniles in the wintering population, the calculations for "20%" seem most realistic. In such years about 40–50% of the population would produce fledged juveniles, which point to a situation where most of the pairs lay eggs and suffer the "normal" predation and other failure. On the contrary, the calculations for 30% juveniles seem to give unrealistically high predictions of breeding success—at least with a 50% juvenile mortality. Such high juvenile percentages may be due to overrepresentation of juveniles in the winter catches or too high estimates of juvenile mortality.

These examples may support the possibility that red knots have been reproducing poorly at Zackenberg during recent years, thus providing an explanation for the reduced population in some of the last years. Total midwinter population sizes as recorded in Northwest Europe do not indicate a decline on population level, however (Stroud *et al.*, 2004).

VII. BREEDING CONDITIONS IN HIGH-ARCTIC GREENLAND IN A CIRCUMPOLAR PERSPECTIVE AND IN THE FUTURE

As mentioned above, the breeding conditions for waders in high-arctic Greenland are generally more favourable than, for example, in the Siberian High Arctic. The climate is continental, with the exception of a narrow fringe of cool outer coasts. This means that the weather is generally favourable with few spells of inclement weather during the breeding season. Also, few years have so much snow and so late snowmelt that it prevents a large proportion of the waders from breeding. Predation on nests and young is moderate in most years, as opposite to the often severe predation pressure experienced by, for example, Siberian waders (Meltofte *et al.*, 2007a). This means that we do not see the same widely fluctuating numbers of juveniles as is so typical for Siberian populations (Meltofte *et al*, in press).

Furthermore, most of the wader species of high-arctic Greenland have a "conservative" breeding strategy with high site and mate faithfulness. In contrast, the Siberian wader assembly holds a high ratio of "opportunistic" species, which breed one year in one area and the next year in another (Tomkovich and Soloviev, 1994).

Also adult mortality seems to be low in high-arctic Greenland, but there are exceptions to this. During a snowstorm in mid June 2001, a pair of arctic skuas *Stercorarius parasiticus* systematically hunted adult waders, of which many may have been exhausted and several may have died from hunger (Meltofte, 2003). Wader populations are generally supposed to be more sensitive to adult survival than to recruitment (e.g. Hitchcock and Gratto-Trevor, 1997), but recruitment may also be relatively important under certain conditions (Ryabitsev, 1993; Troy, 1996; Boyd and Piersma, 2001; Atkinson *et al.*, 2003).

The prospects for high-arctic Greenland waders under future climate scenarios are uncertain. As dealt with by Stendel *et al.* (2008, this volume), snow precipitation will probably increase, but spring temperatures will also increase, possibly balancing each other out, so that snowmelt on wader breeding habitats may not change much on average (Hinkler *et al.*, 2005). Snow accumulation areas will see even more snow accumulation, making them even more unsuitable as wader breeding habitat. And in a year with much more snow in combination with a cold spring, large areas may clear too late for breeding. On top of this comes the possibility even for a more maritime summer climate with more frequent fog, rain and sleet increasing chick mortality. All of this means that the frequency of late and poor breeding seasons may increase significantly.

At the same time, an increase in plant growth is expected in the Arctic (Callaghan *et al.*, 2005). This is likely to involve at least some of the plant communities in high-arctic Greenland (Elberling *et al.*, 2008a, this volume), and more lush vegetation will reduce the quality of heath habitats for most of the waders, in that most high-arctic waders are dependent on very low vegetation (Meltofte, 1985; Mortensen, 2000). On the other hand, more precipitation may result in presently barren gravel expanses, particularly in high-arctic desert areas, being vegetated (Callaghan *et al.*, 2005), thereby increasing the extent of breeding habitat for most waders in the presently poorly vegetated end of the habitat scale. Also the endemic *arctica* subspecies of the dunlin may benefit from more extensive and lush fens.

VIII. CONCLUSIONS

During 1996–2005, an average of between 260 and 300 pairs of six species of waders bred in the Zackenberg bird census area of 19 km^2, and their inter-annual variability in population density together with timing of reproduction and breeding success were monitored. Early spring food availability turned out to be the most important determinant of timing of egg-laying, followed by snow-cover in years with less than average snow-free land in early June. Mean clutch size decreased during June–July and the total length of the laying period was shortened in years of late snowmelt, meaning that the chances for re-laying in case of failure were limited in such years. All this point to reduced breeding success in late breeding seasons. Events of inclement weather and predation, including availability of alternative prey for the predators, had little effect on breeding success in most years.

Densities of breeding waders in high-arctic Greenland are low both in the desert-like north and in the snow-rich south, while higher densities are found in central Northeast Greenland. Here, the balance between spring snow-cover and vegetation cover available for the waders during the critical pre-laying period is more favourable. Hence, densities at Zackenberg are among the highest recorded in Northeast Greenland, and the populations show relatively limited year-to-year variation. Yet, common ringed plover seems to have decreased significantly.

The results are compared with conditions in other parts of the Arctic, focusing on the effects of the more continental climate in high-arctic Greenland than in most other parts of the Arctic. The future of Greenland's high-arctic waders is uncertain. According to climate scenarios, the frequency of poor breeding seasons may increase, and spread of lusher low-arctic vegetation into the high-arctic may reduce the area of breeding habitat for most of the specialised high-arctic species. On the other hand, vegetation may expand onto presently barren expanses and the secondary productivity of the tundra may increase, benefiting some of the waders including the possibility for more southern species to move into the area.

ACKNOWLEDGMENTS

The monitoring data for this chapter were provided by the BioBasis programme, run by the National Environmental Research Institute, University of Aarhus, and financed by the Danish Environmental Protection Agency, Danish Ministry of the Environment. We further thank the Danish Polar Center for access and accommodation at the Zackenberg Research Station during all the years. Dr. Pavel Tomkovich kindly criticised an earlier draft of the manuscript and provided valuable suggestions for improvements.

REFERENCES

Atkinson, P.W., Clark, N.A., Bell, M.C., Dare, P.J., Clark, J.A. and Ireland, P.L. (2003) *Biol. Conserv.* **114**, 127–141.

Battley, P.F. (2006) *Biol. Lett.* doi:10.1098/rsbl.2006.0535.

Boertmann, D., Meltofte, H. and Forchhammer, M. (1991) *Dansk Orn. Foren. Tidsskr.* **85**, 151–160.

Boyd, H. and Piersma, T. (2001) *Ardea* **89**, 301–317.

Byrkjedal, I. (1980) *Ornis Scand.* **11**, 249–252.

Callaghan, T., Björn, L.O., Chapin, III, F.S., Chernov, Y., Christensen, T.R., Huntley, B., Ims, R., Johansson, M., Riedlinger, D.J., Jonasson, S., Matveyeva, N., Oechel, W., *et al.* (2005) *Arctic Climate Impact Assessment*, chapter 8, pp. 243–352. Cambridge University Press, Cambridge.

CHASM (The Committee for HolArctic Shorebird Monitoring) (2004) *Wader Study Group Bull.* **103**, 2–5.

Cramp, S. and Simmons, K.E.L. (1983) *The Birds of the Western Palearctic. Vol. 3. Waders to Gulls*, Oxford University Press, Oxford.

Drent, R.H. and Daan, S. (1980) *Ardea* **68**, 225–252.

Ganter, B. and Boyd, H. (2000) *Arctic* **53**, 289–305.

Green, G.H., Greenwood, J.J.D. and Lloyd, C.S. (1977) *J. Zool. Lond.* **183**, 311–328.

Hansen, J. and Meltofte, H. (2006) In: *Zackenberg Ecological Research Operations, 11th Annual Report, 2005* (Ed. by M. Rasch and K. Caning), pp. 57–68. Danish Polar Center, Ministry of Science, Technology and Innovation, Copenhagen.

Hinkler, J., Hansen, B.U., Tamstorf, M.P. and Meltofte, H. (2005) In: *From digital cameras to large scale sea-ice dynamics,* pp. 159–173, The Faculty of Science, Unpublished Ph.D. thesis by J. Hinkler, University of Copenhagen, Denmark.

Hitchcock, C.L. and Gratto-Trevor, C. (1997) *Ecology* **78**, 522–534.

Hötker, H. (1995) *J. Ornithol.* **136**, 105–126.

Hötker, H. (1999) *J. Ornithol.* **140**, 57–71.

Kirby, J.S. (1997) *Bird Study* **44**, 97–110.

Klaassen, M. (2003) In: *Avian Migration* (Ed. by P. Berthold, E. Gwinner and E. Sonnenschein), pp. 237–249. Springer-Verlag, Berlin Heidelberg.

Klaassen, M., Lindström, Å., Meltofte, H. and Piersma, T. (2001) *Nature* **413**, 794.

de Korte, J. and Wattel, J. (1988) *Ardea* **76**, 27–41.

Leon, M.T.D. and Smith, L.M. (1999) *Condor* **101**, 645–654.

Masero, J.A. and Pérez-Hurtado, A. (2001) *Condor* **103**, 21–30.

Maher, W.J. (1970) *Arctic* **23**, 112–129.

Meltofte, H. (1976) *Meddr Grønland* **205,** 1, 1–57.
Meltofte, H. (1985) *Meddr Grønland, Biosci.* **16**, 1–43.
Meltofte, H. (1996) *Ardea* **84**, 31–44.
Meltofte, H. (1998) In: *Zackenberg Ecological Research Operations, 3rd Annual Report, 1997* (Ed. by H. Meltofte and M. Rasch), pp. 27–31. Danish Polar Center, Ministry of Research and Information Technology, Copenhagen.
Meltofte, H. (2000) In: *Zackenberg Ecological Research Operations, 5th Annual Report, 1999* (Ed. by K. Caning and M. Rasch), pp. 32–39. Danish Polar Center, Ministry of Research and Information Technology, Copenhagen.
Meltofte, H. (2001a) *Arctic* **54**, 367–376.
Meltofte, H. (2001b) In: *Zackenberg Ecological Research Operations, 6th Annual Report, 2000* (Ed. by K. Caning and M. Rasch), pp. 30–39. Danish Polar Center, Ministry of Research and Information Technology, Copenhagen.
Meltofte, H. (2003) In: *Zackenberg Ecological Research Operations, 7th Annual Report, 2001* (Ed. by K. Caning and M. Rasch), pp. 30–44. Danish Polar Center, Ministry of Science, Technology and Innovation, Copenhagen.
Meltofte, H. (2006a) *Dansk Orn. Foren. Tidsskr.* **100**, 16–28.
Meltofte, H. (2006b) *Wildfowl* **56**, 129–151.
Meltofte, H. and Høye, T.T. (2007) *Dansk Orn. Foren. Tidsskr.* **101**, 109–119.
Meltofte, H. and Lahrmann, D.P. (2006) *Dansk Orn. Foren. Tidsskr.* **100**, 75–87.
Meltofte, H., Sittler, B., and Hansen, J. (in press) *Arctic Birds.*
Meltofte, H., Piersma, T., Boyd, H., McCaffery, B., Golovnyuk, V.V., Graham, K., Morrison, R.I.G., Nol, E., Schamel, D., Schekkerman, H., Soloviev, M.Y., Tomkovich, P.S., *et al.* (2007a) *Meddr Grønland, Biosci.* **59**, 1–48.
Meltofte, H., Høye, T.T., Schmidt, N.M. and Forchhammer, M.C. (2007b) *Polar Biol.* **30**, 601–606.
Morrison, R.I.G. and Hobson, K.A. (2004) *Auk* **121**, 33–344.
Mortensen, C.E. (2000) *Dansk Orn. Foren. Tidsskr* **94**, 29–41 (in Danish, with English summary).
Nol, E., Blanken, M.S. and Flynn, L. (1997) *Condor* **99**, 389–396.
Ntiamoa-Baidu, Y., Piersma, T., Wiersma, P., Poot, M., Battley, P. and Gordon, C. (1998) *Ibis* **140**, 89–103.
Piersma, T., Gudmundsson, G.A. and Lilliendahl, K. (1999) *Physiol. Biochem. Zool.* **72**, 405–415.
Piersma, T., Meltofte, H., Jukema, J., Reneerkens, J., de Goeij, P. and Ekster, W. (2006) *Wader Study Group Bull.* **109**, 83–87.
Robinson, R.A., Clark, N.A., Lanctot, R., Nebel, S., Harrington, B., Clark, J.A., Gill, J.A., Meltofte, H., Rogers, D.I., Rogers, K.G., Ens, B.J., Reynolds, C.M., *et al.* (2005) *Wader Study Group Bull.* **106**, 17–29.
Roudybush, T.E., Grau, C.R., Peterson, M.R., Ainley, D.G., Hirsch, K.V., Gilman, A.P. and Patten, S.M. (1979) *Condor* **81**, 293–298.
Ryabitsev, V.K. (1993) *Territorial Relations and Dynamics of Bird Communities in Subarctic.* Nauka Publ., Ekaterinburg (In Russian).
Scheiffarth, G., Wahls, S., Ketzenberg, C. and Exo, K.-M. (2002) *Oikos* **96**, 346–354.
Schekkerman, H., van Roomen, M.J.W. and Underhill, L.G. (1998) *Ardea* **86**, 153–168.
Shepherd, P. (2001) *Wader Study Group Bull.* **97**, 17.
Soloviev, M.Y., Minton, C.D.T. and Tomkovich, P.S. (2006) In: *Waterbirds Around the World* (Ed. by G.C. Boere, C.A. Galbraith and D.A. Stroud), pp. 131–137. The Stationery Office, Edinburgh, UK.

Stroud, D.A., Davidson, N.C., West, R., Scott, D.A., Haanstra, L., Thorup, O., Ganter, B. and Delany, S. (2004) *Status of migratory wader populations in Africa and Western Eurasia in the 1990s*, International Wader Studies no.15.
Tomkovich, P.S. and Soloviev, M.Y. (1994) *Ostrich* **65**, 174–180.
Tomkovich, P.S. and Soloviev, M.Y. (2001) *Ornithologia* **29**, 125–136.
Thorup, O. and Meltofte, H. (2005) In: *Zackenberg Ecological Research Operations, 10th Annual Report, 2004* (Ed. by K. Caning and M. Rasch), pp. 43–51. Danish Polar Center, Ministry of Research and Information Technology, Copenhagen.
Troy, D.M. (1996) In: *Shorebird Ecology and Conservation in the Western Hemisphere* (Ed. by P. Hicklin, A.J. Erskine and J. Jehl), *International Wader Studies*, **8**, pp. 15–27.
Tulp, I. and Schekkerman, H. (2007) In: *Environmental forcing on the timing of breeding in long-distance migrant shorebirds* I. Tulp, PhD thesis, University of Groningen.
Tulp, I., Schekkerman, H. and Klaassen, R. (2000) *Studies on breeding shorebirds at Medusa Bay, Taimyr, in summer 2000*. Altera report 219, Wageningen, The Netherlands.
van de Kam, J., Ens, B.J., Piersma, T. and Zwarts, L. (2004) *Shorebirds. An Illustrated Behavioural Ecology*, KNNV Publishers, Utrecht.
Zöllner, T. (2002) *Charadrius* **38**, 9–23.

Vertebrate Predator–Prey Interactions in a Seasonal Environment

NIELS M. SCHMIDT, THOMAS B. BERG, MADS C. FORCHHAMMER, DITTE K. HENDRICHSEN, LINE A. KYHN, HANS MELTOFTE AND TOKE T. HØYE

SUMMARY

The High Arctic, with its low number of species, is characterised by a relatively simple ecosystem, and the vertebrate predator–prey interactions in the valley Zackenbergdalen in Northeast Greenland are centred around the collared lemming *Dicrostonyx groenlandicus* and its multiple predators. In this chapter, we examine these interactions in a climatic context through the predator–lemming model developed for the more southerly Greenlandic site, Traill Ø (Gilg *et al.,* 2003, Science **302**, 866–868), parameterised by means of data from the BioBasis monitoring programme to reflect the situation in Zackenbergdalen.

Despite large differences in relative predator densities between these two locations, the two lemming populations exhibit remarkably similar and synchronous population fluctuations. Also, in both lemming populations the annual fluctuations seem primarily driven by a 1-year delay in stoat *Mustela erminea* predation and stabilising predation from the generalist predators, in Zackenbergdalen mainly the arctic fox *Alopex lagopus*. In Zackenbergdalen, however, the coupling between the specialist stoat and the lemming population is relatively weak. During summer, the predation

ADVANCES IN ECOLOGICAL RESEARCH VOL. 40
© 2008 Elsevier Ltd. All rights reserved
0065-2504/08 $35.00
DOI: 10.1016/S0065-2504(07)00015-3

pressure is high, and in most years so high that the lemming population declines during summer. This heavy predation pressure is also reflected in the summer behaviour of the lemmings, and lemmings spend most of their time under ground, and when above, they devote equal amounts of time to being vigilant and foraging. In most winters, predation by the only remaining predator, the stoat, is insufficient to regulate the lemming population.

In the predator–lemming model, seasonality plays an important role in determining the growth rate of the lemming population as well as the density of the various lemming predators. We therefore examined how variation in the relative length of seasons affected the pattern of fluctuation in the lemming population by gradually advancing or delaying the time of onset of winter. In contrast to advanced winter onset, delayed winter onset led to increased periodicity. The climatic conditions, hence, affect not only the seasonal but also the inter-annual dynamics of the collared lemming population in Zackenbergdalen.

I. INTRODUCTION

The Arctic terrestrial vertebrate community is very simple relative to those at lower latitudes, and the trophic interactions are centred around a few species (Callaghan *et al.*, 2004). In the valley Zackenbergdalen (74°30′N, 20°30′W), musk oxen *Ovibos moschatus*, arctic hares *Lepus arcticus*, collared lemmings *Dicrostonyx groenlandicus*, rock ptarmigans *Lagopus mutus* and a number of wader species (see Meltofte *et al.*, 2008, this volume) form the main prey basis for the vertebrate predators. However, whereas several predators consume lemmings, the only true predator on musk oxen is the arctic wolf *Canis lupus*, and musk oxen constitute the bulk of wolf diet in Northeast Greenland, though lemmings and arctic hares also are consumed (Petersen, 1998). Wolves, however, apparently do not assert a significant negative impact on musk oxen in Northeast Greenland (Schmidt, 2006).

In contrast to the other herbivore species, the collared lemming is resident in the valley lowland throughout the year, and it forms the food basis for a number of predators. Zackenbergdalen holds three important lemming predators, namely the arctic fox *Alopex lagopus*, the stoat *Mustela erminea* and the long-tailed skua *Stercorarius longicaudus*. In late summer, however, long-tailed skuas migrate south towards their pelagic winter areas. Also, the large amount of snow during winter impedes the arctic fox from accessing the lemmings (see e.g., Hansson and Henttonen, 1985). During winter, the predator community is thus even more impoverished, and only the stoat is residing in the valley year-round, hunting lemmings in their subnivean runways (Sittler, 1995). Although all three predators feed extensively on collared lemmings (Gilg *et al.*, 2006), and thus may be regarded as more or less strict lemming specialists, the nomadic and migratory nature of all predators but the stoat makes them *de facto* generalists (Gilg *et al.*, 2006). Hence, the only true year-round specialist is the stoat.

The interplay between the collared lemming and its predators in high-arctic Greenland is highly influenced by the inter-annual changes in climatic conditions, but the intra-annual variation in climate may also be important for the shaping of the long-term population dynamics of this simple vertebrate predator–prey system.

In this chapter, we describe and analyse the intricate predator–prey dynamics representative of the terrestrial ecosystem in Zackenbergdalen. Using comprehensive, temporally replicated data on the collared lemming and its predators, we focus on the predator–lemming dynamics embracing a period of 10 years. Pivotal for our description is the predator–lemming model developed by Gilg and co-workers (Gilg *et al.*, 2003, 2006), specifically modified to reflect the observed predator–lemming system in Zackenbergdalen (Box 1).

Box 1

Modelling the Predator–Prey Dynamics in Zackenbergdalen

The day-to-day changes in the densities of the collared lemming and its predators in Zackenbergdalen were examined using a simple predator–prey model approach (Gilg *et al.*, 2003). The density of lemmings was modelled as continuous, unlimited growth, with different growth rates in summer (low) and winter (high), respectively. The impact of all the major lemming predators on the lemming population was modelled by combining the functional response (i.e., the rate by which a predator consume its prey in relation to prey density) and the numerical response (i.e., the number of a given predator in relation to prey density) of all predatory species, hence equalling the total impact of predators on the prey population. The model thus resembles the traditional Lotka-Volterra equations for coupled predator–prey systems (e.g., Begon *et al.*, 1996), but also includes seasonality in both the reproductive potential of the prey and seasonal changes in the predator community.

The annual recordings of winter nests conducted by the BioBasis programme (Meltofte and Berg, 2006) yielded estimates of collared lemming density within a 2 km^2 designated area. To get a more representative estimate of lemming density, we extrapolated the habitat-specific nest density within the monitoring area onto an *c.* 17 km^2 area below 300 m a.s.l., in which the vegetation types have been mapped in detail (Bay, 1998). The slope of the correlation between the corrected and the uncorrected density was 0.67, thus indicating that the density estimates

(continued)

Box 1 *(continued)*

are lower in the valley in general than in the monitoring area. These corrected density estimates were used in the modelling, and combined with the annual records of predators allowed for the estimation of the numerical response of the lemming predators. Combining these numerical responses with the functional responses adapted from Gilg *et al.* (2003, 2006), we can parameterise the predator–lemming model developed for the Traill Ø population of collared lemmings (Gilg *et al.*, 2003, 2006). Hereafter, we can calculate the number of lemmings and the number of lemmings consumed per predator per day. All model-specific parameters are given in Box Table 1 below. Please refer to Gilg *et al.* (2003, 2006) for full exposition of the predator–prey model and the abbreviations used.

Box Table 1 Model parameters for the numerical and functional responses of all lemming predators in Zackenbergdalen

		Zackenberg	Traill Ø
Numerical responses (adults)			
B_o	Maximum owl density (individuals per hectare)	0.0006[a]	0.00366
Y_o	Slope of the numerical response for adult owls	[a]	2.86
P_l	Skua density (individuals per hectare)	0.0253	0.02
B_f	Maximum fox density (individuals per hectare)	0.000822[b]	0.0008
Y_f	Slope of the numerical response for adult foxes	[b]	11
Numerical responses (young)			
b'_o	Maximum density for young owls (fledglings per hectare)	0.0011[a]	0.0011
Y'_o	Slope of the numerical response for young owls	[a]	4
b'_l	Maximum density for young skuas (fledglings per hectare)	0.016[c]	0.016
Y'_l	Slope of the numerical response for young skuas	6[c]	6

b'_f	Maximum density for young foxes (weaned young per hectare)	0.002592	0.0028
Y'_f	Slope of the numerical response for young foxes	1.59356	5.3
Functional responses			
W_o	Maximum predation rate for the owl (lemmings per day)	4.7[c]	4.7
D_o	Slope of the type III functional response ($e = 2$) for the owl	1.08[c]	1.08
W_l	Maximum predation rate for the skua (lemmings per day)	4.4[c]	4.4
D_l	Slope of the type III functional response ($e = 4$) for the skua	2.2[c]	2.2
W_f	Maximum predation rate for the fox (lemmings per day)	3.8[c]	3.8
D_f	Slope of the type III functional response ($e = 2$) for the fox	0.13[c]	0.13

Parameters for the Traill Ø model are given for comparison (see Gilg *et al.*, 2003, 2006).
[a]Owls are present (and breed) on average every fourth year (random function) if the lemming density at snowmelt is above two individuals per hectare.
[b]Adult fox density was modelled as a constant.
[c]Values adapted from the Traill Ø model (Gilg *et al.*, 2003, 2006).

II. THE COLLARED LEMMING AND ITS PREDATORS IN ZACKENBERGDALEN

In Greenland, collared lemming population fluctuations have been monitored in detail in Zackenbergdalen since 1996 (Meltofte and Berg, 2006), and on Traill Ø since 1988 (Sittler, 1995; Gilg *et al.*, 2003, 2006), about 220 km south of Zackenberg. These two locations host the same lemming predator species, but differ markedly with respect to the relative predator densities: both predator communities include arctic fox, stoat and long-tailed skua. However, while snowy owls *Nyctea scandiaca* in some years are present in very high numbers on Traill Ø, this species is essentially absent in Zackenbergdalen (see Table 1). Nevertheless, the pattern of fluctuations in the two

Table 1 Data from the annual density estimates of the collared lemming and its predators in Zackenbergdalen 1996–2005. All densities are number of individuals per hectare

	Size of area censused (ha)	Year									
		1996	1997	1998	1999	2000	2001	2002	2003	2004	2005
Collared lemming (winter nests)	208	0.7854	1.7854	3.5171	1.6146	0.9366	1.5902	1.4000	0.4634	2.1024	1.1415
Stoat winter nests depredated)	208	0.0049	0.0146	0.1262	0.0194	0.0485	0.0388	0.0437	0.0049	0.0000	0.0000
Arctic fox (adults)	≈5000	0.0008	0.0006	0.0008	0.0010	0.0012	0.0008	0.0004	0.0010	–	0.0008
Arctic fox (weaned young)	≈5000	0.0018	0.0000	0.0024	0.0000	0.0017	0.0024	0.0000	0.0012	0.0026	0.0000
Long-tailed skua (adults)	1930	0.0284	0.0242	0.0232	0.0221	0.0253	0.0242	0.0253	0.0284	0.0232	0.0326
Long-tailed skua (young)	1930	0.0000	0.0026	0.0032	0.0005	0.0000	0.0026	0.0021	0.0011	0.0116	0.0005
Snowy owl (adults)	≈5000	0.0000	0.0004	0.0000	0.0000	0.0000	0.0008	0.0000	0.0000	0.0000	0.0000
Snowy owl (young)	≈5000	0.0000	0.0006	0.0000	0.0000	0.0000	0.0016	0.0000	0.0000	0.0000	0.0000

separated collared lemming populations are remarkably similar and synchronised (see below), and both characterised by a ≥4-year periodicity (Figure 1). The amplitude, however, differs between the two sites, being smallest in Zackenbergdalen (Figure 1).

The causality behind these multi-annual lemming fluctuations are widely debated, and several causative agents have been suggested in both empirical and theoretical studies (reviewed by Stenseth and Ims, 1993; Lindström *et al.*, 2001; see also Berg *et al.*, 2008, this volume, for the plant–herbivore interactions in Zackenbergdalen). However, although Turchin *et al.* (2000) stated that no single mechanism is likely to cause the microtine cycles, the hypothesis that the fluctuations are driven by predators has gained strongest support (Turchin and Hanski, 2001). Additionally, on the basis of empirical data, Gilg *et al.* (2003) were able to create multi-annual cycles resembling the observed fluctuations for the collared lemming population on Traill Ø in a simple predator–prey model. We adopted this predator–prey model as a tool to elaborate the interactions between the collared lemming and its predators in Zackenbergdalen in the context of climate-driven seasonality (Box 1). Direct extrapolation from Gilg *et al.* (2003) is of course not recommendable (see Hudson and Bjørnstad, 2003), but the BioBasis monitoring programme

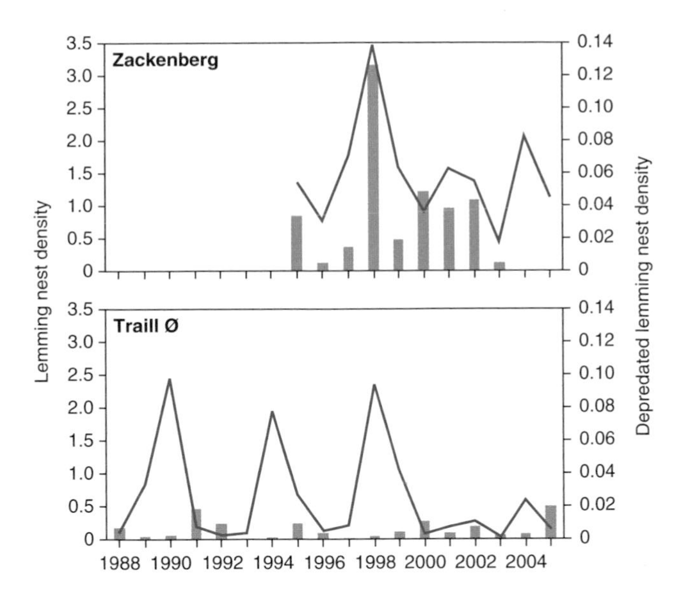

Figure 1 Density (numbers per hectare) of collared lemming winter nests (line) and winter nests depredated by stoats (bars) in Zackenbergdalen (top) and on Traill Ø (bottom). Data from Traill Ø by courtesy to Benoit Sittler. The fluctuations of the two collared lemming populations are highly correlated ($R^2 = 0.77$, $F_{1,9} = 29.76$, $p < 0.001$).

at Zackenberg has enabled us to parameterise the model to reflect the situation in Zackenbergdalen (Box 1).

As the only rodent species in arctic Greenland, the collared lemming plays a potentially central role in the terrestrial community. In the BioBasis monitoring of mammals at Zackenberg, the population fluctuations of the collared lemming are monitored by annual records of winter nests within an *c.* 2 km^2 designated monitoring area in the valley lowland (Meltofte and Berg, 2006). The number of winter nests is a good proxy for the actual density of lemmings at the time of snowmelt (Gilg *et al.*, 2003, 2006). During 1996–2005, the density of nests within the lemming monitoring area has fluctuated markedly, and ranged between *c.* 1 and 4 nests per hectare (Figure 1 and Table 1).

The number and the breeding success of the predatory species in Zackenbergdalen are also monitored as part of the BioBasis monitoring programme (Table 1). Arctic foxes are encountered frequently all over the valley, and 14 dens have been mapped. Breeding has been verified in five of these dens. During summer, dens are checked on a weekly basis to look for signs of pups. Later, observations around-the-clock verify the minimum number of young. The most conspicuous predator in the valley during summer is the long-tailed skua, and around 25 pairs arrive in the valley each year (Meltofte and Høye, 2007). The stoat exerts its strongest impact on the lemming population during winter where it hunts under the snow, but it is difficult to monitor directly as individuals are rarely encountered. We used the number of lemming winter nests taken over by stoats as a proxy for the stoat density during winter (Figure 1). A winter nest taken over by a stoat is easily recognisable, as the stoat uses the lemming fur to line the nests. Additionally, sporadic observations of gyrfalcon *Falco rusticolus*, polar bear *Ursus maritimus* and arctic wolf are made in the valley and surrounding areas. These species were not included in the model because of their insignificant numbers.

III. PATTERNS OF PREDATION ON COLLARED LEMMINGS

The predator–prey model parameterised for Zackenbergdalen was indeed able to reproduce the pattern of fluctuations in the lemming population observed in the valley. Hence, the model strongly suggested that the observed fluctuations in the collared lemming populations in Zackenbergdalen are attributable to interactions between the collared lemming as prey and its multiple predators as reported for Traill Ø (Gilg *et al.*, 2003, 2006). Collared lemming densities as inferred from the model were, however, much lower than what was expected from the records of winter nests. In this respect, it is noteworthy that the lemming monitoring area is located in the rich valley lowland, and densities estimated here, thus, may be higher than the average for the whole valley. We tried to extrapolate lemming densities based on relative densities of winter nests in different habitats, and these corrected densities were lower, but still larger than model predictions (Box 1). Nonetheless, the model lemming densities are within the range of densities reported from other collared lemming populations (e.g., Ehrich *et al.*, 2001; Gilg, 2002; Gilg *et al.*, 2003). The sporadic breeding of owls in Zackenbergdalen may also be indicative of low lemming densities.

From the predator–prey model, we can estimate the relative importance of the various predators through the lemming cycle. As is evident from Figure 2, the model suggests that the arctic fox is responsible for the vast majority of predated lemmings throughout the lemming cycle. Long-tailed skuas assert their largest impact during the increase phase, whereas the snowy owl affects the lemming population only in the peak phase, and here still only have limited impact. The importance of stoats increases steadily during the lemming phase, and stoats assert their largest impact on the lemming population in the decrease phase (Figure 2). The pattern of predation in Zackenbergdalen is, thus, consistent with that found on Traill Ø (Gilg *et al.*, 2003), where the lemming fluctuations are driven by the 1-year delay in the numerical response of the stoat and stabilised by strongly density-dependent predation by arctic fox, snowy owl and long-tailed skua. In Zackenbergdalen, however, long-tailed skua and especially arctic fox seemingly take over the lead role of snowy owls. Also, the 1-year delay in stoat predation is less pronounced in

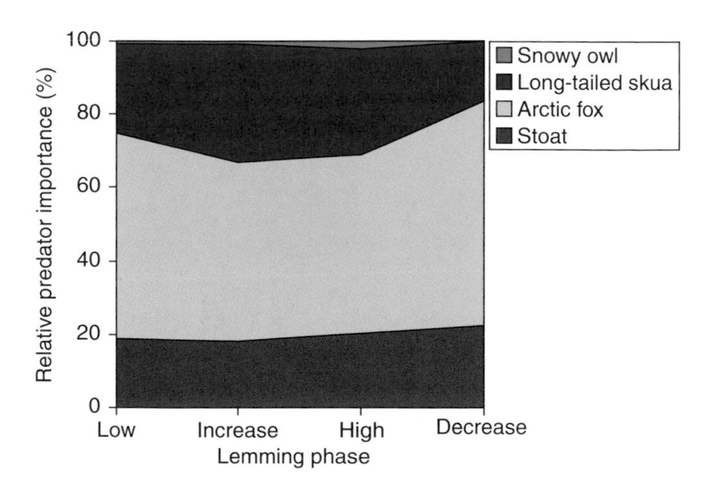

Figure 2 The relative importance of the four predators present in Zackenbergdalen during four collared lemming population phases. Percentage predation estimated from simulations (100 years) in the predator–prey model with daily time steps.

Zackenbergdalen (Figure 3; see also Figure 1) compared to Traill Ø (Gilg *et al.*, 2006), and the interactions between the collared lemming and the only true specialist predator less tightly coupled. The causality behind the less strict interactions between collared lemmings and stoats is currently unknown, but may at least in part be due to the occasional lack of stoat predated nests within the lemming monitoring area (Figure 1). Additionally, in the winter 1997–1998 the stoat population responded instantly to the high lemming density (Figure 1), which also weakens the link between the specialist predator and the lemming prey.

The role of the arctic fox as the dominating lemming predator throughout the lemming cycle may be due to its ability to switch between available food sources (type 3 functional response; Holling, 1959; see also Gilg *et al.*, 2006). Compared to Traill Ø and surroundings, Zackenbergdalen and Wollaston Forland holds more musk oxen (Boertmann and Forchhammer, 1992), and in Zackenbergdalen musk ox carcasses are found in relatively high numbers each year [between 1 and 13 within Zackenbergdalen (*c.* 50 km^2), mean = 4.6 per year]. Also, the arctic fox in Zackenbergdalen has been observed feeding heavily on arctic charr *Salvelinus alpinus* trapped in shallow parts of the river. Both carcasses and arctic charr may constitute an important alternative food source for arctic foxes when lemmings are scarce, as also reported elsewhere (Roth, 2003), and may also explain the current lack of numerical response of arctic fox to variation in lemming density in Zackenbergdalen (Box 1). Past hunting statistics from Northeast Greenland, however, indicates a 4-year periodicity in arctic fox numbers in the first half of the twentieth century

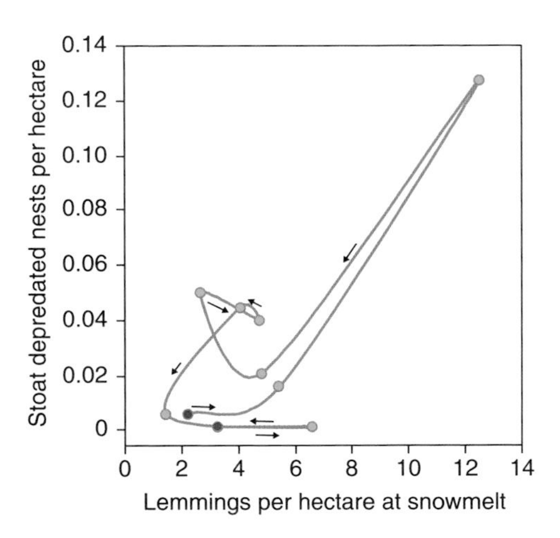

Figure 3 Stoat numerical response to variation in lemming density at snowmelt in Zackenbergdalen during 1996–2005. The blue dot indicates the year 1996 and the red dot indicates 2005.

(Figure 4), which may have been linked to the annual lemming fluctuations (see Macpherson, 1969). Stoats presumably also take advantage of musk ox carcasses, and on Traill Ø the observed change in the stoat–lemming dynamics in winter 2000–2001 may be attributed to increased stoat survival due to high musk ox carcass availability (Gilg *et al.*, 2006). In Zackenbergdalen, however, we have no evidence of stoat numbers responding positively to musk ox carcasses. In fact, although musk ox carcasses have been found every year, the stoat population seemingly has been reduced to a minimum during the last years (Figure 1).

Many microtine populations are characterised by declining population densities during the summer season (Reid *et al.*, 1995, 1997; Schmidt, 2000; Gilg, 2002). By combining the numerical and functional responses for each of the lemming predators, we can calculate their summed impact on the lemming population in summer (see Gilg *et al.*, 2006). Taking the very conservative estimate of collared lemming reproductive potential in summer of 1% per day (see Gilg *et al.*, 2006), we can, by comparing the reproductive potential with the calculated daily predation rates, examine the likelihood that the collared lemming population is able to increase during summer. Figure 5 shows the daily predation rates of the individual predators along with their combined impacts on the lemming population in summer. In most cases, the daily predation rates exceed the lemming reproductive potential, indicating that the collared lemming population in Zackenbergdalen is unable to increase, except at very low densities or just prior to the peak phase (Figure 5). At most densities, arctic fox, long-tailed skua and stoat alone seem capable of reducing the summer population of collared lemming. Including the snowy owl

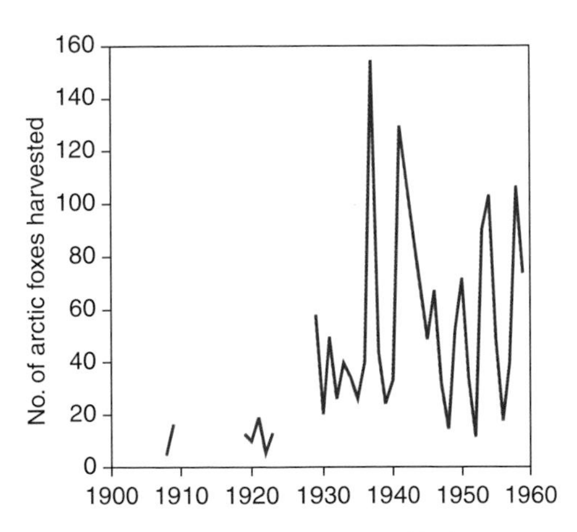

Figure 4 Numbers of arctic foxes harvested annually (mean per trapper) in Northeast Greenland during the entire period of trapping 1908–1960. Total catch was 19,287 arctic foxes. Data from Mikkelsen (1994).

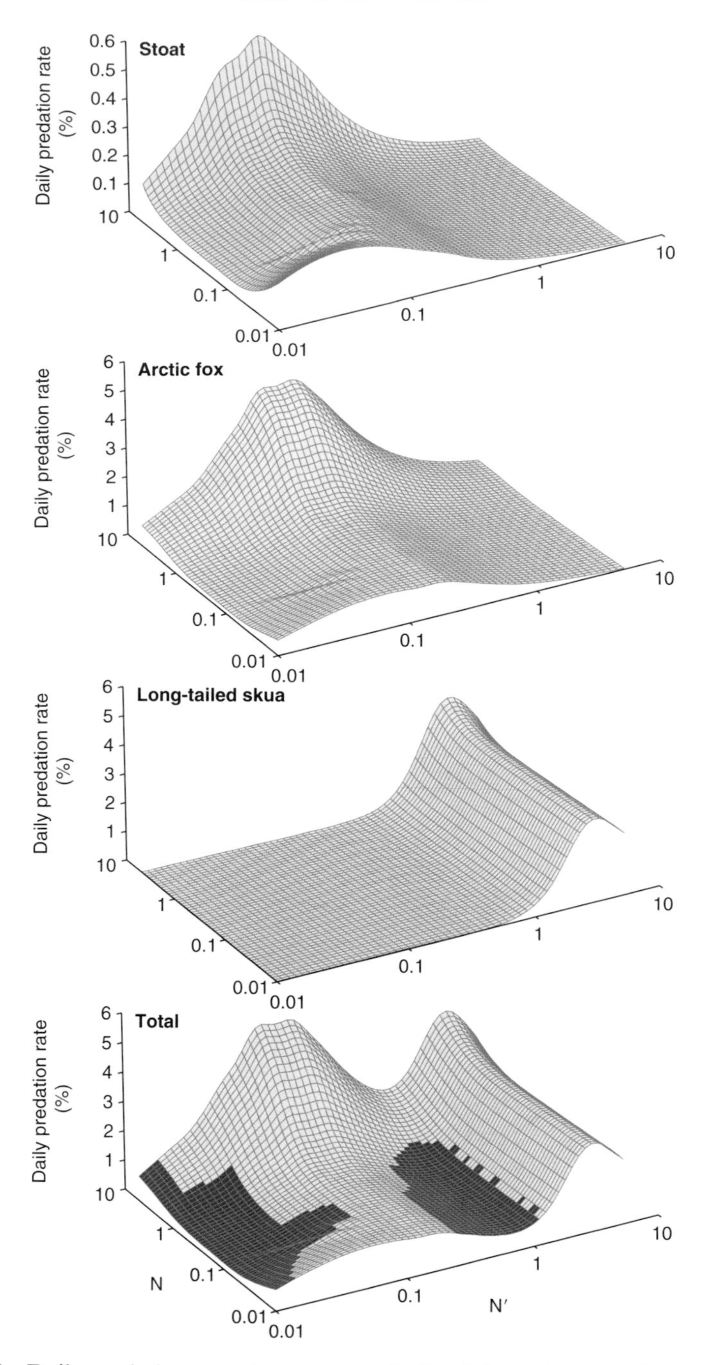

Figure 5 Daily predation rates in summer calculated from the numerical and functional responses of the three main predators on collared lemmings in Zackenbergdalen in relation to the log-transformed daily density of lemmings (N) and the

only increases the total daily predation marginally, because of the sporadic breeding of this species in Zackenbergdalen (see Box 1).

These impacts of predation on the Zackenberg collared lemming population were also supported by empirical data from the valley. In a radio-telemetry study (Schmidt *et al.*, 2002), we found that at least 76% of the lemming population was predated during summer following a winter peak. Similar predation rates have been documented from Traill Ø (Gilg, 2002), as well as for other collared lemming populations (Krebs *et al.*, 1995; Wilson *et al.*, 1999). In summers following winter lows, however, predation rates may be much lower (Berg, 2003; Gilg *et al.*, 2006). It is likely that collared lemmings are particularly susceptible to predation in early spring when moving from their subnivean winter nests to their summer habitat concurrent with the melting of the snow. Lemming tracks on the snow in early June showed that *c.* 30% of these ended in tracks of either arctic fox or long-tailed skua, whereas 50% of the collared lemmings made it to the barren grounds, which constitute the summer habitat. Remaining tracks were between holes in the snow. Mortality data from radio-collared lemmings (Schmidt, 2000; Schmidt *et al.*, 2002) also showed that the most rapid decline in survival is observed in early spring, concurrent with the habitat shift (Figure 6; see also Gilg, 2002). Radio-tracking data from summer also points at the long-tailed skua and the arctic fox as the main lemming predators (Schmidt, 2000).

The collared lemming has several enemies during summer and has adjusted its behaviour accordingly. When in their summer habitat, mainly on dry *Dryas* heath (Schmidt, 2000; Berg, 2003), collared lemmings spent most of their time inside their burrows (Brooks, 1993; Schmidt, 2000; Kyhn, 2004). When outside the burrow, collared lemmings spent about 50% of their time being vigilant, while about 40% is spent in eating (Figure 7). At all times, lemmings stay in close proximity of the burrow entrance (Figure 8), and retract instantly when feeling threatened. Such short-term anti-predator responses seem the most beneficial in fluctuating populations of short-lived animals (Norrdahl and Korpimäki, 2005). However, access to receptive females may force male collared lemmings to move long distances (Schmidt *et al.*, 2002), thereby increasing the risk of predation (e.g., Norrdahl and Korpimäki, 1998). Predators hunting on visual cues, such as avian predators, are known to kill more males than females, whereas mammalian predators hunting by means of olfactory clues may have a female-biased prey composition (Norrdahl and Korpimäki, 1998). From observations of arctic fox foraging behaviour in Zackenbergdalen, we know that the species generally

---

log-transformed density of lemmings at snowmelt (N'). Note the difference in scale. Dark blue in the lower panel (combined daily predation of all three predators) indicates daily predation rates below 1%. Functional responses were obtained from Gilg *et al.* (2006).

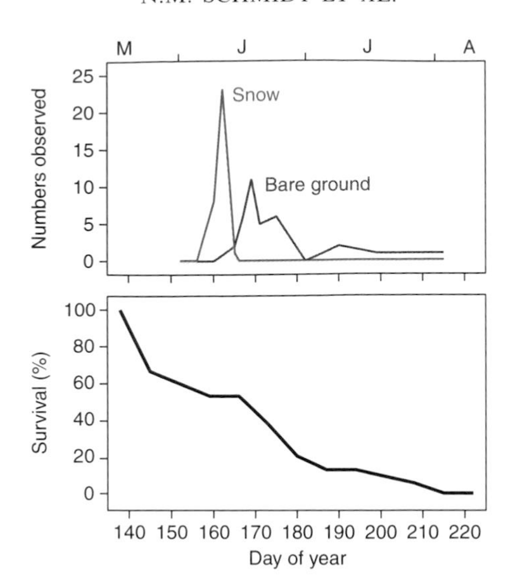

Figure 6 Observations of collared lemmings on snow and on bare ground (upper panel) in relation to day of year (May 15–August 3) in 1998. The corresponding survival is given in the lower panel. The shift from winter habitat to summer habitat around day 165 (June 14) onwards is associated with a marked decline in survival. The earlier steep decline in survival is mainly caused by small sample size.

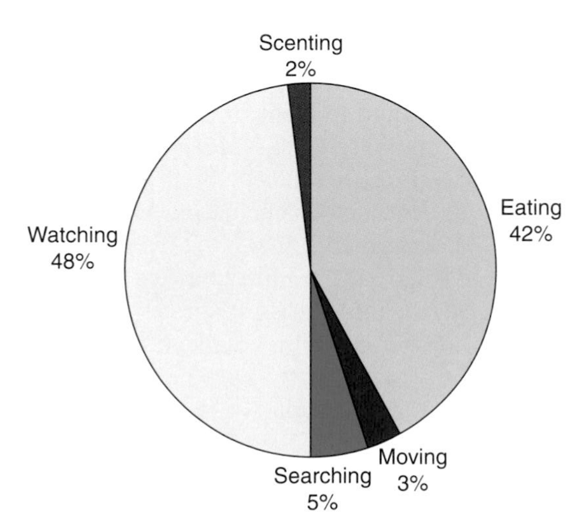

Figure 7 The percentage time spent on five behavioural categories by collared lemmings when outside their summer burrows in Zackenbergdalen (modified from Kyhn, 2004).

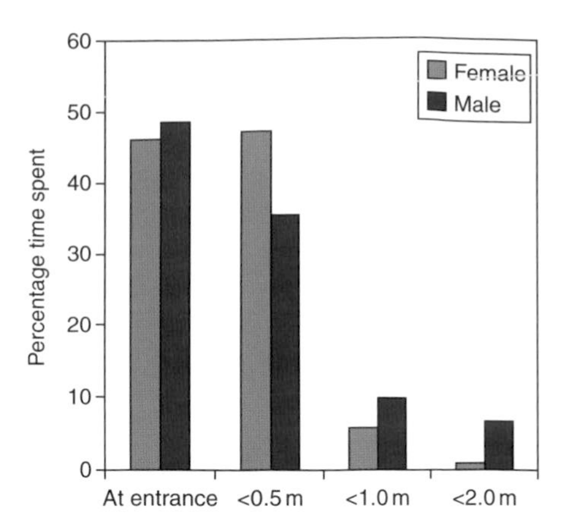

Figure 8 The mean percentage time spent by female and male collared lemmings at four distances to the summer burrow entrance (modified from Kyhn, 2004).

moves along straight lines across large distances, and only with occasional sallies when an olfactory cue is detected. This foraging behaviour probably reflects the generally low prey availability in the arctic region (Box 2).

During winter, the stoat predation may have substantial negative impact on the collared lemming population in Zackenbergdalen, and especially following high lemming densities (see Figure 1). Unfortunately, the actual impact of stoats onto the lemming population is difficult to assess directly, but the predator–prey model enables us to estimate the daily predation rates of stoats year-round. In the model, stoat daily predation in winter was never higher than 1.2%, indicating that in most winters collared lemmings are indeed able to escape the predator pit of the summer, thereby allowing the population to increase (e.g., Fuller *et al.*, 1977; Reid and Krebs, 1996). The summer season in Zackenbergdalen, and probably in most other high-arctic areas, must therefore be regarded as a bottleneck to the population (Schmidt, 2000), a bottleneck attributable to severe predation when lemmings are forced to aggregate in the summer habitat, which covers a much smaller area than the winter habitat (Berg, 2003; Berg *et al.*, 2008, this volume).

IV. INDIRECT EFFECTS OF COLLARED LEMMING FLUCTUATIONS

One important side-effect of the presence of lemmings in high-arctic Greenland is that they sustain a much higher density of arctic foxes than would otherwise have been the case, and that these foxes exert a considerable

Box 2

Arctic fox hunting

The arctic fox has a food search strategy distinctly different from the red fox *Vulpes vulpes*. Where a red fox often will move slowly through the vegetation in a non-oriented direction, nosing and zigzagging its way (Meia and Weber, 1995), arctic foxes cover extensive areas of land, constantly galloping in a more or less straight direction (Box Figure 1). This discrepancy in behaviour between the two fox species is most likely a function of food availability and distribution, with food density being much higher and less patchy distributed in the temperate region compared to the Arctic. Hence, densities of breeding arctic foxes are typically in the order of 8–12 per 100 km^2 (e.g., Smits and Slough, 1993; Wiklund *et al.*, 1999; this study), while red fox densities in temperate areas may be more than 10 times as high (Meia and Weber, 1995; Dekker *et al.*, 2001).

While galloping across the tundra, the arctic fox keeps its nose 10–15 cm above the ground. When detecting an olfactory cue, it slows down and searches the area with the nose at the ground. Afterwards it most often continues in the once chosen direction. In Zackenbergdalen, this strategy is used both on snow and on bare ground in summer. Fox tracks on snow showed that foxes move between snow-free spots, sometimes continuing for kilometres along almost straight lines of snow-free spots (max. recorded distance in summer: 4.5 km). By the time the fjord ice has gone (July), the foxes make extensive use of the littoral zone, where they search for food in straight lines following the contours of the beach. Only few detours were observed from these routes. Individually marked foxes have been observed up to 5 km from their natal den.

Probably as an anti-predator strategy, collared lemmings use several burrows during summer (Schmidt *et al.*, 2002), of which only one is used for nesting (Berg, 2003). Most burrows are used for retreat and expansion of the potential forage areas of the individual lemming as they forage within half a meter from the burrow entrance (Figure 8). Lemmings seemingly defecate only at latrine sites below ground (Boonstra *et al.*, 1996; Berg, 2003), thereby minimising the risk of predation from visual hunters capable of detecting microtine faeces and urine in UV light, but probably also from the arctic fox, since defecating in many burrows may reduce the chances of a fox to encounter a lemming inside a burrow by the use of olfactory cues. In Zackenbergdalen, burrows dug out by arctic foxes have been observed only in a few cases (Berg, 2003).

Ground-nesting birds use a wide variety of camouflage and distraction behaviours to reduce predation. Also, the waxes some of these birds use for preening undergo changes to less volatile chemical composition during incubation, thereby leaving less olfactory cues on the nest and eggs for mammalian predators (Reneerkens *et al.*, 2005).

Box Figure 1 In search of food, arctic foxes cover large expanses of land running in relatively constant directions. Photo: Aurora Photo.

predation pressure, not only on lemmings themselves but in particular also on eggs and young of ground-nesting birds. On Traill Ø, collared lemmings constitute an important fraction of arctic fox diet irrespective of lemming density (Dalerum and Angerbjörn, 2000; Gilg *et al.*, 2006). Birds, including their eggs and young, constituted a smaller but still important fraction of the diet. Moreover, there was a close correlation between the frequency of consumed bird prey and bird density around the individual fox dens (Dalerum and Angerbjörn, 2000). In Zackenbergdalen, arctic foxes depredate about 50–60% of the wader nests in most years (Table 2), while the predation on, for example, long-tailed skua nests have varied markedly inter-annually between 6% and 100% (Table 2). Also, predation, presumably by arctic foxes, on goslings of barnacle geese *Branta leucopsis* is heavy, and *c.* 0.6% of the goslings disappear per day (Meltofte, 2006). The inter-annual variation in breeding success of most ground-nesting birds was, however, not significantly correlated with variation in the lemming population (Table 2). Thus, the typical Siberian situation where high abundance of lemmings results in high breeding success for ground-nesting birds, succeeded by very poor breeding success the following year when lemmings are few and arctic foxes abundant (Summers and Underhill, 1987; Underhill *et al.*, 1993;

Table 2 Breeding success of ground-nesting bird species in Zackenbergdalen 1996–2005: Red-throated diver *Gavia stellata*, waders Charadrii, long-tailed duck *Clangula hyemalis* and long-tailed skua *Stercorarius longicaudus*

	1996	1997	1998	1999	2000	2001	2002	2003	2004	2005
All wader species										
Nest success %	33–63	52–100	32–37	38–39	44	43	43	42–44	87–90	18
Red-throated diver										
Population (pairs)	4–5	5	5	4–6	5	5	6	6	6–7	5–7
Chicks recorded	1	5	8	0	2	2	3	7	2	0
Long-tailed duck										
Population (pairs)	5–8	4–6	6–8	7–8	5–8	5–7	6–7	7–9	5–7	6–8
Chicks recorded	0	2	4	0	5	0	7	0	29–48	0
Long-tailed skua										
Population (pairs)	25–29	22–25	21–24	19–24	21–28	22–25	23–26	25–29	21	24–29
Nest success %	0	81	27	18	17	40	44	76	94	52

Data from Meltofte (2006) and Meltofte and Høye (2007).

Please refer to Table 1 for data on the densities of collared lemmings and all predators.

Blomqvist *et al.*, 2002), was not found in Zackenbergdalen, and has only been reported few times in high-arctic Greenland (Sittler *et al.*, 2000; Meltofte, 2006). Instead, we see a complex pattern of good breeding success in years with relatively few foxes—with or without many lemmings—while high lemming density may result both in good and poor breeding success (Table 2). Taken together, the lemming fluctuations give rise to a very dynamic and unpredictable predation scenario for ground-nesting birds, and an intricate interplay exists between lemming density and arctic fox predation on birds' nests. The breeding success of several ground-nesting bird species in high-arctic Greenland can therefore be characterised as "eruptive," rather than solely dependent upon collared lemming density (Meltofte and Høye, 2007).

V. PREDATOR–PREY INTERACTIONS IN A CHANGING CLIMATE

At northern latitudes, snow depth and snow-cover may have marked effects on predators and their prey as well as their interactions. For instance, early onset of winter may be detrimental to vertebrate reproduction in the Arctic (Mech, 2000), and Stenseth *et al.* (2004) reported that unfavourable snow conditions may act as a barrier for the Canadian lynx, thereby affecting the well-known lynx–hare interactions there. Also, climatic effects may be mediated through predator–prey interactions (e.g., Post *et al.*, 1999).

The expected climatic changes in Zackenbergdalen include changes especially in winter precipitation (Stendel *et al.*, 2008, this volume), and dendroclimatological analyses from the valley indicate that the snow-cover extent in Zackenbergdalen in early spring is indeed increasing (Schmidt *et al.*, 2006), at least in the *Salix* snow-beds (see Bay, 1998). Both data on reproduction and mortality (Reid *et al.*, 1995, 1997; Schmidt, 2000; Schmidt *et al.*, 2002; Gilg, 2002; this study) suggests that the winter season is the prime time for collared lemmings in Zackenbergdalen, which in the predator–prey model is reflected in a higher winter than summer growth rate of the lemming, as well as the absence of long-tailed skua, snowy owl and arctic fox. These differences between seasons have a profound effect on collared lemming dynamics, and we therefore expect the structural dynamics of the lemming population fluctuations to be highly sensitive to changes in the relative length of seasons. To examine this, we ran the predator–prey model with varying time of onset of winter (Box 1). In Zackenbergdalen, the onset of winter (defined as a snow depth of >10 cm) is much more variable than the onset of summer (see Sigsgaard *et al.*, 2006), and we gradually (with 5-day intervals) advanced the onset of winter from the initial model onset of winter on September 20 to September 6, or delayed the onset of winter to November 15, respectively. Remaining model parameters were kept unchanged, except for the arctic fox whose access to the lemmings was set to stop 26 days after the (varying) onset of winter as in the original model. After extracting time series on lemming density at snowmelt, and plotting the second-order autoregressive structure of these time series in the parameter plane (Royama, 1992), we evaluated the potential structural changes in the collared lemming population dynamics (see Bjørnstad *et al.*, 1995; Stenseth *et al.*, 1996a). When looking at Figure 9, advancing the onset of winter did not affect the pattern of cyclicity in the collared lemming population markedly (Figure 9, bottom), and the population fluctuations remained characterised by a 4–5-year periodicity (see Bjørnstad *et al.*, 1995, for details). Delaying the onset of winter, however, changed the dynamics profoundly, and the periodicity increased steadily with later onset of winter, and the pattern of fluctuation moved towards more chaotic fluctuations (Figure 9). In Fennoscandia, decreasing winter snow-cover and an increasing frequency of mild winters have been suggested as being responsible for the apparent collapse of the cyclicity in the microtine populations there (Hörnfeldt *et al.*, 2005), a pattern of cyclicity that prior to its collapse was believed to be driven mainly by predators (Hanski *et al.*, 1991, 2001), partly mediated through the effects of snow-cover (Hansson and Henttonen, 1985). A parallel gradient in rodent dynamics has been observed on Hokkaido, Japan, and has also been ascribed to a climatic gradient (Stenseth *et al.*, 1996b). The rodent populations from Fennoscandia and Japan are influenced by seasonal

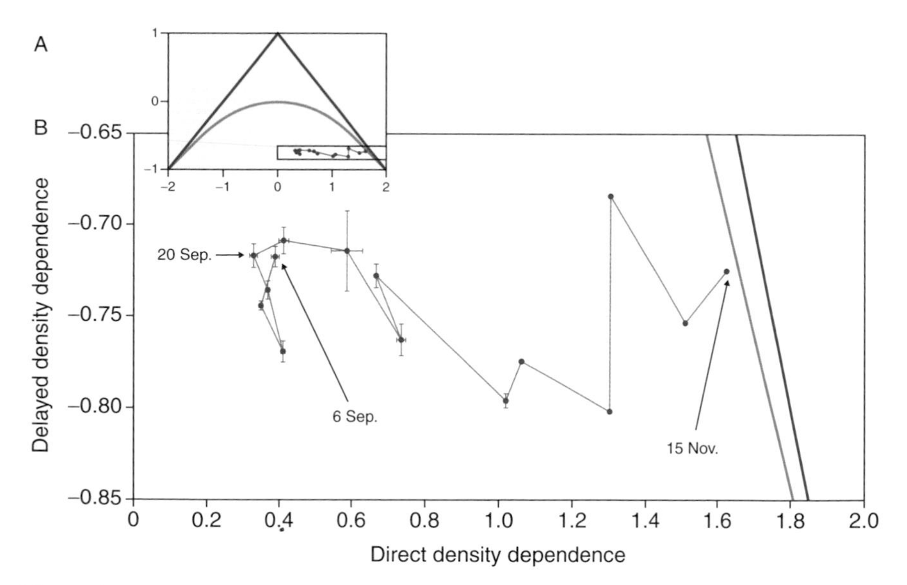

Figure 9 The pattern of collared lemming fluctuations evaluated in the parameter plane. The cycle length increases from left to right along the x-axis (A) presents estimates of direct and delayed density dependence in collared lemming density at snowmelt with varying time of onset of winter are plotted. In the initial model, onset of winter was September 20 (B) shows an enlarged view of the initial manipulations. Bars are the SE (standard error) for five replicate runs.

forcing where a short summer implies regular multi-annual fluctuations, whereas an increase in summer length leads to a breakdown of the regularity of fluctuations (Stenseth *et al.*, 1998, 2002). With decreasing length of winter, the Japanese populations shift from multi-annual cyclic dynamics to stable dynamics, with changes in both direct and delayed density dependence (Stenseth *et al.*, 2002). The model predictions of the Zackenberg data, on the other hand, display a somewhat different pattern with only changes in the direct density dependence, from weakly to strongly positive. Delayed density dependence, usually interpreted as inter-specific interactions, however, does not change, suggesting that the change in dynamics with decreasing winter length is caused primarily by factors working within the lemming population alone. These differences between Fennoscandia and Hokkaido gradients on the one hand and the model predictions from Zackenberg on the other hand are difficult to explain. These, however, may potentially be caused by the differences in rodent living conditions in the High Arctic versus lower latitudes, primarily the fact that wintertime at high latitudes is the main breeding season, whereas at lower latitudes breeding takes place mainly in summer.

The predicted increasing amounts of snow (Stendel *et al.*, 2008, this volume), especially if combined with earlier onset of winters as shown above, hinder the access of, in particular, arctic foxes to their lemming prey in larger parts of the year than seen today. Hence, a shift in the relative importance of lemming predators is likely to occur, and the stoat especially may become more important. It is, however, important to stress that in this predator–prey model, though the shift from summer to winter (and *vice versa*) has a profound effect on several model components, it is modelled in a very simplistic manner. Snow and ice have numerous facets, and, for instance, thaw events during winter may lead to the formation of ice-crust, which may affect rodent populations negatively, and potentially change or disrupt the cyclic nature of their population dynamics (Aars and Ims, 2002; Korslund and Steen, 2006). Also, shallow snow-cover combined with low temperatures may assert a strong negative effect on collared lemmings and confine lemming populations to low numbers for several years (Reid and Krebs, 1996). Finally, both the numerical and the functional responses of the predators were kept unchanged when modelling the importance of timing of onset of winter. To what extent changes in either one may amplify or dampen the impact of winter length is currently unknown, but it may indeed have a profound effect on the response of the lemming population towards inter-annual variation on onset of winter. Additionally, changes in snow conditions also affect the distribution of vegetation types, as well as the build up of biomass available for the herbivores, which in turn may affect the lemming population fluctuations. Hence, the changes in the pattern of fluctuations observed in our model with earlier winter onset, and the change in relative predator importance, are likely to occur in the short run. In the long run, however, there may be a snow threshold beyond which the entire lemming–predator system changes non-linearly. A situation that resembles the situation in Southeast Greenland, that is, with large amounts of snow, large plant biomass availability, but no lemmings, may be the long-term outcome of future climate changes in this region.

VI. COLLARED LEMMING POPULATION SYNCHRONY

The population fluctuations of microtines are often synchronised over vast geographical areas (Krebs *et al.*, 2002; Korpimäki *et al.*, 2004; Sundell *et al.*, 2004), and fluctuations may also be synchronised temporally across species (Krebs *et al.*, 2002; Korpimaki *et al.*, 2005). While the latter can be explained by the alternative prey hypothesis by which predators induce a synchronous low phase in small mammal populations by reducing the densities of both main and alternative prey after the peak phase (Korpimaki *et al.*, 2005), the explanation for spatial synchrony may be more complicated. Geographical synchrony may arise from shared predators, dispersal between populations, and/or from common climatic influences (Post and Forchhammer, 2002). Sundell *et al.* (2004) argued that large-scale spatial synchrony is more likely to be caused by movements of predators and/or by climate rather than by dispersal. Also, Ims and Andreassen (2005) found that dispersal was insufficient to synchronise vole and lemming populations. However, as pointed out by Huitu *et al.* (2005), no single mechanism is likely to explain the synchronisation of populations, and rather the combined influence of abiotic (climate) and biotic (predation) conditions explain microtine population synchrony (see also Norrdahl and Korpimäki, 1996).

In case of the two synchronous collared lemming populations in Zackenbergdalen and on Traill Ø (Figure 1), it is evident that dispersal among the two populations is impossible. Moreover, the two lemming populations do not share predator populations, as the otherwise nomadic predator, the snowy owl, is almost absent in Zackenbergdalen. Climate is therefore expected to play an important role in shaping not only the pattern of fluctuations within populations (see above) but also across populations (see e.g., Moran, 1953).

VII. CONCLUSIONS

Despite relatively different predator communities, the collared lemming populations in Zackenbergdalen and on Traill Ø exhibit remarkably similar and synchronous population fluctuations. In both populations, predators seem to be the main driver of these fluctuations. Hence, the pattern of lemming fluctuations driven by a 1-year delay in stoat predation and stabilising predation from the generalist predators (long-tailed skua and especially arctic fox) found on Traill Ø (Gilg *et al.*, 2003, 2006) was also observed in Zackenbergdalen, though there was a much weaker link between the specialist stoat and its lemming prey.

In spring and summer, predation pressure is heavy, and the lemming predators are capable of reducing the density of collared lemmings in most summers. In winter, however, stoat predation alone is insufficient to regulate the collared lemming population. These marked effects of predators are also visible in the behaviour of collared lemmings in summer. Collared lemmings spent most of their time below ground, and while above ground they divide their time almost equally between foraging and being vigilant.

Climate is not only contributing to the shaping of the seasonal dynamics of the lemming population, but the onset of winter also affects the pattern of fluctuations markedly. Earlier onset of winter has no impact on the periodicity, whereas the later the onset of winter the longer the periodicity of the lemming fluctuations, moving towards less stable population fluctuations. Climate may, thus, be important for shaping not only the pattern of fluctuations within populations but also across populations, ultimately synchronising collared lemming populations in Greenland.

ACKNOWLEDGMENTS

Monitoring data for this chapter were provided by the BioBasis programme, run by the National Environmental Research Institute, University of Aarhus, and financed by the Danish Environmental Protection Agency, Danish Ministry of the Environment. Logistic facilities and support at Zackenberg Research Station were provided by the Danish Polar Center. Furthermore, we sincerely thank Benoit Sittler for allowing us to use the Traill Ø data, Olivier Gilg for discussions and feedback on the Zackenberg predator–prey model and Mikkel Tamstorf for preparation of vegetation coverage data.

REFERENCES

Aars, J. and Ims, R.A. (2002) *Ecology* **83**, 3449–3456.

Bay, C. (1998) *Vegetation Mapping of the Zackenberg Valley, Northeast Greenland.* Danish Polar Center & Botanical Museum, University of Copenhagen, Copenhagen, Denmark.

Begon, M., Harper, J.L. and Townsend, C.R. (1996) *Ecology: Individuals, Populations and Communities.* Blackwell Science.

Berg, T.B.G. (2003) *The Collared Lemming (Dicrostonyx Groenlandicus) in Greenland: Population Dynamics and Habitat Selection in Relation to Food Quality*. PhD thesis, National Environmental Research Institute, Denmark.

Bjørnstad, O.N., Falck, W. and Stenseth, N.C. (1995) *Proc. R. Soc. Lond. B* **262**, 127–133.

Blomqvist, S., Holmgren, N., Akesson, S., Hedenstrom, A. and Pettersson, J. (2002) *Oecologia* **133**, 146–158.

Boonstra, R., Krebs, C.J. and Kenney, A. (1996) *Can. J. Zool.* **74**, 1947–1949.

Boertmann, D. and Forchhammer, M.C. (1992) *A Review of Muskox Observations from North and Northeast Greenland. Greenland Environmental Research Institute. Report series* No. **4**, 1–36.

Brooks, R.J. (1993) *Lin. Soc. Symp. Ser.* **15**, 355–386.

Callaghan, T.V., Bjorn, L.O., Chernov, Y., Chapin, T., Christensen, T.R., Huntley, B., Ims, R.A., Johansson, M., Jolly, D., Jonasson, S., Matveyeva, N., Panikov, N., Oechel, W., Shaver, G. and Henttonen, H. (2004) *Ambio* **33**, 436–447.

Dalerum, F. and Angerbjörn, A. (2000) *Arctic* **53**, 1–8.

Dekker, J.J.A., Stein, A. and Heitkönig, M.A. (2001) *Can. J. Zool.* **225**, 505–510.

Ehrich, D., Jorde, P.E., Krebs, C.J., Kenney, A.J., Stacy, J.E. and Stenseth, N.C. (2001) *Mol. Ecol.* **10**, 481–495.

Fuller, W.A., Martell, A.M., Smith, R.F.C. and Speller, S.W. (1977) In: *Truelowe Lowland, Devon Island, Canada: A High Arctic Ecosystem* (Ed. by L.C. Bliss), pp. 437–459. University of Alberta Press, Edmonton.

Gilg, O. (2002) *Oikos* **99**, 499–510.

Gilg, O., Hanski, I. and Sittler, B. (2003) *Science* **302**, 866–868.

Gilg, O., Sittler, B., Sabard, B., Hurstel, A., Sané, R., Delattre, P. and Hanski, I. (2006) *Oikos* **113**, 193–216.

Hanski, I., Hansson, L. and Henttonen, H. (1991) *J. Anim. Ecol.* **60**, 353–367.

Hanski, I., Henttonen, H., Korpimaki, E., Oksanen, L. and Turchin, P. (2001) *Ecology* **82**, 1505–1520.

Hansson, L. and Henttonen, H. (1985) *Oecologia* **67**, 394–402.

Holling, C.S. (1959) *Can. Entomol.* **XCI**, 385–398.

Hörnfeldt, B., Hipkiss, T. and Eklund, U. (2005) *Proc. R. Soc. Lond. B* **272**, 2045–2049.

Hudson, P.J. and Bjørnstad, O.N. (2003) *Science* **302**, 797–798.

Huitu, O., Laaksonen, J., Norrdahl, K. and Korpimaki, E. (2005) *Oikos* **109**, 583–593.

Ims, R.A. and Andreassen, H.P. (2005) *Proc. R. Soc. Lond. B* **272**, 913–918.

Korpimäki, E., Brown, P.R., Jacob, J. and Pech, R.P. (2004) *Bioscience* **54**, 1071–1079.

Korpimaki, E., Norrdahl, K., Huitu, O. and Klemola, T. (2005) *Proc. R. Soc. Lond. B* **272**, 193–202.

Korslund, L. and Steen, H. (2006) *J. Anim. Ecol.* **75**, 156–166.

Krebs, C.J., Boonstra, R. and Kenney, A.J. (1995) *Oecologia* **103**, 481–489.

Krebs, C.J., Kenney, A.J., Gilbert, S., Danell, K., Angerbjörn, A., Erlinge, S., Bromley, R.G., Shank, C. and Carriere, S. (2002) *Can. J. Zool.* **80**, 1323–1333.

Kyhn, L.A. (2004) *A Study on the Behaviour of Wild Collared Lemmings (Dicrostonyx Groenlandicus Traill, 1823)*. Report in Behavioural Biology, University of Copenhagen.

Lindström, J., Ranta, E., Kokko, H., Lundberg, P. and Kaitala, V. (2001) *Biological Reviews* **76**, 129–158.

Macpherson, A.H. (1969) *The Dynamics of Canadian Arctic Fox Populations*. Can. Wildlife Serv. Report Series No. 8.

Mech, L.D. (2000) *Arctic* **53**, 69–71.
Meia, J.-S. and Weber, J.-M. (1995) *Can. J. Zool.* **73**, 1960–1966.
Meltofte, H. (2006) *Wildfowl* **56**, 129–151.
Meltofte, H. and Berg, T.B. (2006) *BioBasis—Conceptual Design and Sampling Procedures of the Biological Programme of Zackenberg Basic, 9th edition*. National Environmental Research Institute, Denmark.
Meltofte, H. and Høye, T.T. (2007) *Dansk Orn. Foren. Tidsskr.* **101**, 109–119.
Mikkelsen, P.S. (1994) *Nordøstgrønland 1908–60. Fangstmandsperioden*. Dansk Polarcenter, Copenhagen.
Moran, P.A.P. (1953) *Aust. J. Zool.* **1**, 291–298.
Norrdahl, K. and Korpimäki, E. (1996) *Oecologia* **107**, 478–483.
Norrdahl, K. and Korpimäki, E. (1998) *Ecology* **79**, 226–232.
Norrdahl, K. and Korpimäki, E. (2005) *Evol. Ecol.* **19**, 339–361.
Petersen, U.M. (1998) *J. Mammal.* **79**, 236–244.
Post, E., Peterson, R.O., Stenseth, N.C. and McLaren, B.E. (1999) *Nature* **401**, 905–907.
Post, E. and Forchhammer, M.C. (2002) *Nature* **420**, 168–171.
Reid, D.G. and Krebs, C.J. (1996) *Can. J. Zool.* **74**, 1284–1291.
Reid, D.G., Krebs, C.J. and Kenney, A. (1995) *Oikos* **73**, 387–398.
Reid, D.G., Krebs, C.J. and Kenney, A.J. (1997) *Ecol. Monogr.* **67**, 89–108.
Reneerkens, J., Piersma, T. and Damsté, J.S.S. (2005) *J. Exp. Biol.* **208**, 4199–4202.
Roth, J.D. (2003) *J. Anim. Ecol.* **72**, 668–676.
Royama, T. (1992) *Analytical Population Dynamics*. Chapman & Hall, London.
Schmidt, N.M. (2000) *Spatiotemporal Distribution and Habitat Use of the Collared Lemming, Dicrostonyx Groenlandicus Traill, in High Arctic Northeast Greenland*. Master thesis. Department of Zoology, University of Aarhus, Denmark.
Schmidt, N.M. (2006) *Climate, Agriculture and Density-Dependent Dynamics Within and Across Trophic Levels in Contrasting Ecosystems*. PhD thesis. Department of Ecology, Royal Veterinary & Agricultural University, Denmark.
Schmidt, N.M., Berg, T.B. and Jensen, T.S. (2002) *Can. J. Zool.* **80**, 64–69.
Schmidt, N.M., Baittinger, C. and Forchhammer, M.C. (2006) *Arct. Antarct. Alp. Res.* **38**, 257–262.
Sigsgaard, C., Petersen, D., Grøndahl, L., Thorsøe, K., Meltofte, H., Tamstorf, M. and Hansen, B.U. (2006) In: *Zackenberg Ecological Research Operations, 11th Annual Report, 2005* (Ed. by A.B. Klitgaard, M. Rasch and K. Caning), pp. 11–35. Danish Polar Center, Ministry of Science, Technology and Innovation, Copenhagen.
Sittler, B. (1995) *Ann. Zool. Fenn.* **32**, 79–92.
Sittler, B., Gilg, O. and Berg, T.B. (2000) *Arctic* **53**, 53–60.
Smits, C.M.M. and Slough, B.G. (1993) *Can. Field-Nat.* **107**, 13–18.
Stenseth, N.C. and Ims, R.A. (1993) *Lin. Soc. Symp. Ser.* **15**, 61–96.
Stenseth, N.C., Bjørnstad, O.N. and Falck, W. (1996a) *Proc. R. Soc. Lond. B* **263**, 1423–1435.
Stenseth, N.C., Bjørnstad, O.N. and Saitoh, T. (1996b) *Proc. R. Soc. Lond. B* **263**, 1117–1126.
Stenseth, N.C., Bjørnstad, O.N. and Saitoh, T. (1998) *Res. Popul. Ecol.* **40**, 85–95.
Stenseth, N.C., Kittilsen, M.O., Hjermann, D.O., Viljugrein, H. and Saitoh, T. (2002) *Proc. R. Soc. Lond. B* **269**, 1853–1863.
Stenseth, N.C., Shabbar, A., Chan, K.S., Boutin, S., Rueness, E.K., Ehrich, D., Hurrell, J.W., Lingjærde, O.C. and Jakobsen, K.S. (2004) *PNAS* **101**, 10632–10634.
Summers, R.W. and Underhill, L.G. (1987) *Bird Study* **34**, 161–171.

Sundell, J., Huitu, O., Henttonen, H., Kaikusalo, A., Korpimäki, E., Pietiainen, H., Saurola, P. and Hanski, I. (2004) *J. Anim. Ecol.* **73**, 167–178.
Turchin, P. and Hanski, I. (2001) *Ecol. Lett* **4**, 267–276.
Turchin, P., Oksanen, L., Ekerholm, P., Oksanen, T. and Henttonen, H. (2000) *Nature* **405**, 562–565.
Underhill, L.G., Prysjones, R.P., Syroechkovski, E.E., Groen, N.M., Karpov, V., Lappo, H.G., Vanroomen, M.W.J., Rybkin, A., Schekkerman, H., Spiekman, H. and Summers, R.W. (1993) *Ibis* **135**, 277–292.
Wiklund, C.G., Angerbjörn, A., Isakson, E., Kjellén, K. and Tannerfeldt, M. (1999) *Ambio* **28**, 281–286.
Wilson, D.J., Krebs, C.J. and Sinclair, T. (1999) *Oikos* **87**, 382–398.

Lake Flora and Fauna in Relation to Ice-Melt, Water Temperature and Chemistry at Zackenberg

KIRSTEN S. CHRISTOFFERSEN, SUSANNE L. AMSINCK, FRANK LANDKILDEHUS, TORBEN L. LAURIDSEN AND ERIK JEPPESEN

SUMMARY

The ecology of arctic lakes is strongly influenced by climate-generated variations in snow coverage and by the duration of the ice-free period, which, in turn, affect the physical and chemical conditions of the lakes (Wrona *et al.*, 2005). Most arctic lakes are characterised by a long period (8–10 months) of ice-cover, cold water and low algal biomass. The water temperature and nutrient concentrations, and most probably the nutrient input from the catchments, are closely related to the duration of snow- and ice-cover in the lakes. In years when the ice-out is late,—that is, in late July,—phytoplankton photosynthesis is limited by the lack of light and nutrients. Less food is then available to the next link in the food chain, such as copepods and daphnids, with implication on their growth rates.

ADVANCES IN ECOLOGICAL RESEARCH VOL. 40
© 2008 Elsevier Ltd. All rights reserved
0065-2504/08 $35.00
DOI: 10.1016/S0065-2504(07)00016-5

A more maritime climate has been predicted for Northeast Greenland, and an increasing duration of the ice-free period has been observed during the last decade in the lakes at Zackenberg. If this continues, it may result in lower food availability to the top predator of the lakes, the arctic charr *Salvelinus alpinus*. The results obtained so far indicate that despite the often low abundance of arctic charr, it has a strong regulating impact on the composition and density of the zooplankton community and the abundance of the arctic tadpole shrimp, *Lepidurus arcticus*, which likely regulates the density of benthic cladocerans.

I. INTRODUCTION

A large number of shallow and deep lakes are found in near-coastal areas in Greenland. Some of the lakes have been subjected to limnological investigations in the past, and valuable information has been obtained on the geographical distribution and population dynamics of fish (e.g., Bergersen, 1996; Riget *et al.*, 2000), macrophytes (e.g., Fredskild, 1992) and zooplankton (e.g., Wesenberg-Lund, 1894; Røen, 1977). However, no major investigations of Greenland lakes at ecosystem level have been undertaken so far.

The physical and chemical conditions in arctic lakes leave a very short growing season for aquatic organisms, and the role of climate is therefore very important. Greenland lakes and ponds are generally ice-covered most of the year, typically for 8–10 months (Figure 1). The maximum ice thickness is around 2.5 m (Røen, 1962), and shallow water bodies thus freeze solid in late winter and spring. An additional snow layer of 0.5–1 m may cover the ice, preventing light penetration (Vincent and Hobbie, 2000). The nutrient concentrations are low due to the inflowing water that primarily originates from runoff of melting ice and snow. The melt-water transports silt, inorganic and organic particles, as well as atmospheric deposition of substances into the lakes. This—combined with the fact that most lakes have low average summer temperatures, low nutrient availability, and, with it, low primary production—implies that species richness is limited and biomass often relatively low compared with lakes in temperate regions. Therefore, interactions between organisms are less complex than elsewhere.

Plants and animals are active before the ice has melted (e.g., Hobbie *et al.*, 1999). The increased influx of light in April–May means that sufficient light can penetrate the snow and ice layer, initiating plant photosynthesis and a consequent phytoplankton biomass increase (Figure 2). This shows that light, and not water temperature, is the limiting factor for primary production prior to ice-out. While the ice slowly melts, the phytoplankton growth stagnates or decreases, primarily because the low nutrient concentrations are exploited very quickly and due to enhanced grazing by zooplankton. A later summer (autumn) peak is observed in some lakes (Forsström *et al.*, 2005).

Figure 1 Lake Sommerfuglesø in Zackenbergdalen that has not yet lost all its winter ice in the beginning of August (1999). Photo: Kirsten S. Christoffersen.

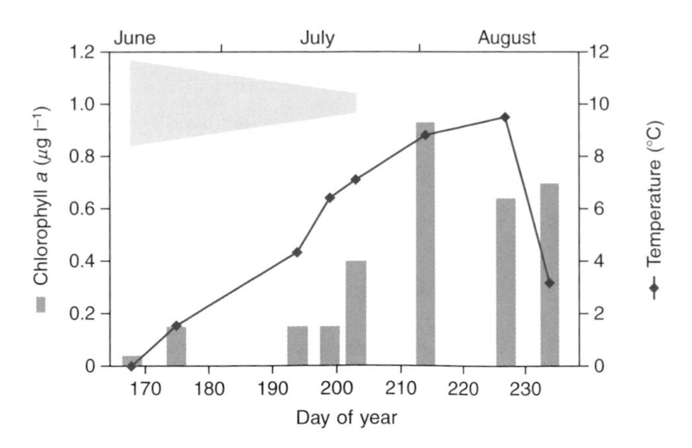

Figure 2 Seasonal variation in phytoplankton chlorophyll (columns), water temperature and ice-cover (blue bar) in Langemandssø during June–August 1998 (day of the year 17 June to 22 August).

The pelagic food web in Northeast Greenland lakes is short, but comprises two important chains (Figure 3) that are well-known from temperate lakes (e. g., Riemann and Christoffersen, 1993). One is the classic grazer food chain

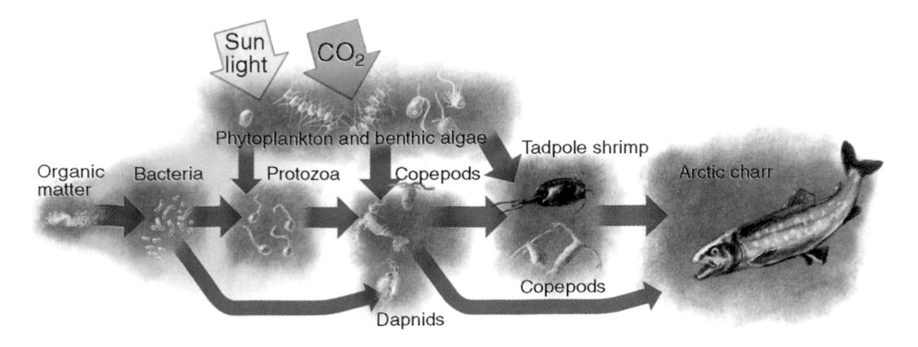

Figure 3 A schematic illustration of the interconnections between the classic, the benthic and the microbial food web in arctic lakes. Tadpole shrimps *Lepidurus arcticus* are generalist predators that consume live and dead organic material and may catch swimming daphnids (modified from Christoffersen, 2006).

where phytoplankton is eaten by zooplankton, such as daphnids and copepods, which themselves serve as prey to benthic invertebrates and fish (Christoffersen, 2001; Jeppesen *et al.*, 2003a). The other link is the "microbial loop" where dissolved organic matter (DOM) from algae and other organisms is lost via excretion and faeces or DOM deriving from the catchment (e. g., humus) is consumed by bacteria. The bacteria, in turn, are eaten by heterotrophic flagellates and ciliates (protozoans), which in turn are preyed upon by daphnids and copepods. These microzooplankton organisms then become prey of daphnids and copepods. Daphnids, however, also consume bacteria and may therefore exploit more sources of food than copepods, which, in turn, may feed on eggs of daphnids (Gliwicz, 2003). Part of the primary production sinks to the bottom, either because it cannot be consumed directly or because it has already been consumed and is transformed into faecal pellets.

A benthic food web is also present and is often very important, as demonstrated recently by Rautio and Vincent (2006, 2007). They quantified the food resources for zooplankton and concluded that the benthic mats were exploited much more intensively than hitherto assumed. Heterotrophic organisms such as bacteria, protozoans, chironomid larvae and often also tadpole shrimps living in the uppermost layers of the lake bottom, are the main consumers of the sediment. Some of these are themselves very important food items for fish and birds. However, the most important component at the lake bottom is the autotrophic organisms represented by epiphytic algae and mosses. Only few vascular plant species exist along the shores and are typically low in abundance.

The results obtained so far indicate that despite the often low abundance of arctic charr, the only fish species present in Northeast Greenland lakes, it has a strong regulating impact on the composition and density of the zooplankton community and the abundance of tadpole shrimps in the lakes; tadpole shrimps, in turn, seem to reduce the density of benthic cladocerans (Jeppesen *et al.*, 2003a,b).

II. RATIONALE AND METHODS

Because of their simplicity and relatively low species diversity, arctic lakes are useful model ecosystems for evaluating the effects of climatic changes. They are also well delimited and heavily influenced by the local climatic conditions in their catchments.

The first sampling of lakes in the Zackenberg area took place in 1997 and included 19 shallow lakes and ponds situated in the western part of the valley Zackenbergdalen (Figure 4), that is, in Morænebakkerne (the moraine hills) and Vestkæret. Additionally, two water bodies situated to the south of the field station were included, namely Lomsø and one of the ponds of Sydkærene (no. 19).

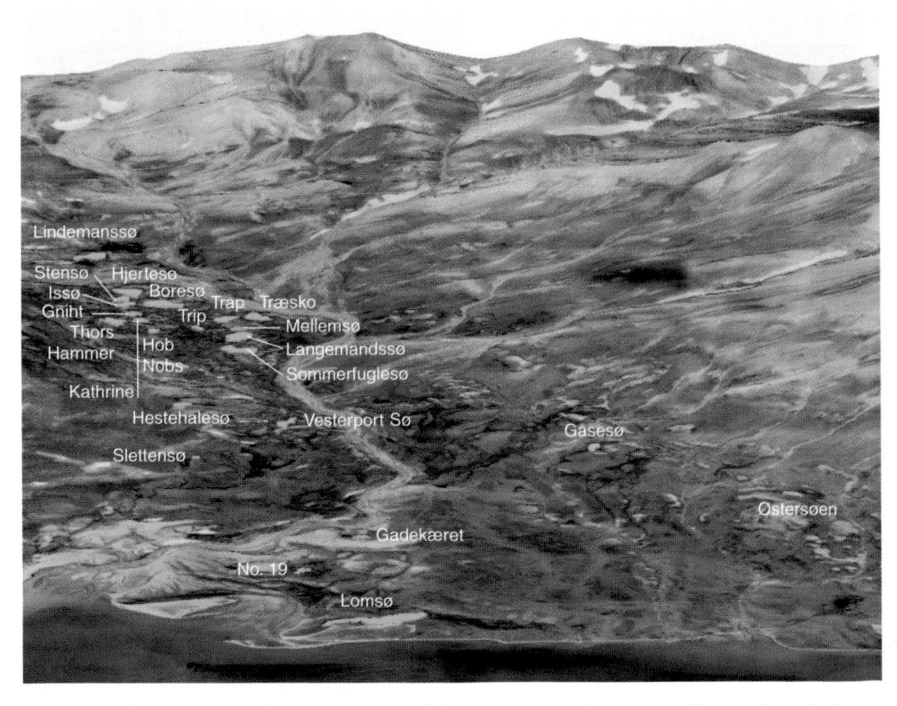

Figure 4 Location of the investigated lakes in the study area in Zackenbergdalen.

Each lake was visited once, and the sampling programme included fish, zooplankton, phytoplankton, ciliates, flagellates, bacterioplankton and a number of physico-chemical parameters (Table 1). Water chemistry and plankton populations were either measured directly in the water from a rubber boat anchored at a mid-lake position or later in the field station's laboratory or in Denmark using water preserved by fixatives or by freezing. All water samples were collected by mixing water from the entire column using a special design sampler. Moreover, the upper 1 cm of the sediment was sampled by a Kajak sampler, and the material was later analysed for

Table 1 Basic characteristics of the lakes investigated in 1997 and 2003

Name	Area (km^2)	Max depth (m)	Conductivity ($\mu S\ cm^{-1}$)	Total P ($\mu g\ l^{-1}$)	Total N ($\mu g\ l^{-1}$)	Chlorophyll ($\mu g\ l^{-1}$)
Vesterport Sø (G)	0.004	0.7	18	27	570	2.2
Sommerfuglesø	0.017	1.8	15	4	250	0.8
Langemandssø (F)	0.011	6.1	8	8	170	1.0
Mellemsø	0.003	0.2	11	2	160	1.1
Trip	0.003	0.9	19	10	830	0.6
Trap	0.008	1.5	17	10	280	3.1
Træsko	0.013	2.7	14	15	300	3.2
Hob Nobs	0.004	0.6	16	13	400	0.7
Kathrine	0.007	0.5	14	2	140	0.5
Gniht Sø	0.005	1.5	25	ND	430	1.0
Boresø (F)	0.025	4.6	9	6	200	1.3
Hjertesø (F)	0.035	6.7	9	7	150	1.5
Stensø	0.003	0.7	12	10	490	1.1
Issø	0.004	1.7	16	15	260	3.3
Thors Hammer	0.008	3.4	10	2	150	0.6
Hestehalesø	0.005	0.3	36	5	270	0.8
Slettensø (G)	0.002	0.3	22	6	460	1.4
Lomsø (G)	0.016	1.5	77	15	580	2.0
Sydkær no. 19	0.004	0.2	44	10	730	4.9
Gåsesø (G)	ND	ND	26	34	920	10.2
Østersøen (G)	ND	ND	10	17	90	3.0
Lindemanssø	ND	ND	33	ND	ND	ND

The area is estimated from aerial photographs and observations on location. The data shown are from the first survey on the particular lakes, and all chemistry data are means for the water column. Abbreviations: ND, no data; F, fish population present; and G, often used by geese. "Max depth" is not necessarily the absolute maximum depth, but denotes the point where the largest depth was recorded and where sampling was undertaken. The two lakes that were monitored every year are marked with bold.

zooplankton remains. Finally, a fish survey was taken using standardised gillnets. The results of the entire survey were presented in Meltofte and Rasch (1998) and will be summarized below.

A second survey was performed in 2003. All the previous 19 lakes studied were revisited to investigate if changes had occurred in basic water chemistry, and 3 additional lakes were included (Table 1, last 3 lakes).

A monitoring programme for two of the shallow lakes in Morænebakkerne, one with and one without arctic charr, was initiated in 1999 (Meltofte and Berg, 2006). Selected physical (ice-cover and water temperature), chemical [pH, conductivity, total nitrogen (TN) and total phosphorous (TP)] and biological parameters (phytoplankton biomass and species composition) are measured at regular intervals during summer (i.e., mid-July to mid-August). Pelagic invertebrates (zooplankton) are sampled in mid-August for determination of species abundance and composition. The monitoring programme also includes investigations of the fish population every fifth year and finally analyses of the surface sediment (micro- and macrofossils) at ~10-year intervals.

The physico-chemical and biological analyses follow conventional procedures used in standard monitoring programmes in Denmark (Jespersen and Christoffersen, 1987; Jeppesen *et al.*, 2003a; Kronvang *et al.*, 2005).

III. GENERAL CHARACTERISTICS OF THE LAKES

The lakes and ponds are generally small and shallow with maximum depths ranging from 0.2 to 6.7 m (Table 1). This means that all ponds and a majority of the lakes freeze solid during winter. While many lowland ponds often have a thick mud layer, lake beds mostly consist of rock, boulders and pebbles as well as coarse and fine sand particles. The deeper lakes also have an organic layer. All lakes have clear waters originating from melted snow and runoff in the catchment area. Thus, there are no obvious effects of humic substances or silt. Several lakes (Sommerfuglesø, Langemandssø, Hjertesø, Boresø and Gåsesø) have extensive coverage of mosses and a few lakes and ponds (e.g., Hestehalesø and Slettensø) have vascular plants mare's tail *Hippuris vulgaris* and high-arctic buttercup *Ranunculus hyperboreus* in the littoral zone.

The rather nutrient-poor conditions are reflected by the low specific conductivity (average 21 μS cm^{-1}), low concentrations of total phosphorus (TP; average is 11 μg TP l^{-1}) and total nitrogen (TN; average 373 μg TN l^{-1}) (Table 1). Such low values are typical for high-arctic lakes and ponds (e.g., Rautio and Vincent, 2006). By contrast, the mean values for Danish lakes studied so far are as high as 210 μg P l^{-1} and 2100 μg N l^{-1} (Søndergaard *et al.*, 2005). In some of the lakes, conductivity and TP are comparatively high (Vesterport Sø, Lomsø and Gåsesø), which may be caused by

Table 2 Average values of temperature, pH, conductivity, total phosphorous and chlorophyll of 19 lakes in Zackenbergdalen during July–August 1997 and in 17 of these lakes during the same period in 2003 (two were dried out in the meantime)

Year	Temperature (°C)	Conductivity (μS cm^{-1})	Total P (μg l^{-1})	Total N (μg l^{-1})	Chlorophyll (μg l^{-1})
1997 (19 lakes)	8.7 (1.9)	21 (17)	10 (6)	314 (182)	1.8 (1.2)
2003 (17 lakes)	11.4 (1.1)	17 (12)	14 (8)	298 (213)	2.5 (1.5)

Mean values are given with the standard deviation in parentheses. *t*-tests show that water temperature, total phosphorous and chlorophyll *a* were significantly higher and that pH and conductivity were significantly lower (two-tailed, *p* 0.01) in 2003 than in 1997.

flocks of geese and pairs of divers that commonly rest and nest here. Thus, birds bring nutrients from land and sea to the lake water via droppings. All lakes have well-oxygenated waters with 80–100% saturation including the bottom waters, and pH ranges between 6.0 and 7.8.

From the revisit to the lakes, it appeared that 2 of these were temporary and had dried out (Mellemsø and Trip), but in the remaining 17 lakes significant differences (2-tailed paired *t*-test) were found in water temperature (higher), pH (lower), conductivity (lower), chlorophyll (higher) and total phosphorous (higher) (Table 2). TN concentrations, in contrast, remained unchanged. These differences were attributed to warmer summers in recent years (see below).

The two lakes monitored each year since 1997, Sommerfuglesø and Langemandssø, are of similar size (0.017 and 0.011 km^2, respectively) with maximum depths of 1.8 m (Sommerfuglesø) and 6.1 m (Langemandssø; Table 1). In both lakes, ice starts melting nearshore in mid-June, but they are usually ice-free only from mid- or late July until the second half of September, as are the other lakes investigated. No precise measure of the ice thickness exists for the monitored lakes, but in the beginning of June 1997, an ice thickness of 1.5–1.7 m was measured in several of the other lakes in the area (Christoffersen and Landkildehus, unpublished data).

IV. BIOLOGICAL STRUCTURE AND FOOD WEB INTERACTIONS

The phytoplankton communities of the intensely studied Greenland lakes include genera of diatoms (Diatomophyceae), dinophytes (Dinophyceae), and chrysophytes (Chrysophyceae), while chlorophytes (Chlorophyceae) and cyanobacteria (Nostocophyceae) are less common. Diatoms are represented by numerous species. A typical and commonly occurring chrysophyte genus is *Dinobryon*; dinoflagellates are represented by *Gymnodinium* spp. and *Peridinium* spp., while *Koliella longista* is dominant among the chlorophytes. Other chlorophytes, as well as several naked flagellates, also occur. The phytoplankton biomass is dominated (often 95% of the biovolume) by chrysophytes in both lakes. The most important taxa in Sommerfuglesø are Chrysophyceae, *Stichogloea* spp. and *Dinobryon bavaricum*, while Chrysophyceae, *D. bavaricum*, *D. boreale*, *Kephyrion boreale* and *Ochromonas* spp. dominate in Langemandssø. In addition, several dinophyceans (*Gymnodinium* spp. and *Peridinium umbonatum* group) as well as nanoflagellates (Sommerfuglesø) and green algae (Langemandssø) occur. Dominance of chrysophytes and dinophytes are typical for high-arctic lakes (e.g., Rautio and Vincent, 2006).

Zooplankton communities are species poor in Greenland lakes (Røen, 1962; Lauridsen *et al.*, 2001) and the species composition is clearly dependent on whether fish are present or not (Jeppesen *et al.*, 1998, 2001; Lauridsen *et al.*, 2001; Jeppesen *et al.*, 2003a,b). The large-sized daphnid *Daphnia pulex* as well as other cladocerans are almost absent in lakes with fish (arctic charr), while cyclopoid copepods (e.g., *Cyclops abyssorum alpinus*) can coexist with fish. Rotifers are almost always present, and their populations proliferate in situations where their greatest competitor, daphnids, is diminished by the fish. Ostracods can also be numerous and are often associated with macrophytes, mosses and other surfaces for attachment. The zooplankton community in Sommerfuglesø consists of the cladoceran *D. pulex*, the copepod *C. abyssorum alpinus* and the rotifers *Polyarthra dolicopthera* and *Keratella quadrata*. The fact that *D. pulex* dominated the community probably explains the low abundances of copepods and rotifers. *C. abyssorum alpinus* and *P. dolicopthera* dominated the community in the fish-containing Langemandssø.

During the initial survey in 1997, fish were sampled with biological multi-mesh-sized gill nets (see Jeppesen *et al.*, 2001 for details). Fish were caught only in the three deepest lakes (maximum depths were 4.6, 6.1 and 6.7 m, respectively). The fish stock consisted solely of dwarf arctic charr that mature sexually at a size of only 11–13 cm. The catch per net (CPUE) was as low as 0.5–1.4 (Jeppesen *et al.*, 2001). In comparison, the fish CPUE typically reaches values as high as 100–400 in nutrient-rich Danish lakes (Jeppesen *et al.*, 2003b).

A new survey of the charr population was undertaken in 2005 in the two intensively studied lakes using "Ella traps." Three traps were set at different heights above the bottom in Langemandssø from 17 July to 4 August. A single trap was simultaneously placed in Sommerfuglesø. The traps caught three dwarf charr in Langemandssø and no fish in Sommerfuglesø, which supports the results from 1997.

V. SIGNIFICANT ANNUAL VARIATIONS IN PHYSICO-CHEMICAL CONDITIONS

During the 9-year monitoring period, the spring date for 50% ice-cover in Sommerfuglesø and Langemandssø has varied by almost a month. The earliest year was 2005 when the ice-cover reached 50% in both lakes already during mid-June (18 and 22 June, respectively). This coincided with a very low snow-cover of 37% on 10 June compared to an average for the study period of 72% (Sigsgaard *et al.*, 2006). By contrast, the snow-cover on 10 June was 92% in 1999, and the snow disappeared very late this year. As a result, the two lakes reached 50% ice-cover as late as 18 and 21 July, respectively.

Water temperatures (averaged for the water column and for the summer) ranged from 4 to 11 °C during the sampling years (Figure 5). This variation was related to hydrological conditions (depth, residence time and inflowing melt-water), water temperature being higher in years with early ice-melt, but was also influenced by the actual weather conditions.

Snow-cower in early summer (as recorded by 10 June) and the timing of the ice-melt apparently have a greater impact on water temperature than has air temperature in summer, as indicated by higher r^2 values of these relationships (Figure 6A and B). Timing of ice-melt and water temperature was also closely related for other locations in Zackenbergdalen for which data are available, the larger Lomsø and pond no. 19 in Sydkærene. The tight coupling between snow-cover, ice-melt and water temperature was reflected in the appearance of the planktonic communities. Increased water temperatures resulted in increases in chlorophyll and in the abundance of key-zooplankton taxa abundance (Figure 6C and D).

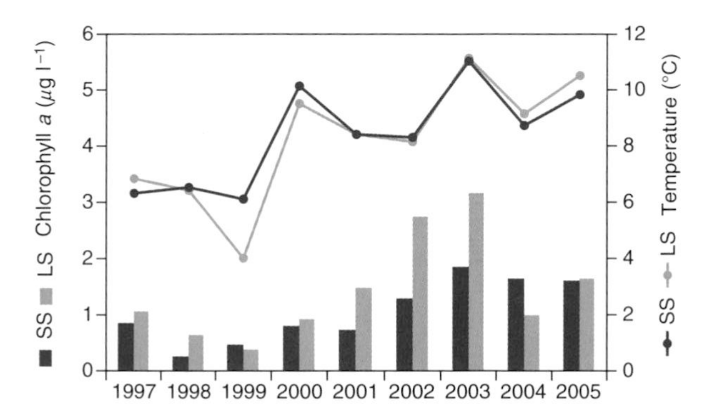

Figure 5 Annual variations in phytoplankton chlorophyll (columns), water temperature and ice-cover in Langemandssø (LS) and Sommerfuglesø (SS) during 1997–2005. The data are either single measurements from one sampling date in the beginning of August (1997) or an average of 2–3 sampling dates in July/August (1998–2005).

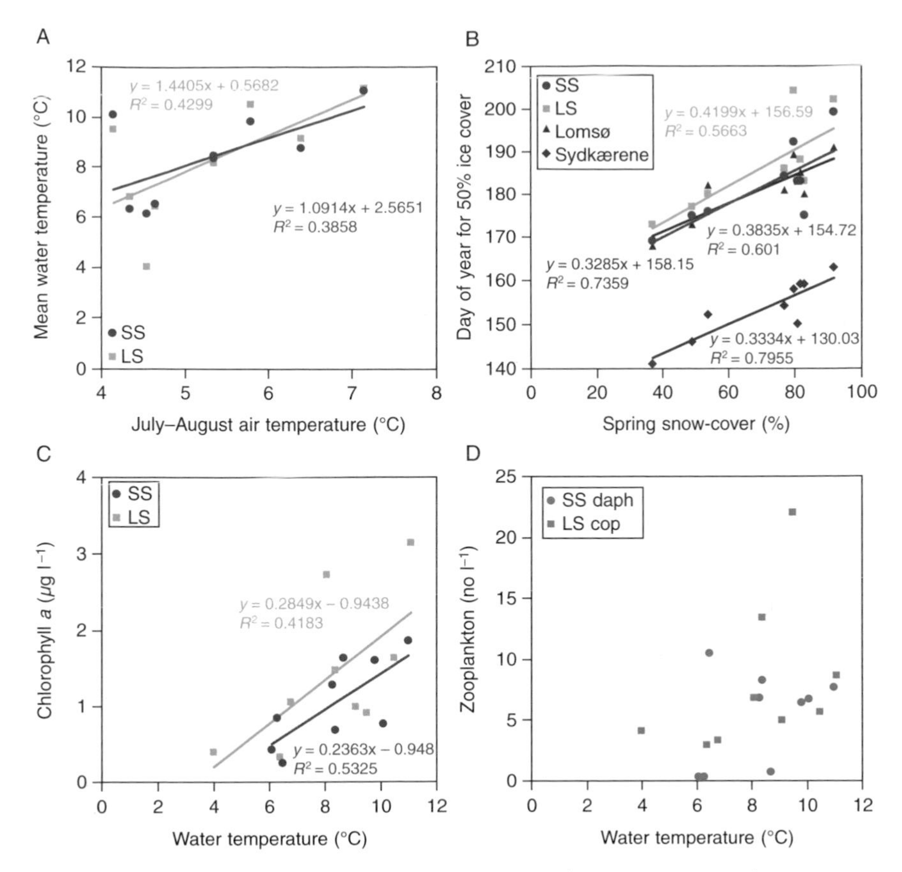

Figure 6 Relationships between (A) July–August air temperature and mean water temperature, (B) the percentage of spring snow-cover (10 June) and day of the year (20 May–29 July) for 50% ice-cover, (C) mean water temperature and chlorophyll and (D) mean water temperature and zooplankton abundance.

VI. WATER TEMPERATURE AFFECTS GROWTH AND ABUNDANCES OF PELAGIC ORGANISMS

The concentration of phytoplankton (expressed as chlorophyll) varies from year to year and between the two lakes (Figure 5), most likely reflecting inter-annual variation in the concentrations of nutrients and water temperature. The low average water temperature and low nutrient concentrations recorded in, for instance, 1999 evidently led to lower phytoplankton abundance and dominance of dinophytes compared with the warmer season of, for example, 2001, with chrysophyte dominance. In 1999, ice-melt did not occur until late July, and the average water temperature was therefore low.

Chrysophytes and dinophytes together made up 93% of total phytoplankton abundance in Sommerfuglesø, while dinophytes constituted 89% of the phytoplankton in the deeper and colder Langemandssø. The average phytoplankton biomass reached its highest level in 2003.

The redundancy analyses (RDA) plots (Figure 7) suggest decreasing importance of Nostocphyceae and Diatomphyceae and increasing importance of Dinophyceae and Chrysophyceae along with rising levels of chlorophyll *a*,

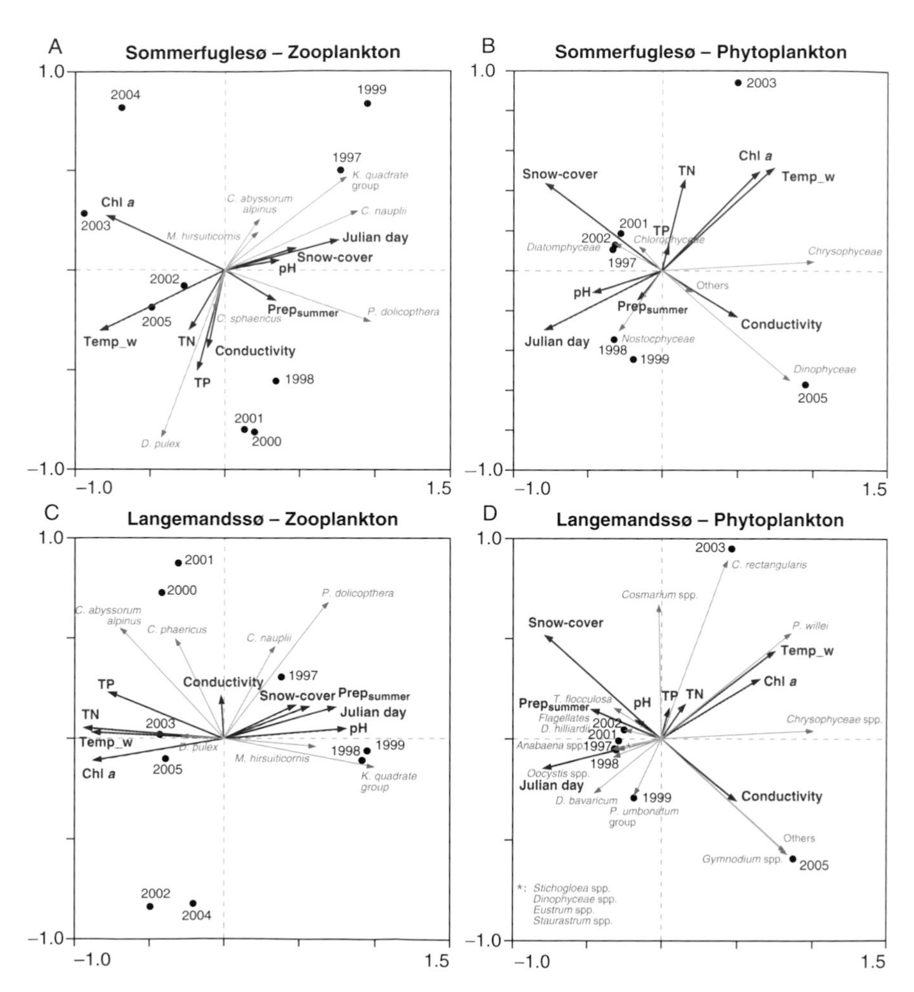

Figure 7 Redundancy analyses (RDA) showing trends and relationships between the community structure of zooplankton and phytoplankton, respectively, and environmental variables of Sommerfuglesø (A, B) and Langemandssø (C, D) during the study period of 1997–2005.

temperature and conductivity in Sommerfuglesø. In Langemandssø, *Gymnodium* spp. seemingly increased in importance along with similar shifts in environmental conditions as those occurring in Sommerfuglesø.

Also the abundances of cladocerans and advanced stages of copepods in the fish-free Sommerfuglesø and advanced stages of copepods in the fish-containing Langemandssø increased in years with warm water temperatures (Figures 6D, 7 and 8). The explanation of higher abundances is probably improved growth conditions in warm years when both food and water temperatures are more favourable. Apparently, rotifers react negatively to high temperatures, most probably reflecting that they are inferior competitors to daphnids and copepods and also sensitive to predation by advanced stages of copepods. The RDA analyses (Figure 7) indicate a change in the zooplankton community structure from reduced importance of the *K. quadrata* group, *Cyclops nauplii* and *P. dolicopthera* towards increasing importance of *D. pulex* along with increasing levels of chlorophyll *a*, temperature and TN; and towards higher dominance of advanced stages of *C. abyssorum alpinus* and *Chydorus sphaericus* in Langemandssø.

The dominance of small-sized specimens in Langemandssø suggests a strong predation pressure on zooplankton (Jeppesen *et al.*, 2001) despite that the low abundances of fish and phytoplankton likely are primarily regulated by nutrients.

The differences between the relatively warm (i.e., early ice-out) and cold years become evident when the 9 years of data are divided into years with cold (1997, 1998 and 1999), warm (2000, 2003 and 2005) and intermediate

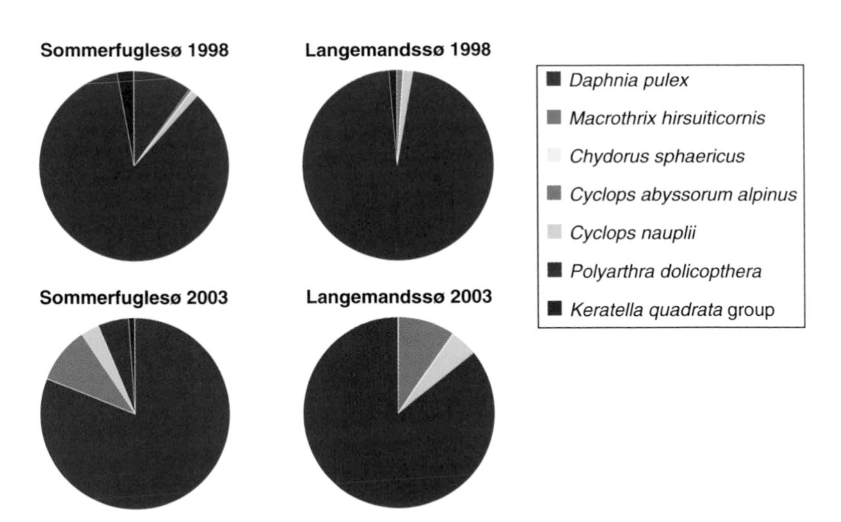

Figure 8 Composition of zooplankton abundances in Langemandssø and Sommerfuglesø in 1998 and 2003 representing a cold and a warm season, respectively.

(2001, 2002, and 2004) seasons, respectively (Figure 9). Despite the short data series, it appears that cold years differ from intermediate and warm years with regard to the outcome of nutrient concentrations (especially nitrogen) as well as phytoplankton and zooplankton populations. Warmer years generally lead to higher abundance of phytoplankton and crustaceans and alter taxon composition. The higher nutrient concentrations in warm years are probably caused by increased loading of nutrients and humus from the catchment when the active layer melts.

However, no significant relationship was found between any of the physico-chemical or biological variables and the winter North Atlantic Oscillation (NAO) index, indicating that local climate variations are more important

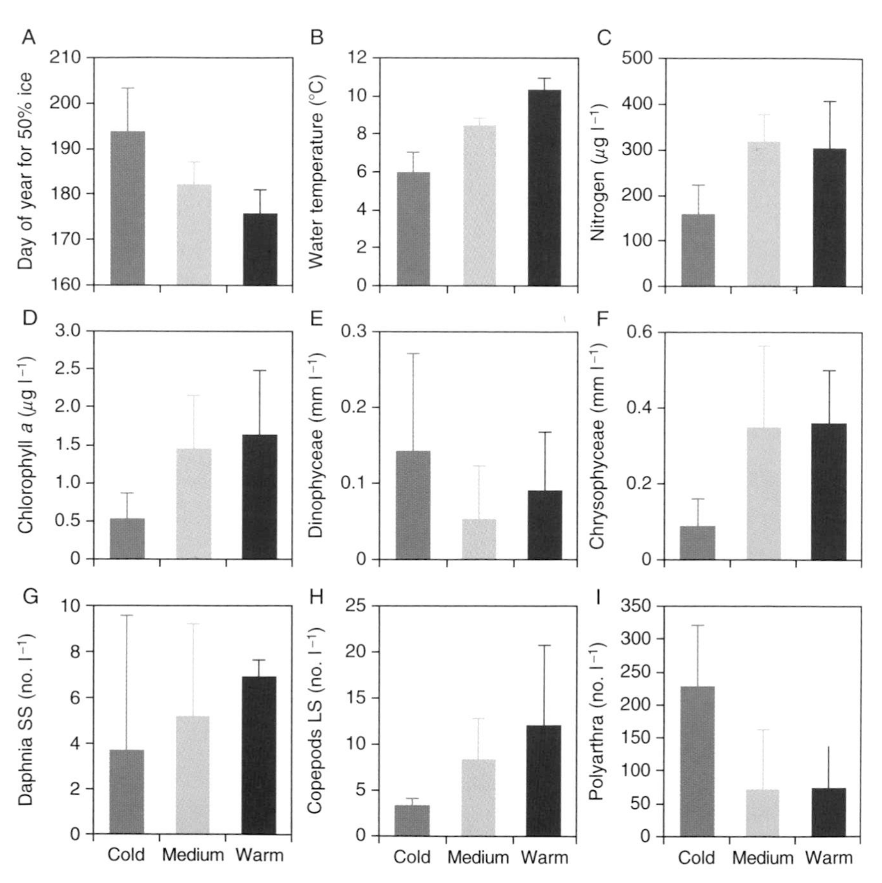

Figure 9 Calculated mean values for a number of parameters in three warm (2000, 2003 and 2005), cold (1997 1998 and 1999) and intermediate (2001, 2002 and 2004) years (day of the year 9 June–29 July).

driving factors than is the NAO as otherwise seen in European lakes (Livingstone and Dokulil, 2001; Straile *et al.*, 2003; George *et al.*, 2004). Another explanation may be that the time series is simply too short to show such relationships.

VII. AN ANCIENT INVERTEBRATE SEEMS TO PLAY A ROLE IN THE FOOD WEB

The arctic tadpole shrimp, *L. arcticus*, occurs in a number of lakes and ponds, especially those with soft sediments, whereas it is completely absent from lakes with arctic charr. Knowledge about the role of tadpole shrimps in lake ecosystems is poor. *Lepidurus* is described in the literature as a scavenger eating settled material and presumably small organisms living in the sediment, but its quantitative role in decomposition has not been studied.

Observations and experimental studies during several summers provided new insight into the life cycle of *L. arcticus*. The animals hatch from resting eggs in the bottom as soon as the ice melts and pass quickly through a number of development stages. During the first few weeks the larvae are frequently seen swimming in the water, but spend more time in and on the sediments as they grow larger. Newly hatched individuals are a few millimetres in length and full-grown specimens reach several centimetres by the end of the summer.

While *L. arcticus* most likely is mainly a benthic feeder, several observations of the behaviour of *L. arcticus* in Gadekæret and ponds in Sydkærene in 1997 indicated that it was able to catch the crustacean *D. pulex* when swimming in the water. It was therefore hypothesised that *Lepidurus* had several feeding strategies and a high food intake (judged from its high growth rate), which potentially could affect benthic and planktonic microorganisms. Experiments to elucidate this potential were carried out. *L. arcticus* and *D. pulex* were sampled in ponds around Zackenberg and placed in small containers (0.5 litre) at ambient temperature. A number of repeated feeding trials demonstrated that *L. arcticus* was able to catch and consume *D. pulex* and easily consumed up to six individuals per hour. Although this estimate is clearly biased due to the manipulated conditions (increased encounter rate), it indicates that *Lepidurus* is an active and effective predator on large-sized prey items from the water column (Christoffersen, 2001).

Abundances of *L. arcticus* in a series of the ponds in the area were estimated from specially designed samplers placed randomly in the littoral zone in areas with different sediment types (sand, pebbles, mud and mosses) for 24 h. On the basis of these samplings, it was estimated that up to several hundred individuals may occur per square meter of lake bottom, with a typical density of 50–100 specimens per square meter (Christoffersen, unpublished data).

VIII. BIOLOGICAL REMAINS IN THE SEDIMENT—WHAT CAN THEY TELL?

Biological remains are stored in the sediment and might be valuable indicators of changes in trophic dynamics related to changes in nutrient loading from the catchments and climate (Battarbee *et al.*, 2005; Smol *et al.*, 2005; Bennike *et al.*, 2008, this volume). Remains in the surface sediment (upper 1 cm) provide a spatial and temporal integrated signal of biological communities in the lake covering a few years to decades, depending on the sedimentation rate. They therefore add additional information to single, snapshot water column data as well as to the relative importance of benthic and pelagic production.

Samples were taken from the upper 1 cm of the sediment in lakes and ponds in Zackenbergdalen and analysed for remains of cladocerans and *Lepidurus*. Resting eggs of *D. pulex* and a number of skeleton fragments of *Daphnia*, *Lepidurus*, *Chydorus*, *Macrothrix* and *Alona* were observed (Figure 10). *Daphnia* remains were highly abundant in fishless lakes, while *Chydorus* remains were dominant in lakes with fish. Sediment investigations thus yield a similar picture of the zooplankton community as the open water point samplings in August.

Apart from planktonic forms, the sediment also contained remains of benthic and plant-associated cladocerans, which were otherwise only

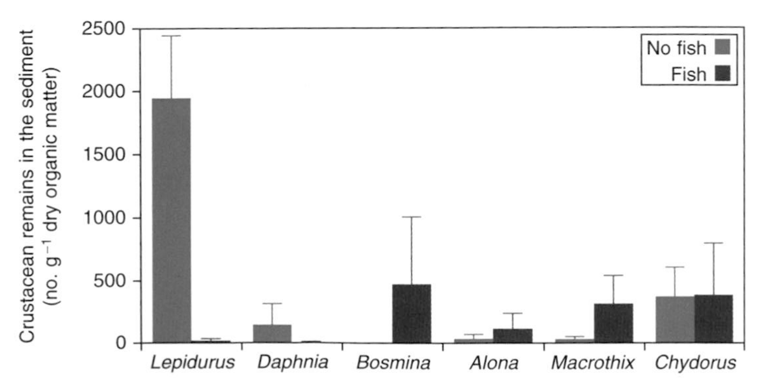

Figure 10 Remains of various cladocerans and *Lepidurus* in the surface sediment in lakes in Zackenbergdalen and Store Sødal with fish (f) and without fish (nf) (arctic charr) (from Jeppesen *et al.*, 2001).

sporadically recorded in the water samples. The numbers of primarily benthic-living forms such as *Macrothrix* and *Alona* were considerably less dense in lakes without fish, probably as a result of predation by *Lepidurus* in the fishless lakes. These preliminary results suggest that changes in the predation pressure from fish may be estimated from changes in the abundance of *Daphnia* and *Lepidurus* remains and that variation in *Lepidurus* predation may be mirrored by the benthic cladoceran community (Jeppesen *et al.*, 2001). These results have also been used to interpret changes in fish predation during the Holocene in Langemandssø (Bennike *et al.*, 2008, this volume).

IX. CONSEQUENCES OF CLIMATE CHANGES

The results from the first 9 years of monitoring show major inter-annual differences in the physico-chemical and biological variables in the investigated lakes, which may be attributed to variations in local climate. How the expected climatic changes (more maritime climate with more precipitation; Stendel *et al.*, 2008, this volume) will influence lakes and ponds in the Zackenberg area is difficult to predict. Several environmental variables seem to be directly coupled with changes in temperature and precipitation and appear highly sensitive to variations in these variables.

A larger amount of snow and the resulting later ice-melt will lead to lower water temperatures, which in turn will result in a shorter growth season and lower nutrient input. Lower water temperature is hardly a limiting factor per se, as most limnic animals and plants are adapted to a life at very low temperatures, but a shorter growth season means that less time is available for plants and animals to fulfil their life cycle. Moreover, zooplankton growth will be lower due to the reduced abundance of food, but the time will probably suffice for reproduction.

By contrast, increased precipitation during summer will result in higher runoff and thus, higher nutrient loading of organic and inorganic matter, and of silt (clay) in the lakes receiving glacial water. Higher nutrient loading may lead to increased phytoplankton production, which in the long term may lead to oxygen depletion under ice in winter and kill of arctic charr—not least in shallow lakes where the water volume under ice is modest. However, input of silt in the glacial lakes means lower light penetration of the water and thus lower primary production. It has been shown that the inter-annual growth of arctic charr populations can be influenced by fluctuations in annual mean temperatures and precipitation (snow depth) and it was concluded that climatic conditions affected short-term and inter-annual growth as well as the long-term shifts in age-specific growth patterns in arctic charr populations (Kristensen *et al.*, 2006).

Increased precipitation in summer or early ice-melt will also lead to increased input of humus. This stimulates bacteria and other microzooplankton, which may fuel the larger animals. However, higher humus content will result in improved protection of fish and microzooplankton against ultraviolet (UV) radiation, since the amount of light penetrating the water declines (the water is less clear). The result may be a decreased level of pigment production by crustaceans, which again will reduce the predation risk (Sægrov *et al.*, 1996). Less UV radiation will probably increase the number of hatched fish eggs and more fish fry will survive, not least in shallow lakes, which due to the low depth are particularly sensitive to UV radiation. Additionally, a short ice-free season and increased cloudiness will reduce both the extent and the duration of UV radiation.

ACKNOWLEDGMENTS

Monitoring data for this chapter were provided by the BioBasis programme run by the National Environmental Research Institute (NERI), University of Aarhus, in corporation with the Freshwater Biological Laboratory, University of Copenhagen and financed by the Danish Environmental Protection Agency, Ministry of the Environment. We are grateful to the Danish Polar Center staff, who contributed with valuable logistic support, and to the National Research Council, The Commission for Scientific Research in Greenland, the Nordic Council, the Carlsberg Foundation, NERI and the Freshwater Biological Laboratory, University of Copenhagen, who made it possible to perform the present study. NERI was also supported by the EU Eurolimpacs project. Thanks to Anne Mette Poulsen for editorial assistance.

REFERENCES

Battarbee, R., Anderson, N.J., Jeppesen, E. and Leavitt, P. (2005) *Freshwater Biol.* **50**, 1772–1780.

Bergersen, R. (1996) *J. Fish Biol.* **48**, 799–801.

Christoffersen, K. (2001) *Hydrobiologia* **442**, 223–229.

Christoffersen, K. (2006) In: *Arktisk Station 1906–2006* (Ed. by L. Bruun), pp. 298–303, University of Copenhagen, Arktisk Station.

Forsstöm, L., Sorvari, S., Korhola, A. and Rautio, M. (2005) *Polar Biol.* **28**, 846–861.

Fredskild, B. (1992) *Acta Bot. Fennica* **144**, 93–113.

George, D.G., Maberly, S.C. and Hewitt, D.P. (2004) *Freshwater Biol.* **49**, 760–774.

Gliwicz, Z.M. (2003) In: *Excellence in Ecology 12* (Ed. by O. Kinne), 379 pp. Oldedorf/Luhe, Germany.

Hobbie, J.E., Bahr, M. and Rublee, P.A. (1999) *Arch. Hydrobiol. Special Issues Adv. Limnol.* **54**, 61–76.

Jeppesen, E., Christoffersen, K. and Landkildehus, F. (1998) In: *Zackenberg Ecological Research Operations, 3rd Annual Report, 1997* (Ed. by H. Meltofte and M. Rasch), pp. 53–56. Danish Polar Center, Ministry of Research and Information Technology, Copenhagen.

Jeppesen, E., Christoffersen, K., Landkildehus, F., Lauridsen, T. and Amsinck, S.L. (2001) *Hydrobiologia* **442**, 329–337.

Jeppesen, E., Jensen, J.P., Jensen, P., Faafeng, B., Hessen, D.O., Søndergaard, M., Lauridsen, T., Brettum, P. and Christoffersen, K. (2003a) *Ecosystems* **6**, 313–325.

Jeppesen, E., Jensen, J.P., Lauridsen, T.L., Amsinck, S.L., Christoffersen, K. and Mitchell, S.F. (2003b) *Hydrobiologia* **491**, 321–330.

Jespersen, A.-M. and Christoffersen, K. (1987) *Arch. Hydrobiol.* **109**, 445–454.

Kristensen, D.M., Jørgensen, T.R., Larsen, R.K., Forchhammer, M.C. and Christoffersen, K.S. (2006) *BMC Ecology* **6**, 10, DOI: 10.1186/1472–6785–6–10.

Kronvang, B., Jeppesen, E., Conley, D.J., Søndergaard, M., Larsen, S.E., Ovesen, N.B. and Carstensen, J. (2005) *J. Hydrol.* **304**, 274–288.

Lauridsen, T., Jeppesen, E., Landkildehus, F., Christoffersen, K. and Søndergaard, M. (2001) *Hydrobiologia* **442**, 107–116.

Livingstone, D.M. and Dokulil, M.T. (2001) *Limnol. Oceanogr.* **46**, 1220–1227.

Meltofte, H. and Berg, T.B. (2006) *Zackenberg Ecological Research Operations. BioBasis: Conceptual design and sampling procedures of the biological programme of Zackenberg Basic. 9th edition.* National Environmental Research Institute, Department of Arctic Environment, Also on http://www2.dmu.dk/1_Viden/2_Miljoe-tilstand/3_natur/biobasis/biobasismanual.asp.

Meltofte, H. and Rasch, M. (1998) *Zackenberg Ecological Research Operations, 3rd Annual Report, 1997.* Danish Polar Center, Ministry of Research and Information Technology, Copenhagen.

Rautio, M. and Vincent, W.F. (2006) *Freshwater Biol.* **51**, 1038–1052.

Rautio, M. and Vincent, W.F. (2007) *Ecography* **30**, 77–87.

Riemann, B. and Christoffersen, K. (1993) *Mar. Microb. Food Webs* **7**, 69–100.

Riget, F., Jeppesen, E., Landkildehus, F., Lauridsen, T.L., Geertz-Hansen, P., Christoffersen, K. and Sparholt, H. (2000) *Polar Biol.* **23**, 550–558.

Røen, I.U. (1962) *Meddr. Grønland* **170**(2), 1–249.

Røen, I.U. (1977) *Fol. Limnol. Scand.* **17**, 107–110.

Sigsgaard, C., Petersen, D., Grøndahl, L., Thorsøe, K., Meltofte, H., Tamstorf, M.P. and Hansen, B.U. (2006) In: *Zackenberg Ecological Research Operations, 11th Annual Report, 2005* (Ed. by A.B. Klitgaard, M. Rasch and K. Caning), pp. 11–35. Danish Polar Center, Ministry of Science, Technology and Innovation, Copenhagen.

Smol, J.P., Wolfe, A.P., Birks, H.J.B., Douglas, M.S.V., Jones, V.J., Korhola, A., Pienitz, R., Rühland, K., Sorvari, S., Antoniades, D., Brooks, S.J. Fallu, M.F., *et al.* (2005) *Proc. Natl. Acad. Sci. USA* **102**, 4397–4402.

Straile, D., Livingstone, D.M., Weyhenmeyer, G.A. and George, D.G. (2003) In: *The North Atlantic Oscillation: Climatic Significance and Environmental Impact.* Geophysical Monograph 134. American Geophysical Union.

Sægrov, H., Hobæk, A. and Lábe-Lund, H.H. (1996) *J. Plankton Res.* **18**, 1213–1228.

Søndergaard, M., Jeppesen, E. and Jensen, J.P. (2005) *J. Appl. Ecol.* **42**, 616–629.

Vincent, W.F. and Hobbie, J.E. (2000) In: *The Arctic: A Guide to Research in the Natural and Social Sciences* (Ed. by M. Nuttall and T.V. Callaghan), pp. 197–232. Harwood Academic Publishers, U.K.

Wesenberg-Lund, C. (1894) *Vid. Meddr. Naturhist. Foren. Kjøbenhavn* **56**, 82–143.

Wrona, F.J., Prowse, T.D., Reist, J.D., Beamish, R., Gibson, J.J., Hobbie, J., Jeppesen, E., King, J., Koeck, G., Korhola, A., Levêsque, L., Macdonald, R., *et al.* (2005) In: *ACIA: Arctic Climate Impact Assessment* (Ed. by C. Symon, L. Arris and B. Heal), pp. 354–452. Cambridge University Press, New York.

Population Dynamical Responses to Climate Change

MADS C. FORCHHAMMER, NIELS M. SCHMIDT,
TOKE T. HØYE, THOMAS B. BERG, DITTE K. HENDRICHSEN
AND ERIC POST

SUMMARY

It is well established that climatic as well as biological factors, in concert, form the mechanistic basis for our understanding of how populations develop over time and across space. Although this seemingly suggests simplicity, the climate–biology dichotomy of population dynamics embraces a bewildering number of interactions. For example, individuals within a population may compete for space and other resources and, being embedded in an ecosystem, individuals in any population may also interact with individuals of competing species as well as those from adjacent trophic levels.

In principal, the effects of climate change may potentially extend through any of these interactions. In this chapter, we focus on the extent to which evolutionarily distinct species at different trophic levels respond to similar changes in climate. By using a broad spectrum of statistically and ecologically founded approaches, we analyse concurrently the influence of climatic

ADVANCES IN ECOLOGICAL RESEARCH VOL. 40 0065-2504/08 $35.00
© 2008 Elsevier Ltd. All rights reserved
DOI: 10.1016/S0065-2504(08)00017-7

variability and trophic interactions on the temporal population dynamics of species in the terrestrial vertebrate community at Zackenberg.

We describe and contrast the population dynamics of three predator species (arctic fox *Alopex lagopus*, stoat *Mustela erminea* and long-tailed skua *Stercorarius longicaudus*), two herbivore species (collared lemming *Dicrostonyx groenlandicus* and musk ox *Ovibos moschatus*) and five wader species (common ringed plover *Charadrius hiaticula*, red knot *Calidris canutus*, sanderling *Calidris alba*, dunlin *Calidris alpina* and ruddy turnstone *Arenaria interpres*) with respect to intra-specific density dependence, consumer–resource interactions and direct as well as indirect inter-trophic level mediated effects of varying snow-cover.

We found that the temporal population dynamics of all three predators, both herbivores and three out of five wader species, displayed significant direct density dependence. Only two species (sanderling and long-tailed skua) displayed dynamics characterised by delayed density dependence. The direct effects of previous winter's snow were related to over-wintering strategies of resident and migrating species, respectively. The dynamics of all four resident species were significantly affected by variations in snow-cover and explained up to 65% of their inter-annual dynamics.

The three predators differed in their numerical response to changes in prey densities. Whereas the population dynamics of arctic fox were not significantly related to changes in lemming abundance, both the stoat and the breeding of long-tailed skua were mainly related to lemming dynamics. The predator–prey system at Zackenberg differentiates from previously described systems in high-arctic Greenland, which, we suggest, is related to differences in the compositions of predator and prey species. The significant inter-trophic interactions are centred on the collared lemming as a result of which there is a significant potential for indirect climate effects mediated across the established consumer–resource interactions.

I. INTRODUCTION

Describing and understanding the inter-annual fluctuations in population size is central for our perception of and ability to predict how climate changes affect species. In essence, populations vary from year to year because of concurrent changes in reproduction, survival and migration (May, 1981), and it is through these vital rates that climate affects population dynamics (Begon *et al.*, 2002). However, climate is not the only factor that affects populations. Individuals within a population may compete for space and other resources and, being embedded in an ecosystem, individuals in any population may also interact with species, which belong to other trophic levels. Obvious examples include predator–prey and herbivore–plant interactions (Begon *et al.*, 2002). Hence, to evaluate population dynamical responses to climate change, we need to embrace simultaneously the combined effects of biological and climatic factors (Forchhammer and Post, 2004; Box 1). Indeed, biological processes may be important in some populations, whereas

Box 1

Integrating Climatic Effects in the Analysis of Population Dynamics

One of the key questions in the study of population dynamics is to what extent do inter-annual variations in population sizes (X) reflect climatic influence and biological interactions (Box Figure 1A)? This is the key question

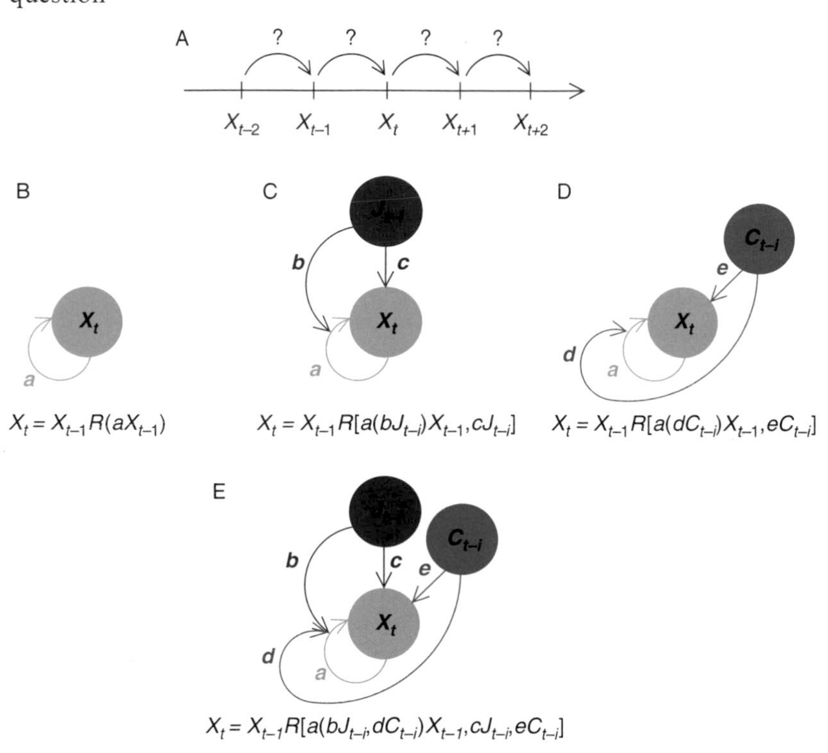

Box Figure 1 (A) Understanding the temporal dynamics of populations revolve around asking what factors determine the change in population size from one year to the next ($X_{t-n-1} \rightarrow X_{t-n}$). (B) Simple population model considering density dependence (X_{t-1}) only. (C) Population model integrating the effects of another trophic level (J_{t-i}) either direct (arrow c) or indirect on the degree of density dependence (arrow b). (D) Model with effects of climate (C_{t-i}),

(continued)

Box 1 *(continued)*

direct (arrow e) and indirect (arrow d). Population model combining trophic and climate influence simultaneously. Equations below each sub-panel give the general population model where the inter-annual change in population abundance is related by the function $R[.]$

approached in this chapter. Obviously, the need for integrating both climatic and biological influence mirrors an increasing analytic complexity in associated population models: $X_t = X_{t-1}R[\cdot]$, that is, which factors should be integrated in the function R that relates consecutive changes in population abundances.

A good approach is a stepwise procedure integrating potential density-dependent as well as density-independent factors (Stenseth *et al.*, 2002). For example, at any given year t, the population size (X_t) will be affected by previous years' population sizes (Box Figure 1B: interaction arrow *a*). In addition, effects of consumers (J_{t-i}) may also be significant. Typically this will be direct (interaction arrow *c*) but can also affect competitive interactions within the population (interaction arrow *b*) (Box Figure 1C). Similarly, climatic influences may be direct (interaction arrow *e*) as well as indirect (interaction arrow *d*) (Box Figure 1D). However, since both climate and biological interactions occur concomitantly, they need to be considered simultaneously in analysing the causal drivers of population time series (Box Figure 1E). It is obvious from the associated population models given under each sub-panel in Box Figure 1, as the number of factors integrated increases, so does model complexity.

other populations may be influenced primarily by climatic conditions. Then again, climate may be important but only under specific biological conditions (Grenfell *et al.*, 1998; Ellis and Post, 2004; Tyler *et al.*, 2007). For example, in a high-arctic reindeer *Rangifer tarandus* population on Svalbard, both mortality and fecundity were significantly affected by ablation (melting of snow during winter), but only when the population was increasing and during prolonged periods with severe winter climate (Tyler *et al.*, 2007).

Notwithstanding the complexity of climatic effects on arctic populations (e.g., Vibe, 1967; Forchhammer and Boertmann, 1993; Post and Stenseth, 1999; Post and Forchhammer, 2002; Schmidt 2006), they may, on a micro-evolutionary scale and in the simplest sense, be divided into direct and indirect effects (Forchhammer, 2001; Forchhammer and Post, 2004; see also Box 1 in Berg *et al.*, 2008, this volume). Direct climatic effects often incur population dynamical changes without time lags. For example, increased winter severity has been directly associated with inter-annual variation in survival in several northern and arctic ungulate species (Milner *et al.*, 1999; Post and Stenseth, 1999). On the contrary, indirect climatic effects may be temporally delayed, and often involve interactions between organisms on different trophic levels. A good example is a northern tri-trophic community involving wolves *Canis lupus*, moose *Alces alces* and balsam fir *Abies balsamea* (Post *et al.*, 1999; Post and Forchhammer, 2001). In this system, snowy winters increased the hunting success of wolves through formation of larger packs, negatively affecting the moose population one year later and eventually causing a two-year delayed increase of growth in the fir population (Post *et al.*, 1999).

The BioBasis monitoring programme at Zackenberg is a rare example of a study that embraces a range of organisms across trophic levels simultaneously in this high-arctic ecosystem (Meltofte and Berg, 2006). This provides an excellent opportunity to analyse and contrast the variability of direct and indirect climatic effects for the population dynamics of evolutionary different but ecologically interrelated species characteristic of a high-arctic ecosystem. In this chapter, we describe and contrast the population dynamics of a range of terrestrial species, which live at Zackenberg, in relation to inter-annual changes in winter climatic conditions. We focus on winter climate for two reasons. First, it is during winter that the predicted warming of the northern environment is most pronounced. Some projections of increases of 4–5 °C with concomitant reduction in the extent and duration of snow-cover have been made (McBean *et al.*, 2005; Walsh *et al.*, 2005; but see Stendel *et al.*, 2008, this volume). Secondly, inter-annual variability in spring snow-cover is central for the functioning of the entire ecosystem at Zackenberg (Meltofte, 2002) and affects responses ranging from the reproductive phenology of plants, insects and waders (Høye *et al.*, 2007a,b; Meltofte *et al.*, 2007) and the spatial distribution of large herbivores (Forchhammer *et al.*, 2005) to earth–atmosphere gas-flux dynamics (Grøndahl *et al.*, 2007).

The species included in our analyses embrace the most important predator–prey and herbivore–plant dynamics at Zackenberg (Figure 1). The predators include arctic fox *Alopex lagopus*, stoat *Mustela erminea* and long-tailed skua *Stercorarius longicaudus*; the herbivores include musk ox *Ovibos moschatus* and collared lemming *Dicrostonyx groenlandicus* and the waders include common ringed plover *Charadrius hiaticula*, red knot *Calidris canutus*, sanderling *C. alba*, dunlin *C. alpina* and ruddy turnstone *Arenaria interpres*. All these species are relatively long lived and breed throughout their adult lives. Yet they display rather different life histories, in particular with respect to over-wintering strategies. The arctic fox, stoat, musk ox and lemming are all resident species; they remain in high-arctic Greenland throughout winter (Muus *et al.*, 1990). In contrast, all the bird species migrate south during winter, the waders to Europe and West Africa, and the long-tailed skua to the open waters of the South Atlantic (Cramp and Simmons, 1983). Hence, whereas the resident species have to cope with changes in winter climate, such as variation in the thickness, hardness and extent of snow-cover, the migrants face entirely different climatic conditions

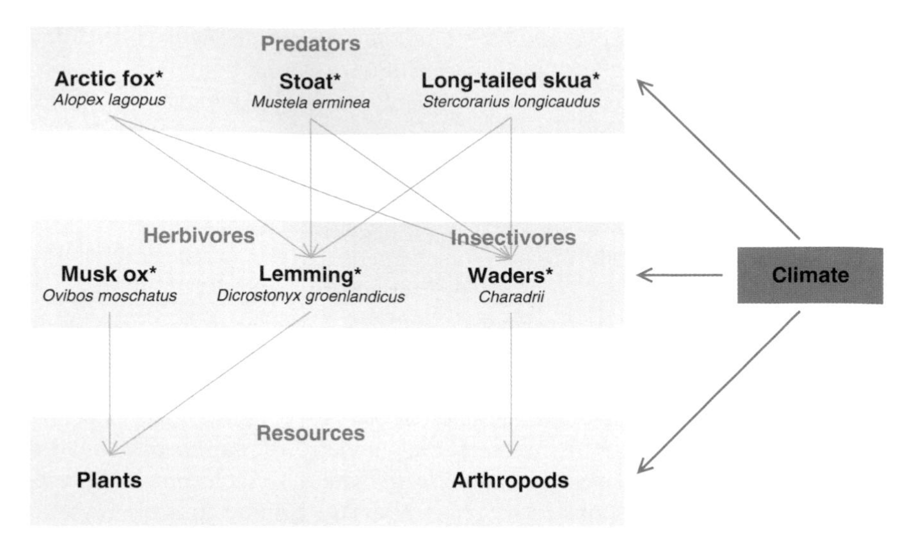

Figure 1 The trophic location of the species (marked with *), whose population dynamics were analysed in relation to changes in winter climate indexed by spring snow-cover (dark blue arrows). Light blue arrows indicate consumer–resource trophic interactions.

during their non-breeding season south of the Arctic. The contrasting strategies in resident and migrant species are important to bear in mind because local winter climate conditions may strongly influence the over-winter survival and breeding success of resident species (e.g., Post and Stenseth, 1999; Forchhammer and Post, 2004). In contrast, although winter conditions do not directly affect the migrants, winter conditions are important because of their impact on snowmelt, and hence, the concomitant spring conditions which the migrant species heavily depend upon to initiate their breeding (e.g., Meltofte, 1985, 2006; Klaassen *et al.*, 2001).

II. CONCEPTS OF POPULATION DYNAMICS

Changes in numbers of individuals present in a population are typically recorded by annual long-term monitoring like the BioBasis programme at Zackenberg. Because of cost–benefit considerations related to maximising the length of study as well as optimising the number of species and trophic levels to be monitored, demographic changes in the age and sex composition are not recorded in detail, although demography is one of several important factors shaping the dynamics of populations (e.g., Caswell, 2001). Nevertheless, autoregressive models, which have been used extensively to describe the auto-covariance in time series of population numbers (Royama, 1992; Bjørnstad and Grenfell, 2001), have by the inclusion of climatic co-variance proven to be highly valuable in pinpointing the temporal importance of biological and climatic population drivers (e.g., Royama, 1992; Forchhammer *et al.*, 1998). For example, it has been demonstrated in both experimental and natural consumer–resource systems that the dimension (i.e., order of density dependence) of autoregressive models for the resource organism depends on the specialisation of the consumer (Forchhammer and Asferg, 2000; Bjørnstad *et al.*, 2001). Similarly, variations in the temporal lags of climatic effects in autoregressive models have for various organisms in the Northern Hemisphere been related specifically to variations in survival, reproduction and cohort quality (Forchhammer, 2001). Hence, being central for describing time series generated by climate-related long-term monitoring of individuals, and because we make extensive use of the method here, we will in this section shortly outline the climate–biological concepts of the autoregressive population models (Royama, 1992).

Basically, any species may be considered as being part of a consumer–resource system affected by climate (Figure 2). If a species functions as a resource in a community (as, e.g., the lemming does for its predators), interactions influencing its population dynamics will include competition among individuals within a population (α_{NN}, Figure 2), predation (α_{PN}, Figure 2) and climate (α_{CN}, Figure 2). The relative importance of each set of interactions acting on the resource population may be determined by the

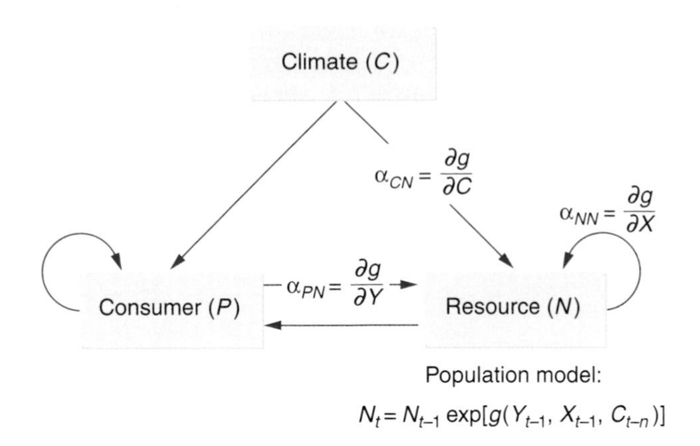

Figure 2 Conceptual representation of consumer–resource interactions in relation to climatic effects. Focusing on the resource species, its population dynamics can be described by a general Gompertz model function $g(\cdot)$ of $Y (= \log_e P)$, $X (= \log_e N)$ and climate (C) (Dennis and Taper, 1994), where P and N denote the population abundances of consumer and resource species, respectively. The relationship between the interaction coefficients (α_{PN}, α_{CN}, α_{NN}) and the population dynamics of the resource species is defined by the partial derivatives of $g(\cdot)$ ($\partial g/\partial Y$, $\partial g/\partial C$, $\partial g/\partial X$).

partial derivatives of the population model (Figure 2). Combining and summing up the information in these climate–biological interactions for the resource species as depicted in Figure 2, the population dynamics of the focal species can be described on a log-linear basis by direct (X_{t-1}) and delayed (X_{t-2}) autoregressive processes (i.e., Gompertz, 1825) with an additive climatic co-variance (Forchhammer *et al.*, 1998),

$$N_t = N_{t-1} \exp(\beta_1 X_{t-1} + \beta_2 X_{t-2} + \omega_1 C_{t-n}) \quad (1)$$

where N_t is the population size in year t, $X_t = \log_e(N_t)$, and C the climatic conditions, for example, winter precipitation. The variables β_1, β_2 and ω_1 are the autoregressive and climate co-variate coefficients, respectively. Although the autoregressive coefficients are purely statistical, they display variation related to changes in the biological interactions outlined in Figure 2 (Forchhammer and Asferg, 2000; Box 2). Whereas the direct density dependence (β_1) embraces intra-trophic interactions only, such as competitive interactions for resources and territories, the delayed density dependence (β_2)

Box 2

Merging Single-Species Population Models with Time Series Analyses

The population dynamics of single species have been previously described by a range of conceptually related discrete time models (May, 1981). A characteristic representative of these, the Maynard Smith–Slatkin (Maynard Smith and Slatkin, 1973) model, provides, despite its simplicity, an impressive general description of single-species population dynamics embracing monotonic damping, damped oscillations and stable limit cycles as well as chaotic dynamics (Bellows, 1981). The Maynard Smith–Slatkin population model unites logistic population growth with concomitant changes in density (N), carrying capacity (K), degree (a) and type (b) of competition:

$$N_t = \frac{N_{t-1} R}{1 + a N_{t-1}^b}, \quad (2.1)$$

where R is the fundamental net reproductive rate and $a = (R–1)/K$. Of particular interest here is how the Maynard Smith–Slatkin population model relates to the analyses of time series, the basic product of long-term monitoring? The answer is found by a simple reformulation taking the natural logarithm ($\log_e$) on both sides of Eq. 2.1:

$$X_t = X_{t-1} + r - \log_e (1 + a N_{t-1}^b), \quad (2.2)$$

where $X = \log_e(N)$ and $r = \log_e(R)$. For sufficiently large N_{t-1}, Eq. 2.2 may be rearranged and reduced to:

$$X_t = \left(r - b \log_e(a)\right) + (1 - b)X_{t-1}. \tag{2.3}$$

Comparing Eqs. 2.1 and 2.3, we see that the Maynard Smith–Slatkin population model can be expressed as a log-linear autoregression linking consecutive changes in population sizes ($X_{t-1} \rightarrow X_t$) through direct density dependence $(1-b)$. Hence, variations in direct density dependence portray which type of competition the population is exposed to. Specifically, $(1-b) < 0$ leads to over-compensation, $(1-b) > 0$ to under-compensation, $(1-b) = 0$ to perfect compensation and $(1-b) = 1$ to even density dependence (Bellows, 1981). Whereas r expresses the fundamental reproductive rate, $[r-b \log_e(a)]$ is the realised reproductive rate for individuals in a population under competition characterised by a and b. Climatic influence on population time series may be additive and/or through competitive density-dependent interactions (see Box 1).

also includes inter-trophic interactions exemplified by predator–prey or herbivore–plant interactions (Forchhammer *et al.*, 1998). Hence, climate-related changes in β_1 suggest an intra-population impact, whereas changes in β_2 may portray inter-trophic impacts by climate (see Post and Forchhammer 2001 for a detailed exposition of this).

The autoregressive population model specified in Eq. 1 assumes log-linearity in density dependence of population growth (Royama, 1992). However, any climatic effect mediated through any seasonal characteristic in an arctic ecosystem may be nonlinear in relation to species dynamics and behaviour (e.g., Forchhammer *et al.*, 2005; Høye and Forchhammer, 2008, this volume). Although linearity was not rejected for any of the $\log_e$-transformed time series analysed here ($p > 0.21$; Tong, 1992), thereby providing the basis for log-linear autoregressive population analyses, we also investigated to what extent climate effects and consumer–resource relations displayed nonlinearity.

III. THE DYNAMICS OF CONTRASTING SPECIES AT ZACKENBERG

Below we describe the inter-annual dynamics of predator, herbivore and wader species at Zackenberg over 10 years from 1996 to 2005 (Figure 3A,C,E). We make extensive use of the autoregressive population model given in

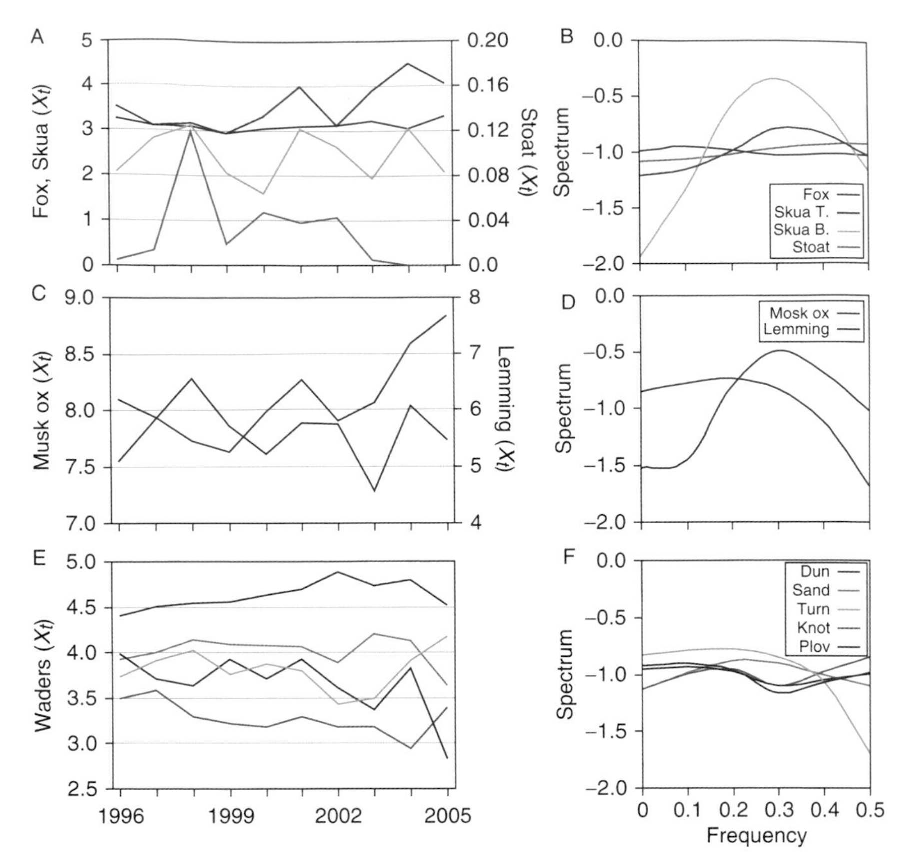

Figure 3 The $\log_e$-transformed (X_t) 10-year time series of the annual density indices (see below) of the selected species (A, C, E) and associated smoothed periodograms (B, D, F). (a) Arctic fox and stoat together with territorial (T) and breeding (B) long-tailed skua, (C) musk ox and collared lemming, and (F) dunlin (Dun), ruddy turnstone (Turn), sanderling (Sand), red knot (Knot) and common ringed plover (Plov). The colour of the species-specific smooth periodograms corresponds to those given in the time series. The smoothed periodograms were calculated using the spec.pgram function in *S–Plus* with spans = 2 (for details, cf. Venables and Ripley, 2002). The following measures of annual population densities were used: Arctic fox: total number of adults and juveniles encountered during fieldwork in Zackenbergdalen June to August; Skua.T: total number of territorial pairs (including those breeding) recorded in the 19.3 km^2 bird census area in Zackenbergdalen; Skua.B: total number of skua nests found; Stoat: indexed as the number of lemming winter nests depredated by stoat during the previous winter (year t–1 to t); Lemming: number of winter nests recorded from the previous winter (year t–1 to t); the accumulated number of musk oxen recorded per day in the 40 km^2 census area in Zackenbergdalen; Waders: estimated number of territorial pairs in the 19.3 km^2 census area (see Meltofte and Berg, 2006 and Meltofte, 2006 for details).

Eq. 1, where the effects of direct and delayed density dependence are estimated simultaneously (Table 1). The fluctuating behaviour of the dynamics of all three predators was estimated by spectral analyses using smoothed periodograms (Figure 3B, D, F) (Venables and Ripley, 2002).

A. Predators

The three predator species, the arctic fox, the stoat and the long-tailed skua, displayed distinctly different temporal population dynamics (Figure 3A). Whereas the annual number of territorial long-tailed skuas varied very little across years, the number of pairs producing eggs each year was highly variable (Figure 3A). The latter had distinct 3–4 year fluctuations (Figure 3B), and only during years with high lemming densities did the entire population of long-tailed skua breed (Figure 3A; Meltofte and Høye, 2007). In contrast, there was no significant 3–4 year fluctuating pattern in the population index of the arctic fox. This is in accordance with the model dynamics where no numerical response to changes in lemming number was observed (Schmidt *et al.*, 2008, this volume). Interestingly, the dynamics of stoats did not display 3–4 year cycles delayed one year with respect to the lemming (Figure 3A, B) as reported from Traill Ø, 220 km south of Zackenberg (Gilg *et al.*, 2006). This may be related to the fact that the index used for stoat density (i.e., number of lemming winter nests depredated by stoat) may be biased due to the differential functional response by stoat in years with different lemming densities (Gilg *et al.*, 2006), that is, population number of stoat may not be represented by the number of lemming nests depredated per se. The population index of stoat has decreased significantly ($p = 0.02$) since 1998, and in 2004–2005 no stoat depredated lemming nests were found (Figure 3A), although the lemming population, during these years, displayed large fluctuations (Figure 3C). In contrast, the population index for arctic fox increased significantly ($p = 0.03$) over the same period, whereas there was no temporal trend for the population of long-tailed skua ($p > 0.50$).

All three species of predators exhibited negative direct density dependence ($\beta_1 < 1$; Table 1), suggesting long-term stability mediated by density dependence through population intrinsic competitive interactions (i.e., $-2 < \beta_1 < 1$; Royama, 1992). The breeding population of only long-tailed skuas displayed significant autoregressive dynamics, with delayed density dependence (β_2; Table 1). The latter suggests an influence of inter-trophic interactions on the breeding population (e.g., Forchhammer and Asferg, 2000), which is corroborated by the breeding population's high dependence on the occurrence of collared lemmings (see below). For all species, density dependence explained between 6% (arctic fox dynamics) and 60% (breeding population of long-tailed skua) of the inter-annual variation in population size (Table 1).

Table 1 Analyses of the population dynamics of the selected predator, herbivore and wader species using the autoregressive population model (Eq. 1) with spring snow-cover as an index for the amount of snow precipitated during the previous winter year t–1 to year t (S_{t-1}) as an additive co-variate: $X_t = \beta_0 + \beta_1 X_{t-1} + \beta_2 X_{t-2} + \omega_1 S_{t-1}$

Species	$\beta_1 \pm$ S.E.M.[a]	$\beta_2 \pm$ S.E.M.[b]	$\omega_1 \pm$ S.E.M.[c]	R^2_{AR}	R^2_{Clim}	AIC$_c$	dAIC$_c$	AR2
Predators								
Arctic fox, *A. lagopus*	**−0.18 ± 0.48**	0.16 ± 0.48	**−0.027 ± 0.009**	0.06	0.33	17.56	−3.38	No (1)
Stoat, *M. erminea*	**−0.62 ± 0.46**	−0.23 ± 0.42	**0.001 ± 0.0005**	0.18	0.07	−33.46	−1.66	No (1)
Long-tailed skua (territorial), *S. longicaudus*	**−0.17 ± 0.43**	0.18 ± 0.52	−0.002 ± 0.002	0.15	0.03	−9.83	−1.8	No (1)
Long-tailed skua (breed), *S. longicaudus*	**−0.49 ± 0.27**	**−0.74 ± 0.31**	0.008 ± 0.007	0.60	0.03	15.17	0	Yes (2)
Herbivores								
Musk ox, *O. moschatus*	**−0.22 ± 0.40**	0.24 ± 0.40	**−0.025 ± 0.004**	0.06	0.65	3.71	−1.45	No (1)
Collared lemming, *D. groenlandicus*	**−0.29 ± 0.39**	−0.38 ± 0.44	**0.003 ± 0.001**	0.19	0.17	21.77	−0.64	Yes (2)
Waders								
Common ringed plover, *C. hiaticula*	**−0.60 ± 0.42**	−0.27 ± 0.67	**0.012 ± 0.006**	0.20	0.30	7.31	−1.61	No (1)
Red knot, *C. canutus*	0.37 ± 0.56	−0.51 ± 0.53	0.004 ± 0.004	0.17	0.04	−0.32	−2.81	No (0)
Sanderling, *C. alba*	**−0.53 ± 0.41**	**−0.84 ± 0.310**	**0.004 ± 0.002**	0.48	0.26	−11.95	0	Yes (2)
Dunlin, *C. alpina*	**0.23 ± 0.35**	0.55 ± 0.52	0.003 ± 0.003	0.38	0.01	−6.62	−1.21	No (1)
Ruddy turnstone, *A. interpres*	0.61 ± 0.37	−0.55 ± 0.38	**−0.003 ± 0.001**	0.13	0.31	−1.60	−0.78	Yes (2)

Regression coefficients are given with standard error of mean (S.E.M.) and significant ($p < 0.05$) coefficients are in bold. R^2_{AR} and $\pm R^2_{Clim}$ are the partial R^2 of density-dependent (X_{t-1}, X_{t-2}) and climatic (S_{t-1}) effects on X_t, respectively. The corrected Akaike information criterion (AICc; Hurvich and Tsai, 1989) is given for each model, and dAIC$_c$ gives the difference between model AIC$_c$ and the AIC$_c$ value for the most parsimonious model. AR2 denotes whether the most parsimonious autoregressive population model is two-dimensional (i.e., includes X_{t-1} and X_{t-2}) with the dimension of the most parsimonious model given in brackets.

Two-tailed t-tests:

[a]H$_0$: $\beta_1 = 1$.

[b]H$_0$: $\beta_2 = 0$.

[c]H$_0$: $\omega_1 = 0$.

B. Herbivores

Collared lemmings are well known to display cyclic, multi-annual fluctuations (Stenseth and Ims, 1993; Gilg *et al.*, 2006). As expected, the Zackenberg population showed marked fluctuations with a periodicity of about 3–4 years (Figure 3C, D; Schmidt *et al.*, 2008, this volume). Numbers of the only other resident mammalian herbivore, the musk ox, were considerably less variable (Figure 3C), although spectral analysis did indicate long-term (>5 years) fluctuations in this species (Figure 3D), as previously observed in other musk ox populations in Greenland (Forchhammer and Boertmann, 1993; Forchhammer *et al.*, 2002). The chief response, however, was that whereas the number of musk oxen at Zackenberg increased dramatically ($p = 0.005$) over the last 10 years, no temporal trend was observed in the lemming population ($p > 0.50$).

Both the musk ox and lemming populations displayed strong direct density dependence ($\beta_1 < 1$), but no delayed density dependence ($\beta_2 = 0$). For the lemming, this contrasts with the autoregressive analyses in Schmidt *et al.* (2008, this volume) where delayed density dependence was recorded. However, our analyses were performed on observational data, whereas those in Schmidt *et al.* (2008, this volume) were done on simulated lemming data with *a priori* build-in model assumptions of the significant influence of predators. The stronger the influence of predators is in such prey population models, the stronger delayed density dependence (D.K. Hendrichsen, unpublished). Also, the length of time series analysed may be partly responsible. As previously reported, the length of time series is highly positively correlated with the ability to statistically detect true delayed density dependence (Saitoh, 1998). For musk oxen and lemmings, respectively, only 6 and 19% of the inter-annual variation in population size was explained by pure density-dependent interactions (Table 1).

C. Waders

All the wader species are long-distance migrants, returning to the High Arctic of Greenland to breed each year (Meltofte, 1985). Although their breeding population dynamics obviously are also affected by conditions at their wintering areas in Europe and West Africa (e.g., Rehfisch *et al.*, 2004; Austin and Rehfisch, 2005), banding studies suggest that individuals from several wader species return to the same breeding location (Cramp and Simmons, 1983). None of the five wader species breeding at Zackenberg showed much annual variation in numbers of territories (Figure 3E, F; Meltofte, 2006). Furthermore, no significant temporal trends were detected in the dynamics of red knot, sanderling and ruddy turnstone (p's > 0.10) (Figure 3E). The dunlin showed a marginal increase in breeding numbers ($p = 0.05$), whereas the ringed plover population decreased over the same period ($p = 0.04$) (Figure 3E).

Significant negative direct density dependence ($\beta_1 < 1$) was recorded in common ringed plover, sanderling and dunlin, but not in red knot and ruddy turnstone (Table 1) indicating across-species differences in intra-population competition for resources. This corroborates previous estimates that only knot and turnstone have dynamics characteristics of populations at carrying capacity (Meltofte, 1985). Only the sanderling displayed significant delayed density dependence ($\beta_2 < 0$).

IV. EFFECTS OF CLIMATE AND INTER-TROPHIC INTERACTIONS

A. Direct Climatic Effects

The direct effects of climate on predators, herbivores, and waders were investigated using the autoregressive population model in Eq. 1, where the additive climatic co-variate, C_{t-n}, winter weather conditions year t–1 to year t indexed by the observed spring snow-cover, S_{t-1} (Table 1). Thus, the additive influence of snow on population dynamics reported here is corrected for the aforementioned statistical density dependence. The index S_{t-1} primarily constitutes changes in the amount of snow precipitated during winter (i.e., recorded snow depth). Whereas the relative influence of changes in snow depth explained 48% of the inter-annual variation in S_{t-1}, positive degree days in spring (1 April–10 June) only explained about 1–2% (GLM: $R^2_{total} = 0.50$, $p < 0.05$). Increased snow depth was associated with an increase in S_{t-1} ($r_{partial} = 0.69$).

Species differences in the response to direct effects of inter-annual variation in snow may be related to differences in life history strategies typical for resident (arctic fox, stoat, musk ox and collared lemming) or migrant (long-tailed skua and waders) species at Zackenberg. Obviously, winter survival of the resident species is potentially closely linked to changes in the amount of snow at Zackenberg only, whereas winter survival of the long-tailed skua and the waders reflects local conditions in their southern winter quarters and/or those *en route* to their high-arctic breeding grounds (e.g., Insley *et al.*, 1997). However, sharing the same summer climatic conditions, the breeding success of both resident and migrating species may be highly influenced by climatic conditions in previous winters. Examples of such influences might include, but are not limited to, the influence of snow on forage availability, such as plant growth, for most of the species in focus here (Ellebjerg *et al.*, 2008, this volume), as well as temporal emergence patterns and abundance of invertebrates (Meltofte *et al.*, 2007; Høye and Forchhammer, 2008, this volume).

This differential effect of previous winter's snow on resident and migrant species may be observed among the predators. Significant direct effects of snow were recorded for the two resident species, arctic fox and stoat, but not for the migratory long-tailed skua (Table 1). Specifically, increased S_{t-1} had a large negative effect on the arctic fox population ($R^2_{\text{partial}} = 0.33$; Table 1), which probably relates to reduced hunting success on lemmings during winters with increased amount of snow and, consequently, increased negative effects on vital rates observed elsewhere (e.g., Angerbjörn *et al.*, 1991) as well as increased migration by foxes out of Zackenbergdalen (Schmidt *et al.*, 2008, this volume). On the contrary, increased snow had a significant but minor ($R^2_{\text{partial}} = 0.07$) positive effect on the stoat population (Table 1), which may seem contradictory. However, in contrast to the arctic fox, the stoat hunts its main prey, the collared lemming, under the snow and exerts its strongest impact on its prey during winter (Sittler, 1995). Since decreased snow reduces the lemming population (Table 1), this may affect the stoat population negatively through the same mechanisms. An effect of increased snow-cover was also expected to influence long-tailed skua negatively, since delayed spring snowmelt may increase predation on nests of ground-nesting birds by arctic fox (Byrkjedal, 1980). Although we found no effects in our log-linear analyses (Table 1), nonlinear analyses indicated that whereas variation in spring snow-cover following the previous winter (i.e., S_{t-1}) in the mid-range exerts little effect, variations during extreme years negatively affect the number of territorial long-tailed skuas (Figure 4A). This, however, was not recorded in the proportion of birds breeding (Figure 4B).

The two resident herbivore species, musk ox and collared lemming, both were affected significantly by previous winter climatic conditions (S_{t-1}, Table 1) but in exactly opposite ways. Increased snow-cover had a considerable negative effect on the musk ox population at Zackenberg and explained 65% of the inter-annual variation in numbers of musk oxen observed (Table 1). Snow represents a severe constraint on foraging in musk oxen (e.g., Forchhammer, 1995), and deep or hard snow results in increased rates of mortality (Forchhammer *et al.*, 2002; Schmidt, 2006). Snow had a positive effect on the lemming population (17%; Table 1), which was probably a result of increased protection from winter predation by arctic fox (Reid and Krebs, 1996; see also Berg *et al.*, 2008, this volume, and Schmidt *et al.*, 2008, this volume).

Like for the long-tailed skua, we found that increased spring snow-cover (S_{t-1}) had a negative effect on ruddy turnstone where 31% of the inter-annual variation in the number of turnstone territories was explained by S_{t-1} (Table 1). In contrast, increased snow-cover was positively associated with numbers of sanderling and common ringed plover territories (Table 1). However, this was

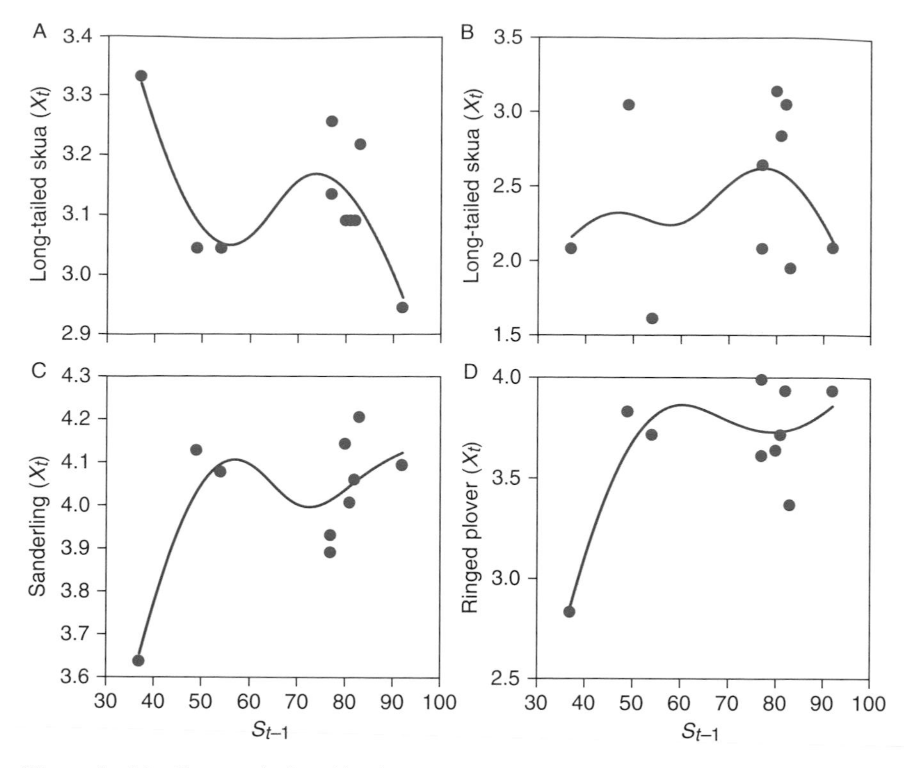

Figure 4 Nonlinear relationships between percent snow-cover on 10 June (S_{t-1}) and the log$_e$-transformed annual abundance (X_t) of (A) territorial and (B) breeding long-tailed skuas, (C) sanderlings and (D) common ringed plovers. The nonlinear regression lines are second-order generalised additive models (Venables and Ripley, 2002). For panels (A), (C) and (D), these described the relationships better than generalised linear models ($p < 0.01$).

due to a single outlier in 2005 for both species (Figure 4C, D), and it remains to be seen whether there is a genuine indirect effect of snow-cover across trophic levels on the local population size of these two wader species, and/or the higher numbers in snow-rich years are caused by more pre-breeders in the census area awaiting snowmelt in adjacent areas (Meltofte, 2006).

B. Inter-Trophic Interactions

In addition to the direct climatic effects of snow, climate may exert its influence indirectly through inter-trophic interactions (Forchhammer and Post, 2004; Box 1 in Berg *et al.*, 2008, this volume), and it is therefore essential to establish whether significant relations occur across trophic levels (Figure 1). Although the dimension (i.e., temporal lag) of density dependence in the autoregressive population model (Eq. 1) indicates the presence of inter-trophic effects (Forchhammer and Asferg, 2000), detection of delayed density dependence requires relatively long time series (Saitoh, 1998). Owing to the relative shortness of the present time series, we investigated consumer–resource interactions in separate analyses by replacing the delayed density-dependent term (X_{t-2}) with a trophic consumer–resource interaction term (Table 2). The specific sign, + or −, of a consumer–resource interaction (Figure 2) indicates whether consumer dynamics are controlled by resource abundance or vice versa (May, 1981).

Dynamics of the three predator species responded differently to changes in the abundance of their prey and, hence, may have been affected differently by indirect influences of climate. The dynamics of arctic fox displayed no numerical response to changes in collared lemming densities (Table 2). Instead, they were found to be positively associated with changes in numbers of dunlins and common ringed plovers, where increased number of territories in these species was followed by increased arctic fox abundance the next year (Table 2). Although it remains to be investigated further, these results suggest that the arctic fox at Zackenberg may display the generalist predator behaviour (here on bird eggs and young) reported elsewhere for this species (e.g., Eide *et al.*, 2005; Gilg *et al.*, 2006). As in the single-species analyses (Table 1), the direct negative effect of increased winter snow was significant (Table 2). Therefore, since negligible effects of snow-cover were detected in the quantitative population dynamics of these wader species (but see Meltofte *et al.*, 2007 for responses in breeding phenology), the direct effect of snow seems to be the main climatic effect on the arctic fox population at Zackenberg.

In contrast to the arctic fox, both stoat and long-tailed skua were significantly influenced by current-year variations in the abundance of collared lemming. Both species increased when there were plenty of lemmings (Table 2; Meltofte and Høye, 2007). However, for the long-tailed skua, this effect was observed in the breeding population only (Table 2). As previously reported in other long-tailed skua populations (e.g., Andersson, 1976; Gilg *et al.*, 2006), the observed numerical response of the breeding population (Figure 3A) suggests density-dependent predation by the skuas at Zackenberg, that is, the skua population displays an increasing dependence on lemming concurrent with increasing lemming densities. Indeed, it has been shown that the functional response of long-tailed skua involves a shift from food consisting mainly of berries and insects to lemming when the latter becomes abundant (Cramp and Simmons, 1983; de Korte and Wattel, 1988; Gilg *et al.*, 2006).

The recent field study by Gilg *et al.* (2003, 2006) on the dynamics of the collared lemming and its predators in a natural consumer–resource system in

Table 2 Inter-trophic analyses of predator–prey interactions

Species	Population model	$\beta_1 \pm$ S.E.M.[a]	$g_1 \pm$ S.E.M.[b]	$l_1 \pm$ S.E.M.[b]	$d_1 \pm$ S.E.M.[b]	$p_1 \pm$ S.E.M.[b]	$\omega_1 \pm$ S.E.M.[b]	R^2_{AR}	R^2_{COV}
Arctic fox	$X_t = \beta_1 X_{t-1} + l_1 L_t + \omega_1 S_{t-1}$	**−0.43 ± 0.42**	–	−0.25 ± 0.30	–	–	**−0.023 ± 0.008**	0.18	0.35
Arctic fox[c]	$X_t = \beta_1 X_{t-1} + d_1 D_{t-1} + \omega_1 S_{t-1}$	**−0.66 ± 0.29**	–	–	**1.45 ± 0.61**	–	**−0.025 ± 0.006**	0.23	0.40
Arctic fox[c]	$X_t = \beta_1 X_{t-1} + p_1 P_{t-1} + \omega_1 S_{t-1}$	**−0.66 ± 0.29**	–	–	–	**6.69 ± 3.01**	**−0.024 ± 0.006**	0.22	0.40
Stoat[d]	$X_t = \beta_1 X_{t-1} + l_1 L_t + \omega_1 S_{t-1}$	**−0.65 ± 0.29**	–	**0.05 ± 0.01**	–	–	**0.001 ± 0.0003**	0.26	0.40
Long-tailed skua (territorial)[e]	$X_t = \beta_1 X_{t-1} + l_1 L_t + \omega_1 S_{t-1}$	−0.01 ± 0.52	–	−0.09 ± 0.07	–	–	−0.002 ± 0.002	0.01	0.13

Long-tailed skua (breed)[e]	$X_t = \beta_1 X_{t-1} + l_1 L_t + \omega_1 S_{t-1}$	**0.12 ± 0.35**	–	**0.82 ± 0.22**	–	–	0.007 ± 0.007	0.07	0.63
Musk ox	$X_t = \beta_1 X_{t-1} + g_1 G_{t-1} + \omega_1 S_{t-1}$	**−0.92 ± 0.17**	**0.02 ± 0.01**	–	–	–	**−0.014 ± 0.005**	0.20	0.71

Because of the shortness of the time series, predator responses (X, $\log_e$-transformed densities; see Figure 3 for details) to changes in prey abundance were analysed for each prey separately, why the R^2 values for each model do not convey any information of a given prey's influence relative to other prey species. Log_e-transformed prey abundance variables integrated in the population models were L, lemming; D, dunlin; P, common ringed plover. The variable S_{t-1} denotes percent spring snow-cover and G the length of growth season (number of days with positive ecosystem assimilation; Grøndahl, 2006). The temporal delays of model variables are specified under population model. Regression coefficients are given with standard error of mean (S.E.M.), and significant ($p < 0.05$) coefficients are bold. R^2_{AR} and R^2_{COV} are the partial R^2 of density-dependent (X_{t-1}) and co-variate (prey abundance and S_{t-1}) effects on X_t, respectively. The sign "–" indicates exclusion of the variable from the population model.

Two-tailed t-tests,

[a]H_0: $\beta_1 = 1$.

[b]H_0: g_1, l_1, d_1, or $p_1 = 0$.

[c]There was no significant influence of the other wader species (red knot, sanderling and ruddy turnstone; see further in the text).

[d]No significant influence of the wader species except for ruddy turnstone ($t-1$): 0.09 ± 0.03 (±S.E.M.; see further in the text). No significant delay ($t-1$) of lemming density.

[e]Neither waders nor arctic fox and stoat influenced significantly the population dynamics of long-tailed skua.

Northeast Greenland represents a classic textbook example of the specialist hypothesis (Bjørnstad *et al.*, 1995; Hudson and Bjørnstad, 2003), which predicts that prey populations undergo periodic fluctuations in numbers in response to predation by specialised predators. Indeed, in their study, Gilg and coworkers found that the periodic fluctuations in collared lemming abundance on Traill Ø was driven by a one-year delay in predation by stoat and stabilised by density-dependent actions of the generalist predators including arctic fox and long-tailed skua (Gilg *et al.*, 2003). On the basis of current field data from Zackenberg, however, we did not find a significant, one-year delayed predation by stoat here. Instead, our analyses suggest that the stoat population at Zackenberg responds to increased lemming abundance in the current year (Table 2). In corroboration, Schmidt *et al.* (2008, this volume) found only a weak, delayed effect of stoat on lemmings when modelling the system at Zackenberg. The interaction between stoat and collared lemming is probably less tightly coupled at Zackenberg than recently described for the specialist predator at the Traill Ø locality, as suggested by its statistical association with the dynamics of ruddy turnstone (Table 2). This, however, remains to be investigated in detail.

The reported linear response of the skua as well as the stoat populations to changes in lemming abundance may disguise potential nonlinearity (Figure 5). Apparently, the stoat population responds primarily to high numbers of collared lemming (Figure 5A) corroborating the suggested

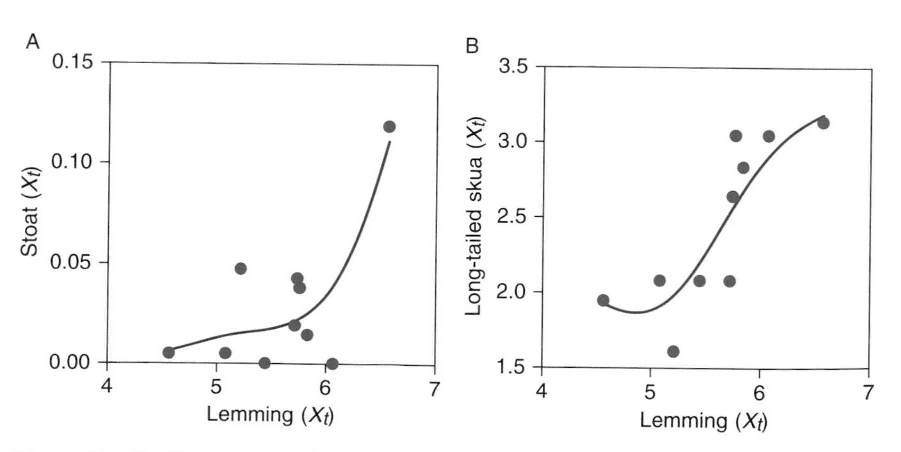

Figure 5 Nonlinear regressions of the lemming abundance year t on (A) stoat abundance year t (B) on breeding long-tailed skuas year t. Abundances were $\log_e$ transformed. The nonlinear regression lines are second-order generalised additive models (Venables and Ripley, 2002). These described the relationships better than generalised linear models ($p < 0.05$).

"generalist" behaviour of stoat at Zackenberg. On the contrary, the response by long-tailed skua occurred at intermediate lemming densities (Figure 5B). The latter is also corroborated by the predator–prey model of Schmidt *et al.* (2008, this volume). However, longer time series would be necessary to fully evaluate the influences of nonlinearity and thresholds. Nevertheless, our predator–prey analyses of stoat and long-tailed skua (Table 2) confirmed the direct climatic effects of snow detected in our single-species analyses (Table 1), indicating that whereas the long-tailed skua population is affected primarily by indirect effects of snow mediated through its impact on the lemming population, the stoat population is potentially exposed to both direct and indirect prey-mediated climatic effects of snow.

A number of common predators at Zackenberg have the collared lemming as an important part of their diet (Figure 1), and the species is often referred to as the key prey species of the system (Gilg *et al.*, 2006; Schmidt *et al.*, 2008, this volume). Key species in ecological communities are potentially key indicators for climate effects. The dynamics of the collared lemming potentially represent the cumulative response of direct and indirect, intertrophically mediated climate effects in the community. Given this multitude of both biological and climatic effects on a single species, it is no simple task to separate single effects or even predict their consequences. The multi-trophic level data collected at Zackenberg have enabled us to establish a better basis for investigating the climatic effects on collared lemmings and their environment. Schmidt *et al.* (2008, this volume) specifically address this by combining field data with different model scenarios. Here, we focus on the extent to which dynamics of collared lemmings over the last 10 years at Zackenberg were associated with the dynamics of their predators. Correcting for the density-dependent and direct climatic effects (Figure 6), our analyses suggest that through predation, the arctic fox and long-tailed skua significantly constrained growth of the collared lemming population, accounting for an additional 25–28% of its variance. The dynamics of the stoat, on the contrary, were positively associated (Figure 6) with lemming dynamics, suggesting that the former are controlled by the latter and not vice versa (May, 1981; Turchin *et al.*, 2000). Hence, climatic effects on the lemming population seem to be perpetuated through direct as well as indirect, predator-mediated interactions.

The dynamics of predator–lemming relations described above clearly suggest that the system at Zackenberg deviates from the system on Traill Ø described by Gilg *et al.* (2003, 2006). This probably relates to the different food webs in which the predator–lemming systems at Zackenberg and Traill Ø are embedded. The low abundance of snowy owl *Nyctea scandiaca* and the high density of musk oxen found at Zackenberg but not on Traill Ø, for example, probably strongly influence the consumer and/or competitive interactions within the system. Lemmings and other voles display quite different

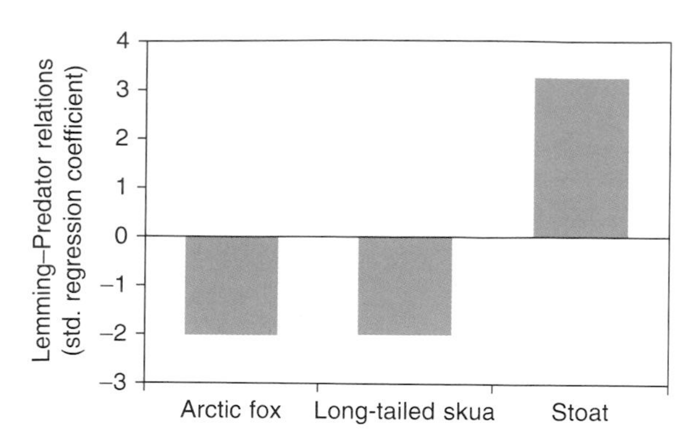

Figure 6 Lemming–predator relations expressed by the standardised regression coefficients (*b*-primes) from the effect of arctic fox, long-tailed skua, (territorial) and stoat on the lemming population, respectively. Because of the relative shortness of time series, predator regression coefficients were estimated in three separate generalised linear models with lemming abundance (X_t) as the response variable, and direct density dependence (X_{t-1}), amount of winter snow indexed by spring snow-cover (S_{t-1}) and predator abundance (Y_{t-1}) as predictor variables.

dynamics across populations, from clear periodic to fluctuations with no clear statistical patterns (Stenseth *et al.*, 1996; Reid *et al.*, 1997; Bjørnstad *et al.*, 1998). The causal mechanisms behind these are often multiple. For example, in Fennoscandia and on Hokkaido, Japan, microtine populations display very similar clinal patterns in their dynamics. However, whereas the Fennoscandian cline has been related to the occurrence of generalist predators (Bjørnstad *et al.*, 1995), the observed geographic cline in microtine dynamics on Hokkaido probably is a result of variations in the interaction between snow-cover and the presence of specialist predators (Stenseth *et al.*, 1996). Which of these factors might account for the differences observed between Zackenberg and Traill Ø remain unknown. It would be premature, however, to embark on integrating delayed responses in population dynamics until the time series at both sites has been substantially extended (e.g., Saitoh, 1998).

As with the aforementioned interactions between the collared lemming and its predators, the interaction between musk oxen and their forage is likely to also embrace indirect, inter-trophic-mediated effects of climate. In addition to the direct negative effects of snow (Table 2), the musk ox population is significantly dependent on the length of the growing season: the longer the growing season, the more musk oxen occur at Zackenberg the following year (Table 2). Since the length of the growing season is negatively associated with the extent of snow-cover ($r_{10} = -0.77$, $p < 0.01$; Grøndahl, 2006), significant climatic influences on plant phenology may be mediated through the plant–musk ox interactions to musk ox population dynamics

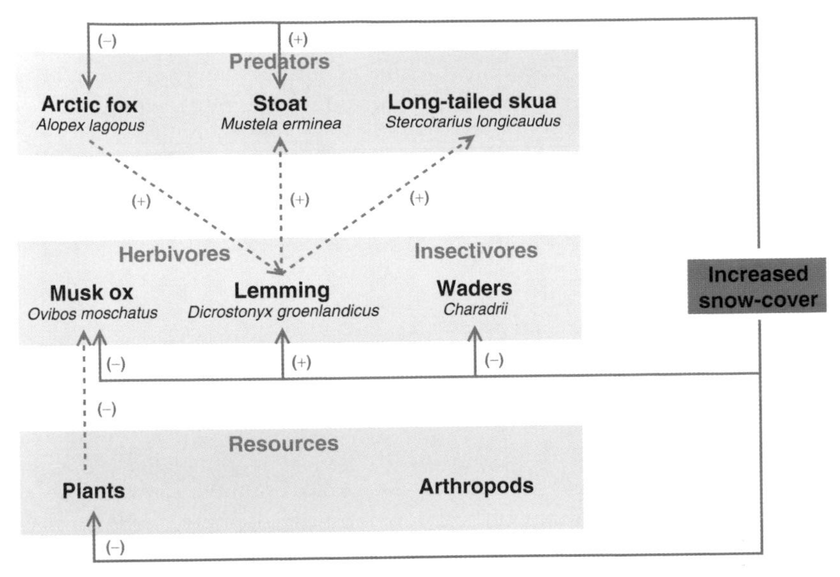

Figure 7 Schematic view of the consequences of increased snow-cover on the terrestrial vertebrate community at Zackenberg. Solid and dashed arrows indicate direct and indirect effects, respectively. Associated signs indicate the direction of influence.

(Berg *et al.*, 2008, this volume). Figure 7 summarises the direct and potential indirect effects of variations in snow-cover for all species analysed and contrasted in this chapter.

V. INTRA-ANNUAL POPULATION DYNAMICS IN RESPONSE TO CLIMATE

So far, we have approached population dynamical responses to climate change on an inter-annual basis, because the life histories of the species described above involve annual reproductive cycles (Begon *et al.*, 2002). However, significant climatic effects on long-lived, iteroparous species and their environment are often considered to occur on multi-annual or even decadal timescales, hence necessitating the use of temporal trend analyses to estimate effects of climate change. However, such approaches lose important short-term responses. In fact, the monitoring programme at Zackenberg has demonstrated the ability of the ecosystem to respond quickly not only on a year-to-year basis but even across seasons (e.g., Meltofte, 2002; Grøndahl, 2006; Høye and Forchhammer, 2008, this volume). We end this chapter by demonstrating how inter-annual variations in snow conditions can exert considerable influence on the seasonal spatio-temporal population dynamics of a single species, the musk ox. In this context, changes in the local occurrence are the result of behaviour rather than changes in vital rates (Forchhammer *et al.*, 2005). Indeed, understanding the short-term, behavioural dynamics of a population is essential because the distribution of the animals may affect ecosystem feedback mechanisms. Population level responses to climate change are ultimately the sum of all individual behavioural decision-making in the population (Sutherland, 1997).

The musk ox is the only large-bodied herbivore inhabiting high-arctic Greenland and the terrestrial system at Zackenberg. It is well known that ungulates like reindeer and musk ox exert, especially in larger herds, considerable influence on the growth and diversity of the terrestrial vegetation through their foraging, defecation and other physical activities (e.g., Post and Klein, 1996; Raillard and Svoboda, 2000; Klein *et al.*, 2008, this volume). Since vegetational changes with respect to species composition recently have been related to concomitant changes in land-atmosphere carbon gas-flux exchange (Ström *et al.*, 2005; Ström and Christensen, 2007), climate-mediated changes in spatio-temporal, short-term population dynamics of the musk ox population at Zackenberg may lead to considerable changes in the feedback from land to atmosphere. Although this remains to be investigated, the data from Zackenberg clearly show that marked climate-mediated shifts in vegetation growth have significant consequences for the spatial usage of the Zackenberg landscape by the musk ox population (Forchhammer *et al.*, 2005). Specifically, current year plant growth is highly dependent on variations in previous winter's snow (S_{t-1}). Winters with large amounts of snow are followed by summers characterized by lower and later biomass production (Tamstorf *et al.*, 2007; Berg *et al.*, 2008, this volume). These climate-mediated changes in plant biomass have important effects on the musk oxen. Increased biomass of forage reduced density-dependent movements of musk oxen in and out of the monitoring area (Figure 8A). The density-dependent usage of the landscape occurred primarily in female musk oxen (Forchhammer *et al.*, 2005). In years with increased plant biomass, the animals gathered in larger herds (Figure 8B), with less distance between them (Figure 8C) and foraged at lower altitudes (Figure 8D) where they concentrated on lush vegetation types, such as fen.

VI. CONCLUSIONS

Notwithstanding the impressive and meticulous field work carried out by the staff at Zackenberg, month after month and year after year, 10 years of time series data may be considered as relatively short when focusing on the inter-annual dynamics of long-lived, iteroparous vertebrate species

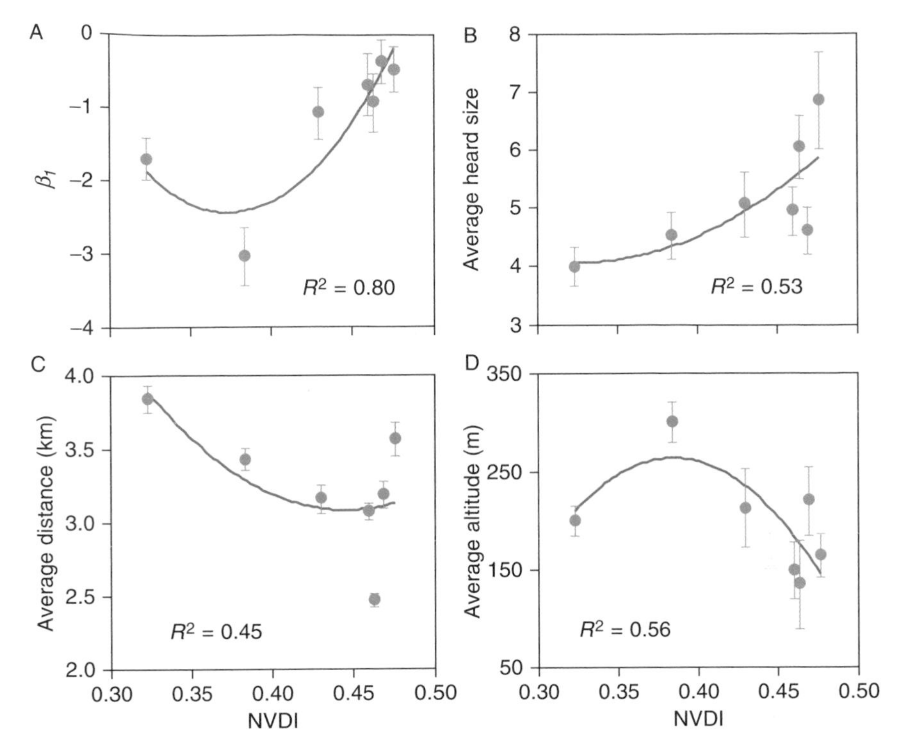

Figure 8 Current-year influence of changes in plant biomass expressed by the Normalised Difference Vegetation Index (NDVI; Todd *et al.*, 1998) derived from satellite images at the peak of summer greening (data from Sigsgaard *et al.*, 2006), on (A) average strength of within-year density-dependent migration in and out of the Zackenberg study area (adapted from Forchhammer *et al.*, 2005), (B) average herd size, (C) average distance between herds and (D) average altitude of observed herds (m a.s.l.). The nonlinear regression lines are second-order generalised additive models (Venables and Ripley, 2002). All R^2 values are significant ($p < 0.05$).

(e.g., Saitoh, 1998). Therefore, we have purposely avoided complex multivariate analyses with interaction terms and, instead, adopted a stepwise approach in our analyses of the population dynamical responses to changes in climate.

Despite these considerations, we have documented a consistent pattern of, we presume, causal factors of (1) intra-specific density dependence, (2) consumer–resource interactions and (3) direct as well as indirect, inter-trophic-mediated climate effects of varying snow-cover underlying the population dynamics of the terrestrial vertebrate community at Zackenberg. Although ecologically interrelated, the species involved displayed differences in their responses to climate, reflecting over-wintering strategies of residence versus migration.

Out of the 10 species described, 8 displayed significant direct density dependence, suggesting an overall density-dependent stability in their dynamics (Royama, 1992; Tong, 1992). However, in five species, equilibrium densities were (including all four resident species) influenced by changes in the amount of snow in the previous winter, introducing climatic stochasticity into their population dynamics. Although we did not directly document inter-trophic-mediated climatic effects, the significant consumer–resource relationships presented suggest a clear potential for this, in particular between the collared lemming and its predators (Figure 7). There is, however, still an urgent need to establish the relative importance of both direct and indirect climatic effects to provide a proper model skeleton for future climate scenarios (but see Schmidt *et al.*, 2008, this volume).

ACKNOWLEDGMENTS

The monitoring data used in this chapter were provided by the BioBasis programme, run by the National Environmental Research Institute, University of Aarhus, and financed by the Danish Environmental Protection Agency, Ministry of the Environment. The Danish Polar Center provided access and accommodation at the Zackenberg Research Station during all the years. We extend our sincere thanks to Nick Tyler who contributed significant improvements to an earlier version of the manuscript.

REFERENCES

Andersson, M. (1976) *J. Anim. Ecol.* **45**, 537–559.
Angerbjörn, A., Arvidson, B., Norén, E. and Strömgren, L. (1991) *J. Anim. Ecol.* **60**, 34–49.
Austin, G.E. and Rehfisch, M.M. (2005) *Global Change Biol.* **11**, 31–38.
Begon, M., Harper, J.L. and Townsend, C.R. (2002) *Ecology: Individuals, Populations and Communities* 4th ed. Blackwell Science, Oxford.
Bellows, T.S., Jr. (1981) *J. Anim. Ecol.* **50**, 139–156.
Bjørnstad, O.N. and Grenfell, B.T. (2001) *Science* **293**, 638–643.
Bjørnstad, O.N., Falck, W. and Stenseth, N.C. (1995) *Proc. R. Soc. Lond. B* **262**, 127–133.
Bjørnstad, O.N., Stenseth, N.C., Saitoh, T. and Lingjærde, O.C. (1998) *Res. Popul. Ecol.* **40**, 77–84.
Bjørnstad, O.N., Sait, S.M., Stenseth, N.C., Thompson, D.J. and Begon, M. (2001) *Nature* **409**, 1001–1006.
Byrkjedal, I. (1980) *Ornis Scand.* **11**, 249–252.
Caswell, H. (2001) *Matrix Population Models,* 2nd ed. Sinauer Associates, Sunderland, Massachusetts.

Cramp, S. and Simmons, K.E.L. (1983) *The Birds of the Western Palearctic. Vol. 3. Waders to Gulls.* Oxford University Press, Oxford.
de Korte, J. and Wattel, J. (1988) *Ardea* **76**, 27–41.
Dennis, B. and Taper, M.L. (1994) *Ecol. Monogr.* **64**, 205–224.
Eide, N., Eid, P.M., Prestrud, P. and Swenson, J.E. (2005) *Wildl. Biol.* **11**, 109–121.
Ellis, A.M. and Post, E. (2004) *BMC Ecol.* **4**, 2.
Forchhammer, M.C. (1995) *Can. J. Zool.* **73**, 1344–1361.
Forchhammer, M.C. (2001) In: *Climate Change Research–Danish Contributions* (Ed. by A.M.K. Jørgensen, J. Fenger and K. Halsnæs), pp. 219–236. Copenhagen, Gad.
Forchhammer, M.C. and Asferg, T. (2000) *Proc. R. Soc. Lond. B* **267**, 779–786.
Forchhammer, M.C. and Boertmann, D. (1993) *Ecography* **16**, 299–308.
Forchhammer, M.C. and Post, E. (2004) *Popul. Ecol.* **46**, 1–12.
Forchhammer, M.C., Stenseth, N.C., Post, E. and Langvatn, R. (1998) *Proc. R. Soc. Lond. B* **265**, 341–350.
Forchhammer, M.C., Post, E., Stenseth, N.C. and Boertmann, D.M. (2002) *Popul. Ecol.* **44**, 113–120.
Forchhammer, M.C., Post, E., Berg, T.B., Høye, T.T. and Schmidt, N.M. (2005) *Ecology* **86**, 2644–2651.
Gilg, O., Hanski, I. and Sittler, B. (2003) *Science* **302**, 866–868.
Gilg, O., Sittler, B., Sabard, B., Hurstel, A., Sané, R., Delattre, P. and Hanski, I. (2006) *Oikos* **113**, 193–216.
Gompertz, B. (1825) *Phil. Trans. R. Soc. Lond. B* **115**, 513–585.
Grenfell, B.T., Wilson, K., Finkenstädt, B.F., Coulson, T.N., Murray, S., Albon, S.D., Pemberton, J.M., Clutton-Brock, T.H. and Crawley, M.J. (1998) *Nature* **394**, 674–677.
Grøndahl, L. (2006) *Carbon Dioxide Exchange in the High Arctic–Examples from Terrestrial Ecosystems.* PhD thesis, National Environmental Research Institute and University of Copenhagen.
Grøndahl, L., Friborg, T. and Soegaard, H. (2007) *Theor. Appl. Climatol.* **88**, 111–125.
Hudson, P.J. and Bjørnstad, O.N. (2003) *Science* **302**, 797–798.
Hurvich, C.M. and Tsai, C.-L. (1989) *Biometrika* **76**, 297–307.
Høye, T.T., Post, E., Schmidt, N.M., Meltofte, H. and Forchhammer, M.C. (2007a) *Curr. Biol.* **17**, R449–R451.
Høye, T.T., Ellebjerg, S.M. and Philipp, M. (2007b) *Arct. Antarct. Alp. Res.* **39**, 412–421.
Insley, H., Peach, W., Swann, B. and Etheridge, N. (1997) *Bird Study* **44**, 277–289.
Klaassen, M., Lindström, Å., Meltofte, H. and Piersma, T. (2001) *Nature* **413**, 794.
May, R.M. (1981) *Theoretical Ecology: Principles and Applications.* Blackwell Scientific, Oxford, UK.
Maynard Smith, J. and Slatkin, M. (1973) *Ecology* **54**, 384–391.
McBean, G., Alekseev, G., Chen, D., Førland, E., Fyfe, J., Groisman, P.Y., King, R., Melling, H., Vose, R. and Whitfield, P.H. (2005) In: *Arctic Climate Impact Assessment* (Ed. by C. Symon, L. Arris and B. Heal), pp. 21–60. Cambridge University Press, Cambridge.
Meltofte, H. (1985) *Meddr. Grønland, Biosci.* **16**, 1–43.
Meltofte, H. (ed.) (2002) *Sne, is og 35 grades kulde. Hvad er effekterne af klimaændringer i Grønland.* National Environmental Research Institute, Ministry of Environment.
Meltofte, H. (2006) *Dansk Orn. Foren. Tidsskr.* **100**, 16–28.

Meltofte, H. and Berg, T.B. (2006) *BioBasis–Conceptual Design and Sampling Procedures of the Biological Programme of Zackenberg Basic*, 9th ed. National Environmental Research Institute, Denmark.
Meltofte, H. and Høye, T.T. (2007) *Dansk Orn. Foren. Tidsskr.* **101**, 109–119.
Meltofte, H., Høye, T.T., Schmidt, N.M. and Forchhammer, M.C. (2007) *Polar Biol.* **30**, 601–606.
Milner, J.M., Elston, D.A. and Albon, S.D. (1999) *J. Anim. Ecol.* **68**, 1235–1247.
Muus, B., Salomonsen, F. and Vibe, C. (1990) *Grønlands Fauna*. Gyldendal, Copenhagen.
Post, E. and Forchhammer, M.C. (2001) *BMC Ecol.* **1**, 5.
Post, E. and Forchhammer, M.C. (2002) *Nature* **416**, 389–395.
Post, E. and Klein, D.R. (1996) *Oecologia* **107**, 364–372.
Post, E. and Stenseth, N.C. (1999) *Ecology* **80**, 1322–1339.
Post, E., Peterson, R.O., Stenseth, N.C. and McLaren, B.E. (1999) *Nature* **401**, 905–907.
Raillard, M. and Svoboda, J. (2000) *Arct. Antarct. Alp. Res.* **32**, 278–285.
Rehfisch, M.M., Austin, G.E., Freeman, S.N., Armitage, M.J.S. and Burton, N.H.K. (2004) *Ibis* **146**, 70–81.
Reid, D.G. and Krebs, C.J. (1996) *Can. J. Zool.* **74**, 1284–1291.
Reid, D.G., Krebs, C.J. and Kenney, A.J. (1997) *Ecol. Monogr.* **67**, 89–108.
Royama, T. (1992) *Analytical Population Dynamics*. Chapman & Hall, London.
Saitoh, T. (1998) *Popul. Ecol.* **40**, 61–76.
Schmidt, N.M. (2006) *Climatic, Agriculture and Density-Dependent Dynamics Within and Across Trophic Levels in Contrasting Ecosystems*. PhD thesis, Department of Ecology, Royal Veterinary and Agricultural University, Denmark.
Sigsgaard, C., Petersen, D., Grøndahl, L., Thorsøe, K., Meltofte, H., Tamstorf, M. and Hansen, B.U. (2006) In: *Zackenberg Ecological Research Operations, 11th Annual Report, 2005* (Ed. by A.B. Klitgaard, M. Rasch and K. Caning), pp. 11–35, Danish Polar Center, Ministry of Science, Technology and Innovation, Copenhagen.
Sittler, B. (1995) *Ann. Zool. Fenn.* **32**, 79–92.
Stenseth, N.C. and Ims, R.A. (1993) *The Biology of Lemmings*. Academic Press.
Stenseth, N.C., Bjørnstad, O.N. and Saitoh, T. (1996) *Proc. R. Soc. Lond. B* **263**, 1117–1126.
Stenseth, N.C., Mysterud, A., Ottersen, G., Hurrell, J.W., Chan, K.-S. and Lima, M. (2002) *Science* **297**, 1292–1296.
Ström, L. and Christensen, T.R. (2007) *Soil Biol. Biochem*, in press.
Ström, L., Mastepanov, M. and Christensen, T.R. (2005) *Biogeochemistry* **75**, 65–82.
Sutherland, W.J. (1997) *From Individual Behaviour to Population Ecology*. Oxford University Press, Oxford.
Tamstorf, M.P., Illeris, L., Hansen, B.U. and Wisz, M. (2007) *BMC Ecol.*, in press.
Todd, S.W., Hoffer, R.M. and Milchunas, D.G. (1998) *Int. J. Remote Sens.* **19**, 427–438.
Tong, H. (1992) *Non-Linear Time Series. A Dynamical System Approach*. Oxford University Press, Oxford.
Turchin, P., Oksanen, L., Ekerholm, P., Oksanen, T. and Henttonen, H. (2000) *Nature* **405**, 562–565.
Tyler, N.J.C., Forchhammer, M.C. and Øritsland, N.A. (2007) *Ecology*. in press.
Venables, W.N. and Ripley, B.D. (2002) *Modern Applied Statistics with S*, 4th edition. Springer, New York.
Vibe, C. (1967) *Meddr. Grønland* **170**(5), 1–227.

Walsh, J.E., Anisimov, O., Hagen, J.O.M., Jakobsson, T., Oerlemans, J., Prowse, T.D., Romanovsky, V., Savelieva, N., Serreze, M., Shiklomanov, A., Shiklomanov, I. and Solomon, S. (2005) In: *Arctic Climate Impact Assessment* (Ed. by C. Symon, L. Arris and B. Heal), pp. 183–242. Cambridge University Press, Cambridge.

Solar Ultraviolet-B Radiation at Zackenberg: The Impact on Higher Plants and Soil Microbial Communities

KRISTIAN R. ALBERT, RIIKKA RINNAN, HELGE RO-POULSEN, TEIS N. MIKKELSEN, KIRSTEN B. HÅKANSSON, MARIE F. ARNDAL AND ANDERS MICHELSEN

SUMMARY

Depletion of the ozone layer and the consequent increase in solar ultraviolet-B (UV-B) radiation may impact living conditions for arctic plants significantly. In order to evaluate how the prevailing UV-B fluxes affect the heath ecosystem at Zackenberg (74°30′N, 20°30′W) and other high-arctic regions, manipulation experiments with various set-ups have been performed.

Activation of plant defence mechanisms by production of UV-B-absorbing compounds was significant in ambient UV-B in comparison to a filter treatment reducing the UV-B radiation. Despite the UV-B screening response, ambient UV-B was demonstrated to decrease photosynthesis and shift carbon allocation from shoots to roots. Moreover, ambient UV-B increased plant stress with detrimental effects on electron processing in the photosynthetic apparatus. Plant responses did not lead to clear changes in the amount of fungal root symbionts (mycorrhiza) or in the biomass of microbes in the soil of the root zone. However, the composition of the soil microbial

ADVANCES IN ECOLOGICAL RESEARCH VOL. 40 0065-2504/08 $35.00
© 2008 Elsevier Ltd. All rights reserved
DOI: 10.1016/S0065-2504(07)00018-9

community was different in the soils under ambient and reduced UV-radiation after three treatment years.

These results provide new insight into the negative impact of current UV-B fluxes on high-arctic vegetation. They supplement previous investigations from the Arctic focussing on other variables like growth and so on, which have reported no or minor plant responses to UV-B, and the presented synthesis clearly indicates that UV-B radiation is an important factor affecting plant life at high-arctic Zackenberg. However, long-time experiments are needed in order to see whether the observed changes are transient or whether they accumulate over years. Such experiments are especially important for valid determination of below-ground responses, which potentially lead to feedbacks on the ecosystem functioning.

I. INTRODUCTION

The ecosystem responses to ultraviolet-B (UV-B) radiation (280–315 nm; CIE, 1999) in the Arctic are a research area of growing interest (Callaghan *et al.*, 2004a, 2005), motivating investigations focused to identify the targets and the relative importance of UV-B alone and in interaction with other global change factors. The increase in UV-B radiation is the result of stratospheric ozone depletion (Webb, 1997; Madronich *et al.*, 1998). The ozone-destroying chemical reactions are caused by chlorine and bromide released from emitted chlorofluorocarbons and halons (Farman *et al.*, 1985) and the very cold stratospheric temperatures (below −78 °C). The process is highly temperature dependent, it is most pronounced during spring with cold stratospheric conditions and it also varies from year to year (Weatherhead *et al.*, 2005). Currently, the UV-B irradiance level in the arctic region is considered to be near its maximum, and the ozone column is estimated to recover towards the middle of the century, but the rate of ozone recovery is uncertain in the northern hemisphere (WMO, 2003). Increased cloudiness decreases the amount of UV-B radiation reaching the ground (Madronich *et al.*, 1998), but the predictions of future cloud cover and cloud types are uncertain (Weatherhead *et al.*, 2005). The UV-B fluxes at Zackenberg (74°30′N, 20°30′W), where this study was performed, peak in late May and early June, and high doses still prevail during July and August (see Figure 1 in Rinnan *et al.*, 2005). This means that changes in snow-cover and length of the growing season may affect the UV-B exposure dose of the vegetation. Moreover, vegetation located in the snow-free patches will receive greatly increased UV-B dose because of irradiance reflected from the surrounding snow (Jokela *et al.*, 1993; Gröebner *et al.*, 2000).

High-arctic plants are "living on the edge" because they are growing on the limit of their distribution in an extreme environment with a short growing season, low temperatures and often nutrient limitation. Therefore, acclimation is of special importance especially when the plants face environmental changes, such as increased UV-B radiation (Caldwell *et al.*, 1980; Robberecht *et al.*, 1980), which can cause additional stress under ambient conditions (Bredahl *et al.*, 2004; Albert *et al.*, 2005a, 2007a). In addition, because the vegetation in the Arctic is evolutionary adapted to low UV-B levels (Robberecht *et al.*, 1980; Caldwell and Flint, 1994), the potential impact on the vegetation is expected to be pronounced (Björn *et al.*, 1999; Paul, 2001). This leads to the hypothesis that arctic plants are negatively influenced by the current UV-B levels. Thus, if the present UV-radiation affects the vegetation significantly, then reduction of the irradiance load would improve the photosynthetic performance of the plants. Therefore, an experimental approach where ambient UV-irradiance is screened off by means of filters was chosen in this study.

The previous knowledge of polar ecosystem responses to UV-irradiance is the result of field experiments with various experimental approaches: UV-radiation has been elevated by various lamp setups (e.g., Johanson *et al.*, 1995a,b; Björn *et al.*, 1999; Gwynn-Jones *et al.*, 1997; Phoenix *et al.*, 2001 and others), transplants have been set up along latitudinal gradients (Lehrner *et al.*, 2001), or UV-radiation has been reduced by means of filters (Xiong and Day, 2001; Phoenix *et al.*, 2002; Robson *et al.*, 2003; Albert *et al.*, 2005a). In principle, the studies supplementing UV-B relate closely to scenarios with future increased UV-B levels, whereas experimental UV-reduction relates to the impacts of the current level of solar UV-radiation.

UV-exclusion experiments by means of filters are attractive in several ways. They are simple and do not require electrical power or any special technical maintenance, which is an advantage in remote areas. Further, differences in spectral ratios, which are a problem in the UV supplementation experiments (Caldwell and Flint, 1994), can be avoided. However, reduction of the UV-B irradiance by 60% or more implies a higher relative change in the UV-B load than is predicted to take place in nature. Anyhow, the clear advantage to emphasise here is that the interpretation of results from UV-exclusion experiments directly relates to the impact of current level of UV-radiation, and that the exposure includes the variability during the growing season and from year to year.

The prevalent view is that—although UV-radiation induces increased production of phenolics and berries and alters the below-ground processes (Gwynn-Jones *et al.*, 1997; Searles *et al.*, 2001a; Johnson *et al.*, 2002)—arctic plants are more or less tolerant to enhanced UV-B in the long term (Phoenix *et al.*, 2001; Callaghan *et al.*, 2004b; Rozema *et al.*, 2006). Moreover, as discussed by Phoenix *et al.* (2001) and Phoenix and Lee (2004), other climatic changes, such as increased CO_2, temperature or changes in precipitation, may further negate the detrimental effects of enhanced UV-B in the sub-arctic.

II. VEGETATION AND PLANT ECO-PHYSIOLOGICAL RESPONSES AT ZACKENBERG

The experiments at Zackenberg approached the effects of ozone depletion on ecosystems by comparing the responses to prevailing UV-B fluxes to responses to reduced UV-levels obtained by different filter arrangements, which covered plots of the ecosystem (see Box 1 for an outline of the conducted

Box 1

Overview of UV-B experiments at Zackenberg

UV-exclusion experiments were initiated in 2001, and until 2003 intensive monitoring of the microclimate in the experimental plots was done. The vegetation composition was mapped, and the hypothesised response parameters, such as photosynthesis, plant stress, leaf content of UV-B-absorbing compounds, carbon, nitrogen and soil characteristics, were investigated. The project is unique in the High Arctic and will be continued in cooperation with the ZERO programme.

2001 Establishment of two permanent sites differing in inclination (“level” and “sloping” sites) with four treatments in four groups: Filter treatments with UV-AB-reducing Lexan, UV-B-reducing Mylar, UV-transparent Teflon and a treatment without filter being an open control (Box Figure 1). Two independent climatic stations continuously logged microclimate: Air and soil temperatures, and in one group soil humidity, under and outside the filters as well as air humidity, air temperature and irradiance of photosynthetically active radiation (PAR) and UV-B at each site. Weekly measures of chlorophyll *a* fluorescence induction curves were conducted on bog blueberry *Vaccinium uliginosum* and arctic willow *Salix arctica* and end season harvest of leaves enabled analysis of UV-B-absorbing compounds. Gas exchange was measured on *S. arctica*. Results were published by Bredahl *et al.* (2004).

2002 Investigations on the permanent experimental plots were continued as the previous year. Moreover, two new experiments with UV-B-reducing Mylar and UV-transparent Teflon lasting one growth season were conducted: (1) Maximum irradiance experiment, where the irradiation doses were homogenised by controlling leaf angle on *S. arctica*, and intensive measures of chlorophyll fluorescence were conducted; (2) Robust measurements of photosynthesis were achieved by chamber measurements of gas exchange on whole canopies of *V. uliginosum* supplemented by chlorophyll fluorescence measurements. Results were published by Albert *et al.* (2005a, in press).

2003 Investigations on the permanent experimental plots were continued as in previous years, and in addition soil samples for root biomass and microbial community analyses were taken. Continued experimentation with homogenised irradiation doses on *S. arctica* were performed on new plants shoots designed to achieve two different irradiance levels. Here, intensive measurements of both gas exchange and of chlorophyll fluorescence were conducted. Results were published by Albert *et al.* (2005b), Rinnan *et al.* (2005) and Håkansson (2006).

2004 No manipulations and no observations.

2005 Re-establishment of the permanent experimental plots and continuation of the measurements on *S. arctica* and *V. uliginosum* as done in previous years. Moreover, dwarf birch *Betula nana* was included on a new site with the same setup and measurement campaign. The experimentation with effects of homogenised irradiation doses on *S. arctica* was continued and measures of simultaneous chlorophyll *a* fluorescence and gas exchange were performed. Reflectance measures in the range 325–1250 nm were performed on all species, and leaves were sampled for analysis of secondary compounds, chlorophyll, C and N. Results presented in 11th Annual ZERO Report, 2005.

2006 Continuation of the measurement campaign on *V. uliginosum*, *S. arctica* and *B. nana*. Results presented in 12th Annual ZERO Report, 2006.

Box Figure 1 The experimental setup on a part of one of the permanent experimental sites (Site 2) comprising open control, filter control, UV-B-excluding Mylar filter and UV-AB-excluding Lexan.
Vegetation below filters is dominated by *Vaccinium uliginosum* and *Salix arctica*. In the background, the Zackenberg Mountain is seen. Photo: K. Albert, 2002.

experiments and their responses). The four main treatments comprised open control plots without any filters, UV-transparent filter controls, filters reducing UV-B (280–315 nm) and filters reducing both UV-B and UV-A (315–400 nm). These treatments exposed the vegetation to *c.*100%, 91%, 39% and 17% of ambient UV-B radiation and 100%, 97%, 90% and 91% of the photosynthetically active radiation (PAR), respectively (Bredahl *et al.*, 2004). Filters also slightly changed the microclimate when compared to the open control. Three succeeding years of measurements during the growing season revealed that the filters increased the mean soil temperature up to 0.6°C, but this did not affect soil humidity. No significant mean air temperature difference was observed within the filter treatments or between the filter treatments and the open control. Between the filter treatments no microclimatic differences were found (K. Albert, unpublished).

From the range of UV-exclusion experiments conducted at Zackenberg, the emerging pattern is that there are significant plant-ecophysiological responses (Table 1). The chosen parameters, that is, photosynthesis and probing of plant stress, are generally expected to respond faster than many other important ecological processes, such as growth, phenology and species composition, which may respond to disturbances on a longer timescale (Callaghan *et al.*, 2004c). However, inter-species differences in plant performance characteristics are important since the traits are likely to influence competition and the resulting plant cover.

Measurements of chlorophyll *a* fluorescence induction curves led to calculation of the much reported parameter, maximal quantum yield (F_V/F_M), which closely relates to photosystem II (PSII) function and is often interpreted as a proxy for plant stress related to photosynthetic performance (Strasser *et al.*, 2004; see Box 2 for more information on photosynthesis and plant performance). Also the so-called performance index, PI (Strasser *et al.*, 2004) was derived from the fluorescence measurements. The PI integrates into one parameter the proportional responses of energy fluxes related to trapping and dissipation within the PSII and also to the energy transport behind PSII. Hereby, the PI expresses the overall effective energy processing through PSII and is believed to sum up the accumulative stress effects on PSII.

The UV-exclusion experiments were initiated in 2001 (see Box 1), and it was found that a reduction in the ambient UV-B level resulted in decreased content of UV-B-absorbing compounds and lower stress level indicated by increased F_V/F_M (Bredahl *et al.*, 2004) and PI in arctic willow *Salix arctica* and Bog blueberry *Vaccinium uliginosum ssp. microphyllum*. Moreover, the analysis of leaf level gas exchange revealed a decreased stomatal conductance and internal CO_2 concentration in *S. arctica* when ambient UV-B was reduced (Bredahl *et al.*, 2004). These findings were initially concluded to indicate the important impact of UV-B in the short term. However, the greatly varying leaf angle results in differences in UV-B doses for the

Table 1 Ecosystem responses to UV-B exclusion at Zackenberg

Response type	Parameters	Short-term effects			Long-term effects	
		Salix	*Vaccinium*	*Betula*	*Salix*	*Vaccinium*
Plant stress	Maximal photochemical efficiency (F_V/F_M)	↓ (~)	↓	↓	↓	↓
	Performance index (PI)	↓	↓	↓	↓	↓
Photosynthetic	Photosynthesis (P_n)	~	↓	#	#	#
	Respiration (R_d)	#	~	#	#	#
	Transpiration (Tr)	Δ	~	#	#	#
	Intercellular CO_2 concentration (C_i)	↓	#	#	#	#
	Stomatal conductance (g_s)	↑	#	#	#	#
Growth	Leaf biomass	~	~	#	~	~
	Stem biomass	~	~	#	~	~
	Root biomass	#	#	#	#	↓
	Leaf area	~	~	#	~	~
	Specific leaf area	~	~	#	~	~
Leaf chemistry	UV-B-absorbing compounds	↑	↑	↑	↑	↑
	Carbon	~	~	~	~	~
	Nitrogen	~	~	~	~	~
	Chlorophyll	#	#	↓	#	#
Plant species composition	Cover	~	~	~	~	~
Mycorrhiza	Mycorrhizal colonisation	#	#	#	#	↑↓
	Root ergosterol concentration	#	#	#	#	~
Microbial biomass	Microbial biomass carbon	#	#	#	~	~
	Microbial biomass nitrogen	#	#	#	~	~
	Microbial biomass phosphorous	#	#	#	~	~
	Soil ergosterol concentration	#	#	#	~	~
Microbial community composition	Phospholipid fatty acid (PLFA) biomarkers	#	#	#	Δ	Δ

Note: Ambient UV-B responses are compared to reduced UV-B after 1 and 3 years labelled short- and long term. Only significant changes are labelled with arrows. Signatures: No effect (~); Negative impact (↓); Positive impact (↑); Changed (Δ); Not investigated (#).

Box 2

Photosynthesis and plant performance

The photosynthetic processes can be separated into energy-producing (sources) and energy-consuming processes (sinks). Source processes are involved in the capture of light and the processing through the photosynthetic apparatus, resulting in available energy equivalents. Sink processes are the energy-demanding processes, primarily CO_2 assimilation in the Calvin cycle. The molecules of CO_2 diffuse into the leaf through stomatal openings into the stomatal cavities and from here further through the internal leaf cells until finally reaching the chloroplast, where the Calvin cycle takes place. Here, the CO_2 molecules assimilated are stored in energy-rich metabolites, that is, sugar, which is allocated to different plant parts for growth and maintenance. Under field conditions, the often used methods to probe both the characteristics and activity of the source and sink sides are to do measurements of chlorophyll *a* fluorescence and CO_2 and H_2O gas exchange. The sink processes are evaluated by parameters as, for example, net photosynthesis (P_n), stomatal conductance (g_s) and transpiration (Tr), while the source processes are evaluated by maximal quantum yield (F_V/F_M), performance indexes (PI), and a range of other parameters related both to the handling of light in the photosynthetic apparatus and partitioning of energy fluxes. The much reported parameter, maximal quantum yield (F_V/F_M), closely relates to photosystem II (PSII) function and is often interpreted as a proxy for plant stress. The PI integrates into one parameter the proportional responses of energy fluxes related to trapping and dissipation within the PSII and also to the energy transport behind PSII. Hereby, the PI expresses the overall effective energy processing through PSII and is believed to sum up the accumulative effects on PSII. Based on such measurements, the targets of environmental stressors can be both identified and quantified and the overall performance of the photosynthetic processes is assessed.

individual leaves, which probably confounded the effects. This hypothesis led to an experimental attempt to homogenise the UV-B dose received by the leaves by manipulative fixation of the plant leaves perpendicular to the Sun.

The second season (2002) included such a setup with fixation of leaves, and this led to clear-cut positive impacts on almost all measured and derived fluorescence parameters on *S. arctica*. The results on F_V/F_M and PI confirmed that ambient UV-B radiation is a significant plant stressor

(Albert *et al.*, 2005a). Also the proportions of energy fluxes per leaf cross section were quantified, and the dissipation of untrapped energy was highest under ambient UV-B, resulting in significantly lower flux of energy beyond the electron intersystem carriers (Albert *et al.*, 2005a). These responses are argued to be specifically due to the UV-B radiation (Albert *et al.*, 2005a), and they demonstrate a less effective energy processing in the photosynthetic machinery. For logistical reasons, no leaf level photosynthetic measurements were done on *S. arctica* that year, but measurements of photosynthesis and respiration were conducted on whole canopies of *V. uliginosum*. The *V. uliginosum* plants showed a decreased photosynthesis in parallel with decreased values of F_V/F_M and PI in ambient UV-B compared to the reduced UV-B treatment (Albert *et al.*, in press). This response was seen through most of the growth season, but in the senescence period in late August the treatment differences disappeared. These results clearly linked the decrease in net photosynthesis to the stress effects on the light-energy harvesting and processing machinery. Also the *V. uliginosum* plants had a higher level of UV-B-absorbing compounds in the leaves under ambient UV-B (Albert *et al.*, in press), but the possible protective screening by these compounds was obviously not sufficient to avoid negative effects on the photosynthetic machinery.

It has been argued by Searles *et al.*, (2001a) and others that changes in UV-B-absorbing compounds as such are not a good indicator of the degree of UV-B impact in plants, although the increase in UV-B-absorbing compounds in response to UV-B is the most consistent and frequent plant response (Searles *et al.*, 2001a). The pool here referred to as UV-B-absorbing compounds includes a range of secondary compounds, and moreover no distinction between wall bound and cellular compounds is generally made. These compounds have also other functions related to antioxidation (Bornman *et al.*, 1998; Rozema *et al.*, 2002) and plant defence against herbivores (Harborne and Grayer, 1993). Hence, the UV-induced alterations in the amount and quality of the UV-B-absorbing compounds may have implications for both herbivory (Ballaré *et al.*, 1996; Rousseaux *et al.*, 2004a) and litter decomposition (Björn *et al.*, 1999), affecting nutrient cycling, although this has not yet been investigated at Zackenberg.

For *S. arctica* plants, physiological responses between male and female plants may differ (Jones *et al.*, 1999; Håkansson, 2006). In a short-term study in 2003 on *S. arctica* at Zackenberg, the sex actually interacted with treatment responses, and surprisingly no significant treatment differences were found on F_V/F_M, photosynthesis or content of UV-B-absorbing compounds. The response was in general ascribed to the particularly high content of UV-B-absorbing compounds, being more than 50% higher than previous years, leading to effective screening against UV-B radiation. If this explanation is correct, then it suggests that plants under some conditions actually are able to cope with the negative impact of UV-B. Håkansson (2006) also investigated the effect of sudden filter removal in filter treatments during peak season. Although no treatment-specific responses *per se* were detected hereafter, the plants actually became more stressed when re-exposed to ambient UV-B. This was indicated by decreased F_V/F_M, whereas the cohort of plants still being treated with filters showed the opposite and decreased their stress level.

Throughout the 3-year period (2001–2003), it was consistently found that *S. arctica* and *V. uliginosum* leaves exposed to current UV-B fluxes had higher content of UV-B-absorbing compounds and were experiencing a higher stress level than when UV-B was reduced (K. Albert, unpublished).

An investigation on dwarf birch *Betula nana* during 2005 clearly demonstrated a similar plant stress release by UV-reduction as earlier observed in *S. arctica* and *V. uliginosum*. F_V/F_M and PI were significantly increased throughout the experimental period in July and the beginning of August in the treatments where large proportions of UV-B and UV-AB were excluded as compared to both filtered and open control (Albert *et al.*, 2006). The stress response was previously hypothesized to be restricted to periods with high irradiance (of both PAR and UV) (Albert *et al.*, 2005a), and this was tested by measurements throughout a day under clear sky conditions. As expected, a midday depression in both F_V/F_M and PI was seen in *B. nana* across treatments in parallel with irradiance doses, which were maximal when the Sun was in Zenith. Surprisingly, the level of the PIs in the UV-reduction treatments stayed higher during all times of the day. This demonstrates that the control plants, which were exposed to the ambient level of UV-radiation, appear to be permanently stressed and do not recover after exposure to the midday high irradiation event by finalising repair processes. This new finding points to the importance of negative impacts of ambient UV-radiation on the photosynthetic apparatus in *B. nana*, which may be rendered as a UV-sensitive plant species, at least in the short term.

III. COMPARISONS OF PLANT RESPONSES THROUGHOUT THE POLAR REGION

Although caution is needed when making generalisations of polar plant responses because of differences between Antarctic and arctic ecosystems (e.g., higher species diversity, more trophic interactions and lower UV-B fluxes in the Arctic) (Rozema *et al.*, 2005), our observations are in agreement with the UV-exclusion studies conducted in the Antarctic ecosystems showing that ambient UV-B can have significant impacts.

Responses from the UV-exclusion studies carried out on the Antarctic Peninsula and in sub-Antarctic Tierra del Fuego point to negative UV-B effects on plant growth (Day *et al.*, 1999; Rousseaux *et al.*, 1999; Ruhland and Day, 2000; Ballaré *et al.*, 2001 and others) and increased phenolic production in most species (Day *et al.*, 2001; Searles *et al.*, 2001a,b). Further, DNA damage (Rousseaux *et al.*, 1999) has been observed particularly during the high springtime UV-B fluxes (Xiong and Day, 2001; Ruhland *et al.*, 2005) and also in the longer term (Robson *et al.*, 2003). The negative impact on plant biomass production, as reported by Xiong and Day (2001), was not associated with reduced photosynthesis per leaf area, but rather with reduced photosynthesis per chlorophyll amount or leaf dry mass. Xiong and Day (2001) interpret this response as that under UV-B, the plants were denser and probably had thicker leaves with a higher amount of photosynthetic and UV-B-absorbing pigments per leaf area. On the other hand, the analysis of chlorophyll *a* fluorescence and photosynthetic light response curves demonstrated that photosynthesis was impaired in the upper cell layers, but this did not translate into changes in photosynthetic rates at the whole leaf level.

There are differences between the responses observed at high-arctic Zackenberg compared to high-arctic Svalbard and the sub-arctic Abisko in northern Sweden, where most UV-B supplementation studies have been conducted.

No effects of 7 years of UV-B supplementation were detected on plant cover, density, leaf weight, leaf area, reproductive parameters, leaf UV-B absorbance and content of total phenolics in plants on arctic Svalbard (Rozema *et al.*, 2006). The absence of responses to enhanced UV-B in Svalbard was discussed to indicate several aspects. First, the differences in UV-B levels posed in supplemental studies are less than in UV-B exclusion studies, where responses were argued to be more difficult to detect. Secondly, the tundra biome in Svalbard originates from latitudes with higher natural solar UV-B fluxes implying a possible higher tolerance to UV-B (Rozema *et al.*, 2006).

Based on long-term studies conducted in the area of Abisko, it has been concluded that the dwarf shrubs there seem tolerant to ambient UV-B (Phoenix *et al.*, 2001, 2002; Callaghan *et al.*, 2004b; Rozema *et al.*, 2006). In some instances, enhanced UV-B radiation reduced plant growth, modified plant—herbivore interactions (Gwynn-Jones *et al.*, 1997), slowed the rate of litter decomposition, altered microbial soil biomass (Johnson *et al.*, 2002) and reduced cyanobacterial nitrogen fixation (Solheim *et al.*, 2002), but did not change plant cover or DNA damage (Rozema *et al.*, 2005, and references herein).

The investigations in sub-arctic Abisko and high-arctic Svalbard have put more weight on traditional parameters such as various measures of growth, phenology and so on, while the Zackenberg research has had a more non-invasive approach by weighting photosynthetic and stress variables, which respond immediately to changes in radiation. To take advantage of both approaches, work focused on linking variables across scales, that is, the photosynthetic response to other measures of growth, should be done. However, because of the difference in variables measured and also a different experimental approach (supplementing UV-B vs UV-B exclusion), direct comparisons are not always possible. In the section V below, these important issues are discussed further.

IV. BELOW-GROUND RESPONSES

While above-ground plant responses have received much attention over the years, possible effects of UV-B radiation on below-ground components of arctic ecosystems are less well understood. Although the presence of creeping tundra plants leads to higher UV penetration to the soil compared to the presence of more shading cushion plants, grasses and mosses (Hughes *et al.*, 2006), UV-radiation mainly affects the soil communities indirectly via effects on plants.

Ambient UV-B radiation at Zackenberg reduced root biomass of *V. uliginosum* as determined by soil core sampling after 3 years of UV-B exclusion (Rinnan *et al.*, 2005). The lower root biomass is well in agreement with the responses in the above-ground plant parts, as ambient UV-B also reduced photosynthesis and induced stress to the photosynthetic machinery (Bredahl *et al.*, 2004; Albert *et al.*, in press) as discussed above. Reductions in below-ground plant components due to ambient UV-B levels have also been reported for southern high latitudes. For instance, root length production of *Carex* spp. at a fen in southern Argentina was significantly lower under near-ambient than under reduced UV-B radiation (Zaller *et al.*, 2002). At Palmer Station, at the Antarctic Peninsula, near-ambient UV-B radiation reduced root biomass of the Antarctic hair grass *Deschampsia antarctica* by 34% compared to the plants under reduced UV-B radiation (Ruhland *et al.*, 2005). However, in this case the above-ground biomass was reduced even more, which led to a higher root-to-shoot ratio under near-ambient UV-B (Ruhland *et al.*, 2005).

Changes in plant photosynthesis and carbon allocation are likely to have an impact on mycorrhizal symbionts living in association with plant roots. At Zackenberg, the response of mycorrhizal fungi to UV-B manipulations was unclear at the level site (see Box 1 for the details of experimental setup). The light microscopical analyses indicated that the roots of *V. uliginosum* were more colonised by ericoid-type mycorrhiza under reduced UV-B radiation, but at the sloping site, the response was nearly opposite (Rinnan *et al.*, 2005). The only other report on effects of UV-B radiation on ericoid mycorrhiza that we are aware of states that 5 years of UV-B enhancement by fluorescent lamp arrays simulating 15% ozone depletion at Abisko had no effects on mycorrhizal colonisation (Johnson, 2003).

Net primary production (Callaghan *et al.*, 2004c) often correlates with the soil microbial biomass (Wardle, 2002). At our sites at Zackenberg, microbial biomass determined by the fumigation–extraction technique was, indeed, significantly associated with the total root biomass per soil volume ($R^2 = 0.23$, $p < 0.01$). However, the UV-B manipulations had no statistically significant effects on the soil microbial biomass or concentrations of nitrogen and phosphorus in the biomass (Rinnan *et al.*, 2005). This is in contrast with the results from a sub-arctic heath at Abisko, where UV-B supplementation for 5 years resulted in lower soil microbial biomass carbon and higher microbial biomass nitrogen concentration (Johnson *et al.*, 2002). However, a similar UV-B supplementation as at Abisko had no effects on the soil microbial biomass in a mesotrophic sub-arctic mire in northern Finland (R. Rinnan, unpublished data).

Potential UV-B-induced changes in the chemical quality and quantity of the labile carbon substances exuded from plant roots (i.e., root exudates) could affect soil microbial community composition. In order to compare the composition of the microbial communities between the UV-B treatments in our experiments at Zackenberg, we extracted phospholipid fatty acids (PLFAs) from the soil, which are biomarkers specific to different bacteria and fungi (Zelles, 1999). The PLFA profiles were indeed different under ambient and reduced UV-B fluxes (Rinnan *et al.*, 2005), which indicates that ambient UV-B radiation in Greenland has indirect effects on the soil microbial communities. This finding is supported by results both from the sub-arctic (Johnson *et al.*, 2002; R. Rinnan, unpublished data) and from Antarctica (Avery *et al.*, 2003), which reported effects of UV-B radiation on the utilisation of different carbon sources by culturable soil bacterial community. As the fungal biomass in the soil from Zackenberg was not affected by UV-B radiation based on the quantity of fungal PLFA biomarkers and ergosterol concentration (Rinnan *et al.*, 2005), the community composition alterations appeared to occur within the bacterial community.

Relating microbial community composition to microbial-driven ecosystem processes such as decomposition and nutrient transformations in the soil is not straightforward. Therefore, it is not possible to extrapolate how UV-B radiation would affect ecosystem functioning based on the observed responses in microbial community composition. Further analyses of microbial community by molecular methods and targeted measurements of ecosystem processes, such as nitrogen transformations, could reveal whether a certain group of bacteria was especially affected. As the indirectly induced below-ground responses can first take place after a strong enough response has occurred in plants, a 3-year-long experiment may not be long enough to show the eventual responses.

V. METHODOLOGICAL CONSIDERATIONS OF THE EXPERIMENTAL APPROACHES

In many cases, the drivers of ecosystem responses can best be identified by an experimental approach. Ideally, this implies well-documented long-term multi-factorial manipulations and comprehensive effects investigations. This approach takes advantage of testing the actual impact of the hypothesised driver and their interactions on ecosystem processes. If the environmental perturbations are realistic and well conducted they not only identify key ecosystem responses but also reveal their strength and relative importance in time and space. The syntheses of such results are the starting point for generation of novel hypotheses, which may be tackled via new experiments and relevant ecosystem modelling.

Concerning the responses to UV-B radiation: Is it possible to extrapolate results from UV-B exclusion experiments to future scenarios of ozone depletion and UV-B radiation climate? This exercise demands a range of premises to be discussed of which the most important are outlined below.

The supplemental studies which are closely simulating future scenarios may be argued to be far more realistic. UV-B exclusions substantially change the total UV-B irradiation to a much higher degree (up to 60%) than supplemental UV-B studies simulating 15–30% enhancement of UV-B do. Hence, UV-exclusion may *per se* be expected to induce greater responses in a dose-dependent context. Also the qualitative spectral differences existing between methodologies may be of importance. This is clearly indicated from studies applying treatments that reduce UV-B (UV-B-absorbing filter), ambient (UV-transmitting filter) and supplemental UV-B (UV-transmitting filter + lamps) in parallel (Gabeřscik *et al.*, 2002; Rousseaux *et al.*, 2004b). Here, a stepwise dose-dependent UV-B response is to be expected on affected parameters, if responses are linear. Further, since the initiation of biological responses also are closely related to the spectral composition of light this adds to complexity.

To approximate the biological effective differences, which may be mediated by such differences in spectral composition, biological spectral weighting functions (BSWS) has been used by Rousseaux *et al.* (2004b). Here, biologically effective UV-B doses were calculated according to widely accepted and much used BSWS in an experiment with reduced UV, ambient UV-B and supplemental UV-B (30%) in parallel. Depending on BSWS, the doses differed by 1.4 to 6.4 times by comparing ambient UV-B to supplemental UV-B, whereas UV-doses differed by 1.5–77 times when comparing reduced UV-B to ambient UV-B. From this, it was concluded that considerable care is needed when comparing studies using the two different methodologies (Rousseaux *et al.*, 2004b).

If plants do not have linear responses to realistic doses of UV-B, are responses then subject to any thresholds? Does this relate to whether responses are equally detectable by either approach? This is an area of dispute, but the directions of responses of parameters, such as UV-B-absorbing compounds, stomatal density, chlorophyll, transpiration and photosynthesis (although depending on species), are the same as with increases in UV-B within those few studies comprising reduced, ambient and enhanced UV-B radiation in parallel (Gabeřscik *et al.*, 2002; Rousseaux *et al.*, 2004b). In addition, most of these responses display stepwise or dose-dependent changes (Gabeřscik *et al.*, 2002; Rousseaux *et al.*, 2004b), although they probably are not universal. Concerning the phenolics, which also function as UV-B-absorbing compounds, UV-B exposure response curves have shown that the production of several phenolics quantitatively are UV-B dose dependent (de la Rosa *et al.*, 2001), but complex contrasting responses have also been seen. An increase in UV-B-absorbing compounds were found in supplemental UV-B (Phoenix *et al.*, 2000; Semerdjieva *et al.*, 2003a, 2003b), but no changes occurred when UV-B was excluded (Phoenix *et al.*, 2002) on sub-arctic *V. uliginosum*. Moreover, along a wide range of UV-B doses no evidence of a possible threshold UV-B dose for UV-B responses has been found (González *et al.*, 1998; de la Rosa *et al.*, 2001). Together these findings provide support that our approach is scientifically acceptable.

The fact that filter treatment may induce important microclimatic differences has to be taken into account. Filters may potentially change temperatures and humidity in air and soil and of course exclude rainwater. At Zackenberg, the filters were shown to only change soil temperature significantly, but this increase did not change the soil humidity. This was probably due to the very low precipitation during 2001–2003 growth seasons (Sigsgaard *et al.*, 2006) and that the filter plots were placed in an angle allowing vegetation to benefit from events of precipitation due to surface runoff from above the filters. Since filter treatments elevated mean soil temperature by 0.6°C, the UV effects may be viewed as a combined effect (warming plus UV-B reduction) compared to warming (filter control) and open control (no filter and no warming).

A special issue concerning the photosynthetic response is that the filter differences in PAR transmission may lead to differences in canopy photosynthesis (Flint *et al.*, 2003). The degree of photosynthesis impact depends on leaf area index (LAI) being increased by increased LAI. According to Flint *et al.* (2003), with an LAI of 1, 5–10% differences in PAR results in 2–4% difference in photosynthesis. Compared to the open control, the transmitted PAR is 97% and 90% in the filtered control and UV-B-reducing treatment, respectively. Thus, between filter treatments, the resulting difference is *c.* 7% less PAR-irradiance in the UV-B-reducing treatments, respectively. Since we observed a stepwise higher photosynthesis in parallel with less PAR-irradiance, the effect may be of little importance here.

In summary, taking the premises above into consideration, we believe that the UV-B exclusion approach is very well suited to identify the impact on ecophysiological processes of current UV-B fluxes, whereas a UV-B supplemental approach may be better suited when evaluating their consequences. Hence, we argue that it is possible to indicate the direction of future ecosystem responses, but it remains speculative to actually quantify the responses within a particular UV-B radiation scenario, primarily because we do not know the UV-B exposure response curves and response to differences in the spectral composition of light.

VI. CONCLUSIONS AND FUTURE DIRECTIONS

The range of significant responses seen in the UV-B exclusion experiments at Zackenberg clearly indicates that ambient UV-B is a plant stress factor in this area. This seems in contrast to the reported robustness towards supplemental UV-B for plants in sub-arctic Abisko and high-arctic Svalbard. However, the results from UV-B exclusion studies in the Antarctic region have demonstrated effects on plants, similar to the results from Zackenberg. There are differences in the chosen response variables, and the contrasting responses may be interpreted to be due to the climatic differences between the areas. Further, the extreme living conditions in the high-arctic Zackenberg and Antarctic region may to a larger degree amplify effects of the stress factors, leading to significant UV-B impacts here. Although the responses from Zackenberg provides new insight and supplements earlier work, more work dedicated to link variables across scales is needed to take full advantage of the earlier findings. Thus, only by making parallel UV-B supplementation and UV-B exclusion field experiments it is possible to exclude the methodological differences and validate the ecosystem responses. Furthermore, the experiments should be conducted over longer time periods and include more traditional parameters (e.g., shoot growth rate and biomass effects) in order to ease comparisons and to elucidate whether the observed changes are transient or whether they accumulate over years. Long-term experiments are especially important for valid determination of below-ground responses, which have the potential to pose great feedbacks on the ecosystem functioning.

If projections from climatic scenarios to future biological responses shall be made, the biological responses and their feedback must be detected in multi-factorial experiments closely resembling the climatic projections. If sufficient reliable biological response functions to climatic parameters can be established then ecosystem modelling shall be possible. Presently, we do not have sufficient knowledge of all responses of importance and their interactions. Concerning ozone layer depletion, a specific UV-B radiation scenario for Zackenberg is needed. What we can state is that ambient UV-B as a single factor affects plant life negatively at high-arctic Zackenberg, and that the methodology developed is very well suited for long-term monitoring.

ACKNOWLEDGMENTS

The work was financially supported by DANCEA (Danish Co-operation for Environment in the Arctic) grant 123/000–0212, the Danish Environmental Protection Board, Climate and Environment Support MST grant 127/01–0205 and the Danish Natural Sciences Research Council grant 272–06–0230 and travel grants from the Svend G. Fiedler Foundation to Kristian Albert in 2005 and 2006. The Danish Polar Center provided excellent logistics and the dedicated staff personnel at the Zackenberg Research Station contributed by making stays in Zackenberg excellent conditions for research. Professor Sven Jonasson is acknowledged for enthusiastic support throughout all phases of the UV project. Finally, the authors thank Esben Vedel Nielsen, Gosha Sylvester, Niels Bruun, Karna Heinsen and Karin Larsen for help on soil and leaf chemical analyses and Svend Danbæk for solving all sorts of IT challenges.

REFERENCES

Albert, K.R., Mikkelsen, T.N. and Ro-Poulsen, H. (2005a) *Physiol. Plantarum* **124**, 208–226.

Albert, K.R., Ro-Poulsen, H., Mikkelsen, T.N., Bredahl, L. and Haakansson, K.B. (2005b) *Phyton* **45**, 41–49.

Albert, K.R., Arndal, M.F., Michelsen, A., Tamstorf, M.F., Ro-Poulsen, H. and Mikkelsen, T.N. (2006) In: *Zackenberg Ecological Research Operations, 11th Annual Report, 2005* (Ed. by M. Rasch and K. Caning), pp. 90–91. Danish Polar Center, Ministry of Science, Technology and Innovation, Copenhagen.

Albert, K.R., Mikkelsen, T.N. and Ro-Poulsen, H. (in press) *Physiol. Plantarum* in press.

Avery, L.M., Smith, R.I.L. and West, H.M. (2003) *Polar Biol.* **26**, 525–529.

Ballaré, C.L., Scopel, A.L., Stapelton, A.E. and Yanovsky, M.J. (1996) *Plant Physiol.* **112**, 161–170.

Ballaré, C.L., Rousseaux, M.C., Searles, P.S., Zaller, J.G., Giordano, C.V., Robson, T.M., Caldwell, M.M., Sala, O.E. and Scopel, A.L. (2001) *J. Photoch. Photobio. B* **62**, 67–77.

Björn, L.O., Callaghan, T.V., Gehrke, C., Gwynn-Jones, D., Lee, J.A., Johanson, U., Sonesson, M. and Buck, N.D. (1999) *Polar Res.* **18**, 331–337.

Bornman, J.F., Reuber, S., Cen, Y.-P. and Weissenböck, G. (1998) In: *Plant and UV-B. Responses to Environmental Change* (Ed. by P.J. Lumsden), pp. 157–170. Cambridge University Press, Cambridge.

Bredahl, L., Ro-Poulsen, H. and Mikkelsen, T.N. (2004) *Arct. Antarct. Alp. Res.* **36**, 363–368.

Caldwell, M.M. and Flint, S.D. (1994) *Climatic Change* **28**, 375–398.

Caldwell, M.M., Robberecht, R. and Billings, W.D. (1980) *Ecology* **61**, 600–611.

Callaghan, T.V., Björn, L.O., Chernov, Y., Chapin, T., Christensen, T.R., Huntley, B., Ims, R.A., Johansson, M., Jolly, D., Jonasson, S., Matveyeva, N., Panikov, N., *et al.* (2004a) *Ambio* **33**, 474–479.

Callaghan, T.V., Björn, L.O., Chernov, Y., Chapin, T., Christensen, T.R., Huntley, B., Ims, R.A., Johansson, M., Jolly, D., Jonasson, S., Matveyeva, N., Panikov, N., *et al.* (2004b) *Ambio* **33**, 418–435.
Callaghan, T.V., Björn, L.O., Chernov, Y., Chapin, T., Christensen, T.R., Huntley, B., Ims, R.A., Johansson, M., Jolly, D., Jonasson, S., Matveyeva, N., Panikov, N., *et al.* (2004c) *Ambio* **33**, 398–403.
Callaghan, T.V., Björn, L.O., Chapin, T., Chernov, Y., Christensen, T.R., Huntley, B., Ims, R.A., Johansson, M., Riedlinger, Jolly, D, Jonasson, S., Matveyeva, N., Oechel, W., *et al.* (2005) In: *ACIA, Arctic Climate Impact Assessment* (Ed. by C. Symon, L. Arris and B. Heal), pp. 243–352. Cambridge University Press, New York.
CIE (1999) *134/1:TC 6–26 report: Standardization of the terms UV-A1, UV-A2 and UVB*. Commission Internationale de l'Eclairage (CIE). Collection in Photobiology and Photochemistry, Vienna, AustriaCIE 134–1999 ISBN 3 900 734 94 1.
Day, T.A., Ruhland, C.T., Grobe, C.W. and Xiong, F. (1999) *Oecologia* **119**(1), 24–35.
Day, T.A., Ruhland, C.T. and Xiong, F.S. (2001) *J. Photoch. Photobio. B* **62**, 78–87.
de la Rosa, T.M., Julkunen-Tiitto, R., Letho, T. and Aphalo, P.J. (2001) *New Phytol.* **150**, 121–131.
Farman, J.C., Gardiner, B.G. and Shanklin, J.D. (1985) *Nature* **35**, 207–210.
Flint, S.D., Ryel, R.J. and Caldwell, M.M. (2003) *Agr. Forest Met.* **120**, 177–189.
Gabeřscik, A., Voňcina, M., Trost, T., Germ, M. and Björn, L.O. (2002) *J. Photoch. Photobiol. B* **66**, 30–36.
González, R., Wellburn, A.R. and Paul, N.D. (1998) *Physiol. Plantarum* **104**, 373–378.
Gröebner, J., Albold, A., Blumthaler, M., Cabot, T., de la Casiniere, A., Lenoble, J., Martin, T., Masserot, D., Müller, M., Philipona, R., Pichler, T., Pougatch, E., *et al.* (2000) *J. Geophys. Res.* **105**, 26991–27003.
Gwynn-Jones, D., Lee, J.A. and Callaghan, T.V. (1997) *Plant. Ecol.* **128**, 242–249.
Harborne, J.B. and Grayer, R.J. (1993) In: *The Flavonoids: Advances in Research since 1986* (Ed. by J.B. Harborne), pp. 589–618. Chapman & Hall, London.
Hughes, K.A., Scherer, K., Svenøe, T., Rettberg, P., Horneck, G. and Convey, P. (2006) *Soil Biol. Biochem.* **38**, 1488–1490.
Håkansson, K.B. (2006) *Påvirker det nuværende niveau af ultraviolet-B stråling Salix arctica? - Et eksklusions forsøg fra Nordøstgrønland.* M.Sc. thesis, Department of Terrestrial Ecology, Biological Institute, University of Copenhagen.
Johanson, U., Gehrke, C., Björn, L.O. and Callaghan, T.V. (1995a) *Funct. Ecol.* **9**, 713–719.
Johanson, U., Gehrke, C., Björn, L.O., Callaghan, T.V. and Sonesson, M. (1995b) *Ambio* **24**, 106–111.
Johnson, D. (2003) *Res. Microbio.* **154**, 315–320.
Johnson, D., Campbell, C.D., Lee, J.A., Callaghan, T. and Gwynn-Jones, D. (2002) *Nature* **416**, 82–83.
Jokela, K., Leszcynski, K. and Visuri, R. (1993) *J. Photoch. Photobiol. B* **58**, 559–566.
Jones, M.H., Macdonald, S.E. and Henry, G.H.R. (1999) *Oikos* **87**, 129–138.
Lehrner, G., Delatorre, J., Lütz, C. and Cardemil, L. (2001) *J. Photoch. Photobiol. B* **64**, 36–44.
Madronich, S., McKencie, R.L., Björn, L.O. and Caldwell, M.M. (1998) *J. Photoch. Photobiol. B* **46**, 5–19.
Paul, N. (2001) *New Phytol.* **150**, 1–8.
Phoenix, G.K. and Lee, J.A. (2004) *Ecol. Res.* **19**, 65–74.

Phoenix, G.K., Gwynn-Jones, D., Lee, J.A. and Callaghan, T.V. (2000) *Plant Ecol.* **146**, 67–75.

Phoenix, G.K., Gwynn-Jones, D., Callaghan, T.V., Sleep, D. and Lee, J.A. (2001) *J. Ecol.* **89**, 256–267.

Phoenix, G.K., Gwynn-Jones, D., Lee, J.A. and Callaghan, T.V. (2002) *Plant. Ecol.* **165**, 263–273.

Rinnan, R., Keinänen, M.M., Kasurinen, A., Asikainen, J., Kekki, T.K., Holopainen, T., Ro-Poulsen, H., Mikkelsen, T.N. and Michelsen, A. (2005) *Global Change Biol.* **11**, 564–574.

Robberecht, R., Caldwell, M.M. and Billings, W.D. (1980) *Ecology* **61**, 612–619.

Robson, T.M., Pancotto, V.A., Flint, S.D., Ballaré, C.L., Sala, O.E., Scopel, A.L. and Caldwell, M.M. (2003) *New Phytol.* **160**, 379–389.

Rousseaux, M.C., Ballaré, C.L., Giordano, C.V., Scopel, A.L., Zima, A.M., Szwarcberg-Bracchitta, M., Searles, P.S., Caldwell, M.M. and Diaz, S.B. (1999) *P. Natl. Acad. Sci. USA* **96**, 15310–15315.

Rousseaux, M.C., Julkunen-Tiitto, R., Searles, P.S., Scopel, A.L., Aphalo, P.J. and Ballaré, C.L. (2004a) *Oecologia* **138**, 505–512.

Rousseaux, M.C., Flint, S.D., Searles, P.S. and Caldwell, M.M. (2004b) *Photoch. Photobiol. B* **80**, 224–230.

Rozema, J., Björn, L.O., Bornman, J.F., Gaberscik, A., Häder, D.-P., Trost, T., Germ, M., Klisch, M., Gröniger, A., Sinha, R.P., Lebert, M., He, Y.-Y., *et al.* (2002) *J. Photoch. Photobiol. B* **66**, 2–12.

Rozema, J., Boelen, P. and Blokker, P. (2005) *Environ. Pollut.* **137**, 428–442.

Rozema, J., Boelen, P., Solheim, B., Zielke, M., Buskens, A., Doorenbosch, M., Fijn, R., Herder, J., Callaghan, T., Björn, L.O., Gwynn-Jones, D., Broekman, R., *et al.* (2006) *Plant. Ecol.* **182**, 121–135.

Ruhland, C.T. and Day, T.A. (2000) *Physiol. Plantarum* **109**, 244–351.

Ruhland, C.T., Xiong, F.S., Clark, W.D. and Day, T.A. (2005) *Photoch. Photobiol. B* **81**, 1086–1093.

Searles, P.S., Flint, S.D. and Caldwell, M.M. (2001a) *Oecologia* **127**, 1–10.

Searles, P.S., Kropp, B.R., Flint, S.D. and Caldwell, M.M. (2001b) *New Phytol.* **152**, 213–221.

Semerdjieva, S.I., Sheffield, E., Phoenix, G.K., Gwynn-Jones, D., Callaghan, T.V. and Johnson, G.N. (2003a) *Plant Cell Environ.* **26**, 957–964.

Semerdjieva, S.I., Phoenix, G.H., Hares, D., Gwynn-Jones, D., Callaghan, T.V. and Sheffield, E. (2003b) *Physiol. Plantarum* **117**, 289–294.

Sigsgaard, C., Petersen, D., Grøndahl, L., Thorsøe, K., Meltofte, H., Tamstorf, M. and Hansen, B.U. (2006) In: *Zackenberg Ecological Research Operations, 11th Annual Report, 2005* (Ed. by M. Rasch and K. Caning), pp. 11–35, Danish Polar Center, Ministry of Science, Technology and Innovation, Copenhagen.

Solheim, B., Johanson, U., Callaghan, T.V., Lee, J.A., Gwynn-Jones, D. and Björn, L.O. (2002) *Oecologia* **133**, 90–93.

Strasser, R.J., Tsimilli-Michael, M. and Srivastava, A. (2004) In: *Chlorophyll a Flourescence: A signature of photosynthesis*, Advances in Photosynthesis and Respiration (Ed. by G.C. Papageorgiou and Godvinjee), pp. 321–363. Volume 19.

Wardle, D.A. (2002) *Monographs in Population Biology 34*. Princeton University Press, Princeton, New Jersey.

Webb, A.R. (1997) In: *Plants and UV-B: Responses to Environmental Change* (Ed. by P.J. Lumsden), pp. 13–30. Cambridge University Press, United Kingdom.

Weatherhead, B., Tanskanen, A., Stevermer, A., Andersen, S.B., Arola, A., Austin, J., Bernhard, G., Browman, H., Fioletov, V., Grewe, V., Herman, J.,

Josefsson, W., *et al.* (2005) In: *ACIA, Arctic Climate Impact Assesment* (Ed. by C. Symon, L. Arris and B. Heal), pp. 151–182. Cambridge University Press, Cambridge.
WMO (2003) *Scientific Assessment of Ozone Depletion: 2002. Global Ozone Research and Monitoring Project*. Report no. 47, World Metrological Organization, Geneva.
Xiong, F.S. and Day, T.A. (2001) *Plant Physiol.* **125**, 738–751.
Zaller, J.G., Caldwell, M.M., Flint, S.D., Scopel, A.L., Osvaldo, E.S. and Ballaré, C.L. (2002) *Global Change Biol.* **8**, 867–871.
Zelles, L. (1999) *Biol. Fert. Soils* **29**, 111–129.

High-Arctic Soil CO_2 and CH_4 Production Controlled by Temperature, Water, Freezing and Snow

BO ELBERLING, CLAUS NORDSTRØM, LOUISE GRØNDAHL, HENRIK SØGAARD, THOMAS FRIBORG, TORBEN R. CHRISTENSEN, LENA STRÖM, FLEUR MARCHAND AND IVAN NIJS

SUMMARY

Soil gas production processes, mainly anaerobic or aerobic soil respiration, drive major gas fluxes across the soil–atmosphere interface. Carbon dioxide (CO_2) effluxes, an efflux which in most ecosystems is a result of both autotrophic and heterotrophic respiration, in particular have received international attention. The importance of both CO_2 and methane (CH_4) fluxes are emphasised in the Arctic because of the large amount of soil organic carbon stored in terrestrial ecosystems and changes in uptake and release due to climate changes.

This chapter focuses on controls on spatial and temporal trends in subsurface CO_2 and CH_4 production as well as on transport and release of gases from the soil observed in the valley Zackenbergdalen. A dominance of near-surface temperatures controlling both spatial and seasonal trends is shown

ADVANCES IN ECOLOGICAL RESEARCH VOL. 40 0065-2504/08 $35.00
© 2008 Elsevier Ltd. All rights reserved
DOI: 10.1016/S0065-2504(07)00019-0

based on data obtained using closed chamber and eddy-correlation techniques as well as in manipulated field plots and in controlled incubation experiments. Despite variable temperature sensitivities reported, most data can be fairly well fitted to exponential temperature-dependent equations. The water content (at wet sites linked to the depth to the water table) is a second major factor regulating soil respiration processes, but the effect is quite different in contrasting vegetation types. Dry heath sites are shown to be periodically water limited during the growing season and respond therefore with high respiration rates when watered. In contrast, water saturated conditions during most of the growing season in the fen areas hinder the availability of oxygen, resulting in both CO_2 and CH_4 production. Thus, water table drawdown results in decreasing CH_4 effluxes but increasing CO_2 effluxes.

Additional controls on gas production are shown to be related to the availability of substrate and plant productivity. Subsurface gas production will produce partial and total pressure gradient causing gas transport, which in well-drained soils is mainly controlled by diffusion, whereas gas advection, bubbles and transport through roots and stems may be important in more saturated soils.

Bursts of CO_2 gas have been observed during spring thaw and confirmed in controlled soil thawing experiments. Field observations as well as experimental work suggest that such bursts represent partly on-going soil respiration and a physical release of gas produced during the winter. The importance of winter soil respiration is emphasised because of the fact that microbial respiration in Zackenberg samples is noted down to a least $-18°C$. Hence, the importance of winter respiration and burst events in relation to seasonal and future climate trends requires more than just summer measurements. For example, the autumn period seems important as snowfall prior to low air temperature may insulate the soil, keeping soil temperatures high. This will extend the period of high soil respiration rates and thereby increase the importance of the winter period for the annual carbon balance.

Because of the complexity of factors controlling subsurface gas production, we conclude that different parts of the landscape will respond quite differently to the same climate changes as well as that short-term effects are likely to be different from long-term effects.

I. INTRODUCTION

Arctic soil organic carbon (SOC) cycling is important because these landscapes hold about 14% of the world's organic carbon (Post *et al.*, 1982), and these soils are part of one of the most sensitive ecosystems with respect to climatic changes (Oechel *et al.*, 1993, 1998). The decomposition of SOC results in a release of several greenhouse gases to the atmosphere (e.g., Coyne and Kelley, 1974; Oberbauer *et al.*, 1996; Burkins *et al.*, 2001) of which the release of carbon dioxide (CO_2) and methane (CH_4) are considered most important. CO_2 is primarily being produced by near-surface respiration in living roots (Billings *et al.*, 1977) and by heterotrophic soil microorganisms (Buchmann, 2000). CH_4 is produced where the oxygen availability is limited and anaerobic microbes control the decomposition, which under most conditions is below the water table or in isolated pockets in the soil (Figure 1).

The decomposition of soil organic matter is sensitive to several site-specific environmental conditions including the quality, abundance and distribution of substrates (Van Cleve, 1974; Nadelhoffer *et al.*, 1991; Fahnestock *et al.*, 2000) and climatic factors, in particular soil temperature (Fang and Moncrieff, 2001), water content (Howard and Howard, 1993; Kirschbaum, 1995) and snow (Welker *et al.*, 2000). Availability of oxygen, types of microbes present, as well as faunal abundance and activity are additional factors influencing soil microbial respiration processes (Lomander *et al.*, 1998). The spatial and temporal variability of such conditions in the field results in substantial variations in soil respiration rates and gas effluxes within and between tundra vegetation types (e.g., Evans *et al.*, 1989; Oberbauer *et al.*, 1991, 1992; Hobbie, 1996; Grogan and Chapin, 1999).

A direct positive feedback on climate changes is possible through an increase in soil respiration with increasing soil temperature, which long has been described using simple temperature sensitivity equations (Arrhenius, 1889) assuming that the rate of carbon input (photosynthesis) stays the same. The increase in reaction rate per 10 °C (Q_{10}) is often used to characterise the

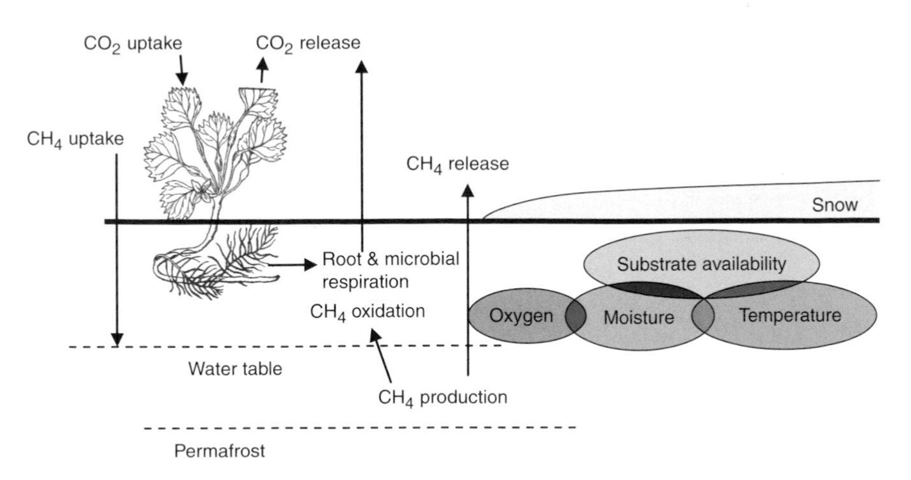

Figure 1 Schematic drawing of the processes and pathways of CO_2 and CH_4 in the landscape at Zackenberg.

temperature dependence of soil respiration on various soil types and temperature ranges above freezing temperatures (Kirschbaum, 1995). However, interacting effects include changes in the availability of substrate and liquid water as a function of soil temperatures (Davidson and Janssens, 2006). Feedback mechanisms of soil processes with respect to climate changes are furthermore complicated by the influence of climate changes on subsurface temperatures, which in arctic ecosystems include the content of unfrozen water and both thickness and duration of snow-cover (Figure 1). In addition, different functional groups in the soil, including bacteria and fungi, have different sensitivities to warming (Pietikainen *et al.*, 2005), which complicates the interpretation and extrapolation of experimentally determined temperature—respiration relationships, especially if warming also modifies the structure of the soil community. For CH_4 production at least, emissions on an annual scale have been shown to be closely related to primary production, which also is directly affected by climate changes (Whiting and Chanton, 1993; Christensen *et al.*, 2000). In wetlands with a high water table and, hence, relatively high rates of CH_4 emissions, these have been found also to be rather closely correlated with mean seasonal temperature across a wide range of sites including Zackenberg (Christensen *et al.*, 2003).

Quantifying subsurface gas production is complicated by the interacting processes of gas production and gas transport. Gas production results in changes in gas composition of the gas-filled pores in soil, which typically consists of between 10 and 100 times higher CO_2 concentrations than in the atmosphere (Welles *et al.*, 2001). The transport of near-surface gases such as CO_2 has long been considered mainly as a result of diffusion (e.g., Lundegärdh, 1927; De Jong and Schappert, 1972) driven by concentration gradients and limited by the decrease in continuous air-filled soil pores with increasing water content. But produced gas may partly be trapped in the frozen soil (Elberling and Brandt, 2003) or in the snow pack (Zimov *et al.*, 1996). In addition, transport through roots and stems, as well as through bubbles of soil air, has been found to be important at waterlogged soils (Sebacher *et al.*, 1985). The fact that both living roots and heterotrophic soil microorganisms produce subsurface CO_2 and that these two processes are coupled add to the complexity of evaluating each of them separately.

Striking differences in arctic plant communities and concurrent soil CO_2 and CH_4 effluxes have been observed over small distances due to microtopography, which seemed predominantly related to hydrology (Oberbauer *et al.*, 1996; Christensen *et al.*, 2000). Consequently, changes in plant community structure, composition and distribution in the landscape will be an additional feedback mechanism for long-term climate changes (Zimov *et al.*, 1996). This feedback is not limited to plants, but also involves the complex interactions between plants, the soil organic matter pools, and the soil organic matter mineralisation, which, again, is closely related to the availability of nutrients for plants and microorganisms competing in nutrient-limited arctic ecosystems (Oberbauer *et al.*, 1991).

Therefore, the assessment of present carbon cycling and long-term trends of net carbon balance in arctic regions requires improved understanding of not only specific controls on subsurface CO_2 dynamics at appropriate spatial and temporal scales (Grogan and Chapin, 1999) but also the interacting processes at a landscape scale including soil types, soil water distribution, vegetation types, soil organic matter quality and soil gas transport. While a mechanistic understanding of how temperatures and other environmental factors affect ecosystem respiration and soil element cycling is still lacking, attempts to establish robust site-specific semi-empirical relationships are crucial in order to scale up from plot level to landscapes and to longer periods.

This chapter aims to evaluate spatial and temporal trends and environmental controls of soil respiration processes during the growing season in characteristic vegetation types found within Zackenbergdalen in Northeast Greenland. CO_2 and CH_4 exchange rates between soil, vegetation and the atmosphere as well as ecosystem carbon budgets are reported elsewhere in this publication (Grøndahl *et al.*, 2008, this volume).

The study sites are situated in the valley Zackenbergdalen near the Zackenberg Research Station in Northeast Greenland (74°30'N, 20°30'W). The valley is generally flat and dominated by non-calcareous sandy fluvial sediments. The present soil development is weak and has been classified using Soil Taxonomy as Typic Psammoturbels (Elberling and Jakobsen, 2000). A relict A-horizon (buried old surface layer, A_b) of 1–5 cm thickness was found in most soil profiles at depths between 15 and 30 cm and was in places associated with a well-developed Podzol. Dating of these buried horizons has previously shown that they represent a soil development during the Holocene Climate Optimum starting at least 5000 years ago (Christiansen *et al.*, 2002).

II. ENVIRONMENTAL CONTROLS ON GAS PRODUCTION

A. Gas Production and Dynamics in Well-Drained Soils

The sites were selected to represent major plant communities, which again reflected the hydrological regimes present in the valley. Vegetation types were identified in the field by topographic position, vegetation and soils and are described in detail by Bay (1998), Soegaard *et al.* (2000) and Elberling *et al.* (2008a, this volume). The abrasion plateaus represent the most exposed and driest areas in the valley. Here, the snow-cover is thin or absent throughout winter, resulting in early and fast development of the active layer during spring. The sparse vegetation cover is dominated by mountain avens *Dryas* sp. Flat low-lying areas receiving the average amount of snow (0.5–1 m $year^{-1}$) typically became free of snow during June and remained moist but aerated during most of the growing seasons (1998–2003). White arctic bell-heather *Cassiope tetragona* dominates these areas, although other heath species such as bog blueberry *Vaccinium uliginosum* are found in patches. The variety of heath vegetations covers more than 30% of the valley below 200 m a.s.l. and is the most dominant vegetation type found in the valley. Areas covered by long-lasting snow (snow-beds) remain moist or wet throughout the growing season and are dominated by arctic willow *Salix arctica*. These sites are all characterised by being partly drained and that oxygen is present in the soil most of the time during the growing season. Thus, CO_2 is the dominating gas being produced and released to the atmosphere.

1. *Observed Soil CO_2 Concentrations*

Measured subsurface concentrations of CO_2 during the 2000 and 2001 growing seasons were about 10–20 times that of atmospheric concentrations, and concentrations generally increased with depth (Figures 2 and 3; Elberling *et al.*, 2004). However, towards the permafrost boundary, CO_2 concentrations tended to decline. This decline, which was found throughout the growing seasons of 2000 and 2001 (Figure 3), indicates that observed concentrations may not represent true steady-state conditions. However, steady state seems to be a reasonable assumption in the upper part of the profiles. Concurrent oxygen profiles (data not shown) indicated that near-atmospheric O_2 concentrations always existed throughout the entire active layer (19–21%). These gas concentrations represent the bulk concentrations in the larger air-filled pores, whereas smaller pores may locally have much higher or lower concentrations. Variations in CO_2 concentrations between vegetation types (Figure 2) are linked to contrasting water contents, temperatures and the distribution of organic substrate to be decomposed. Attempts by Elberling *et al.* (2004) to simulate the entire subsurface concentration profiles using a steady-state production-diffusion model, PROFILE, failed (Figure 2). However, predicted gas concentrations within the upper 30 cm agreed well with observations.

2. *Observed Soil CO_2 Effluxes*

Significant variations in soil CO_2 effluxes between plant communities have been reported by Elberling *et al.* (2004) and are summarised in Table 1. Minimum effluxes were observed at *Cassiope* heath sites (on average 0.52 ± 0.1 μmol CO_2 m^{-2} s^{-1}) and maximum effluxes at cottongrass *Eriophorum* fen sites (on

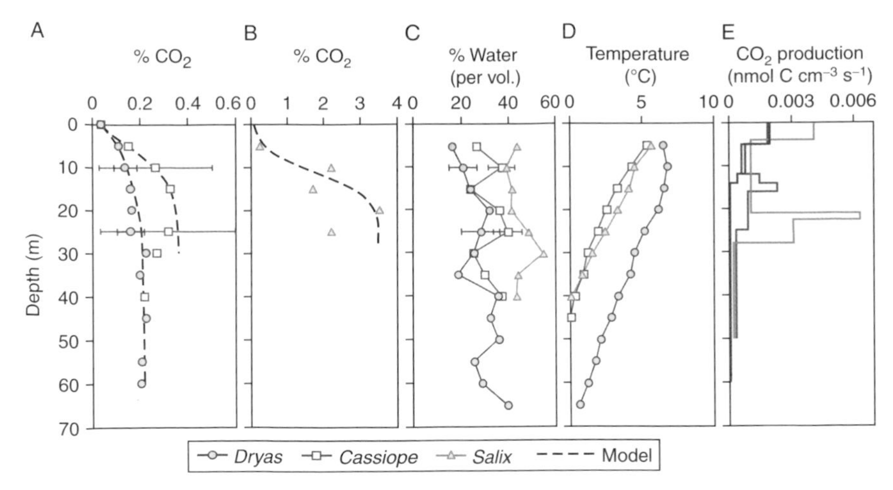

Figure 2 Field measurements of soil CO_2 concentration profiles (A and B), water contents (C) and soil temperatures (D) at *Dryas*, *Cassiope* and *Salix* sites as observed on August 2, 1999. CO_2 production profiles (E) as observed in the laboratory have been used as input to a diffusion model and the resulting model outputs (using PROFILE) are shown as solid lines for *Dryas* and *Cassiope* (A) and *Salix* (B). One standard deviation (±) based on 16 replicates of CO_2 concentrations and water contents observed at two depths at *Dryas* and *Cassiope* sites is shown as horizontal lines (modified from Elberling *et al.*, 2004).

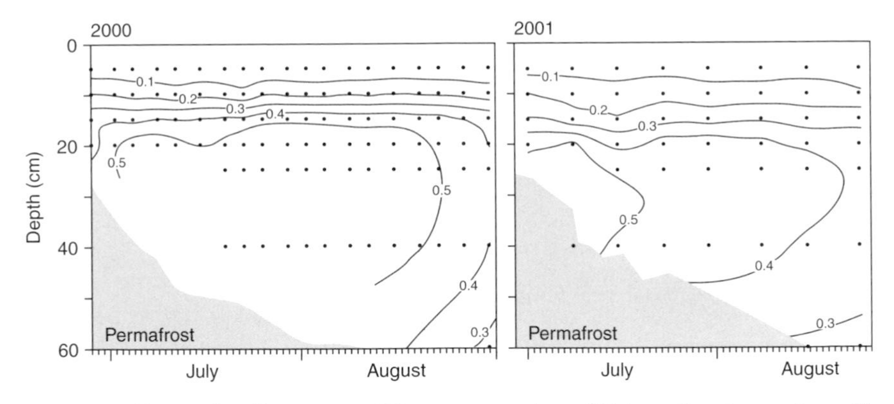

Figure 3 Observed soil pore gas CO_2 concentrations (%) in a *Cassiope* soil profile during the 2000 and 2001 growing seasons (measurements are shown as dots) (modified from Elberling *et al.*, 2004).

average 2.48 ± 0.5 μmol CO_2 m^{-2} s^{-1}). The range of these CO_2 fluxes is comparable with previously reported soil CO_2 effluxes along a Eurasian transect (Christensen *et al.*, 1998; Jones *et al.*, 2000) and with the CO_2 effluxes observed at Zackenberg by Marchand *et al.* (2004a) in mixed sites dominated by *Salix arctica* and wideleaf polargrass *Arctagrostis latifolia*. Comparing

Table 1 Soil biogeochemical properties of major soil/vegetation types at Zackenberg, Greenland (± 1 standard deviation)

Vegetation type	*Dryas* heath	*Cassiope* heath	*Salix* snow-bed	*Eriophorum* fen
Percent water saturation range (0–5 cm)	40–60	60–80	65–90	90–100
Area below 200 m a.s.l. (km^2)	4.36 ± 0.5	4.09 ± 0.6	3.17 ± 0.3	4.87 ± 0.6
Total soil C 0–20 cm (kg C m^{-2})	4.1 ± 0.4	6.3 ± 3.0	10.5 ± 1.8	5.6 ± 3.2
Total soil C 0–50 cm (kg C m^{-2})	6.1 ± 1.1	8.5 ± 2.6	21.2 ± 4.1	11.3 ± 3.8
Vegetation (C:N)	35	43	28	31
Plant cover (%)	65	65	80	88
Litter–C (g C m^{-2})	13 ± 7	21 ± 6	36 ± 10	nd[b]
Root–C (g C m^{-2})	155 ± 25	123 ± 10	nd[b]	nd[b]
Belowground debris-C (g C m^{-2})	265 ± 150	423 ± 121	877 ± 156	nd[b]
Above-below ground biomass ratio	0.81	0.70	nd[b]	nd[b]
Maximum thaw in August (cm)	80	65	45	40
[a]Soil temperature (5 cm) (°C)	9.13 ± 3.8	5.49 ± 1.8	7.59 ± 3.8	7.59 ± 3.8
[a]Soil CO_2 efflux (μmol m^{-2} s^{-1})	0.836 ± 0.2	0.52 ± 0.1	1.22 ± 0.3	2.48 ± 0.5

[a]Average of hourly readings in July and August 2001.
[b]Not determined.

effluxes at the same temperature, for example, 10 °C (Figure 4), shows that effluxes from *Cassiope* and *Dryas* heath are approximately the same, but a factor of 2 less than effluxes at *Salix* snow-bed sites. In turn, effluxes at *Salix* snow-bed sites are a factor of 2 less than effluxes from *Eriophorum* fen sites (Table 1). Figure 4 illustrates that ambient rates at *Dryas* heath sites are consistently higher than at *Cassiope* heath sites, which is in accordance with drier and, therefore, warmer soil temperatures at *Dryas* heath sites (Figure 2D). In contrast, higher rates at 10 °C in *Salix* snow-beds must be explained by other factors. As noted in Table 1, *Salix* snow-bed sites consist of both significantly more near-surface soil organic C (0–20 cm) and receive more easily degradable litter (more N per unit C) as compared to *Cassiope* and *Dryas* heath sites. This illustrates the complexity of factors in play for describing vegetation-specific variations.

Temporal trends in observed soil CO_2 effluxes, active layer depths, water contents and soil temperatures during the 2001 growing season (Figure 5)

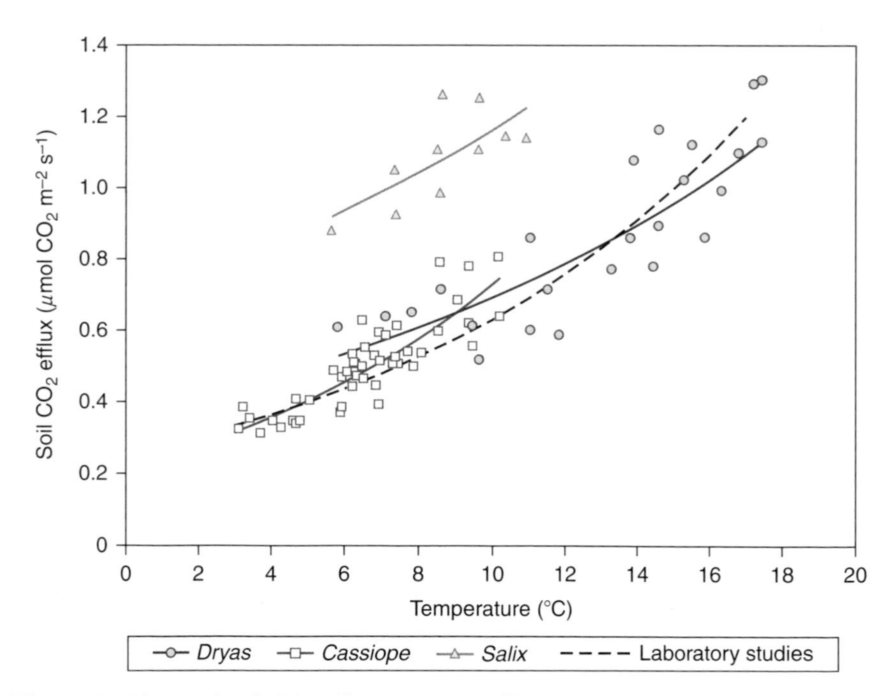

Figure 4 Observed soil CO_2 effluxes versus soil temperatures (at a depth of 2 cm). Exponential regression lines are shown as solid lines for *Cassiope* (solid squares; $0.2203e^{0.1196x}$; $R^2 = 0.8$; $Q_{10} = 3.3$; $R_{10} = 0.73$), *Dryas* (open squares; $0.3553e^{0.0662x}$; $r^2 = 0.7$; $Q_{10} = 1.9$; $R_{10} = 0.69$) and *Salix* (solid triangles; $0.4648e^{0.1013x}$; $R^2 = 0.7$; $Q_{10} = 2.7$; $R_{10} = 1.28$). The dashed line is based on laboratory studies previously reported for *Cassiope* (Elberling and Brandt, 2003) (modified from Elberling *et al.*, 2004).

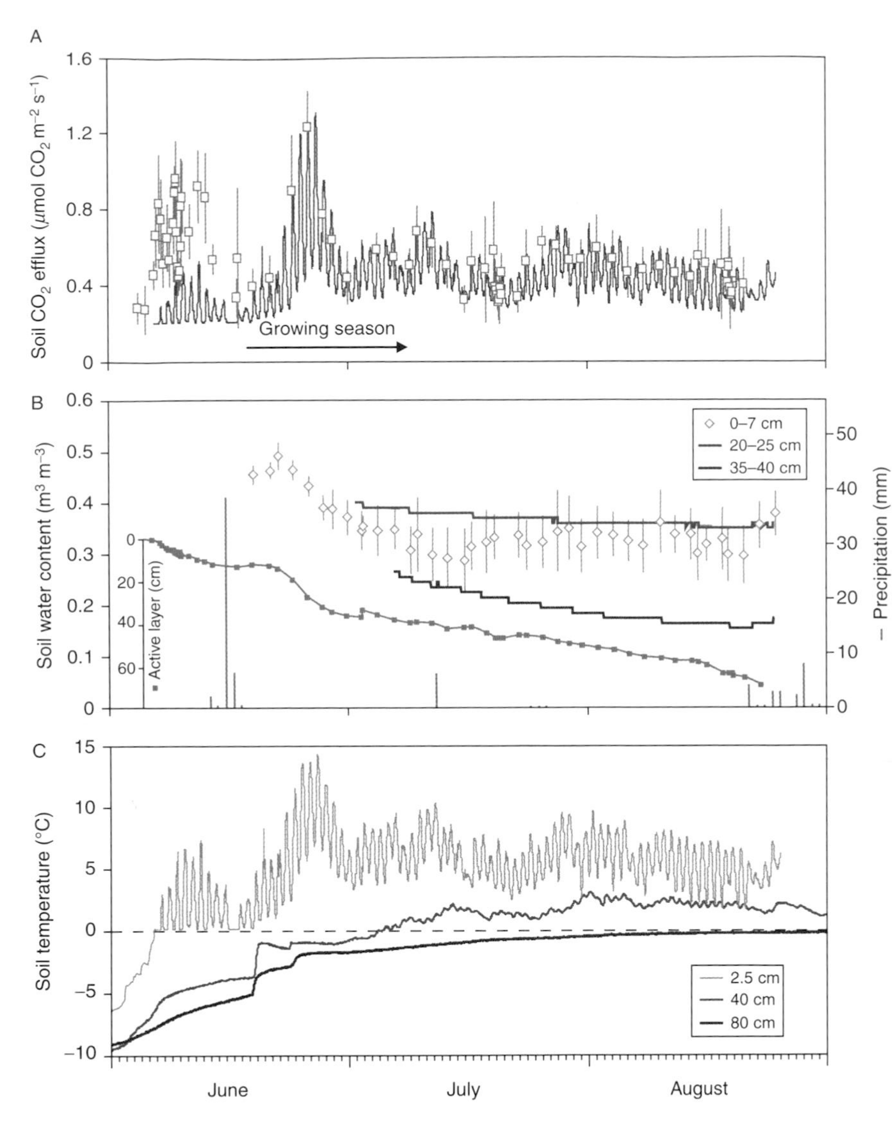

Figure 5 Field observations during June, July and August 2001 including (A) soil CO_2 effluxes shown as solid squares, with ±1 standard deviation shown as vertical lines (n=17), (B) soil water contents at three depths with ±1 standard deviation shown for manual near-surface measurements (n = 3) as vertical lines, precipitation and active layer thickness and (C) soil temperatures. Calculated soil CO_2 effluxes are based solely on temperature variations observed at 2.5 cm depth and a Q_{10} of 3.6.

indicate that efflux variations over the 2-month period correlated relatively well with near-surface soil temperature (Figures 4, 5A and 5C). Q_{10} estimates based on field-observed soil temperatures at 2 cm and observed CO_2 effluxes were equal to 3.3 ($R^2 = 0.8$) for *Cassiope* heath sites, 1.9 ($R^2 = 0.7$) for *Dryas* heath sites and 2.7 ($R^2 = 0.7$) for *Salix* snow-bed sites. Data from the pre-growing season were excluded in this analysis, as they could not be described by the same set of equations (Figure 6). For sites dominated by *Salix arctica* and *Arctagrostis latifolia*, soil effluxes during the 1999 growing season (Marchand *et al.*, 2004a) were likewise explained best by temperature (Q_{10} values amounting to 2.9) and with additional effects of time of the year (probably reflecting senescence) and soil moisture content.

3. Estimated Soil Respiration Rates Based on Eddy Covariance Measurements

The eddy covariance (EC) technique was used to obtain the exchange of CO_2 between the soil and the atmosphere at two sites during the period 1996–2005, a well-drained heath and a wet fen, and has been described in more detail by Grøndahl *et al.* (2008, this volume). Unlike the chamber method, which usually provides information on soil fluxes only, the EC technique does not provide separate information on the ecosystem respiration and

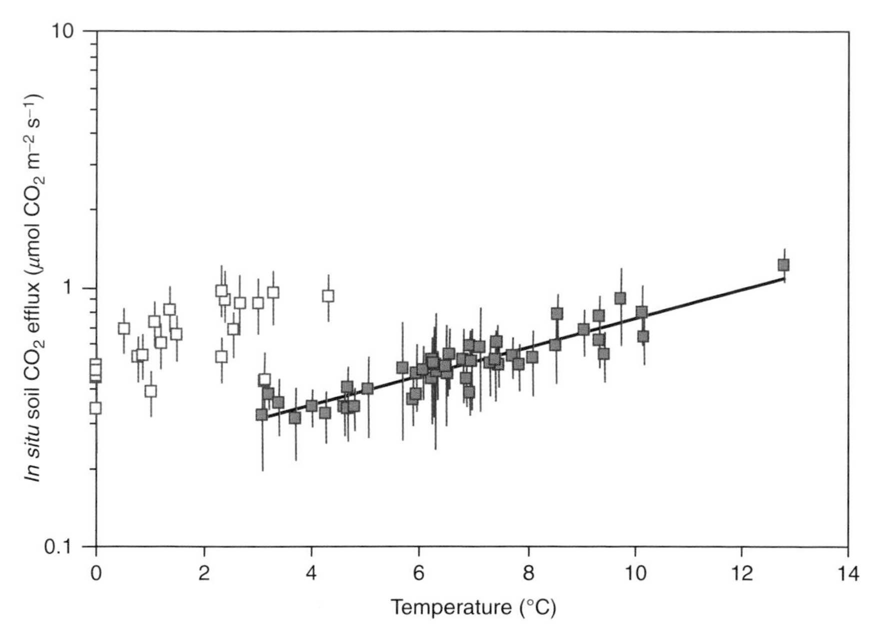

Figure 6 Observed soil CO_2 effluxes versus soil temperatures at 2.5 cm depths with ±1 standard deviation shown as vertical lines ($n = 17$). Open squares represent observations during spring (4–20 June), whereas solid squares represent subsequent data (until August 21). The exponential regression line shown is based on the latter data ($r^2 = 0.82$) and gives a Q_{10} equal to 3.6.

photosynthesis. Consequently, the soil respiration components have to be modelled. This has been done for the heath and the fen ecosystems assuming that EC measurements during night-time as well as the snow-free pre-leaf period after the frost table had attained a certain depth (to avoid physical release of CO_2 trapped below the frost table) represent the soil respiration flux from the ecosystem. Extrapolation of these EC data can be made subsequently assuming an exponential increase in respiration with increasing soil temperatures as mentioned earlier. The reference soil respiration at 10 °C (R_{10}) is subsequently reported (according to Lloyd and Taylor, 1994) based on 3 cm depth temperatures. Because of the northern latitude of the research site, solar radiation is received 24 h a day during most of the growing season. The night-time was defined as the period when incoming photosynthetically active radiation (PAR) was below 50 μmol m^{-2} s^{-1}. The night-time is also characterised by periods of high atmospheric stability during calm nights when the turbulent transport is suppressed. To avoid unreliable flux measurements, a friction velocity threshold of 0.06 m^{-2} s^{-1} was used at the heath site. Ecosystem Respiration (ER) was modelled using the fluxes measured during the period following the spring burst until the flux again turns positive, that is, only the period during the growing season with uptake is used to estimate ER. Usually rain causes the ecosystem to turn into a source of CO_2, and such days are not representative of the growing seasonal ER flux and were therefore neglected. The Q_{10} and R_{10} values for the heath and fen sites are presented in Table 2 and for the heath site comparable to the reported soil respiration chamber values (Figure 4) taking into account that the heath represents a composite of *Cassiope*, *Salix* and *Dryas* vegetation.

B. Gas Production and Dynamics in Poorly Drained Soils

Fen vegetation dominated by graminoids is found in poorly drained landscape depressions and other areas receiving substantial amounts of water from snowmelt throughout the summer. *Arctagrostis latifolia* and tall cottongrass *Eriophorum triste* are key species found in less wet parts, whereas white cottongrass *Eriophorum scheuchzeri* and Fisher's tundragrass *Dupontia*

Table 2 Estimated temperature sensitivity (Q_{10}) and mean soil respiration rates at 10 °C (R_{10}) based on night-time eddy covariance fluxes between 1996 and 2005

	1996	1997	1999	2000	2001	2002	2003	2004	2005
Heath									
Q_{10}		2.19		2.25	2.60	2.58	2.88	2.20	2.00
R_{10}		0.61		0.50	0.90	0.65	0.77	0.80	0.58
Fen									
R_{10}	1.64	1.84	1.77						

psilosantha dominate the wetter parts. In fen areas, the soil is peaty with ground-water levels over, at, or just beneath the surface throughout the growing season.

1. Estimated Soil Respiration Rates Based on EC Measurements

In addition to measurements at partly drained sites, EC measurements were made over a water logged fen, Rylekærerne, in 3 years (1996, 1997 and 1999). This is a roughly $600 \times 1200\ m^2$ fairly coherent homogenous fen in the central part of the valley with a maximum active layer of about 55 cm. In 1997, measurements began in a period with a snow-covered surface (Soegaard and Nordstroem, 1999). In this period, mean flux values were in the range −0.59 to 1.86 g CO_2 day^{-1} and revealed no diurnal trends (Figure 7). Since the soil temperature in 3 cm depth in the same period varied between −1.6 and −0.5°C, fluxes can be explained by ongoing soil respiration. This is in contrast to the results for the snow-free pre-leaf period, where difference in the diurnal variations between 1997 and 1999 can be related to different rates of snowmelt in the two years. It was predicted that the positive fluxes in the snow-free pre-leaf period are a mixture of physical released stored CO_2 and ongoing respiration. Contributions from physical released CO_2 emission are probably most prominent in the period, when the snow pack and top soil melts, until the active layer reaches a certain depth.

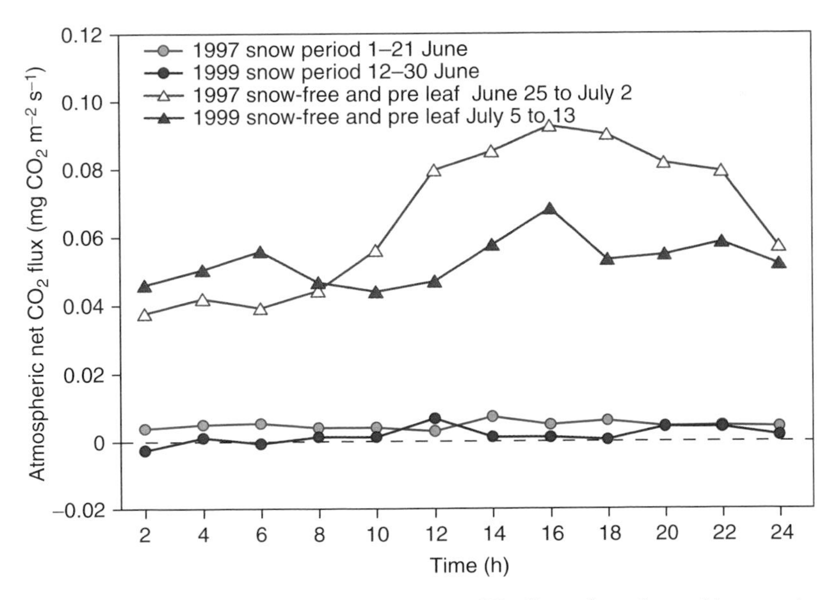

Figure 7 Diurnal variations of 2-h average net CO_2 fluxes based on eddy covariance measurements.

2. *Observed Soil CH_4 Effluxes*

Similar to CO_2 effluxes, significant differences in CH_4 effluxes were reported for various vegetation types at Zackenberg for the summer of 1997 (Christensen *et al.*, 2000). Comparisons between vegetation types have been made based on chamber measurements. Contrasting soil effluxes of both CO_2 and CH_4 (Figure 8) show the overall importance of the presence of a water-saturated zone within the active layer. Major effluxes of CH_4 were noted for hummocky fen, continuous fen and grassland, whereas CH_4 effluxes were absent from more or less well-drained heath sites. Calculated for the

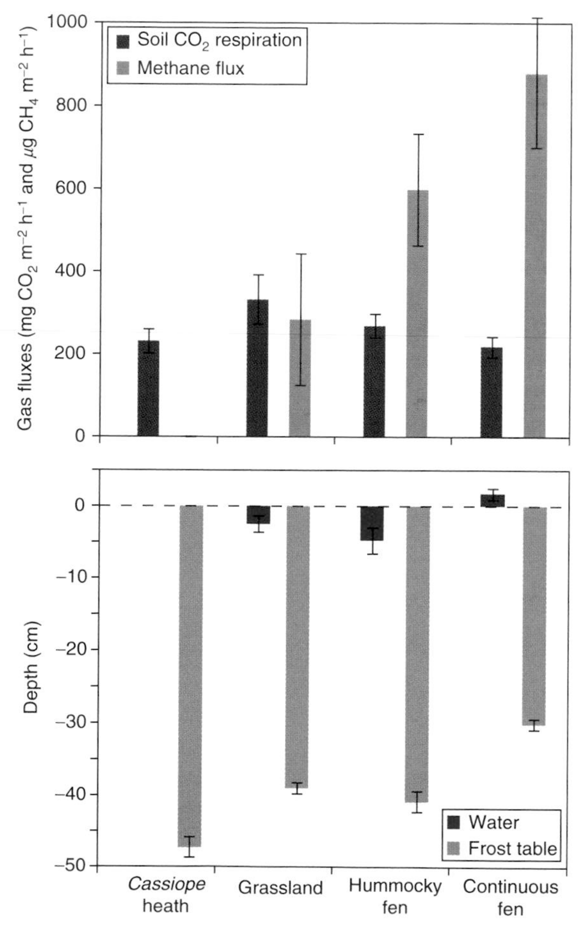

Figure 8 Average daytime CO_2 and CH_4 effluxes (±SE) measured 15 times during the growing season 1997 in four vegetation types along with mean thaw depth and depth to water table (modified from Christensen *et al.*, 2000).

entire valley, the CH_4 emission equalled 2.9 ± 1.0 mmol CH_4 m^{-2} d^{-1} or 0.033 μmol C m^{-2} s^{-1} (Christensen *et al.*, 2000).

Seasonal trends in CH_4 effluxes during the summer of 1997 based on eddy correlation data were reported by Friborg *et al.* (2000), while effluxes on a landscape scale based on chamber measurements were reported by Christensen *et al.* (2000). Chamber measurements made it possible to evaluate vegetation-specific controls on temporal trends. Regression analysis (Christensen *et al.*, 2000) showed that water table fluctuations were the single most important factor controlling seasonal trends. Studies both in the field at Zackenberg and using so-called peat monoliths (whole peat and live plant samples) have documented that vascular plants influence the CH_4 emissions on a species-specific basis. Ström *et al.* (2003) showed for the Zackenberg fen that in particular *Eriophorum scheuchzeri* has a strong positive effect on emissions through root exudation of organic compounds leading to high organic acid concentrations in the root vicinity, which is acting as substrate for CH_4 formation. This interaction was documented on the basis of the amount of organic acids in the peat water combined with analysis of potential rates of CH_4 production (Figure 9). Field measurements of pore water

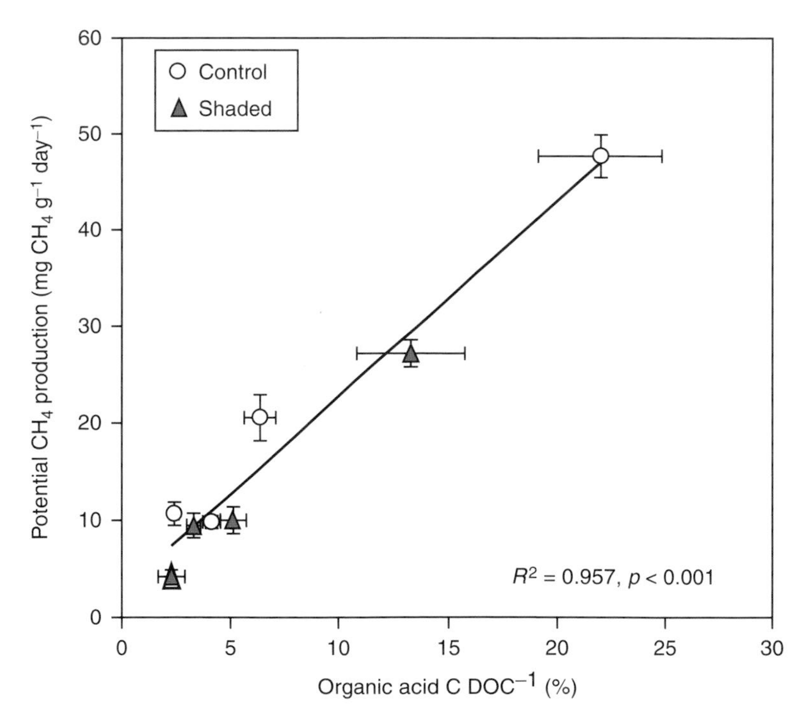

Figure 9 Potential CH_4 production in the soil profile of control (open symbols) and shaded (filled symbols) treatments plotted against the proportion of organic acids in the DOC (modified from Ström *et al.*, 2003).

concentrations of organic acids showed that the proportion of labile organic acid carbon was higher in control than in shaded plots (Table 3), indicating a close linkage between plant photosynthesis and production of these compounds. The plant–microbe linkage was further documented in experiments in the laboratory, where ^{14}C labelled acetate at the same rates and concentrations as measured around the roots in the field was shown to result in ^{14}C labelled CH_4 emissions (Ström *et al.*, 2003).

C. Soil Respiration Rates at Manipulated Field Sites

1. Measurements on Barren Ground and after Litter Removal

Soil CO_2 effluxes observed in the field represent the combined effects of soil microbial and root respiration, whereas CO_2 production observed during laboratory incubations (see section IV) represents only soil microbial respiration. As reported by Buchmann (2000), the ratio between the two respiration components is site-specific and may vary between 1:9 and 9:1. Soil CO_2 effluxes from non-vegetation patches were in the following compared to effluxes from their vegetated counterparts to evaluate the importance of root respiration on observed soil CO_2 effluxes. Measurements were made at *Dryas* and *Cassiope* heath sites, where patches of barren ground were large enough (>20 cm to nearest plant), and values amounted to an average of 94% and 70% of control rates, respectively (Figure 10).

Another set of measurements was made at sites after the litter layer was removed. A litter layer in the proper sense was not detected at *Dryas* heath and *Eriophorum* fen sites, which therefore were excluded. Removing the litter layer at *Cassiope* heath and *Salix* snow-bed sites reduced the soil respiration rates to 70–75% compared to control soils. Elberling *et al.* (2004) concluded that microbial soil respiration at least in some vegetation types dominates the total soil CO_2 efflux. However, the discussion on microbial- versus plant-associated respiration is controversial because estimates of the importance of microbial respiration are partly related to the methodology used. For example, it remains uncertain to what extent root respiration (and plant respiration in the chamber) is actually reduced during dark chamber measurements. Using different methods, Billings *et al.* (1977) found root respiration to account for 66–90% of the total soil respiration in arctic soils dominated by sedge *Carex* sp. and *Eriophorum* spp., whereas Oberbauer *et al.* (1992) concluded that microbial respiration accounted for a major portion of soil respiration at a similar site. In a tree girdling experiment in a sub-alpine forest, Scott-Denton *et al.* (2006) showed that "priming" of the soil with carbohydrates during winter, probably due to damaged shallow roots that use sucrose as protection against low-temperature extremes, contributed significantly to the increase in microbial biomass and associated CO_2 efflux

Table 3 Concentrations of organic acids and total dissolved organic carbon at different depths in the fen as subjected to experimental treatments (with sack cloth) reducing the rate of photosynthesis (Joabsson and Christensen, 2001; Ström *et al.*, 2003)

	Depth (cm)	Acetate (μM)	Lactate (μM)	Formate (μM)	Propionate (μM)	Malate (μM)	DOC (μM)	OA/DOC (%)
Control	5	87.0 ± 66.5	9.0 ± 3.8	32.3 ± 4.6	4.8 ± 2.9	0.17 ± 0.07	969 ± 107	23.7 ± 9.4
	10	42.1 ± 10.2	7.9 ± 0.6	28.7 ± 2.9	5.9 ± 1.9	0.16 ± 0.01	695 ± 67	22.0 ± 2.9
	15	12.4 ± 3.6	12.5 ± 1.3	27.9 ± 2.6	1.8 ± 0.6	0.21 ± 0.02	1608 ± 190	6.3 ± 0.7
	20	9.7 ± 1.7	12.8 ± 2.3	30.0 ± 2.6	1.5 ± 0.3	0.23 ± 0.03	2341 ± 219	4.2 ± 0.4
	25	7.2 ± 1.2	12.2 ± 1.6	31.0 ± 2.5	0.7 ± 0.1	0.22 ± 0.02	3719 ± 532	2.4 ± 0.2
Shaded	5	31.9	10.8	21.6	3.0	0.14	1024	12.0
	10	36.7 ± 13.6	9.2 ± 1.3	25.9 ± 1.3	9.2 ± 2.1	0.19 ± 0.01	1295 ± 255	13.3 ± 2.5
	15	9.4 ± 2.0	12.7 ± 2.0	26.9 ± 2.2	1.3 ± 0.3	0.16 ± 0.02	1871 ± 274	5.1 ± 1.1
	20	8.0 ± 1.7	12.2 ± 1.5	26.4 ± 2.1	4.0 ± 0.2	0.20 ± 0.02	2628 ± 176	3.3 ± 0.8
	25	8.6 ± 1.5	12.5 ± 0.6	31.0 ± 4.9	1.7 ± 1.2	0.24 ± 0.02	3497 ± 537	2.8 ± 0.5
Ancova C^1 S		$p = 0.342$	$p = 0.687$	$p = 0.114$	$p = 0.498$	$p = 0.337$	$p = 0.234$	$p = 0.030$

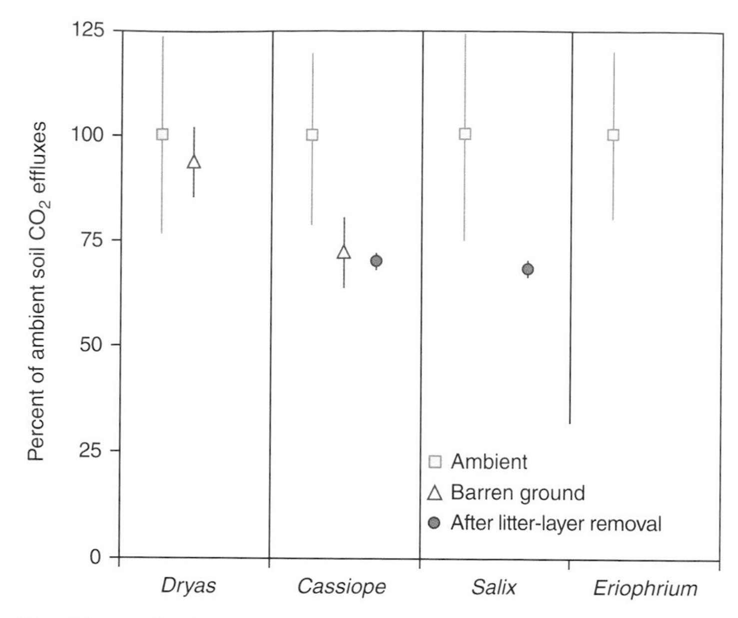

Figure 10 Observed soil CO_2 effluxes at *Dryas* and *Cassiope* heath sites without vegetation (more than 20 cm to the nearest plant) and after the removal of the litter layer (0–2 cm) as percentage of ambient effluxes (modified from Elberling *et al.*, 2004).

during the growing season. This is in line with Grogan *et al.* (2001) showing that soil respiration derived from recently fixed plant C (i.e., plant respiration and respiration associated with rhizosphere exudates and decomposition of fresh litter) was the principal source of CO_2 efflux. This illustrates the linkage between the autotrophic and the heterotrophic component.

The effect of litter layer is uncertain as the layer may act as a diffusion barrier. Also the litter layer, consisting of newly plant-derived C substrate, may decompose rapidly without being incorporated in the soil. Litter layers at Zackenberg have not been shown to represent a diffusion barrier. On the contrary, the importance of litter layer decomposition at *Cassiope* heath and *Salix* snow-bed sites were surprisingly similar, despite *Salix* litter being considered easier degradable than *Cassiope* litter. Thus, results suggest that decomposition of organic matter on top of the soil is an important component of the ecosystem element cycling at Zackenberg.

2. *Effects of Growing Season Warming on Soil Respiration*

To compare the effects of continuous, moderate warming with the effects of heat extremes, Marchand *et al.* (2004a,b, 2005) exposed plots of "wet" tundra vegetation at Zackenberg to both a limited temperature increase during an entire growing season in 1999 (+2.5 ± SD 0.5°C above the mean of 8.4 °C) and to an experimental heat wave during several days in 2001 (+9.2 ± SD 3.3°C above the mean of 7.7°C). The experimental site was a lower grassland plateau dominated by *Salix arctica* and *Arctagrostis latifolia*, with Bigelow's sedge *Carex bigelowii*, alpine bistort *Polygonum viviparum*, chestnut rush *Juncus castaneus* and *Dryas* sp. as subdominants. One of the objectives of the experiments was to detect possible changes in soil CO_2 efflux, using a Free Air Temperature Increase-system designed to homogeneously heat limited areas of short vegetation (Nijs *et al.*, 1996, 2000). Effects of warming during the entire growing season (Figure 11) included increasing thawing depth, green cover and soil respiration, while soil moisture was not significantly affected. Soil respiration was enhanced 33.3%, mainly through direct warming impact (Marchand *et al.*, 2004a), and in spite of lower Q_{10} in the heated plots (2.37 ± SE 0.20 relative to 2.86 ± SE 0.28 in the control plots). Contrary to the findings of Johnson *et al.* (2000), the unheated plots responded more strongly to temperature increase than the heated plots. A possible reason for this is that Q_{10} itself is temperature dependent (Schleser, 1982; Fang and Moncrieff, 2001). Additional factors controlling soil respiration were day of the year and soil moisture (Marchand *et al.*, 2004a). During the entire growing season, this high-arctic tundra ecosystem acted as a relatively small net sink both under current (0.86 mol CO_2 m^{-2}) and heated (1.24 mol CO_2 m^{-2}) conditions. Nevertheless, C turnover was increased, which was best explained by a combination of direct and indirect temperature effects.

When the same vegetation (but different plots) was exposed to the extreme temperature event during 2001, below- and above-ground respiration was likewise stimulated by the instantaneous warmer soil and canopy, respectively, but this time these components of the carbon balance outweighed the increased gross photosynthesis (Marchand *et al.*, 2005). As a result, during the heat wave, the heated plots were a smaller sink compared with their unheated counterparts, whereas afterwards in the recovery period the balance was not affected. If other high-arctic tundra ecosystems react similarly, more frequent extreme temperature events in a future climate may shift this biome towards a CO_2 source and stimulate climate warming through positive feedback.

3. *Effects of Water and Nutrient Addition on Soil Respiration*

At least two different studies have been performed at Zackenberg aiming to describe effects of addition of water and nutrients on measured soil CO_2 effluxes. Søndergaard (2004) focused on short-term effect (up to 12 h) following addition of water or glucose in water at sites dominated by *Cassiope*, *Dryas* and *Salix* during the growing season 2001. Significant increases in CO_2

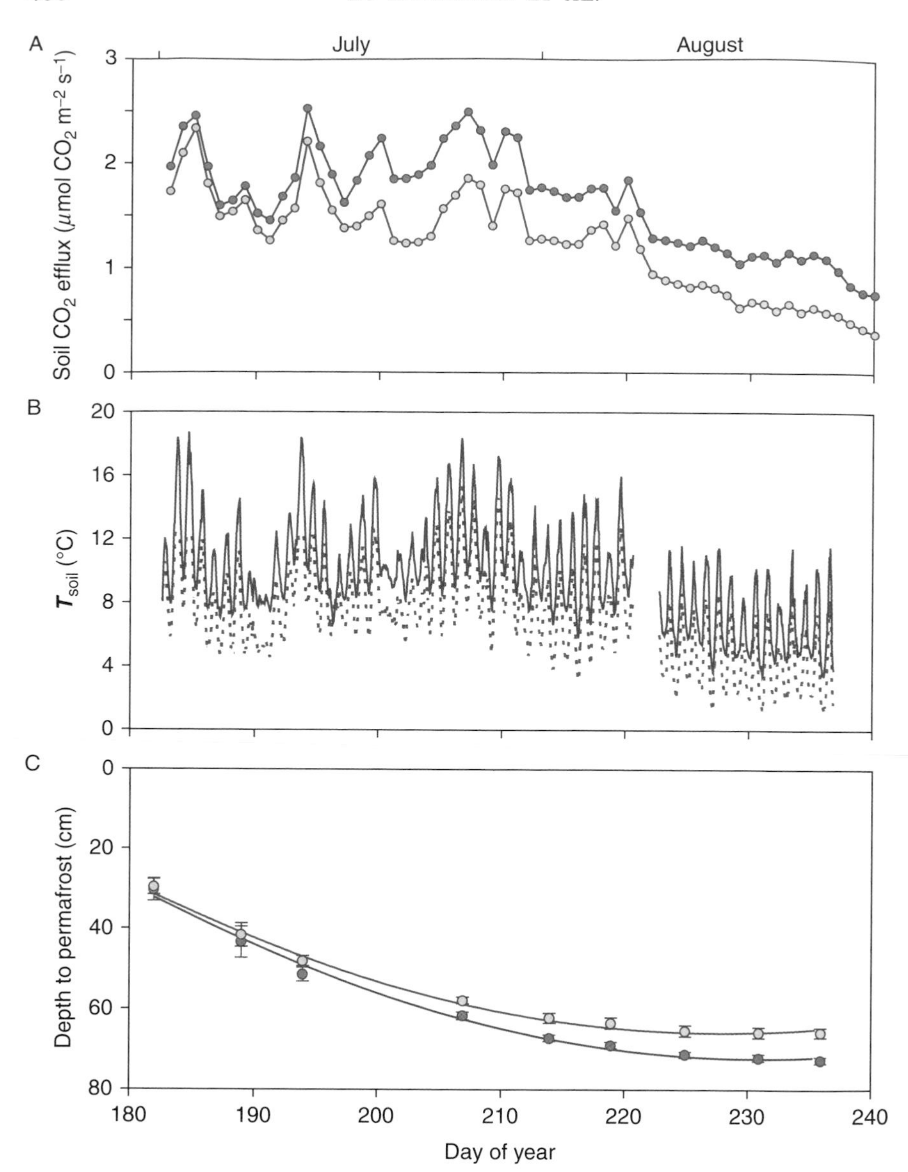

Figure 11 Seasonal trends of (A) daily soil respiration rates, (B) average soil temperature (T_{soil}) and (C) time course of active layer depth (means ± 1 SE on different days of the year and fitted polynomial curve, second order), during the entire growing season warming experiment June 29–August 28, 1999, for the unheated (blue line) and unheated (orange line) treatment.

effluxes following glucose addition were noted for all vegetation types; a factor of 3, 4 and 5 at *Dryas*, *Salix* and *Cassiope* sites, respectively. Surprisingly, only *Cassiope* and *Salix* showed a significant response to water addition (a factor of 2) and not the driest site characterised by *Dryas*. In all experiments, maximum effects of treatment were noted immediately, and effects decline markedly during the experimental time (12 h). In contrast, Illeris *et al.* (2003) studied long-term effects on soil CO_2 effluxes and microbial biomass 1 and 3 years after water, nitrogen and phosphorus application at sites similar to the *Dryas* site studied by Søndergaard (2004). Significant effects of water addition included up to 24% increase in biomass after 1 year and up to 47% increase in soil CO_2 effluxes 3 years after additions. Effects of nitrogen and phosphorus were not significant—perhaps as a result of the realistic but low quantities of nutrient added: 3.75 g N m^{-2} (as NO_3NH_4) and 0.81 g P m^{-2} (as Na_2HPO_4).

4. Early Season Snow Removal and Diurnal Variation of CO_2 Effluxes

Short-term efflux measurements (Figure 12) indicated that daily variations in soil CO_2 effluxes are strongly influenced by soil temperatures. However, daily variations in CO_2 effluxes during a period in June 2001 dominated by fast soil thawing (Figures 6 and 12) could not be modelled using temperature variations observed, whereas similar modelling could explain measurements in July. Model results from June 9 to 10, 2001 (shown as a solid line in Figure 12), indicated that only 75% of the total flux observed during 24 h could be explained by on-going soil respiration. Over the same period, the active layer was observed to increase about 2–3 cm in thickness. Using a simple mass balance for the day, the remaining 25% (equal to 0.16 g C m^{-2} d^{-1}) corresponded to the release of CO_2 from a 2.7 cm thick soil layer containing 10% CO_2 in the air-filled pore space. Unfortunately, subsurface gas concentrations have not been measured in frozen soil, but the total release of CO_2 after rapid thawing of frozen soil collected below more than 60 cm snow in June 2001 was measured and reported by Elberling (2003). Results indicated that subsurface pore gas CO_2 concentrations were between 2% and 18% CO_2 (in average 6.8%, $n = 42$) and therefore significantly higher than measurements made at the end of August 2000 just prior to ground freezing (Figure 3). It was concluded that the burst observed in 2001 (Figures 5 and 6) was mainly a result of a built-up of CO_2 produced prior to the thaw period but released on thawing and in line with similar CO_2 burst reported by Soegaard and Nordstroem (1999) based on ecosystem net fluxes (EC). It is likely that these observations for CO_2 are relevant for understanding similar pulses of CH_4 noted by Friborg *et al.* (1997) during early spring thaw for a sub-arctic mire. However, bursts of CH_4 have not been documented at Zackenberg.

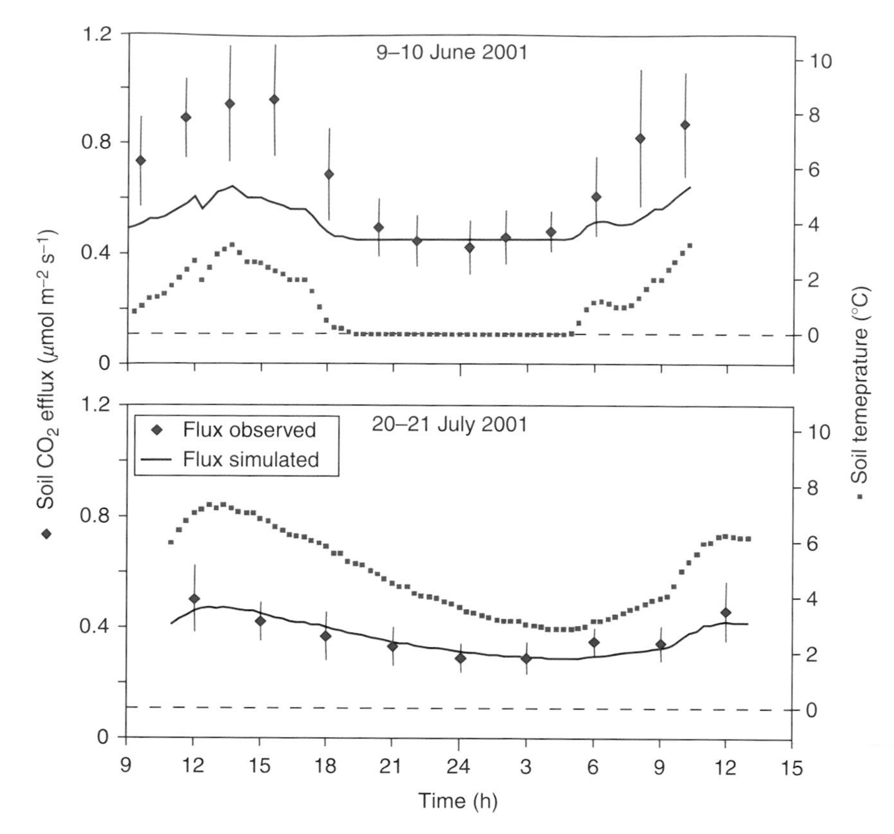

Figure 12 Diurnal variations in soil temperature at 2.5 cm depth as well as observed and simulated soil CO_2 effluxes as observed during 2 days with ±1 standard deviation shown as vertical lines ($n = 10$). Data from June 9 to 10, 2001, represented a non–steady-state condition, whereas data from July 20 to 21, 2001, represented a steady-state condition. The solid lines are the calculated soil respiration rates based on temperature variations observed at 2.5 cm depth and a Q_{10} of 3.3 (modified from Elberling, 2003).

III. CONTROLS OF CO_2 RELEASE FROM INCUBATED SOIL SAMPLES

A. Temperature and Water-Limited CO_2 Release

The temperature dependence of microbial soil respiration was investigated by Elberling (2003) in soil samples representing the four main horizons identified (A, B/C, A_b, C) with a focus on the A-horizon in *Cassiope* heath (Figure 13). An exponential increase in soil respiration with increasing

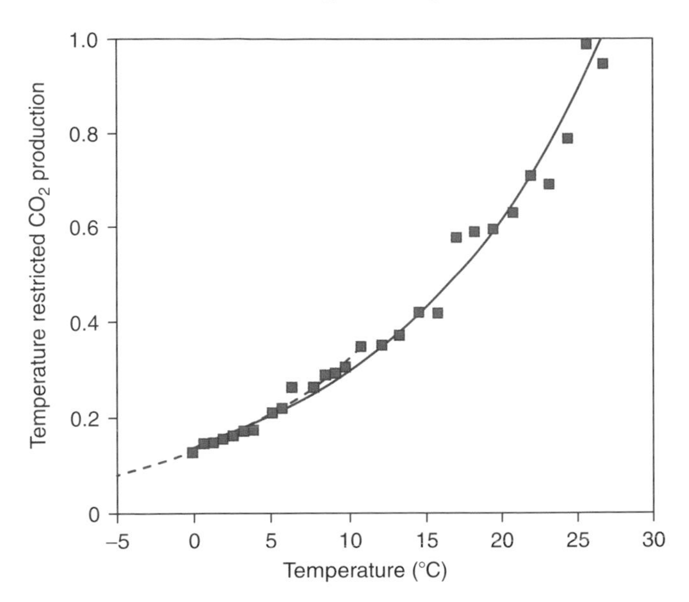

Figure 13 Experimental data on CO_2 production from soil collected from the A_1-horizon versus temperature. Data are normalised with the average observation at 26 °C to show the restriction of CO_2 production relative to production at 26 °C. The exponential fit shown as a solid line is based on all data from the A_1-horizon ($Q_{10} = 2.1$, $r^2 = 0.99$). The fit shown as a dashed line is based on data from 0 to 10 °C ($Q_{10} = 2.5$, $r^2 = 0.98$) (modified from Elberling, 2003).

temperature was observed, consistent with most other studies in the field. Q_{10} equalled 2.1 ($R^2 = 0.99$, based on 29 temperatures) for the A-horizon for the entire temperature range. For the temperature range most relevant for this study (0–10 °C), Q_{10} equalled 2.5. Q_{10} estimates based on four temperatures indicated that Q_{10} decreased from 4 around 3°C to 2.4 around 5–6 °C and became stable between 1.7 and 2 at temperatures above 7°C. Although this trend is based on only a few data points for each Q_{10}, the change in Q_{10} with temperature is consistent with previous observations (Fang and Moncrieff, 2001). Deviation from these Q_{10} values depending on the season and the depth of sampling could not be documented based on soil samples collected over the growing season (June–August) and soil samples collected from the four horizons. The same conclusion has been made previously based on studies of other arctic soils (Oberbauer *et al.*, 1996).

In contrast to the temperature dependency, soil respiration studies on the dependence of microbial respiration at 7 °C and 26 °C on water content (expressed as percentage of water per volume to match field observations) revealed distinctive differences between horizons. Results for the four horizons at 26°C (Figure 14) indicated a lower threshold of activity at a water

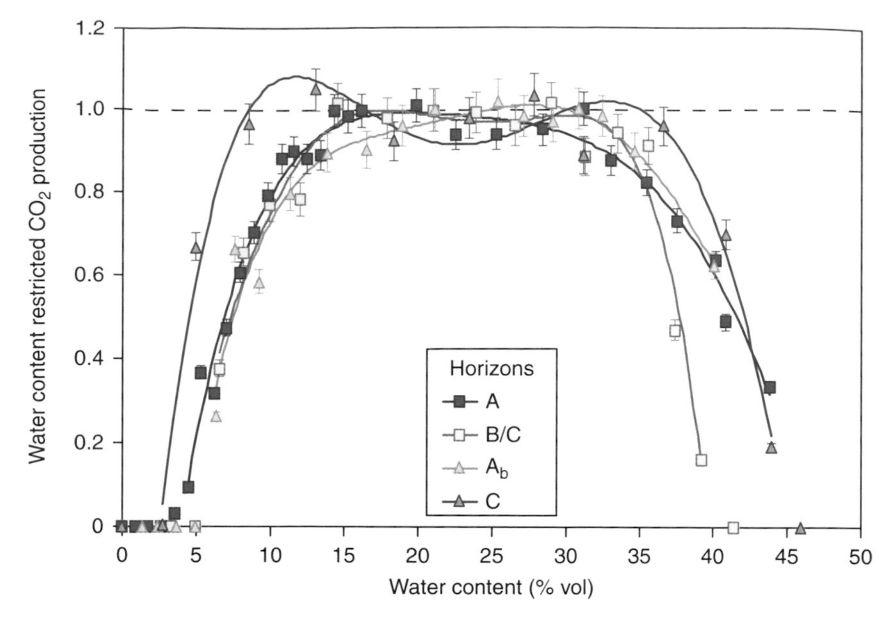

Figure 14 Experimental data on CO_2 production from soil collected at four horizons as a function of the water content at 26 °C. Data are normalised with respect to the average rate of observations, which are within one standard deviation (shown as vertical lines) of the maximum rate observed (modified from Elberling, 2003).

content of about 4% and almost no reduction in activity between 12% and 35% of water. Above 35% water, activity gradually decreased, probably due to the lack of continuing air-filled pores and thereby limiting the renewal of oxygen. Fitted equations describing the CO_2 production as a function of the water content were found to explain more than 95% of the variations observed at both temperatures, although at the lower temperature high water contents tended to have a less pronounced effect on the activity.

The combined effects of temperature and water on soil respiration are in line with a number of similar laboratory investigations (Howard and Howard, 1993; Lomander *et al.*, 1998) and were used by Elberling (2003) to simulate seasonal and spatial trends in subsurface CO_2 production and effluxes.

1. Subzero CO_2 Production and Release on Thawing

To investigate the importance of trapped CO_2 for uncoupling the actual CO_2 production by soil microbes from the subsequent release of the produced gas from frozen soil, Elberling and Brandt (2003) performed a laboratory experiment with soils incubated in a stable temperature gradient thermoblock. The results revealed an abrupt change in the temperature sensitivity of CO_2 release from soil around 0°C. Q_{10} equalled 2.0 ($R^2 = 0.94$) for temperatures above 0°C and 21.7 ($R^2 = 0.96$) for temperatures below 0°C. In order to identify the primary controls responsible for the higher Q_{10} values below 0°C, additional measurements were performed on samples manipulated by water and salt amendments (Figure 15). The results demonstrated that the actual water or ice content (varying from below 18 to 39% water by weight) to a large extent controlled the CO_2 release below 0 °C. In relatively dry soil samples

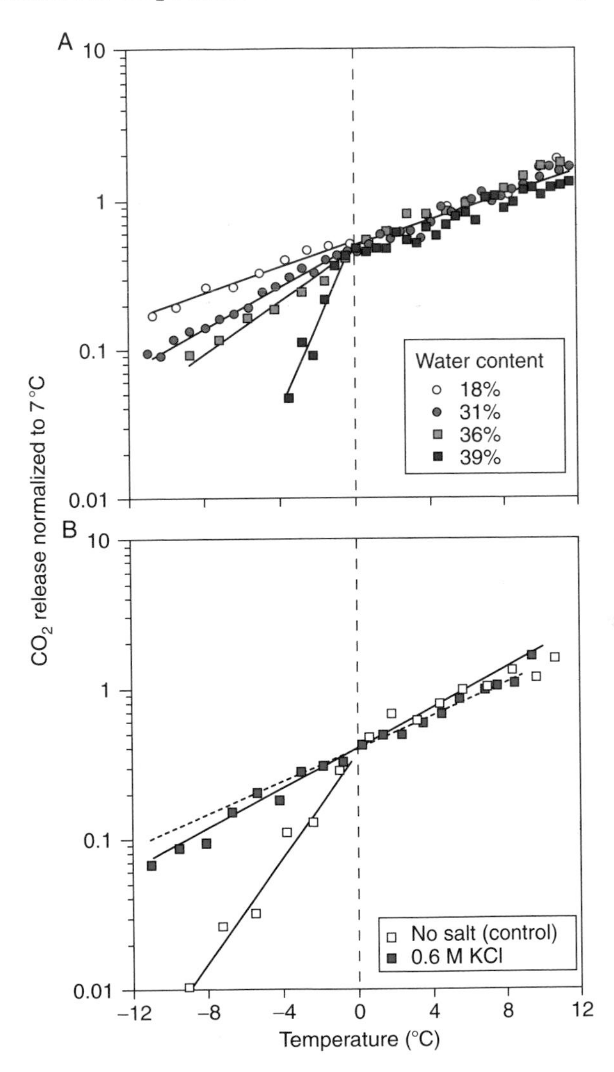

Figure 15 Thermo-block data on CO_2 release from manipulated soil samples at various water (A) and salt contents (B). Exponential fits above and below 0°C are shown as solid lines. The dashed line is a fit for the KCl experiment for temperatures above −3 °C (modified from Elberling and Brandt, 2003).

(18% water), the Q_{10} values below and above 0 °C were not significantly different, whereas a Q_{10} of 430 was observed for a water content of 39%. To lower the freezing point of pore water, up to 0.6 M KCl was added in solution. Simultaneous water content measurements and temperature readings indicated that freezing of water started at temperatures around −3°C, and more than 10% unfrozen water still existed at temperatures near −10°C. In these experiments, the shift in Q_{10} at 0 °C was eliminated (Figure 15B). However, Q_{10} based on observations below −4°C differed significantly from Q_{10} based on observations from −3 to 9 °C as illustrated with a dashed line (Figure 15B). In all experiments, the degree of observed CO_2 trapping explained more than 95% of the variation between Q_{10} below and above 0°C.

A minimum temperature threshold for basal soil respiration (BSR) for Zackenberg soil could not be identified in the present study as BSR was observed even at −18°C (Elberling and Brandt, 2003).

B. Substrate-Induced Respiration and Controls on Depth-Specific CO_2 Release

Production of CO_2 from depth-specific soil samples collected from *Dryas* and *Cassiope* heath and *Salix* snow-bed sites at Zackenberg have been reported by Elberling (2003). Results on production rates reported per unit dry weight show that the release from a near-surface A-horizon and a buried A-horizon at *Cassiope* sites (A and A_b, respectively) was not significantly different ($p < 0.05$), and that these A-horizons produced about 3–5 times as much CO_2 as B and C-horizons. The same was noted for the other vegetation types. Calculated CO_2 production rates per dry weight soil-C at optimum conditions represent a measure of the relative differences in the quality of soil organic matter (Figure 16A). Surprisingly, the soil organic C in a buried A-horizon at *Cassiope* heath sites was easier to decompose than organic C stored in the present A-horizon. However, this was not the case at *Salix* snow-bed sites. CO_2 production rates per unit soil organic-C in near-surface layers (A-horizons) were the same for *Dryas* and *Salix* but were 40% higher than for A-horizon material at *Cassiope* sites.

The production of CO_2 increased significantly ($p < 0.05$) in all samples after glucose addition, suggesting that soil respiration is also limited by the availability of easily degradable carbon substrates (Figure 16B) in line with substrate experiments (Søndergaard, 2004). A maximum increase in CO_2 production after glucose addition was observed for A-horizon material from the *Cassiope* site, which increased by a factor of 2. The lack of increase at A_b in *Cassiope* heath sites may be explained by a limited microbial biomass.

C:N ratios have previously been used to relate carbon substrate quality to decomposition. Van Cleve (1974) showed that C:N ratios were negatively correlated with organic matter decomposition (measured as percentage weight

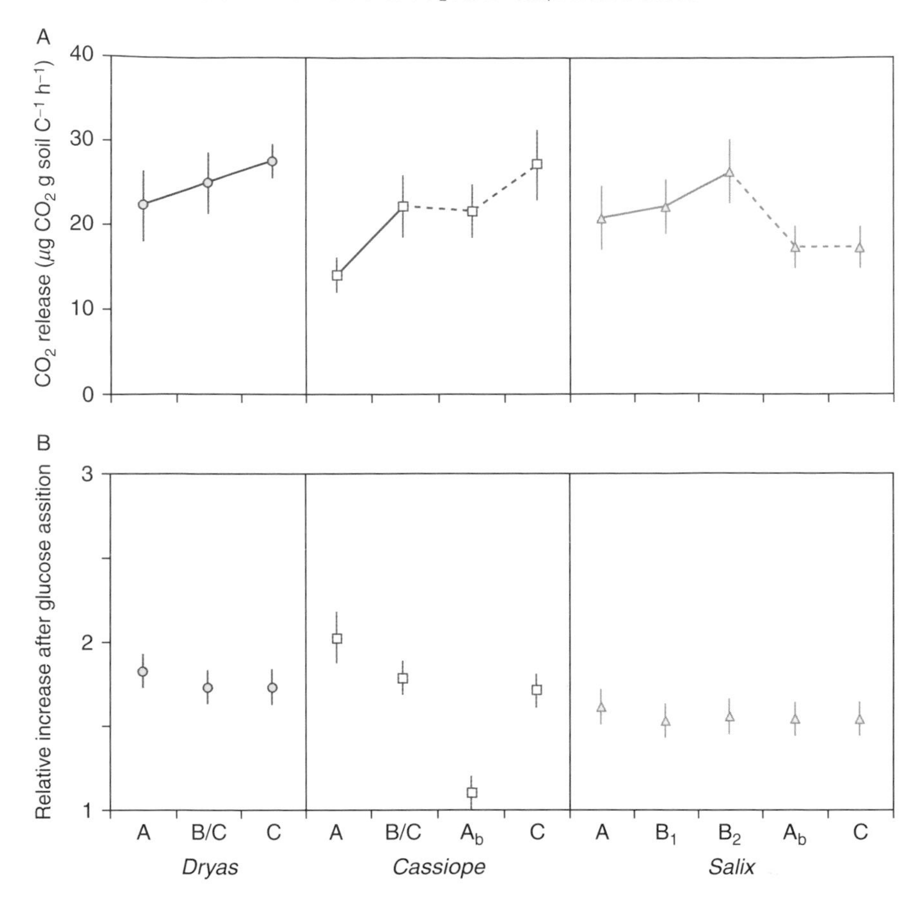

Figure 16 CO_2 production rates per unit organic soil-C (at 26°C) as observed in individual soil horizons at *Dryas* heath, *Cassiope* heath and *Salix* snow-bed sites (A) and the relative increase in CO_2 production rate after glucose addition (B). One standard deviation (±) of three replicates per horizon is shown as vertical lines (modified from Elberling *et al.*, 2004).

loss over 2 years) for a range of circumpolar tundra sites. C:N ratios in A-horizon material collected at *Dryas*, *Cassiope* and *Salix* sites (Table 1) were similar (16.2–17.3), with the lowest value being observed at *Cassiope* sites (Figure 4). Thus, the C:N ratios alone cannot explain the lower activity of carbon sources at *Cassiope* sites. In contrast, C:N ratios of vegetation and litter at *Cassiope* sites were the highest detected of all vegetation types investigated (Figure 4). Plant species control on litter decomposition has previously been reported from Alaskan tundra (Hobbie, 1996), showing that decomposition rates were more related to carbon quality than to the content of nitrogen.

IV. CONCLUSIONS AND FUTURE PERSPECTIVES

Several biogeochemical and physical factors and their interactions affect SOC reservoirs and mineralisation. This study has focused on the interactions between environmental drivers as temperature and water content, depth-specific distribution and quality of C sub-strate, and vegetation types. From laboratory and field observations as well as modelling, the following conclusions can be made:

1. Consistent with most related field investigations, sub-surface temperatures are a major environmental driver for temporal variations in subsurface gas productions and consequently at least during the growing season the main factor controlling soil gas emissions. During winter, freezing temperatures decrease rates, but experimental results suggest that microbial soil respiration continues at least as low as $-18°C$. Freezing moist to saturated soil may furthermore capture winter produced gases, which during soil thaw have been seen to be released in non–steady-state bursts.
2. Soil water condition is another important environmental driver which in heath areas seems not to be a dominant factor controlling gas productions, as the water content has to be very low in order to limit soil respiration rates substantially. However, a positive response to watering has been noted at exposed *Dryas* heath sites in a dry year. In contrast, the location of the water table is crucial at wet fen sites, where the CH_4 production is highly sensitive to both spatial and temporal trends in water table fluctuations.
3. Sampling in order to quantify total C reservoirs requires a procedure which takes into account site-specific active layer development, for example, the occurrence of buried surface layers. The simulated CO_2 activity profiles suggest that buried old surface layers (A_b) are presently important for the overall soil CO_2 dynamics.
4. Decomposition within the litter layer influences strongly total soil CO_2 fluxes observed, contributing up to 30% of the total CO_2 fluxes to the atmosphere but without having the equivalent importance on subsurface CO_2 concentrations and associated soil weathering processes. Furthermore, the spatial distribution of litter material seems to be responsible for most of the spatial variations of observed soil CO_2 effluxes.
5. Studies of CH_4 emissions from the fen in Zackenbergdalen have shown substantial seasonal emission rates for a high-arctic setting. The seasonal course and the landscape pattern of emissions is strongly controlled by the water table position, while in the wet areas the spatial variation in emission has been shown to be closely linked with the vascular plant species composition.

6. The release of trapped CO_2 from thawing soil layers seemed to account for unpredictable CO_2 bursts observed. These events are considered non–steady-state situations and partly to be a result of CO_2 being produced prior to soil thawing. Winter soil respiration has not been measured at Zackenberg, but based on extrapolating simulations it is suggested that soil respiration during winter may account for more than 25% of the annual CO_2 production.

Taken together, it appears that temperature, water content and substrate quality are all crucial parameters controlling the observed spatial variation in subsurface gas dynamics at Zackenberg. Changes in these parameters will rather quickly and directly affect the active soil carbon pool and therefore the soil carbon dynamics. Indirect and more slowly responding effects may also have a significant impact on soil carbon pools. Such effects are likely to be the result of changes on a landscape scale, for example, changing wind pattern, snow distribution, draining and formation of thermokarsts (Evans *et al.*, 1989; Hobbie *et al.*, 2000).

Based on long-term scenarios for climate changes in the Zackenberg region (see Stendel *et al.*, 2008, this volume) it is expected that short-term effects resulting from a longer growing season will enhance all sub-surface respiration processes. The exception may be the driest sites, which may become water-limited during the growing season (e.g., *Dryas* heath sites). Increasing winter temperatures alone are unlikely to enhance soil effluxes markedly as a minor increase in extremely low temperatures will not affect soil respiration rates. More important are the effects of increasing active layer thickness and the potential exposure of peat layer or other permanently frozen layers holding reactive carbon. This effect will be most pronounced at well-drained sites, while smaller increase in the active layer is expected at wet sites, which hold the largest stock of buried SOC.

Changes in the hydrological regime are closely related to changes in winter precipitation and snow-cover (Zimov *et al.*, 1996; Fahnestock *et al.*, 1998, 1999). Increasing water contents because of more snow and thawing of permafrost layers will decrease the oxygen availability and potentially increase CH_4 production rates significantly (Chapman and Thurlow, 1996; Blodau and Moore, 2003). On the contrary, more snow in parts of the landscape will limit the growing season as well as the period of respiration. At Zackenberg, these effects are probably the most important direct effects of predicted climate changes on the greenhouse gas budget at Zackenberg—and effects, which remain unsolved.

A more slowly responding climatic effect is the shift/transition in plant community structure and area distribution of vegetation types (Oberbauer *et al.*, 1996; Grogan and Chapin, 2000). As species respond differently to warming and water supply (Henry and Molau, 1997), successional trajectories are likely to vary with vegetation type (see Elberling *et al.*, 2008a, this volume).

ACKNOWLEDGMENTS

Thanks are extended to students taking part in the collection of field data, in particular J. Søndergaard and C. Sigsgaard of the GeoBasis programme; my colleague K. Brandt, who have been involved in thermo-block measurements; the Danish Polar Centre (M. Rasch) for access to climate data; colleagues who participated in soil sampling and discussions, including K.K. Brandt, R. Sletten and L. Greenfield as well as Jeff Welker, University of Alaska, Anchorage, for reviewing the paper. Monitoring data for this chapter were provided by the GeoBasis program, run by the Geographical Institute, University of Copenhagen, and the National Environmental Research Institute, University of Aarhus, and financed by the Danish Environmental Protection Agency, Danish Ministry of the Environment.

REFERENCES

Arrhenius, S. (1889) *Zeitschrift für Physikalische Chemie.* **4,** 226–248 [as translated in M.H. Back and K.J. Laidler (eds.) *Selected Readings in Chemical kinetics.* Oxford: Pergamon, 1967].

Bay, C. (1998) *Vegetation mapping of Zackenberg valley Northeast Greenland.* Danish Polar Center and Botanical Museum, University of Copenhagen, 29 pp.

Billings, W.D., Peterson, K.M., Shaver, G.R. and Trent, A.W. (1977) *Arctic and Alpine Res.* **9**, 129–137.

Blodau, C. and Moore, T.R. (2003) *Soil Biol. Biochem.* **35**, 535–547.

Buchmann, N. (2000) *Soil Biol. Biochem.* **32**, 1625–1635.

Burkins, M.B., Virginia, R.A. and Wall, D.H. (2001) *Global Change Biol.* **7**, 113–125.

Chapman, S.J. and Thurlow, M. (1996) *Agr. and Forest Meteorol.* **79**, 205–217.

Christensen, T.R., Jonasson, S., Michelsen, A., Callaghan, T.V. and Havström, M. (1998) *J. Geophys. Res.* **103**, 29,015–29,021.

Christensen, T.R., Friborg, T., Sommerkorn, M., Kaplan, J., Illeris, L., Soegaard, H., Nordstroem, C. and Jonasson, S. (2000) *Global Biogeochem. Cy.* **14**, 701–713.

Christensen, T.R., Joabsson, A., Ström, L., Panikov, N., Mastepanov, M., Öquist, M., Svensson, B.H., Nykänen, H., Martikainen, P. and Oskarsson, H. (2003) *Geoph. Res. Lett.* **30**, 1414.

Christiansen, H.H., Bennike, O., Böcher, J., Elberling, B., Humlum, O. and Jakobsen, B.H. (2002) *J. Quaternary Sci.* **17**, 145–160.

Coyne, P.I. and Kelley, J.J. (1974) *Nature* **234**, 407–405.

Davidson, E.A. and Janssens, I.A. (2006) *Nature* **440**, 165–173.

De Jong, E. and Schappert, H.J.V. (1972) *Soil Sci.* **113**, 328–333.

Elberling, B. (2003) *J. Hydrol.* **276**, 159–175.

Elberling, B. and Brandt, K.K. (2003) *Soil Biol. Biochem.* **35**, 263–272.

Elberling, B. and Jakobsen, B.H. (2000) *Can. J. Soil Sci.* **80**, 283–288.

Elberling, B., Jakobsen, B.H., Berg, P., Soendergaard, J. and Sigsgaard, C. (2004) *Arct. Antarct. Alp. Res.* **36**, 509–519.

Evans, B.M., Walker, D.A., Benson, C.S., Nordstrand, E.A. and Petersen, G.W. (1989) *Holarctic Ecol.* **12**, 270–278.

Fahnestock, J.T., Jones, M.H., Brooks, P.D., Walker, D.A. and Welker, J.M. (1998) *J. Geophys. Res.* **103**, 29,023–29,027.
Fahnestock, J.T., Jones, M.H. and Welker, J.M. (1999) *Global Biogeochem. Cy.* **13**, 775–779.
Fahnestock, J.T., Povirk, K.L. and Welker, J.M. (2000) *Ecography* **23**, 623–631.
Fang, C. and Moncrieff, J.B. (2001) *Soil Biol. Biochem.* **33**, 155–165.
Friborg, T., Christensen, T.R. and Soegaard, H. (1997) *Geophys. Res. Lett.* **24**, 3061–3064.
Friborg, T., Christensen, T.R., Hansen, B.U., Nordstroem, C. and Soegaard, H. (2000) *Global Biogeochem. Cy.* **14**, 715–723.
Grogan, P. and Chapin, F.S. (1999) *Ecosystems* **2**, 451–459.
Grogan, P. and Chapin, F.S. (2000) *Oecologia* **125**, 512–520.
Grogan, P., Illeris, L., Michelsen, A. and Jonasson, S.E. (2001) *Climatic Change* **50**, 129–142.
Henry, G.H.R. and Molau, U. (1997) *Global Change Biol.* **3**, 1–9.
Hobbie, S.E. (1996) *Ecological Monographs* **66**, 503–522.
Hobbie, S.E., Schimel, J.P., Trumbore, S.E. and Randerson, J.R. (2000) *Global Change Biol.* **6**, 196–210.
Howard, D.M. and Howard, P.J.A. (1993) *Soil Biol. Biochem.* **25**, 1537–1546.
Illeris, L., Michelsen, A. and Jonasson, S. (2003) *Biogeochem.* **65**, 15–29.
Joabsson, A. and Christensen, T.R. (2001) *Global Change Biol.* **7**, 919–932.
Johnson, L.C., Shaver, G.R., Cades, D.H., Rastetter, E., Nadelhoffer, K., Giblin, A., Laundre, J. and Stanley, A. (2000) *Ecology* **81**, 453–469.
Jones, M.H., Fahnestock, J.T., Stahl, P.D. and Welker, J.M. (2000) *Arct. Antarct. Alp. Res.* **32**, 104–106.
Kirschbaum, M.U.F. (1995) *Soil Biol. Biochem.* **27**, 753–760.
Lomander, A., Kätterer, T. and Andrén, O. (1998) *Soil Biol. Biochem.* **14**, 2017–2022.
Lloyd, J. and Taylor, J. (1994) *Functional Ecol.* **8**, 315–323.
Lundegärdh, H. (1927) *Soil Sci.* **23**, 417–453.
Marchand, F.L., Nijs, I., De Boeck, H.J., Kockelbergh, F., Mertens, S. and Beyens, L. (2004a) *Arct. Antarct. Alp. Res.* **36**, 298–307.
Marchand, F.L., Nijs, I., Heuer, M., Mertens, S., Kockelbergh, F., Pontailler, J.-Y., Impens, I. and Beyens, L. (2004b) *Arct. Antarct. Alp. Res.* **36**, 390–394.
Marchand, F.L., Mertens, S., Kockelbergh, F., Beyens, L. and Nijs, I. (2005) *Global Change Biol.* **11**, 2078–2089.
Nadelhoffer, K.J., Giblin, A.E., Shaver, G.R. and Laundre, J.R. (1991) *Ecol.* **72**, 242–253.
Nijs, I., Kockelbergh, F., Teughels, H., Blum, H., Hendrey, G. and Impens, I. (1996) *Plant Cell Environ.* **19**, 495–502.
Nijs, I., Kockelbergh, F., Heuer, M., Beyens, L., Trappeniers, K. and Impens, I. (2000) *Arc. Antarct. Alp. Res.* **32**, 242–53.
Oberbauer, S.F., Tenhunen, J.D. and Reynolds, J.F. (1991) *Arct. Alp. Res.* **23**, 162–169.
Oberbauer, S.F., Gillespie, C.T., Cheng, W., Gebauer, R., Sala Serra, A. and Tenhunen, J.D. (1992) *Oecologia* **92**, 568–577.
Oberbauer, S.F., Gillespie, C.T., Cheng, W., Sala, A., Gebauer, R. and Tenhumen, J.D. (1996) *Arct. Alp. Res.* **28**, 328–338.
Oechel, W.C., Hastings, S.J., Vourlitis, G.L., Jenkins, M., Riechers, G. and Grulke, N. (1993) *Nature* **361**, 520–523.
Oechel, W.C., Vourlitis, G.L., Brooks, S., Crawford, T.L. and Dumas, E. (1998) *J. Geophys. Res.* **103**, 28,993–29,003.

Pietikainen, J., Pettersson, M. and Baath, E. (2005) *FEMS Microbiol. Ecol.* **52**, 49–58.
Post, W.M., Emanual, W.R., Zinke, P.J. and Stangenberger, A.G. (1982) *Nature* **298**, 156–159.
Sebacher, D.I., Harriss, R.C. and Bartlett, D. (1985) *J. Environ. Qual.* **14**, 40–46.
Schleser, G.H. (1982) *Zeitung Naturforschung* **37**, 287–291.
Scott-Denton, L.E., Rosenstiel, T.N. and Monson, R.K. (2006) *Global Change Biol.* **12**, 205–216.
Ström, L., Ekberg, A. and Christensen, T.R. (2003) *Global Change Biol.* **9**, 1185–1192.
Søndergaard, J. (2004) *Sub-surface CO_2 dynamics in a high-arctic tundra ecosystem.* M.Sc. Thesis, Institute of Geography, University of Copenhagen.
Soegaard, H. and Nordstroem, C. (1999) *Global Change Biol.* **5**, 547–562.
Soegaard, H., Nordstroem, C., Friborg, T., Hansen, B., Christensen, T.R. and Bay, C. (2000) *Global Biogeochem Cy.* **14**, 725–744.
Van Cleve, K. (1974) In: *Soil Organisms and Decomposition in Tundra* (Ed. by A. J. Holding, O.W. Heal, S.F. MacLean, Jr. and P.W. Flanagan), pp. 311–324. Tundra Biome Steering Committee, Stockholm, Sweden.
Welker, J.M., Fahnestock, J.T. and Jones, M.H. (2000) *Climate Change* **44**, 139–150.
Welles, J.M., Demetriades-Shah, T.H. and McDermitt, D.K. (2001) *Chemical Geology* **177**, 3–13.
Whiting, G.J. and Chanton, J.P. (1993) *Nature* **364**, 794–795.
Zimov, S.A., Davidov, S.P., Voropaev, Y.V., Prosiannikov, S.F., Semiletov, I.P., Chapin, M.C. and Chapin, F.S. (1996) *Climatic Change* **33**, 111–120.

Spatial and Inter-Annual Variability of Trace Gas Fluxes in a Heterogeneous High-Arctic Landscape

LOUISE GRØNDAHL, THOMAS FRIBORG,
TORBEN R. CHRISTENSEN, ANNA EKBERG, BO ELBERLING,
LOTTE ILLERIS, CLAUS NORDSTRØM, ÅSA RENNERMALM,
CHARLOTTE SIGSGAARD AND HENRIK SØGAARD

SUMMARY

Summertime measurements of CO_2 and CH_4 fluxes were carried out over a range of high-arctic ecosystem types in the valley Zackenbergdalen since 1996 using both chamber and eddy covariance methodology. The net ecosystem CO_2 exchange and CH_4 flux data presented reveal a high degree of inter-annual variability within the dominant vegetation types in the valley, but also show distinct differences between them. In particular, the wet and dry parts of the valley show distinct differences. In general, the wet parts of the valley, the fens dominated by white cotton grass *Eriophorum scheuchzeri*, show high productivity, also in comparison with other sites, whereas CO_2 uptake rates

0065-2504/08 $35.00
© 2008 Elsevier Ltd. All rights reserved
DOI: 10.1016/S0065-2504(07)00020-7

in the white arctic bell heather *Cassiope tetragona* and mountain avens *Dryas* spp.-dominated heaths are much smaller.

Also within the different ecosystem types, a high degree of spatial variability can be documented. The spatial variability both within and between ecosystem types is especially pronounced for the CH_4 flux and can, at least partly, be related to differences in vegetation composition and water table level. The importance of the CH_4 emission from the various ecosystem types is evaluated both in relation to carbon and greenhouse gas budgets. In both wet and drier ecosystem components, inter-annual variability seems best explained through differences in the amount and distribution of snow in spring and the length of the growing season.

A large number of replicate chamber measurements carried out over various vegetation types in the valley are used to produce a synthesis of 10 years of flux data available on growing season carbon dynamics and CH_4 emission patterns in the individual parts of this high-arctic ecosystem and relates the differences between the ecosystems found in Zackenbergdalen to comparable sites in the circumpolar North.

I. INTRODUCTION

At present, the global atmospheric content of CO_2 increases by nearly 2 ppm/y, which is mainly attributed to the increasing anthropogenic emissions (IPCC, 2007). During the twentieth century, the global average surface air temperature has increased by 0.06 °C/decade, while in the arctic region the rise has been ~0.09 °C/decade. According to ACIA (2005), the current predictions for future warming over the period 2000–2100 range between 3.5 and 7 °C for the arctic region. The cause and effects of the increasing temperature are not fully understood, but there is a general consensus that part of the temperature increase can be explained through the increasing concentration of greenhouse gases (GHG) in the atmosphere.

The arctic region is sensitive to changes in climate. This emphasises the need for an understanding of the processes controlling the GHG exchange rates in the region and the effects of future climatic changes on the exchange rates. The general circulation models (GCMs), which simulate climate change scenarios due to changes in atmospheric GHG concentrations, predict that future global climate change will have the highest impact in the polar regions (ACIA, 2005; IPCC, 2007), due to feedback mechanisms exerted by change in snow-depth and -extent, variations in thawing of permafrost, changes in vegetation patterns and thawing of sea ice.

Rising temperatures cause the thawing of permafrost, resulting in increased active layer depth. Holding ~14% of the global organic carbon

(Post *et al.*, 1982), these regions are highly sensitive to changes in decomposition rates due to changed temperatures. Consequently, the arctic ecosystems are some of the most vulnerable ecosystems in the world to climatic changes (ACIA, 2005).

Few ecosystem level studies on carbon balances have been conducted in the Arctic, and the carbon exchange of these vegetation types is generally poorly represented in the global ecosystem models. Depending on future climate and the response from the arctic ecosystems, the region has a potential to be a net sink (Oechel *et al.*, 2000a) or a source (Oechel *et al.*, 1993; Grogan and Chapin, 2000) for CO_2 exchange with the atmosphere. Hence, in Northern Alaska Oechel *et al.* (1993) have observed that some ecosystems have changed from a net sink of carbon to a net source.

This calls for a synthesis of existing data on the inter-annual variability in the fluxes in relation to climatic differences between years in high-arctic settings as at Zackenberg. The exchange of CO_2 between arctic ecosystems and the atmosphere is a delicate balance between the photosynthetic uptake rates and the respiratory losses. The inter-annual variability is to a large extent dependent on the climate during the growing season. Global climatic change in the form of increasing temperatures, melting permafrost, longer growing seasons and altered precipitation and hydrological drainage patters is leading to dramatic changes in the Arctic, while the full implications of such changes on regional terrestrial carbon cycle dynamics is unknown.

This chapter aims at describing the spatial and inter-annual variability in the observed CO_2 and CH_4 fluxes within the dominating vegetation types during the growing season in the valley Zackenbergdalen and discusses the physical processes that govern the exchange in these particular ecosystems.

II. STUDY SITES AND METHODS

The high-arctic climate at Zackenberg is generally characterised by low temperatures and low precipitation (Hansen *et al.*, 2008, this volume). With respect to vegetation, the valley is representative of the lowland areas in the continental part of Northeast Greenland (Bay, 1998) characterised by a few dominating ecosystem types: wet fens, well-drained heaths, snow-beds and dry abrasion plateaus distributed according to hydrology and morphological properties. The vegetation has been classified into nine plant communities, which can be grouped into five main types. The fen community has a high water table during most of the growing season. The vegetation is composed of sedges and mosses. The wet continuous fens are dominated by white cotton grass *E. scheuchzeri* and Fisher's tundragrass *Dupontia psilosantha*, whereas the hummocky fen also contains elements of arctic willow *Salix arctica*. The intermediate dry grasslands are dominated by russet sedge *Carex saxatilis*, tall cotton grass *Eriophorum triste*, wideleaf polargrass *Arctagrostis latifolia* and *S. arctica*. The heaths are mainly composed of dwarf shrubs and herbaceous vegetation types. Mosses and lichens are also present and form a dry crust in the mesic and arid habitats. Additionally, abrasion plateaus are found.

Data presented in this chapter have been sampled in the period 1996–2005 using approaches to measure the gaseous fluxes of carbon between the surface and the atmosphere. In general, the chamber method is the most commonly used, and it has been applied for measurements of CO_2 and CH_4 for decades in various ecosystem types. The method is excellent for providing measurements from a small well-defined surface area for short time intervals; however, the measurements have great spatial variability within the same vegetation type, due to patchiness of vegetation and spatial variability in soil composition, soil moisture and temperature.

At Zackenberg, the chamber approaches have been used in various vegetation types, where different infra red gas analysers (IRGA) have been used for the CO_2 flux measurements with chambers, including LI-6200 and Environmental Gas Monitor (EGM) (PP systems). CH_4 flux measurements have been carried out using static chambers and Flame Ionization Detector (FID) gas chromatographs.

The other commonly used method for measurements of gas and energy exchange between the surface and the atmosphere is the eddy covariance method. The method is non-intrusive and continuously measures the flux at landscape scale (hectares). The method is ideal when gas exchange budgets for longer time periods for a fairly homogeneous surface are of interest. By not affecting the environment in which measurements are carried out, the fluxes obtained in this way are suitable for comparisons of, for example, carbon budgets or GHG effects (Friborg *et al.*, 2003).

The eddy covariance technique has been used in two different ecosystems; a wet fen in Rylekærene (1996–1999) and a dry *Cassiope*- and *Dryas*-dominated heath (1997 + 2000–2005) north of the climate station (see Groendahl, 2006) during the growing season (early June to late August). Fluxes were measured using a LI-6262 IRGA and a sonic anemometer recording half-hourly exchange of CO_2 between the atmosphere and the surface. Additionally, in 1997 measurements were performed at an *S arctica*-dominated site in the valley. The eddy covariance towers were located ~1 km apart, yet represent areas of contrasting moisture conditions and vegetation composition, which are typical representatives for the major ecosystem types in the High Arctic. Eddy correlation measurements of CH_4 exchange were performed during the growing season in 1997 over the wet fen using a tunable diode laser (Friborg *et al.*, 2000).

III. SPATIAL VARIABILITY

Environmental conditions have been shown to control the variability in the CO_2 fluxes at Zackenberg (Rennermalm *et al.*, 2005; Groendahl *et al.*, 2007). In vegetated areas, the net ecosystem CO_2 exchange (NEE) is controlled by (a) the photosynthetic uptake and (b) the respiration from soil and plants. NEE therefore expresses the balance between CO_2 assimilated by the vegetation and CO_2 emitted through the respiratory loss from the vegetation and soil. If vegetation is not under nutrient or water stress, photosynthesis often shows larger diurnal variations than respiration.

Spatial variability of fluxes among vegetation types relies upon the vegetation type itself and upon variability in soil texture and composition, soil temperature and soil moisture. In addition the variability in nutrient availability might influence the fluxes (Callaghan *et al.*, 2004). At Zackenberg, fluxes have been measured from five vegetation types: *Cassiope* heath, *Dryas* heath, *Salix* snow-bed, grassland and fen (see Elberling *et al.*, 2008b, this volume, for details). In total, they cover 79.2% of the valley below 200 m a.s.l. (Soegaard *et al.*, 2000).

A. Spatial Variability in CO_2 Exchange

The long-term measurement series from the two eddy covariance towers reveals that overall the fen site has a higher CO_2 sequestration rate than the heath (Figures 1 and 2). Comparing the fen and the heath, it is seen that the period dominated by assimilation of CO_2 by the vegetation starts at different times during the growing season. At the heath, the plant growth is initiated rapidly after snowmelt mainly because of the dominance of wintergreen and evergreen shrubs, whereas the fen is dominated by sedges, which need to develop leaves before assimilation can take place.

Net uptake (negative NEE) is initiated in mid or late June, whereas the sedge-dominated fen starts assimilating CO_2 in late June or early July. The late start of uptake is, however, more than compensated by the magnitude of the uptake during the middle of the growing season. The fen sequesters ~50% more carbon during the summer than the heath, which is the result of the fen being characterised by dense vegetation with leaf area index (LAI) reaching 1.2 during the peak of the growing season, whereas LAI at the heath reaches about 0.2 during the peak of the growing season (Soegaard *et al.*, 2000).

During the 2004 growing season, a 9-week field campaign was conducted, where measurements with the chamber method were undertaken in the five most frequently occurring vegetation types in the valley. The sites were selected to represent the spatial variability between the major plant

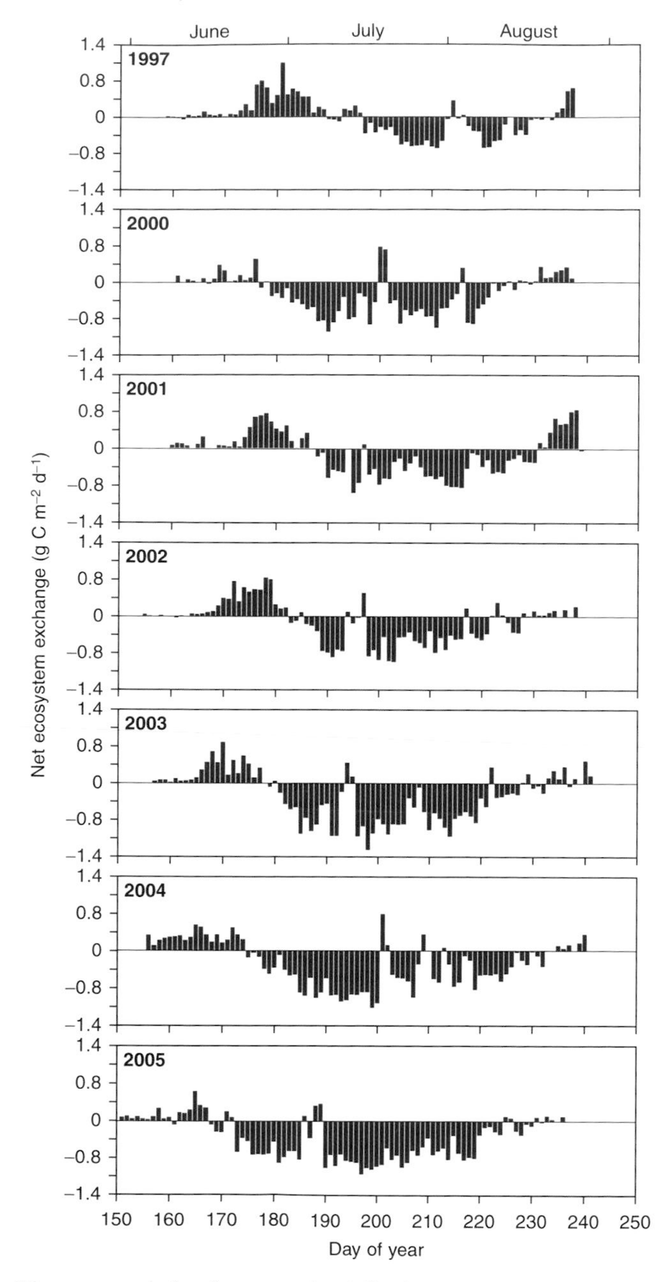

Figure 1 The seasonal development in daily integrated net ecosystem exchange (NEE, May 30–September 7) during 7 years of measurements at a heath at Zackenberg (modified from Groendahl, 2006).

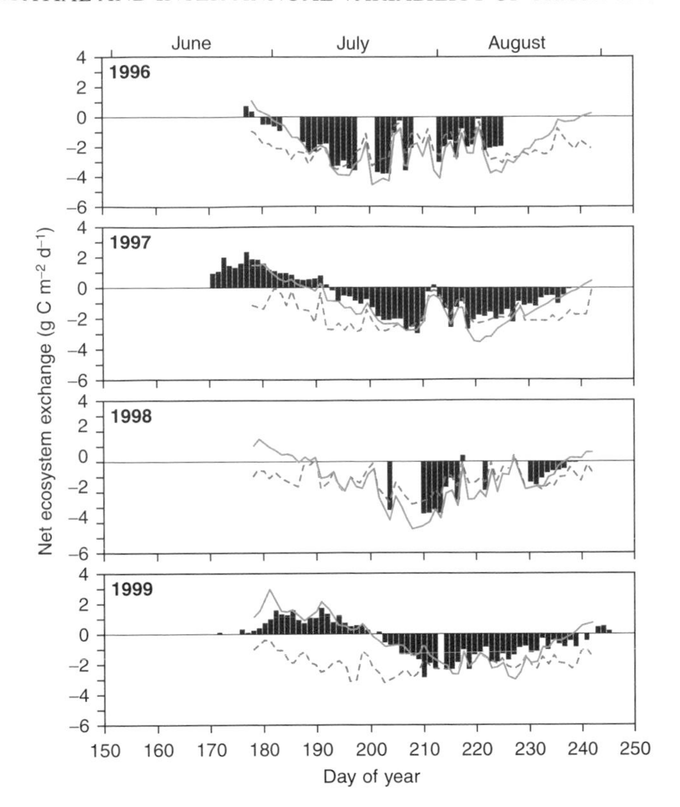

Figure 2 Daily net ecosystem exchange (NEE) measured in the fen during May 30–September 7, 1996–1999. The bars represent measurements and the solid line shows the modelled NEE based on leaf area index (LAI) and V_{max}, which varied individually for each year. The dashed line shows the modelled NEE based on a constant LAI and V_{max} (modified from Rennermalm *et al.*, 2005 by permission from Arctic, Antarctic, and Alpine Research).

communities, which also reflects the hydrological regimes present in the valley (L. Groendahl unpublished; L. Illeris, unpublished). Significant differences in NEE between the plant communities are seen in Figure 3. The wet fens, covering about 12% of the valley, showed the highest uptake rates during the summer season (–916 mg CO_2 m^{-2} h^{-1}), whereas the dry heaths (*Cassiope* and *Dryas*), covering ~30% of the valley, had the lowest uptake rates (~–300 mg CO_2 m^{-2} h^{-1}). The mesic *Salix* snow-bed areas and the grasslands had uptake rates of about −400 mg CO_2 m^{-2} h^{-1}. The shrub-dominated sites were characterised by losses of CO_2 at the beginning of the 2004 growing season, whereas the sedge-dominated sites (fen and grassland) had uptake during the same period.

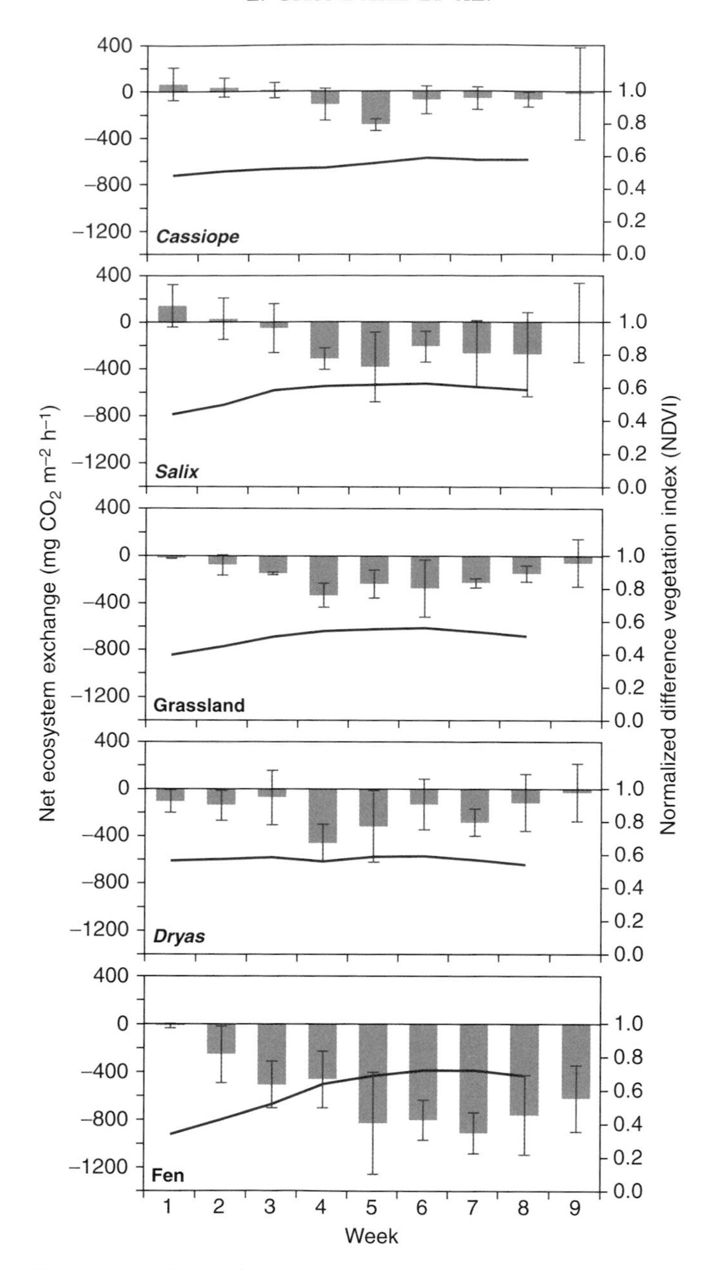

Figure 3 Temporal variation in daytime (10:00–16:00 h) average net ecosystem CO_2 exchange (NEE) for each plant community during the 9 weeks of measurements from June 23 to August 19, 2004. The solid line shows the development in normalised difference vegetation index (NDVI), which can be used as a surrogate for leaf area index (LAI).

Integrating fluxes from the dominant vegetation types at Zackenberg, Soegaard *et al.* (2000) found that during the 1997 growing season, the area functioned as a sink of CO_2 with an uptake rate of ~-10.4 g C m^{-2} season^{-1} from June to August. This was confirmed by Groendahl *et al.* (submitted), who found that when the measured fluxes were scaled up to regional level, the area had a daytime average uptake rate of -102.8 mg CO_2 m^{-2} h^{-1} during the 2004 growing season.

In 2004, there were significant differences in growing season integrated net ecosystem production (NEP), gross ecosystem production (GEP) and ecosystem respiration (ER) among the five vegetation types (Table 1) with a trend towards increasing GEP and NEP from dry to wet ecosystems (L. Illeris, unpublished). However, only the fen had significantly higher fluxes than the other vegetation types. The very productive fen showed GEP values almost twice as high as those in the second most productive vegetation type, the *Salix* snow-bed (Table 1). In contrast, ER differed less among vegetation types and was only slightly higher in the fen than in the *Salix* snow-bed (Table 1) even though the permafrost table and the amount of soil organic matter was highest in the fen. The anaerobic conditions here limited ER to almost the same level as in the other vegetation types. This resulted in NEP and carbon accumulation being highest in the fen. *Cassiope* heath turned out to be the least productive vegetation type among the five investigated types (Table 1).

Since both this study (L. Illeris, unpublished) and earlier investigations of fluxes in the valley (Christensen *et al.*, 2000) found that the anaerobic conditions in the fen limit ER to almost the same level as in some of the other vegetation types, this is probably a general phenomenon for the High Arctic.

Table 1 Means ± standard error (SE) of integrated growing season (9 weeks) net ecosystem production (NEP), gross ecosystem production (GEP) and ecosystem respiration (ER) (g CO_2 m^{-2}) from the five most common vegetation types in Zackenbergdalen as measured by the chamber method

Vegetation type	NEP	GEP	ER
Cassiope heath	4.32 ± 0.80	17.45 ± 2.68	−13.12 ± 3.39
Dryas heath	10.04 ± 6.31	22.69 ± 6.71	−12.65 ± 1.34
Salix snow-bed	9.91 ± 6.73	29.50 ± 5.88	−19.49 ± 1.69
Grassland	8.08 ± 1.74	24.34 ± 3.46	−16.26 ± 1.92
Fen	33.84 ± 5.97	56.32 ± 4.21	−22.50 ± 1.78

Positive values indicate uptake.

B. Spatial Variability in CH_4 Exchange

Significant variations in CH_4 effluxes have been reported for different vegetation types at Zackenberg for the summer of 1997 using chambers measurements (Christensen *et al.*, 2000). Contrasting effluxes of both CO_2 and CH_4 (Figure 4) emphasise the overall importance of the presence of a water-saturated zone within the active layer. Major effluxes of CH_4 were noted for hummocky fen, continuous fen and grassland, whereas CH_4 effluxes were absent from more or less well-drained heath sites.

In 1997, measurements of CH_4 emission were conducted using both eddy covariance technique (Friborg *et al.*, 2000) and static chambers over different ecosystem types (Christensen *et al.*, 2000) followed by more chamber measurements during 1998–2000 as part of an experiment on the impact of altered light intensity (Joabsson and Christensen, 2001). The measurements of CH_4 flux from the valley have shown that the production and thus emission can be related to environmental factors, for example, water table

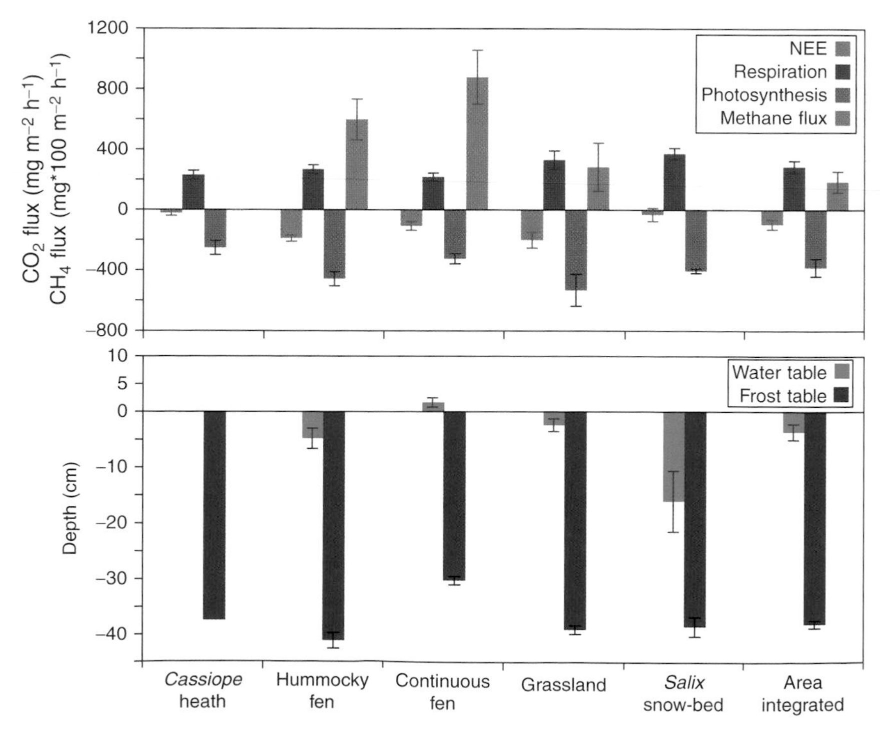

Figure 4 Average daytime CO_2 and CH_4 effluxes [± standard error (SE)] measured 15 times during the growing season 1997 in five vegetation types along with the mean thaw depth and depth to water table (modified from Christensen *et al.*, 2000).

level, soil temperature and development of the active layer over the summer. Friborg *et al.* (2000) found from eddy covariance measurements from Rylekærene that a combination of these three factors could explain 82% of the variation in the measured flux over the course of the growing season (June 1–August 26). Fluxes ranged from 0 in mid June to 120 mg CH_4 m^{-2} d $^{-1}$ around August 1 with an integrated emission of 3720 mg CH_4 m^{-2} over the season, giving an average daily flux of 43 mg CH_4 m^{-2}.

The eddy covariance measurements from Rylekærene covered different vegetation types ranging from the wettest continuous fen over hummocky fen (as continuous fen with elements of *S arctica*) to the intermediate dry grassland where the two fen types covered ~75% of the Rylekærene site. Corresponding CH_4 fluxes were found in 1997 by Christensen *et al.* (2000) with static chambers. When fluxes from these three different "wet" vegetation classes were integrated, a mean flux of 6.3 mg CH_4 m^{-2} h^{-1} was found over the season, which is slightly higher than fluxes measured with the eddy covariance technique during the middle of the summer. Continuous fen showed the highest emission over the season (10–20 mg CH_4 m^{-2} h^{-1}) and grasslands (2–6 mg CH_4 m^{-2} h^{-1}) the lowest of the wet classes. The hummocky fen showed intermediate emission rates between 0 and 15 mg CH_4 m^{-2} h^{-1} with distinct differences in emission rates according to micro relief (hummocks and hollows) as found in other studies (Kutzbach *et al.*, 2004). Overall very few measurements of CH_4 flux have been carried out in high northern latitudes, and in those studies the variability in fluxes was large, but the observation from Zackenberg seems to be within the range of what has been found in the circumpolar Arctic.

Methane production in the soil is stimulated by NEP or net primary production (NPP) through root exudation and production of litter, acting as a substrate for methanogenesis (Ström *et al.*, 2003). The differences in emission rates of CH_4 between vegetation classes could be related to differences in water table levels and also differences in ecosystem production, NEE. Especially in the hummocky part of the fen, variations in the water table showed a high degree of control over the CH_4 emission, whereas this control was less evident in the wetter, continuously water logged parts of the fen, where CH_4 oxidation can be assumed to be small. For the two drier vegetation types on the valley floor, *Cassiope* heath and *S arctica* snow-bed, no substantial emission was found, which is in accordance with findings from other dry ecosystems types in other high-arctic areas (Christensen *et al.*, 1995).

Also biotic factors like photosynthetic activity and the presence of vascular plants show a high degree of control over the CH_4 emission in Zackenbergdalen (Joabsson and Christensen, 2001). When fen plots were shaded, Joabsson and Christensen (2001) found that especially the production of vascular plants was positively correlated to CH_4 emission and that this relationship was due to vascular plants acting as pathways for CH_4, thus preventing oxidation in the top soil. Ström *et al.* (2003) showed for the Zackenberg fens that in particular *E. scheuchzeri* had a strong effect stimulating emissions through root exudation of sugars, leading to high organic acid concentrations in the rhizosphere acting as substrate for CH_4 formation. This significant plant—microbe interaction was documented both in the field through correlations between rates of photosynthesis in experimentally shaded plots and the amount of organic acids in the peat water combined with analysis of potential rates of CH_4 production (Figure 5). The linkage was further documented in experiments in the laboratory where ^{14}C labelled acetate at the same rates and concentrations as those measured in the field was shown to result in ^{14}C-labelled CH_4 emissions (Ström *et al.*, 2003).

IV. INTER-ANNUAL VARIABILITY

Insight into the inter-annual variability is given by the repeated measurements spanning 4 years at the fen site and 6 years at the heath site using the eddy covariance technique. Fluxes of CO_2 were measured continuously from before snowmelt until senescence in late August. Each year was characterised by different climatic conditions during the summer season in terms of timing of snowmelt, air temperature and precipitation.

A. Inter-Annual Variability in CO_2 Exchange

Abiotic as well as biotic factors have been found to drive the inter-annual variability of the measured NEE (Rennermalm *et al.*, 2005; Groendahl *et al.*, 2007). Significant inter-annual variability in NEE during the growing season at the heath have been reported by Groendahl *et al.* (2007) and is seen from Figure 6. Each year is characterised by substantial losses of carbon after snowmelt followed by increasing uptake rates as assimilation exceeds respiratory losses. While the extremely early and warm summer of 2005 constituted a strong carbon sink (-32 g C m^{-2}), the lowest uptake rates were observed in 1997 (-1.4 g C m^{-2}), an year with a cold summer. The range of these CO_2 fluxes is within the range of the previously reported growing-seasonal NEE ranging from near zero for an arctic desert on Svalbard (Lloyd, 2001) to -96 g C m^{-2} for the Zackenberg fen in 1996 (Soegaard and Nordstroem, 1999).

Inter-annual variability observed at the heath reveals that there is a tendency of increased length of the growing season from 2001 to 2005 (Figure 1), which is caused by increasingly early snowmelt date. Timing of snowmelt and air temperature have been found to be the main processes controlling carbon balance at the heath during the years of eddy covariance

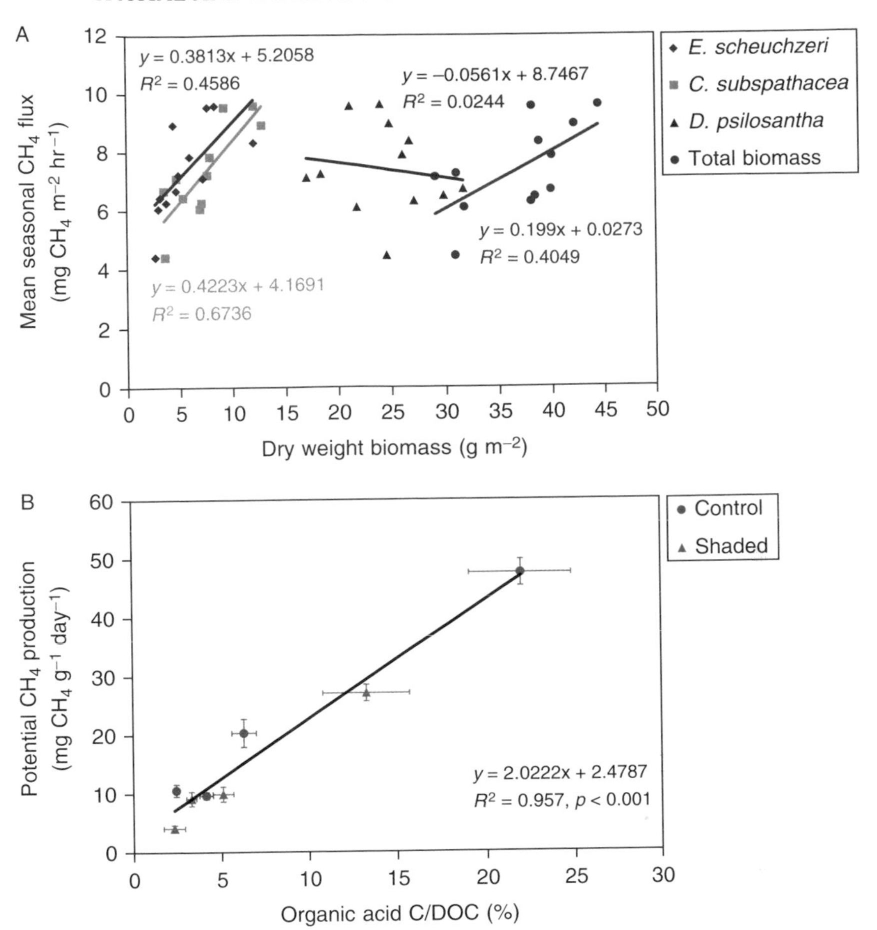

Figure 5 (A) Seasonal (2000) mean CH_4 emission plotted against leaf biomass of white cotton grass *Eriophorum scheuchzeri*, Fisher's tundragrass *Dupontia psilosantha*, Hoppner's sedge *Carex subspathacea* and total leaf biomass of the three species in the plots measured (modified from Joabsson and Christensen, 2001). (B) Potential CH_4 production in the soil profile of control (open symbols) and shaded (filled symbols) treatments plotted against the proportion of organic acids in the dissolved organic C (DOC) (modified from Ström *et al.*, 2003).

measurements (Groendahl *et al.*, 2007). Uptake rates during the summer correlated well with the characteristics of the spring conditions, where early snowmelt tends to increase uptake rates, explaining 97% of the variation in the cumulative NEE over the course of the growing season across the years (Figure 7). Likewise, the temperature during the growing season is positively influencing the uptake rates. Late snowmelt tends to delay canopy development, and consequently photosynthesis tends to start late in years of late

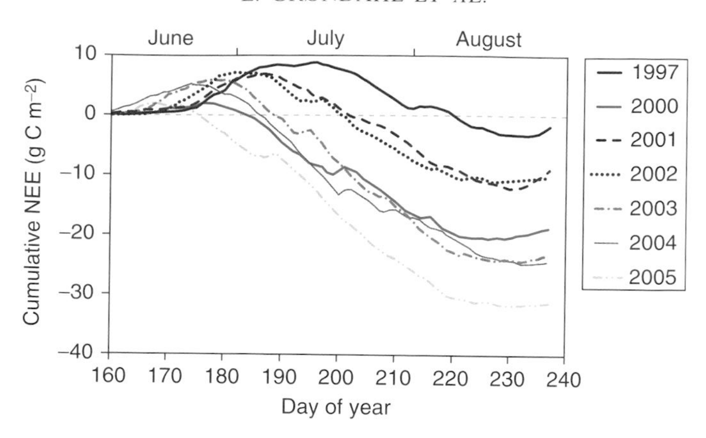

Figure 6 Growing season cumulative net ecosystem exchange (NEE) (g C m^{-2} June 9–August 28) during the 6 years of measurements at the heath site (modified from Groendahl, 2006).

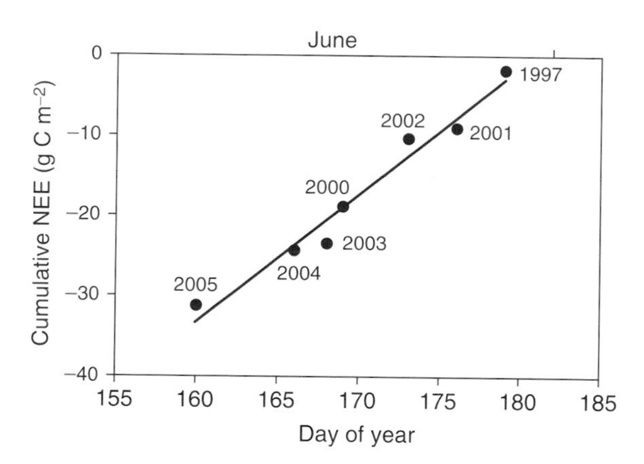

Figure 7 Cumulative net ecosystem exchange (NEE) from June 8 to August 26 for all 7 study years on the heath site plotted against day of snowmelt at the snow sensor. $R^2 = 0.97$, $p < 0.0001$ (modified from Groendahl *et al.*, 2007).

snowmelt (Figure 8). The inter-annual differences in growing season NEE at the heath was explained by variations in GEP, as the photosynthetic component was found to be affected positively by the temperature to a larger extend than the ER (Groendahl *et al.*, 2007). The main explanation for this might be found in the fact that some of the dominant plants in the ecosystems at Zackenberg are at their northern limit (Havstrom *et al.*, 1993) and therefore respond positively to the increment in summertime air temperature seen during the past decade (Groendahl *et al.*, 2007; Ellebjerg *et al.*, 2008, this

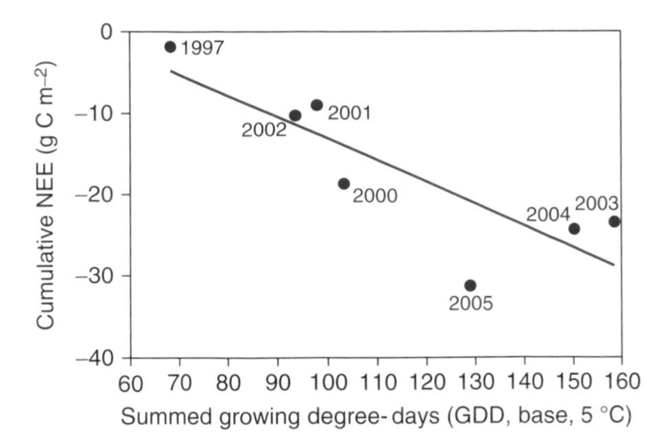

Figure 8 Cumulative net ecosystem exchange (NEE) from DOY 159 to 238 for all 7 study years on the heath site versus summed growing degree days. $R^2 = 0.71$, $p = 0.017$ (modified from Groendahl, 2006).

volume). This leads to the assumption that the ecosystem carbon balance at Zackenberg is strongly controlled by the timing of snowmelt and temperature during the growing season. This is in contrast to previous findings in low-arctic Alaska, where Vourlitis and Oechel (1999) found that respiration was the main explanatory component in inter-annual variation.

During the 4-year study, the fen site showed great variability in uptake rates during the growing season (Figure 2). Uptake rates were constrained by the timing of snowmelt, rubisco capacity (V_{max}) and LAI, resulting in measured growing season uptake rates ranging from 50 g C m^{-2} to 63 g C m^{-2} (Rennermalm *et al.*, 2005). In 1997, the maximum green LAI was not reached until mid August due to late snowmelt, which resulted in low uptake rates during the growing season. This is in contrast to findings in other wet ecosystems in the arctic and boreal regions, where studies have concluded that the inter-annual variability in CO_2 flux was caused by differences in soil moisture and precipitation (e.g., Shurpali *et al.*, 1995; Fahnestock *et al.*, 1999; Griffis *et al.*, 2000; Harazono *et al.*, 2003). The strongest sink activity was found during the dry and warm growing season in 1996, whereas the very late and cold growing season in 1999 had the weakest sink activity. The growing season in 1998 had an intermediate uptake rate of CO_2. This seems to contradict previous findings in the Arctic showing that wet summers are associated with strong CO_2 sink activity (e.g., Shurpali *et al.*, 1995; Lafleur *et al.*, 2003).

The observed uptake rates in the fen are supported by Joabsson and Christensen (2001) who found low uptake rates using static chambers in 1999 when snow melted very late, in contrast to 2000 when snow melted early. The measured exchange rates were comparable to the observations from the eddy covariance towers in the fen.

Comparing inter-annual variability in NEE is difficult due to the difference in length of growing season, as year round measurements are missing. The source/sink strength of the ecosystems may be affected by different climatic conditions during the growing season and the shoulder seasons. Year round measurements would give insight to the wintertime fluxes, whereas the shoulder seasons could be captured by simply prolonging the measuring period by 1 month. The contrasting controls on the NEE at the two sites demonstrate the variable sensitivity of arctic tundra ecosystems to environmental conditions as well as the potential sensitivity and variable response of the regional carbon budget to climate change. The fen site was strongly controlled by LAI (Rennermalm *et al.*, 2005), whereas timing of snowmelt and temperature during the growing season constrained the uptake rates at the heath site. Thus, if snow melts early, the growing season becomes longer. If temperature is high during this period, uptake rates are increased as seen in Figure 8.

B. Inter-Annual Variability in CH_4 Exchange

The CH_4 data from the valley does not allow a robust interpretation of the inter-annual variability in the flux, as measurements have only been conducted over period of three summers (1997, 1999 and 2000), with only the last 2 years over exactly the same plots (continuous fen). The averaged growing season CH_4 emission in 1999 and 2000 was 6.5 and 8.3 mg CH_4 m^{-2} h^{-1}, respectively, and the difference could be related to a late snowmelt in 1999 (Joabsson and Christensen, 2001), but more data from the wet parts of the valley will be needed to determine the effects of climatic variability on CH_4 emission at Zackenberg. A new programme has recently been set up to measure the CH_4 emission from the fen, and this will hopefully enable a more through study of the inter-annual variability, and clarify if environmental parameters such as temperature and water table, which drives the seasonal variation, can also explain the differences in CH_4 fluxes between years.

V. MANIPULATIONS

Various climate change-related manipulations have been performed at Zackenberg in which the exchange of GHG has been studied. Temperature manipulations have been performed with the free-air temperature increase (FATI) technique (Marchand *et al.*, 2004), observing the effects of temperature increment on the heath vegetation. Additionally, effects of increased nutrient availability and precipitation have been studied on an abrasion plateau (Illeris *et al.*, 2003). Shading, simulating increased cloud coverage, has been observed to decrease the photosynthetic responses (Joabsson and Christensen, 2001).

A. Water and Nutrient Manipulations

To investigate the effects of a future changed precipitation pattern in the High Arctic on CO_2 fluxes, water was added at two different experimental sites at Zackenberg. One site was a dry gravel terrace adjacent to the delta banks of Zackenbergelven with a sparse vegetation cover (bare soil covered almost 50% of the surface) belonging to the abrasion plateau vegetation type dominated by bog sedge *Kobresia myosuroides*, mountain avens *Dryas octopetala* × *integrifolia* and *S. arctica*, and with curly sedge *Carex rupestris* as a subdominant. Water was added weekly in an amount corresponding to 8 mm precipitation during the growing season with and without N and P in a factorial experimental design (Illeris *et al.*, 2003). The other site was a white arctic mountain heather *Cassiope tetragona*-dominated heath site, where three levels of watering corresponding to 8, 4 and 2 mm precipitation were added weekly during the growing season (L. Illeris, unpublished). At both sites CO_2 fluxes were measured weekly during the growing season by the closed chamber technique.

At the abrasion plateau, a significant effect of added water was observed in the ER in 1999 [$p = 0.0017$, repeated measurements ANOVA, whereas at the heath no effects on ER of added water were observed during two consecutive growing seasons (repeated measurements ANOVA, $p = 0.9766$ and $p = 0.9962$, respectively, for the years 1998 and 1999, Figure 9). Also, no effects on gross photosynthesis or NEP (which showed a small release of C in both growing seasons) were found at the heath (not shown). The observed difference in response of ER to added water between the two sites is probably caused by the difference in soil types between sites. The abrasion plateau contained 5.4% soil organic matter (SOM, 0–12 cm depth), while SOM at the heath site was 14.1%. Hence, water percolates more readily through the abrasion plateau soil than the heath soil, which causes the soil microbial community to be moisture limited in between precipitation events at the abrasion plateau, whereas this is not the case at the heath (Illeris *et al.*, 2003). Note that 1998 was an exceptionally wet year with a total of 93 mm precipitation during July and August, whereas the growing season in 1999 was slightly dryer than average (34 mm precipitation as opposed to 41 mm in average during the three summer months). Hence, neither high nor low amounts of precipitation during the experiment could explain that the heath did not respond to the water manipulations.

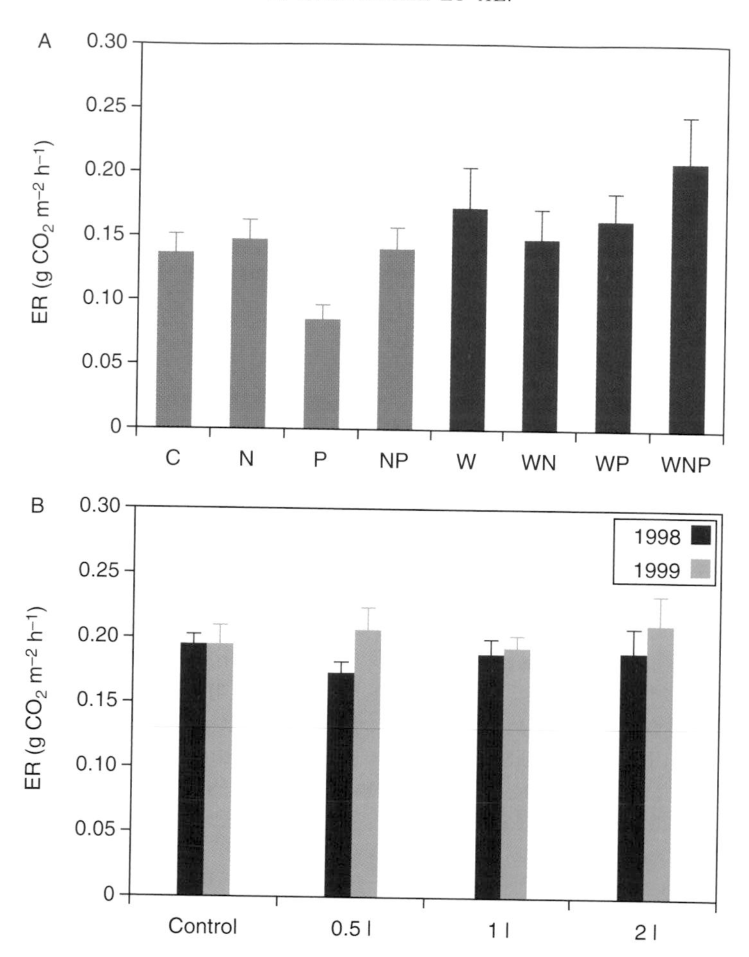

Figure 9 Average seasonal ecosystem respiration (ER) ± standard error (SE) after experimental manipulations at a dry abrasion plateau (A) and a *Cassiope. tetragona*-dominated heath (B) in Zackenbergdalen. At the abrasion plateau, treatments were C = control, N = nitrogen amendment, P = phosphorus amendment, W = water addition and combinations thereof. At the heath sites, treatments were control = no water addition, together with 0.5 l, 1 l and 2 l weekly additions of the mentioned amount of water per experimental plot of 0.25 m^2.

Also, the average seasonal ER in control plots was 42% higher at the heath than at the abrasion plateau in 1999. This is probably caused by higher above-ground plant productivity at the heath site. Illeris *et al.* (2003) found that the major part of ER at the abrasion plateau was due to root exudation of labile C, which decreased at low light levels due to low photosynthesis. Additionally, both the abrasion plateau and a drained fen area at Zackenberg showed no response of ER to addition of nutrients (Christensen *et al.*, 1998;

Illeris *et al.*, 2003), indicating that in some high-arctic ecosystems ER is not nutrient limited, in contrast to results from the sub-arctic (see e.g., Illeris *et al.*, 2004). The result of these experiments indicates that the carbon balance cannot be assessed without differentiating between vegetation types even within limited areas in the arctic landscape.

B. Modelling

Modelling reveals the impact of the variables controlling ecosystem carbon balance. To model the growing season fluxes at the fen site, Rennermalm *et al.* (2005) used the MOSES model to obtain hourly fluxes throughout the growing season. Forcing the model with a parameterisation of the LAI and meteorological variables, the NEE showed large inter-annual variability, with a seasonal uptake ranging from -53 g C m^{-2} in 1999 to -123 g C m^{-2} in 1996 (Figure 2). An objective of the modelling study was to identify the factors responsible for the variability, and it was found that the seasonal variability in LAI largely controlled the inter-annual variability in NEE (Rennermalm *et al.*, 2005).

Using a modelling approach combining carbon assimilation and soil respiration, Soegaard and Nordstroem (1999) evaluated the effects of a warmer climate on the growing seasonal NEE for the fen ecosystem. A clear indication of the ecosystem sensitivity to changes in temperature was found, with a peak uptake rate at + 2 °C above the present temperature level (H. Søgaard, unpublished data). Further increase in the temperature would cause the respiratory loss from the soil to increase faster than the CO_2 assimilation rate, resulting in a weaker summertime sink of the fen (Figure 10).

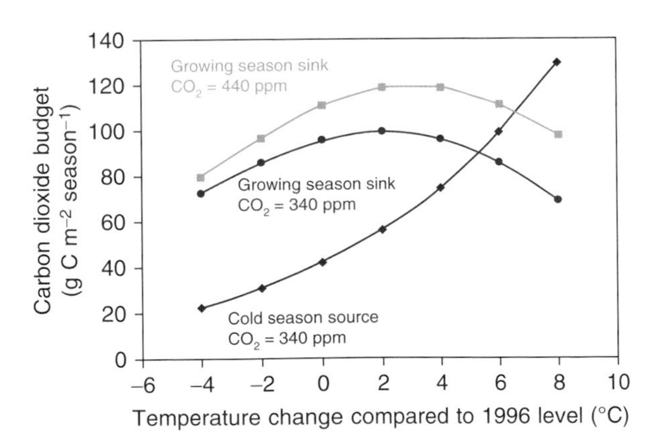

Figure 10 Net ecosystem CO_2 balance at different temperature scenarios compared to the 1996 level. Dotted line and closed circles, growing season total with an atmospheric content = 340 ppm; solid line with open circle, growing season total with an atmospheric CO_2 content = 440 ppm; solid line with square, net emission during cold season, CO_2 content = 340 ppm (modified from Soegaard *et al.*, 1999).

VI. GLOBAL WARMING POTENTIAL

A carbon or CO_2 budget alone does not illustrate the full effect of ecosystem gas exchange on the atmospheric radiative properties, due to different gases (here CO_2 and CH_4) having different greenhouse effect in the atmosphere. To account for this difference in global warming potential (GWP), the combined effect of GHG fluxes between the surface and atmosphere can be calculated. Only few directly comparable measurements of CO_2 and CH_4 fluxes are available from Zackenberg, but the eddy covariance measurements of both gases from the summer of 1997 allow a comparison of the carbon balance and GWP from the fen area (Figure 11).

Despite that the ecosystem showed a strong positive radiative forcing due to the CH_4 emission (Figure 12). As seen, the sink functioning is strongly offset if the CO_2 equivalent of CH_4 is calculated (using the 100-year time horizon) and added to the budget. Using a 100-year time horizon, the ecosystem is still a small sink of CO_2 equivalent at the end of the growing season. If the autumn and winter time fluxes were included, the annual total would probably add up to a source at that time perspective as well.

Although CH_4 only accounts for a minor part (5–10%) of the carbon exchanged between surface and atmosphere in the wet ecosystems in the valley, the greenhouse effect may be significant due to the GWP of CH_4.

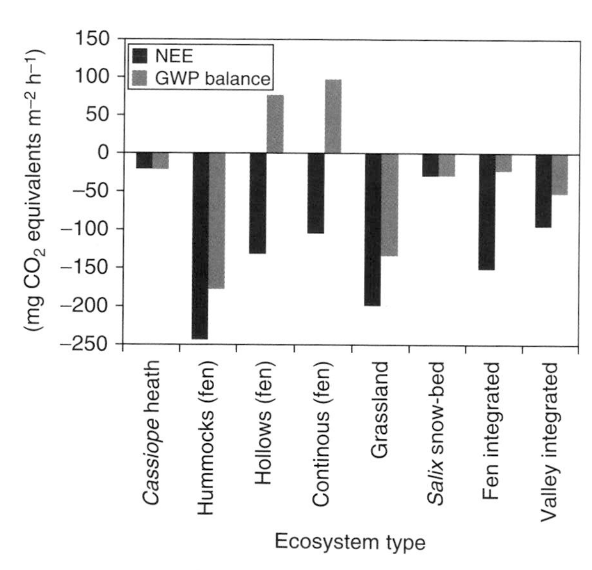

Figure 11 Net ecosystem exchange (NEE) and global warming potential (GWP) of the growing season for the different vegetation types found in Rylekærene 1997. GWP is based on daytime fluxes of CO_2 and CH_4 (data from Christensen *et al.*, 2000).

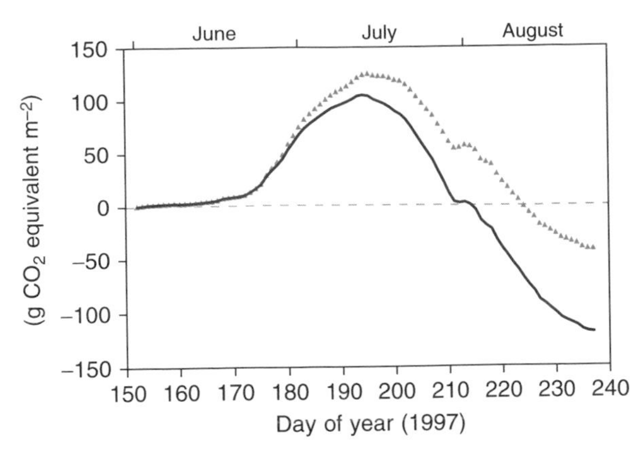

Figure 12 Accumulated carbon balance (solid line) and global warming potential (GWP) in Rylekærene during June 1–August 30, 1997. GWP for CH_4 is calculated as 23 g CO_2 equivalents according to IPCC (2001).

Many natural peat lands in Europe and Russia act as significant emitters of GHG because of high CH_4 emission, despite a substantial CO_2 uptake (Christensen and Friborg, 2004). The eddy covariance measurements from the fen in 1997 show that Rylekærene were a sink of CO_2 of 33 g C m^{-2} (121 g CO_2 m^{-2}) during the period from June to August and a source of CH_4 of 2.9 g C m^{-2} (3.8 g CH_4 m^{-2}), which in total gives a carbon uptake over the period of 30.1 g C m^{-2}. However, if the GWP of CH_4 (GWP = 23 times CO_2) is applied to evaluate the greenhouse effect on the atmosphere, the uptake is only 34 g CO_2 equivalents m^{-2}, $121 - (3.8 \times 23)$, as shown in Figure 12. To evaluate how the different ecosystem types act in a global warming context, daytime fluxes from Christensen *et al.* (2000) have been calculated into mean seasonal NEE and GWP as shown in Figure 11. Only the hollow part (wet- and low-lying parts) of the hummocky fen and the continuous fen acts as net sources of GHG during daytime, whereas the remaining types are all net sinks of GHG in the growing season. Integrated over the full area of the fen or the valley floor, these areas are sinks in terms of both CO_2 and other GHG during the summer season.

This interpretation should be taken with caution as measurements are based on summer values and during daytime only, but it gives an indication on differences between carbon balance, here as NEE, and GWP for especially the wet and moist ecosystem types of the valley. Furthermore, the interpretation of GWP for ecosystems is not unambiguous and depends highly on the time horizon considered. The GWP/NEE comparison in Figures 11 and 12 builds on the widely used 100-year time horizon, but model estimates have recently shown that a much longer time horizon should be considered for peat lands (Frolking *et al.*, 2006).

VII. ZACKENBERG IN A CIRCUMPOLAR CONTEXT

The consistent measurements of carbon fluxes from Zackenbergdalen are quite unique for this latitude and as such an exceptional site with respect to terrestrial ecosystem carbon flux monitoring. Compared to other parts of the circumpolar North, the high-arctic Zackenberg area seems productive with a consistent uptake of CO_2 during the summer period, when vegetation coverage is high (i.e., high LAI). The photosynthetic capacity is highest during mid-summer as a consequence of a high chlorophyll content in the leaves, highradiation levels and relatively high temperatures. The average uptake rates of 0.8–1.1 g C m^{-2} d^{-1} for the integrated valley area (Soegaard *et al.*, 2000; Groendahl, 2006) during summer season are comparable to summer season NEE measurements from moist ecosystems at the north slope of Alaska (69–71° N), which range from uptakes between 0.5 and 1.1 g C m^{-2} d^{-1} (Vourlitis *et al.*, 1999; Harazono *et al.*, 2003). The fen area in Rylekærene on average shows a higher NEE uptake (1.3–2.2 g C m^{-2} d^{-1}, Rennermalm *et al.*, 2005) than found in the moist systems of Northern Alaska. Also, most studies (e.g., Oechel *et al.*, 2000b; Kwon *et al.*, 2006) show that drier shrub systems in arctic Alaska have a small NEE emission, while the heath site at Zackenberg has a small but consistent uptake during summertime (Groendahl *et al.*, 2007). In arctic Alaska, there are indications that the soil moisture status dominates the carbon sequestration process. The moist sites have been net sinks for a period in the 1970s, which shifted to a source for a period in the 1980s. At present the ecosystems are net sinks for CO_2 (Oechel *et al.*, 2000a).

The studies performed at Zackenberg during the past 8 years show that there is a large inter-annual variability in summertime NEE observed both at the well-drained heath and at the wet fen sites. Compared to other artic sites this variability is not unique. The magnitude of the variability for the fen is similar to the observed variability in Alaska (Harazono *et al.*, 2003), whereas the variability at the *Cassiope* heath is larger compared to a polar desert site on Svalbard (Lloyd, 2001). Difference in soil composition and vegetation at the individual sites probably explain the great differences in NEE. Organic rich soils might show larger variability compared to the mineral soils, and the inter-annual variability in LAI has been shown to largely control NEE (Rennermalm *et al.*, 2005).

With respect to CH_4 flux measurements, the comparison between Zackenberg and other sites in the arctic is difficult due to the scarcity of data, and the only comparable data are from a few studies in Northern Alaska and a study in Northern Eurasia. As a daily average for the fen, both Christensen *et al.* (2000) and Friborg *et al.* (2000) found CH_4 emission rates around 30 mg C m^{-2} d^{-1}, which is very much in correspondence with findings from Alaska ranging from 20 to 70 mg C m^{-2} d^{-1} averaged over the summer season (Christensen, 1993; Harazono *et al.*, 2003). In a transect study from Northern Eurasia, Christensen *et al.* (1995) found emission rates from wet ecosystem types of 35 g C m^{-2} d^{-1} and lower rates (2 mg C m^{-2} d^{-1}) from mesic ecosystems. All sites reveal a large variation in emission rates both over the summer season and within apparently similar ecosystem types, which make the comparison of CH4 fluxes between Zackenberg and elsewhere less certain than the good match in rates indicates.

VIII. CONCLUSIONS AND FUTURE PROSPECTS

The research at Zackenberg during the past 10 years has provided a unique insight into the functioning of the high-arctic tundra. Measurements from several dominating ecosystem types have given solid information on the summertime CO_2 and CH_4 exchange between the terrestrial ecosystem and the atmosphere, both on the landscape scale and on plot scale. The heterogeneity in the landscape has been documented through chamber measurements, and these measurements have been validated and upscaled by comparison with eddy covariance measurements, which provide information on landscape level.

Results from Zackenberg indicate that

- Timing of snowmelt and temperature during the growing season along with the resulting LAI are the primary factors controlling the summertime CO_2 sink.
- The spatial variability in measured CO_2 fluxes is highly dependent on the plant composition and surface hydrology, but all vegetated surfaces act as sinks during summer.
- The CH_4 emission rates from the fen areas are comparable to what has been found at lower latitudes, and variations in the individual parts of the fen areas can be related to differences in vegetation composition and the level of the water table.

As also stated by ACIA (2005), the vegetation is likely to change under changing climatic conditions. The question whether the arctic tundra at Zackenberg will remain a sink of CO_2 and carbon will highly depend on the nature of future changes in climate. According to the predictions by the climate models, an increase in both precipitation and spring temperature is expected. On that background, it can be anticipated that fen areas might expand, if permafrost degradation increases as a result of a rise in temperature. For the fen areas, this could lead to an overall increase in GHG emission as a result of an enhanced CH_4 production, as found at lower latitudes (Johansson *et al.*, 2006). Also the grassland areas might be expected to turn into fen-like areas, whereas the heaths in a warmer climate might increase their growth (and LAI). It can further be speculated that the *Salix* snow-bed communities might increase in size, due to increased winter precipitation. If these predictions are true, the overall exchange of GHG from the different ecosystems types of the valley will change in a direction where the CO_2 uptake will increase during the summer season as a result of longer growing seasons. This will likely favour plant photosynthesis and increase NPP.

The overall annual GHG budget for the valley is presently unknown, as all CO_2 and CH_4 measurements are related to 3 months during the summer, and the net effects of future climatic change is therefore highly uncertain.

The next step is to incorporate the information from the variety of different ecosystem types presented here, which eventually would be capable of improving the physical models which have been used to estimate the carbon balance for the arctic biome. Such parameterisation have previously been performed for the Alaskan low-arctic sites, but have until now not been applied to the high-arctic Northeast Greenland.

ACKNOWLEDGMENTS

Monitoring data for this chapter were provided by the BioBasis and GeoBasis programmes, run by the National Environmental Research Institute, University of Aarhus, Denmark, and the Geographical Institute, University of Copenhagen, Denmark, and financed by the Danish Environmental Protection Agency. We also wish to acknowledge Anders Michelsen for a constructive review of the manuscript.

REFERENCES

ACIA (2005) In: *Arctic Climate Impact Assessment.* p. 1042. Cambridge University Press, New York.

Bay, C. (1998) In: *Vegetation Mapping of Zackenberg Valley, Northeast Greenland.* p. 29. Danish Polar Center and Botanical Museum, University of Copenhagen.

Callaghan, T.V., Bjorn, L.O., Chernov, Y., Chapin, T., Christensen, T.R., Huntley, B., Ims, R.A., Johansson, M., Jolly, D., Jonasson, S., Matveyeva, N. Panikov, N., *et al.* (2004) *Ambio* **33**, 459–468.

Christensen, T.R. (1993) *Biogeochemistry* **21**, 117–139.

Christensen, T.R. and Friborg, T. (2004) (http://gaia.agraria.unitus.it/ceuroghg/ReportSS4.pdf).

Christensen, T.R., Friborg, T., Sommerkorn, M., Kaplan, J., Illeris, L., Soegaard, H., Nordstroem, C. and Jonasson, S. (2000) *Global Biogeochem. Cycle* **14**, 701–713.

Christensen, T.R., Jonasson, S., Callaghan, T.V. and Havstrom, M. (1995) *J. Geophys. Res.-Atmos.* **100**, 21035–21045.

Christensen, T.R., Jonasson, S., Michelsen, A., Callaghan, T.V. and Havstrom, M. (1998) *J. Geophys. Res.-Atmos.* **103**, 29015–29021.

Fahnestock, J.T., Jones, M.H. and Welker, J.M. (1999) *Global Biogeochem. Cycle* **13**, 775–779.

Friborg, T., Christensen, T.R., Hansen, B.U., Nordstroem, C. and Soegaard, H. (2000) *Global Biogeochem. Cycle* **14**, 715–723.

Friborg, T., Soegaard, H., Christensen, T.R., Lloyd, C.R. and Panikov, N.S. (2003) *Geophys. Res. Lett.* **30**(3), 715–723.

Frolking, S., Roulet, N.T. and Fuglestved, J. (2006) *J. Geophys. Res.* **111,** G1, Art. No. G01008.

Griffis, T.J., Rouse, W.R. and Waddington, J.M. (2000) *Global Biogeochem. Cycle* **14**, 1109–1121.

Groendahl, L. (2006) Carbon dioxide exchange in the High Arctic—examples from terrestrial ecosystems. PhD thesis,University of Copenhagen.

Groendahl, L., Friborg, T. and Soegaard, H. (2007) *Theor. Appl. Climatol.* **88**, 111–125.

Grogan, P. and Chapin, F.S. (2000) *Oecologia* **125**, 512–520.

Harazono, Y., Mano, M., Miyata, A., Zulueta, R.C. and Oechel, W.C. (2003) *Tellus B* **55**, 215–231.

Havstrom, M., Callaghan, T.V. and Jonasson, S. (1993) *Oikos* **66**, 389–402.

Illeris, L., Michelsen, A. and Jonasson, S. (2003) *Biogeochemistry* **65**, 15–29.

Illeris, L., Konig, S.M., Grogan, P., Jonasson, S., Michelsen, A. and Ro-Poulsen, H. (2004) *Arctic Alpine Res.* **36**, 456–463.

IPCC (2001) In: *Climate Change 2001: The Physical Science Basis* (Ed. by J.T. Houghton, Y. Ding, D.J. Griggs, M. Noguer, P.J. van der Linden and D. Xiaosu). Cambridge University Press, Cambridge, United Kingdom and New York, NY, USA, 944 pp.

IPCC (2007) In: *Climate Change 2007: The Physical Science Basis.* (Ed. by S. Solomon, D. Qin, M. Manning, Z. Chen, M. Marquis, K.B. Averyt, M. Tignor and H.L. Miller). Cambridge University Press, Cambridge, United Kingdom and New York, NY, USA, 996 pp.

Joabsson, A. and Christensen, T.R. (2001) *Global Change Biol.* **7**, 919–932.

Johansson, T., Malmer, N., Crill, P.M., Friborg, T., Akerman, H.J., Mastepanov, M. and Christensen, T.R. (2006) *Global Change Biol.* **12**, 2352–2369.

Kutzbach, L., Wagner, D. and Pfeiffer, E.M. (2004) *Biogeochemistry* **69**, 341–362.

Kwon, H.J., Oechel, W.C., Zulueta, R.C. and Hastings, S.J. (2006) *J. Geophys. Res.-Biogeosci.* **111,** (G3): Art. No. G03014.

Lafleur, P.M., Roulet, N.T., Bubier, J.L., Frolking, S. and Moore, T.R. (2003) *Global Biochem. Cycle* **17**, 5-1–5-14.

Lloyd, C.R. (2001) *Theor. Appl. Climatol.* **70**, 167–182.

Marchand, F.L., Nijs, I., de Boeck, H.J., Kockelbergh, F., Mertens, S. and Beyens, L. (2004) *Arctic Alpine Res.* **36**, 298–307.

Oechel, W.C., Hastings, S.J., Vourlitis, G., Jenkins, M., Riechers, G. and Grulke, N. (1993) *Nature* **361**, 520–523.

Oechel, W.C., Vourlitis, G.L., Hastings, S.J., Zulueta, R.C., Hinzman, L. and Kane, D. (2000a) *Nature* **406**, 978–981.

Oechel, W.C., Vourlitis, G.L., Verfaillie, J., Crawford, T., Brooks, S., Dumas, E., Hope, A., Stow, D., Boynton, B., Nosov, V. and Zulueta, R. (2000b) *Global Change Biol.* **6**, 160–173.

Post, W.M., Emanuel, W.R., Zinke, P.J. and Stangenberger, A.G. (1982) *Nature* **298**, 156–159.

Rennermalm, A.K., Soegaard, H. and Nordstroem, C. (2005) *Arctic Alpine Res.* **37**, 545–556.

Shurpali, N.J., Verma, S.B., Kim, J. and Arkebauer, T.J. (1995) *J. Geophys. Res.-Atmos.* **100**, 14319–14326.
Soegaard, H. and Nordstroem, C. (1999) *Global Change Biol.* **5**, 547–562.
Soegaard, H., Nordstroem, C., Friborg, T., Hansen, B.U., Christensen, T.R. and Bay, C. (2000) *Global Biogeochem. Cycle* **14**, 725–744.
Ström, L., Ekberg, A. and Christensen, T.R. (2003) *Global Change Biol.* **9**, 1185–1192.
Vourlitis, G.L. and Oechel, W.C. (1999) *Ecology* **80**, 686–701.

Zackenberg in a Circumpolar Context

MADS C. FORCHHAMMER, TORBEN R. CHRISTENSEN,
BIRGER U. HANSEN, MIKKEL P. TAMSTORF,
NIELS M. SCHMIDT, TOKE T. HØYE, JACOB NABE-NIELSEN,
MORTEN RASCH, HANS MELTOFTE,
BO ELBERLING AND ERIC POST

SUMMARY

Throughout the Northern Hemisphere, changes in local and regional climate conditions are coupled to the recurring and persistent large-scale patterns of pressure and circulation anomalies spanning vast geographical areas, the so-called teleconnection patterns. Indeed, the atmospheric fluctuations described by the North Atlantic Oscillation (NAO) are closely associated with the last four decades of inter-annual variability in local snow and ice conditions observed in the Arctic. Since the NAO has also been connected

ADVANCES IN ECOLOGICAL RESEARCH VOL. 40 0065-2504/08 $35.00
© 2008 Elsevier Ltd. All rights reserved
DOI: 10.1016/S0065-2504(07)00021-9

with changes in the global climate, the behaviour of species, communities and other ecosystem elements at Zackenberg in relation to the NAO enables us to view these in circumpolar and global contexts.

Large-scale systems like the NAO constitute the link between the global change and local climate variability to which ecosystem components respond. Here, we place selected ecosystem elements from the monitoring programme Zackenberg Basic presented in previous chapters in a circumpolar context related to NAO-mediated climatic changes. We begin by linking the local variability in winter weather conditions at Zackenberg to fluctuations in the NAO. We then proceed by linking the observed intra- and inter-annual behaviour of selected ecosystem elements to changes in the NAO. The functional ecosystem characteristics in focus are landscape gas exchange dynamics phenological patterns at different trophic levels, consumer–resource dynamics and community stability. The influence of the NAO is presented and discussed in a broader perspective based on information obtained from other arctic localities.

The relation between the NAO and the Zackenberg winter weather is non-linear, reflecting differential effects of the NAO as the index moves between high and low phases. The inverse hyperbolic relationship found between the NAO and the amount of winter snow was also evident as non-linear response in organisms and systems to inter-annual changes in the NAO. Responses investigated included growth and reproduction in plants and animals, population dynamics and synchrony, inter-trophic interactions and community stability together with system feedback dynamics.

I. INTRODUCTION

The implementation of the long-term ecosystem monitoring programme at Zackenberg (Zackenberg Basic) 10 years ago marked a paradigm shift in our approach to observe and record how changes in climate influence the environment. By simultaneously recording concurrent changes in the physical, biological and feedback properties of a single ecosystem, Zackenberg Basic provides an unparalleled system-level approach (Figure 1) to describe and analyse the effects of climate change integrating within-year seasonal variability as well as inter-annual long-term changes. The previous chapters in this book portray this unique multi-dimensional approach for a range of physical and biological components in the high-arctic ecosystem at Zackenberg.

Notwithstanding the uniqueness of Zackenberg Basic, ever since Charles Darwin (1859) observed that the "conditions of life may be said, not only to cause variability, but likewise to include natural selection," we have been aware that abiotic factors constitute a central component in forming the evolution and life cycle of organisms. Consequently, many previous studies have embraced, conceptually as well as empirically, the influence of weather

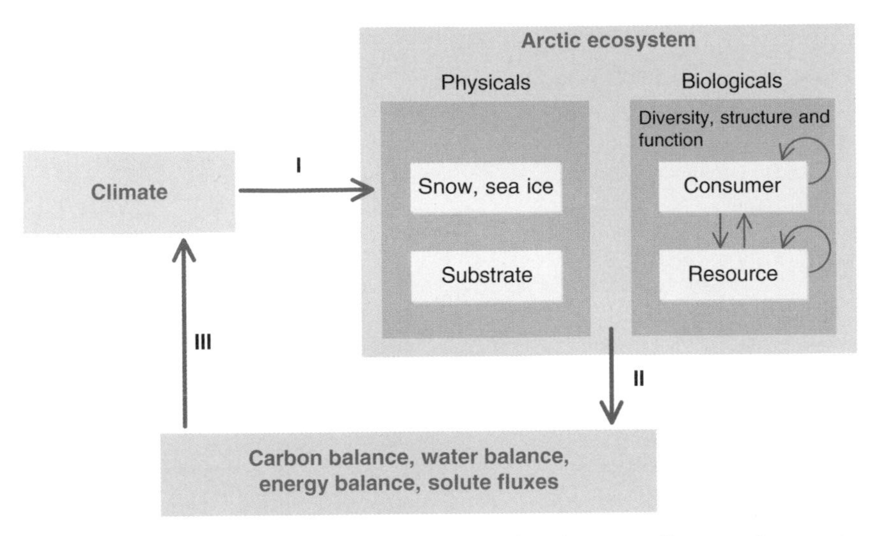

Figure 1 Conceptual visualisation of the interactions between climate and ecosystem responses (arrows I, II) and ecosystem feedback (arrow III).

on a wide range of species. Although many of these focused primarily on the effects of experimentally altered abiotic conditions (e.g., Chapin *et al.*, 1992), the approach of integrating retrospective data of the long-term fluctuations of species was pioneered by the studies of a handful of scientists such as the seminal work by Charles Elton (1958) and, in the arctic regions, by Christian Vibe (1967). In the early 1990s, there was a philosophical and methodological shift in the analytical approach to the study of climate effects, where the focus on local weather parameters changed to the integration of large-scale climate variability such as those described by the El Niño Southern Oscillation, the North Atlantic Oscillation (NAO) and the North Pacific Oscillation (Forchhammer and Post, 2004, and references therein). There are several important features accounted for through the integration of large-scale climate systems (Stenseth *et al.*, 2003; Forchhammer and Post, 2004; see below) of which, the large-scale perspective of these systems enables comparisons of climatic influence on species in physically widely separated systems and, hence, an interpretation in a more global context than would be possible with the use of local weather variables only.

In this chapter, we focus on the single most important atmospheric phenomenon in the Northern Hemisphere, the NAO (Hurrell, 1996; Hurrell *et al.*, 2003; Stendel *et al.*, 2008, this volume). The chapter is divided into two conceptually different parts. The first describes the conceptual and structural perspectives of the monitoring programme Zackenberg Basic in relation to the recent compilation and synthesis of current knowledge by the Arctic Climate Impact Assessment (ACIA, 2005). The second part relates the dynamics of the NAO to the performance of a range of hierarchically selected components of the ecosystem at Zackenberg and how these are reflected in species and systems elsewhere in the Arctic. In particular, we focus on how variations in the NAO are portrayed in species-specific responses, community-level responses and ultimately, ecosystem feedback dynamics.

II. INTEGRATED ECOSYSTEM MONITORING

A. Conceptualising Zackenberg Basic: Climate Effects and Feedback

The specific responses to climate change addressed in the previous chapters may be summarised as a two-way process (Figure 1). First, any change in climate, such as increased variability in large-scale atmospheric–ocean systems or the extension of sea ice cover, will cause changes in the physical characteristics of ecosystems like snow-cover (Figure 1, arrow I). For example, the atmospheric fluctuations described by the NAO/Arctic Oscillation (AO) are closely associated with the last 35 years of inter-annual variability in snow onset, snowmelt and number of snow-free days observed in the Northern Hemisphere (Bamzai, 2003), including Northeast Greenland (Hinkler, 2005; Hansen *et al.*, 2008, this volume; Stendel *et al.*, 2008, this volume). Any climate-mediated changes in the physical characteristics will, in turn, affect the functioning of organisms and their interactions in the system. These effects may be divided into direct and indirect effects (Forchhammer and Post, 2004). Direct climatic effects on the organisms themselves are easily observed with no time lags. For example, from the monitoring at Zackenberg, we have learned that even small annual changes in the amount and extension of snow and sea ice have dramatic influences on, for example, seasonal growth, distribution and production of terrestrial vegetation as well as marine and freshwater plankton the following summer (Christoffersen and Jeppesen, 2002; Mølgaard *et al.*, 2002; Tamstorf and Bay, 2002).

Indirect climatic effects, on the contrary, involve multi-organism interactions often between several trophic levels and are therefore more difficult to monitor using a single-organism monitoring approach alone (Forchhammer and Post, 2004; Forchhammer *et al.*, 2008, this volume). This has been recognised in several temporal ecosystem communities, including Zackenberg. For example, we know that following winters with much snow and prolonged ice cover on lakes, the seasonal production of freshwater zooplankton decreases dramatically in response to the low abundance of their food, phytoplankton, and not ice cover per se (Christoffersen and Jeppesen, 2002).

The second aspect of the two-way interaction between climate and ecosystems is the reciprocal feedback from ecosystem to climate through changes in, for example, carbon, water and energy balances (Figure 1, arrows II and III). Documented from the work at Zackenberg and other studies, we know that changes in the physical characteristics of ecosystems are highly correlated with changes in, for example, the annual flux of carbon from system to atmosphere (e.g., Nordström *et al.*, 2001; Grøndahl *et al.*, 2007).

The monitoring programmes at Zackenberg have been purposely constructed to enable a complete spatial coverage of the general climate–ecosystems interactions portrayed in Figure 1. Indeed, the scientific structure of the monitoring programmes at Zackenberg embraces a total of 14 central themes covering the climatic (climate), physical [snow, soil, ice, sea ice, lakes, hydrology, oceanography and ultraviolet (UV) radiation], as well as biological (soil, vegetation, UV radiation effects, gas flux, lakes, arthropods, birds and mammals) ecosystem compartments (Figure 2).

B. Zackenberg Basic in a Circumpolar Context: Recommendations by ACIA

As exemplified by the previous chapters in this book, persistent climatic changes are likely to cause rather complex and, in many cases, unexpected indirect changes in arctic ecosystems confirming the conceptually visualised dynamics suggested in previous review studies (Forchhammer, 2001; Stenseth *et al.*, 2002; Walther *et al.*, 2002; Forchhammer and Post, 2004). Indeed, our ability to understand, monitor, evaluate and model the consequences of

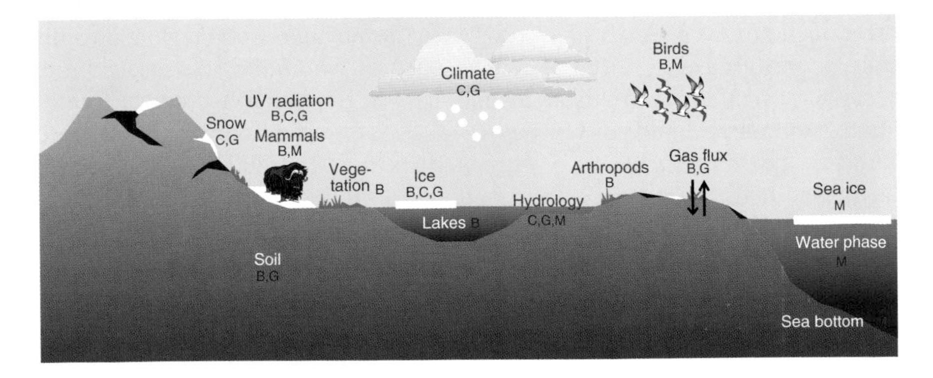

Figure 2 Schematic landscape representation of the major scientific themes in Zackenberg Basic and how these are related to the four basis monitoring programmes: ClimateBasis (C), GeoBasis (G), BioBasis (B) and MarineBasis (M). Under each theme title the capital letters for the basis programmes involved are given.

such changes requires a comprehensive synthesis of current and comprehensive knowledge of information on the observed and projected climatic effects across the Arctic.

Recently, this challenge was taken up by the ACIA, which has provided us with an unparalleled and comprehensive assessment of climate impacts based on previously observed concomitant changes in climate, terrestrial, freshwater and marine systems throughout the Arctic (ACIA, 2005). Founded upon the large amount of information provided by the assessment, ACIA has specified a range of recommendations pivotal for future climate change research in the Arctic (Table 1). These together with those proposed by the Arctic Monitoring and Assessment Programme (AMAP) in their Climate Change Effects Monitoring Programme (AMAP, 2000) and in their follow-up of ACIA (AMAP, 2005) and by the International Conference on Arctic Research Planning II (ICARP II) (Bengtsson, 2005; Callaghan 2005; Prowse *et al.*, 2005) inherently form the objective core of the monitoring at Zackenberg. Indeed, most of the monitoring activities conducted under the basis monitoring programmes at Zackenberg (Rasch *et al.*, 2003) dovetail with the recommendations issued by ACIA embracing the long-term monitoring of cryosphere and hydrology, arctic tundra systems, freshwater systems, marine systems and UV radiation (Table 1).

The recommendations provided by ACIA are formulated in a general context; that is, actions to be taken in future climate change monitoring are not specifically addressed to be carried out within a single ecosystem. Indeed, inherent to ACIA's (ACIA 2005) notion of the need for increased spatial coverage of climate impacts, actions to be taken may be performed at different locations on selected organisms or communities without specifically monitoring the entire system in which these are embedded. In contrast and indeed as a unique additional feature, Zackenberg Basic addresses ACIA recommendations within a single ecosystem in the high-arctic region. Specifically, monitoring is performed on most physical and biological levels of the ecosystem, so that all observed changes can be functionally connected and, hence, summarised and conveyed as a holistic ecosystem response to climate changes. The current overview presented in this book has described state-of-the-art of the monitoring programme and the results it has produced so far; at the same time, to improve its potential strength, the chapter is used to evaluate the programme and tune it to become even better integrated and targeted to grasp future climate change effects in the short and longer terms (Callaghan *et al.*, 2007).

In the remainder of this chapter we proceed by focusing on how central biological responses to climate changes at Zackenberg relate to similar responses throughout the Arctic. This is not trivial, as one of the key objectives of Zackenberg Basic is to provide a conceptual foundation for how biological and physical elements in an ecosystem react to climate,

Table 1 Recommendations by the Arctic Climate Impact Assessment (ACIA) of relevance to the monitoring programmes at Zackenberg and their relation to the specific basis programmes

ACIA recommendations	Programme	Action
Cryosphere and hydrology		
Sea ice: Fine resolution studies of sea ice cover in coastal waters	M	Satellite and photo surveillance
Sea ice: Seasonal, inter-annual, and interdecadal measurements of sea surface albedo		None
Snow-cover: *In situ* measurements of snow water equivalents in high latitude areas	CG	Sonic snow depth and manual density measurements together with daily photo surveillance of snow extent year round
Snow-cover: Measurements of snow albedo over northern terrestrial regions	CG	Point measurements
Snow-cover: Establishment of models to simulate snowmelt process	BCG	Point and spatial monitoring through manual, camera and satellite surveillance
Permafrost: Long-term field data on permafrost–climate interactions and on permafrost–hydrology interactions	G	CALM
River and lake ice: Improve understanding of hydrological and meteorological control of freeze-up and break-up	BCG	Manual hydrological monitoring and camera surveillance
Freshwater discharge: Increase the network of gauge stations for monitoring discharge rates	C	Sonic and manual hydrological monitoring
Freshwater discharge: Better estimation of subsurface flow		None
Arctic tundra and polar desert ecosystems		
Biodiversity changes: Monitor currently widespread species that are likely to decline under climate change	B	Systematic monitoring of species from many taxa
Relocation of species: Measure and project rates of species migration	B	Systematic monitoring of species from many taxa
Vegetation zone redistribution: Improve information about current boundaries of vegetation zones	B	Transect and NDVI monitoring (manual, cameras, satellite)
Carbon sinks and sources: Long-term, annual C monitoring throughout the Arctic	G	Summer and "shoulder" periods (CO_2 and CH_4)

(*continued*)

Table 1 (*continued*)

ACIA recommendations	Programme	Action
Carbon sinks and sources: Models capable of scaling ecosystem processes from plot experiments to landscape scale	G	Spatial modelling
Carbon sinks and sources: Develop observatories to relate disturbance to C dynamics		Not relevant
Carbon sinks and sources: Combine ecosystem carbon flux estimates with C flux from thawing permafrost	G	Summer season CO_2 monitoring over CALM plot
Ultraviolet-B (UV-B) radiation and CO_2 impacts: Long-term impact on ecosystem of increased CO_2 concentrations and UV-B radiation	BC	UV-B and CO_2 monitoring
Increasing and extending the use of indigenous knowledge: Expand use of indigenous knowledge		Not relevant
Monitoring: More networks of standardised, long-term monitoring are required	BCGM	Four comprehensive programmes in operation
Monitoring: Integrated cross-disciplinary monitoring of covarying environmental variables	BCGM	This concept is the fundament for Zackenberg Basic
Monitoring: Long-term and year-round eddy covariance sites and other long-term flux sites for C flux measurements	G	Summer CO_2 by eddy correlation, CH_4 by chamber measurements
Long-term and year-round approach: Long-term observations are required	BCG	ClimateBasis year round, Bio- and GeoBasis seasonal
Long-term and year-round approach: Year-round observations are necessary to understand importance of winter processes	CG	ClimateBasis year round, Bio- and GeoBasis seasonal
Freshwater ecosystems and fisheries		
Freshwater ecosystems: Increase knowledge on long-term changes in physical, chemical and biological attributes	BCG	Physical, chemical and biological monitoring
Freshwater ecosystems: Establish integrated, comprehensive monitoring programmes at regional, national and circumpolar scales	BCG	International co-operation

(*continued*)

Table 1 (*continued*)

ACIA recommendations	Programme	Action
Freshwater ecosystems: Standardise international approach for monitoring	BCG	International co-operation
Freshwater ecosystems: Improve knowledge of synergistic impacts of climate on aquatic organisms	BCG	Possible with existing data
Freshwater ecosystems: Increase understanding of cumulative impacts of multiple environmental stressors on fresh water ecosystems	B	Zackenberg Basic addresses undisturbed ecosystems
Freshwater ecosystems: Increase knowledge of effects of UV radiation–temperature interactions on aquatic biota		None
Freshwater ecosystems: Increase knowledge of linkages between structure and function of aquatic biota	B	Possible with existing data
Freshwater ecosystems: Increase knowledge on coupling among physical/chemical and biotic processes	B	Possible with existing data
Marine systems		
Observational techniques: Increase application of recently developed techniques	M	State-of-the-art equipment and techniques in use
Surveying and monitoring: Undertake surveys that are poorly mapped and whose resident biota has not been surveyed	M	No investigations like these before the establishment of Zackenberg Basic
Surveying and monitoring: Continue and expand existing monitoring programmes	M	A permanent challenge
Surveying and monitoring: Evaluate monitoring data through data analysis and modelling	M	A permanent effort
Data analysis and reconstruction: Reconstruct twentieth-century forcing field		None
Data analysis and reconstruction: Establish database with all available physical and biological data	M	Included—data can easily be provided to other databases
Data analysis and reconstruction: Recover past physical and biological data	M	Included

(*continued*)

Table 1 (*continued*)

ACIA recommendations	Programme	Action
Data analysis and reconstruction: Past climate events to understand physical and biological responses to climate forcing	M	Included
Field programmes: Undertake field studies to quantify climate-related processes	M	Major purpose
Modelling: Develop reliable regional models	M	Included
Approaches: Prioritize ecosystem-based research by integrating multiple ecosystem components in models concerning climate effects	M	The concept of Zackenberg Basic
Ozone and UV radiation		
UV radiation: Address the impact of increased UV irradiance	BCG	Included

B, BioBasis; *G, GeoBasis*; *C, ClimateBasis*; *M, MarineBasis.*

applicable in any system in the Northern Hemisphere (Rasch *et al.*, 2003; Forchhammer *et al.*, 2007). Specifically, we may ask whether the responses observed at Zackenberg are unique or portray any mechanisms common for other systems as well. Indeed, the accumulation of biological and physical time series has recently provided us with the opportunity not only to specifically test whether species do in fact respond to changes in global climate but also, equally important, to what extent such responses vary in a large-scale spatial context (Forchhammer, 2001; Stenseth *et al.*, 2002; Walther *et al.*, 2002; Forchhammer and Post, 2004). Previously, results emerging from species-specific or community-level responses to local weather conditions have often extrapolated to draw conclusions as well as predictions from General Circulation Models (e.g., Oechel *et al.*, 1997). Here, we take a different approach by integrating responses to large-scale climatic variability expressed through the inter-annual variation of climatic indices, a method rooted in an early paper by Turchin *et al.* (1991) in which the stochastic error term of species responses was modified to incorporate climatic conditions (Forchhammer and Post, 2004; Forchhammer *et al.*, 2008, this volume). Before we look at how different organisms at Zackenberg and elsewhere in the Arctic respond to variations in large-scale climate, we shortly introduce the properties of such large-scale climate indices with a special focus on the ecological properties.

III. INDICES OF LARGE-SCALE CLIMATE

A. The NAO and Its Non-Linear Properties

Extensive spatio-temporal changes in our climate are reflected in the recurring and persistent, large-scale inter-annual dynamics of the pressure and circulation anomalies covering extensive geographical areas. These so-called teleconnection patterns express statistically strong relationships with weather conditions across space and time (Stenseth *et al.*, 2003; Forchhammer and Post, 2004). For example, the winter of 1995/1996 was unusually dry in northern Europe and Russia, whereas North America and West Greenland experienced an unusually large amount of snow the same winter. These regional weather conditions could be related to a strong negative phase of the NAO (Kushnir, 1999) (Figure 3).

Although the NAO is the most prominent teleconnection pattern in the Northern Hemisphere (Barnston and Livezy, 1987) and has, during its latest transition from a low to high phase during 1965–1995, been linked to the raise of 0.21 °C in mean temperature in the Northern Hemisphere (Hurrell, 1996), a range of other important large-scale teleconnections have been identified worldwide, including the well-known El Niño Southern Oscillation

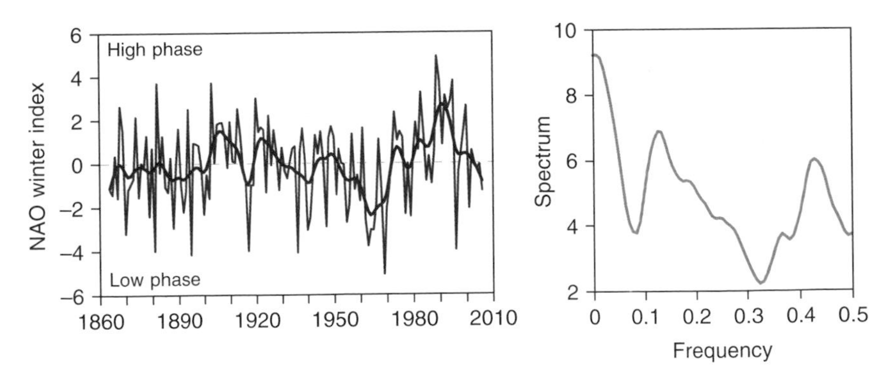

Figure 3 Long-term (1864–2006) variability in the NAO winter index; for definitions see Hurrell (1995) and Stendel *et al.* (2008, this volume). The blue line shows the time series of annual indices, whereas the red line gives the trend of the times series calculated as the smoothed spline of the raw data [i.e., non-parametric cubic B-spline with 30 degrees of freedom; see Hastie and Tibshirani (1990) for details]. Right figure: Spectral analysis of the NAO time series using the smoothed periodogram function in S-plus (spec.pgram). A smoothing window vector of c(7,7) was defined; see Venables and Ripley (1994) for details. The spectral decomposition of the NAO time series displays two peaks in the frequency spectrum at 0.43 and 0.13, respectively. This suggests that variability in the NAO dynamics may be characterised by two sets of multi-annual fluctuations, one of 2–3 years (1/0.43) and one of 7–8 years (1/0.13).

(Stenseth *et al.*, 2003). Here, we focus on the inter-annual variation in the NAO during winter (December–March) when it is most pronounced. The specific definition of the winter index and the meteorological rationale behind the NAO as well as its relation to the Northern Hemisphere Annual Mode (NAM)/AO are given in detail by Stendel *et al.* (2008, this volume). Although the winter NAO displays high inter-annual variations, it may be characterised by alternating periods of low and high phases (Figure 3). Specifically, the temporal dynamics of the winter NAO have two distinct sets of multi-annual fluctuations of 2–3 years and 7–8 years (Figure 3).

The negative and positive phases of the NAO (Figure 3) are important as they often characterise contrasting influences of the NAO on winter weather conditions (Hansen *et al.*, 2008, this volume; Hinkler *et al.*, 2008, this volume). For example, during the low NAO phase, 1965–1983, there was a significant positive linear correlation between winter precipitation and the NAO in Northeast Greenland (Figure 4A; Hinkler *et al.*, 2008, this volume). In contrast, during the high NAO in 1979–1997 (Figure 3), the association between winter precipitation and the NAO became negative (Figure 4B). Note also that during high-NAO periods, the differential effect of the NAO on winter precipitation across the North Atlantic is deepened (Figure 4B). Although long-term linear associations between the NAO and winter precipitation in Northeast Greenland are found to be non-significant (Hinkler, 2005; Hansen *et al.*, 2008, this volume), the contrasting NAO–precipitation relations over time, do suggest a non-linear association between winter weather conditions and the NAO. Indeed, over a 10-year period at Zackenberg, the snow-cover present on June 10 displayed a significant hyperbolic ∩ shaped relation for high NAO indices (Figure 5A). This pattern was corroborated for the inter-annual variations in maximum snow depth (Figure 5B) as well as the date for which the snow is melted to below 0.1 m (Figure 5C) recorded at Zackenberg over the same period. Given the importance of inter-annual variations in winter precipitation and its melt-off found for the ecosystem at Zackenberg reported in previous chapters, we may expect to find similar non-linear relations between NAO and the functioning of organisms, their population dynamics and, potentially, biological interactions in the communities in which they are embedded (Forchhammer, 2001; Stenseth *et al.*, 2002).

The latest increasing phase of the NAO (1965–1995; Figure 3; Stendel *et al.*, 2008, this volume) was associated with a dichotomous response in winter (December through March) weather conditions across the Arctic. Specifically, whereas Northeast Greenland, northern Scandinavia and Russia experienced on average warmer and wetter winters, West and South Greenland, eastern Canada and western Alaska had on average colder and

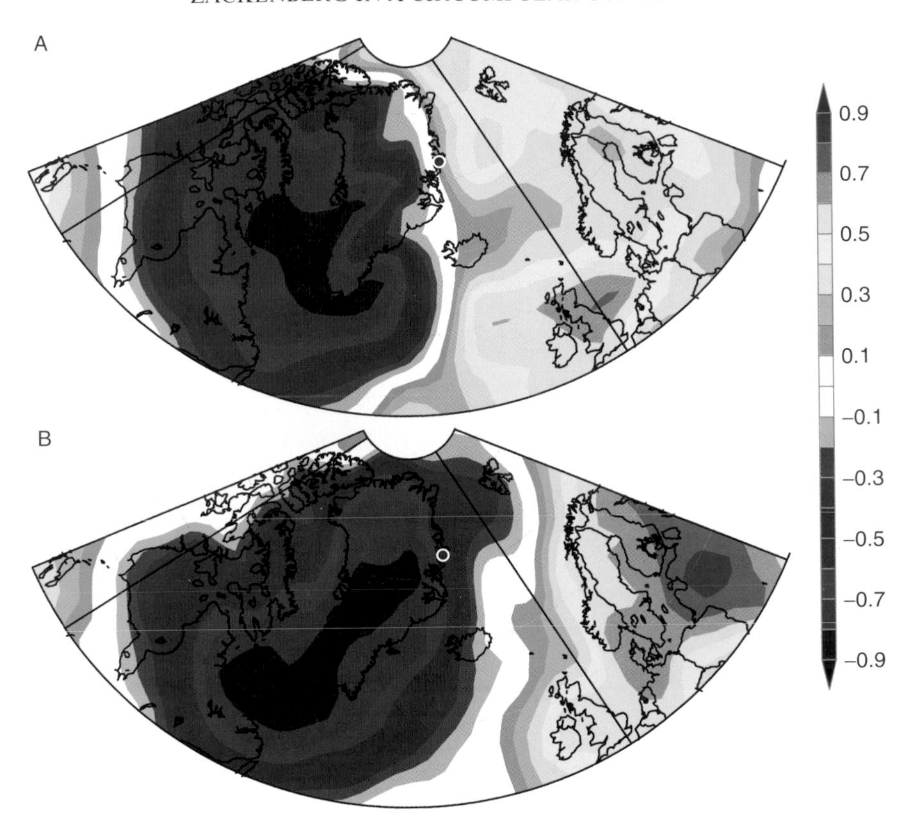

Figure 4 Map of the northern Atlantic region with Greenland centrally placed. Graded colouring indicates the degree of seasonal linear correlation (−1 to 1) between the NAO and winter (December–March) precipitation during (A) 1965–1983 (low NAO phase, mean ± SEM: −0.45 ± 0.45) and (B) 1979–1997 (high NAO phase: 1.32 ± 0.50). The horizontal bar relates colouring and correlation coefficients. Red dot gives the location of Zackenberg on the east coast of Greenland. Analyses were performed by the NOAA-CIRES/Climate Diagnostic Center (http://www.cdc.noaa.gov).

drier winters (Figure 6A, B). As pointed out by Stendel *et al.* (2008, this volume), many of the models discussed in the latest report from the Intergovernmental Panel on Climate Change (IPCC) suggest an increase in the positive phase of the NAO and associated teleconnection patterns (Christensen *et al.*, 2007). Although the direct local effects of the NAO vary considerably, indirect effects through changes in the extension of sea ice may become increasingly important (Stendel *et al.*, 2008, this volume). Hence, we may expect that future NAO-related changes in local winter weather conditions will be as depicted in Figure 6.

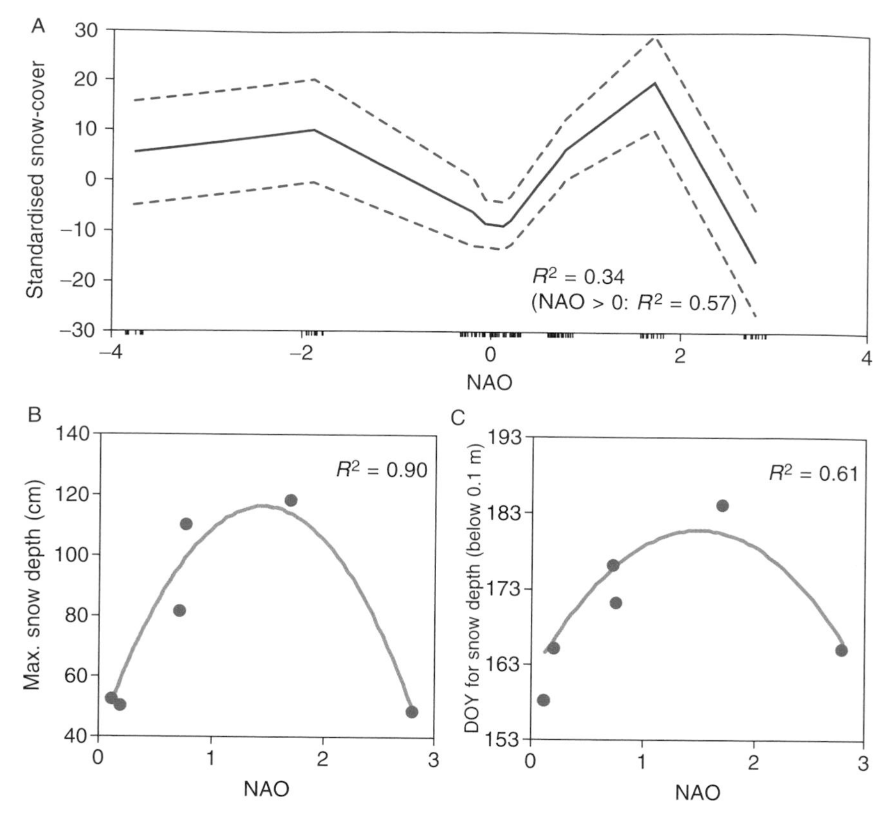

Figure 5 (A) Inter-annual variation in snow-cover on June 10 (standardised) in relation to the NAO winter (December through March) index. The solid line is a non-linear generalised additive model (GAM) with 95% CIs as dotted lines; the rugplot along the x-axis marks the "observed" x values (Venables and Ripley, 1994). (B) Maximum recorded winter snow depth and (C) day of the year for snow depth below 0.1 m at Zackenberg June 2–July 17 in relation to NAO indices above 0. All R^2 vales are significant ($p < 0.05$). Snow data from Sigsgaard *et al.* (2006).

B. Ecological Perspectives of the NAO

Three aspects are central to our perception of how large-scale climate indices are related to the functioning of organisms and their interactions in the ecosystem. First, organisms do not respond to changes in the large-scale indices like the NAO directly but to the variations in local weather that follow changes in the NAO. Their responses embrace horizontal within-generation and vertical across-generation changes (Figure 7A; Forchhammer and Post, 2004). Within-generation responses are related to the individual's ability to respond to short-term climate changes (i.e., phenotypic plasticity)

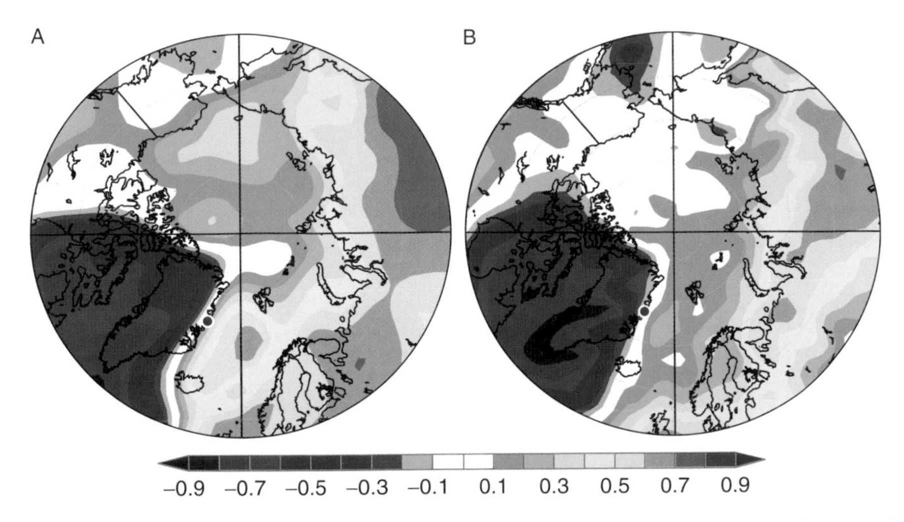

Figure 6 Map of the circumpolar area. Graded colouring indicates the degree of seasonal linear correlation (−1 to 1) between the winter NAO and (A) winter (December–March) temperature 1965–1994 and (B) winter precipitation 1965–1994. The vertical bar relates colouring and correlation coefficients. Red dots give the location of Zackenberg on the east coast of Greenland. Analyses were performed by the NOAA-CIRES/Climate Diagnostic Center (http://www.cdc.noaa.gov).

and involve life history trade-offs such as early reproduction and senescence (Post *et al.*, in press) at the expense of somatic growth. Those individuals responding optimally to climate changes will maximise their reproductive value and bear most offspring (Stearns, 1992).

The across-generation responses of organisms to climate assumes that optimal life history strategies are genetically inherited and passed from one generation to the next through the process of adaptation (Stearns, 1992). The biological responses to changes in the NAO reported below probably integrate both types where the relative importance depends on the evolutionary background of organisms as well as the length of the study. Since the design of the monitoring programmes at Zackenberg include no genetic components, we cannot here differentiate between the two types of responses to climate (but see Høye *et al.*, 2007a).

The second aspect relates to the integrative nature of both organisms and the NAO. Although organisms at any time respond to prevailing weather and not to the indices of the NAO, many studies have demonstrated that the NAO and other similar indices seem to be better predictors than local weather (Hallett *et al.*, 2004). The answer was provided by the zoologist H.D. Picton (1984), who in many ways was ahead of his time. In his early study of climatic effects on ungulates, he realised that he needed an integrative climate index, "since animals are physiological weather integrators the

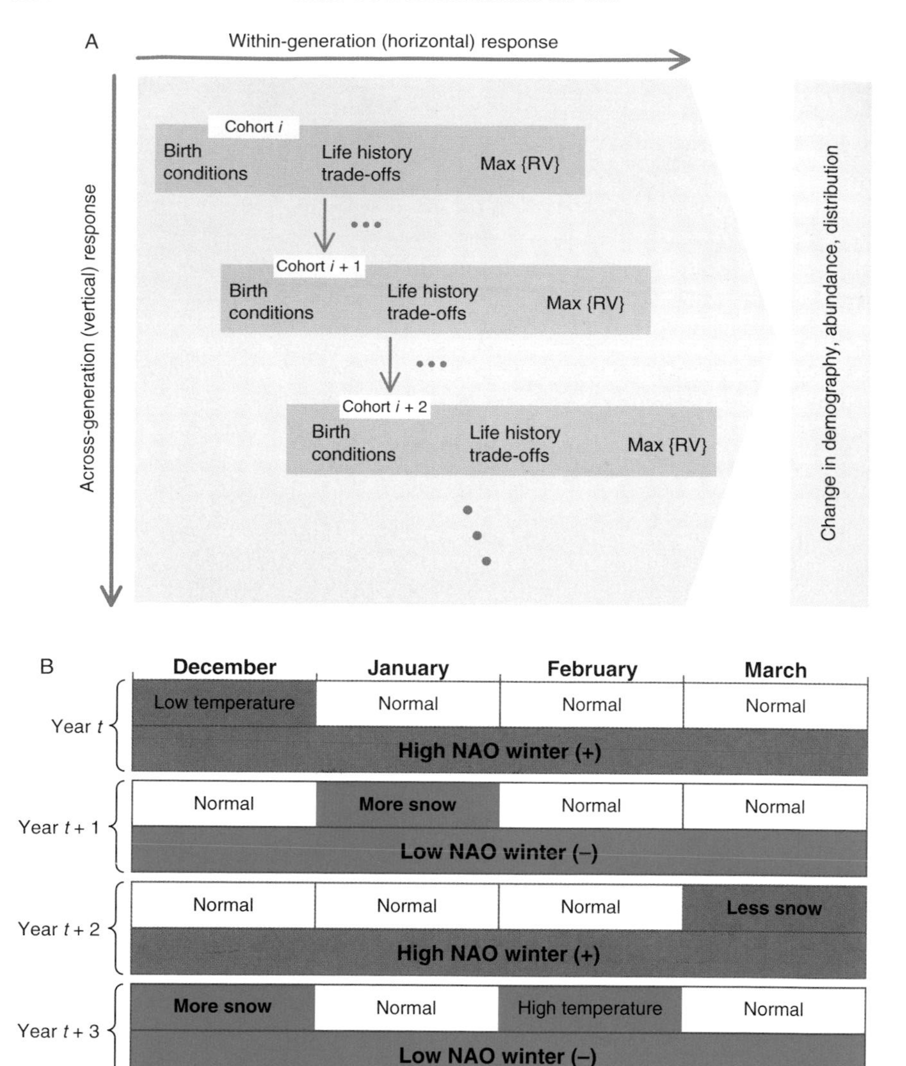

Figure 7 (A) Responses to climate changes visualised in a two-dimensional response diagram: within-generational (horizontal) and across-generation (vertical) responses. RV=reproductive value (Stearns, 1992). Changes in horizontal and vertical responses to climate will lead to changes in demography, abundance and distribution of species. Modified from Forchhammer and Post (2004). (B) The integrative properties of the NAO comparing the temporal monthly scale of mean weather measurements and the 4-months scale of the NAO index December through March. The relationship between monthly means and the NAO mirrors the observed linear correlation between the winter NAO and local winter precipitation and temperature found in western Greenland and eastern Canada (Hurrell, 1996; see also Figure 4). Modified from Stenseth and Mysterud (2005).

index must allow the weather data to be integrated or summed in a fashion which approximates the net result of the manner in which the animals integrate the weather" (Picton, 1984). This was developed further with respect to the NAO by Hallett *et al.* (2004), who pointed out that since precipitation, strong winds and low temperatures at any time during, for example, winter may negatively affect organisms in an additive manner, most measures of local weather previously used by ecologists failed to capture this complexity. In contrast, seasonal indices of large-scale climate such as the winter NAO index integrate both the temporal and compositional aspects of weather affecting organisms (Figure 7B; Stenseth and Mysterud, 2005). For example, if a study of the effects of weather focused on the average amount of winter precipitation during January only, this would disregard any influence of precipitation in other months as well as the influence of other abiotic conditions (Figure 7B). The NAO winter index, however, would capture these variations (Figure 7B) and as such depict a far better measure of how organisms integrate weather changes in their responses.

Finally, the NAO also has the unique feature of embracing spatial variations of climate on a much larger scale than local weather measurements (Stenseth *et al.*, 2003; Forchhammer and Post, 2004; Stenseth and Mysterud, 2005). For example, the influence of the NAO on the dynamics of different populations of Canadian lynx *Lynx canadensis* across Canada varies from negative in eastern Canada to positive in central Canada to negligible in western Canada (Stenseth *et al.*, 1999). This is due to large-scale spatial differences in the atmospheric circulation characteristic of the NAO and, hence, its relation to local weather (Stenseth *et al.*, 1999). This spatial aspect of the NAO makes it particularly useful when merging the biological responses observed at Zackenberg with responses observed elsewhere in the Arctic. Below, we focus selectively on species-specific responses, community-level responses and ecosystem feedback dynamics in relation to the spatio-temporal dynamics of the NAO integrating the dynamics observed by the monitoring at Zackenberg.

IV. SPECIES-SPECIFIC RESPONSES

A. Vegetational Changes

1. NAO Phases Affect Growth and Carbon Fixation in Cassiope Tetragona

The white arctic bell-heather *Cassiope tetragona* is a long-lived dwarf shrub with a wide circumpolar distribution and occurs throughout Greenland, although it is confined to the inland and higher altitude in the southern regions

(Böcher *et al.*, 1968). It is a key species in dry heath communities and forms extensive heaths in the lowland at Zackenberg on mesic ground covered by snow in winter (Bay, 1998). Despite its high abundance, *C. tetragona* is not grazed (Callaghan *et al.*, 1989). Individuals in high-arctic populations of *C. tetragona* have been estimated to be between 30 years old and 60 years old, and genets may live for several hundred years (Havström *et al.*, 1995; Johnstone and Henry, 1997). This high longevity combined with its extensive circumpolar distribution makes *C. tetragona* a suitable species for studying ecophysiological responses in high-arctic plants to large-scale spatio-temporal changes in climate. Indeed, previous retrospective analyses have showed that the annual growth of *C. tetragona* is dependent on variations in temperature, precipitation and thawing degree days, which vary geographically (Callaghan *et al.*, 1989; Havström *et al.*, 1995; Johnstone and Henry, 1997).

Combining the knowledge of annual growth sensitivity to weather conditions with isotope analyses of *C. tetragona* annual growth increments, Welker *et al.* (2005) were able to couple concurrent changes in water sources and leaf gas exchange with changes in large-scale climate mediated by the NAO/AO in a high-arctic population in eastern Canada. Welker *et al.* (2005) estimated the occurrence of two isotope characteristics in annual growth increments: oxygen isotope ratios ($\delta^{18}O$) and carbon isotope ratios (Δ), where the former expresses the ratio of oxygen atoms $^{18}O/^{16}O$ (e.g., Dansgaard, 1964) and the latter expresses the ratio of carbon atoms $^{13}C/^{12}C$ in the plant relative to the $^{13}C/^{12}C$ ratio in the atmosphere (e.g., Farquhar *et al.*, 1989; Sandquist and Ehleringer, 2003). Since the $\delta^{18}O$ values are significantly lower in snowmelt water than in rain water, they were able to trace across years whether annual growth in *C. tetragona* was due primarily to increased uptake of water from snowmelt or rain. Furthermore, as high Δ values reflect increased carbon fixation by plants, and Δ values are higher when *C. tetragona* plants are exposed to increased subsurface snowmelt water (and hence decreased $\delta^{18}O$ values) throughout the summer, Welker *et al.* (2005) could couple the inter-annual variations in growth increments and degree of carbon fixation in *C. tetragona* to concurrent changes in the amount of snow. In eastern high-arctic Canada, low NAO/AO winters result in increased winter precipitation and vice versa (Figures 4 and 6; Hurrell, 1995; Bamzai, 2003). Therefore, here a negative phase of the NAO/AO will increase winter precipitation and snowmelt water use of *C. tetragona*, decrease $\delta^{18}O$ in plant segments, increase carbon fixation by leaves (high Δ values) and, finally, result in higher stem and leaf growth (Figure 8A). Recently, the co-isotopic approach presented above has been tested successfully using local environmental conditions (Sullivan and Welker, 2007) confirming the interaction described by Welker *et al.* (2005).

Although the techniques of isotope characterisation of plant growth are not implemented in the monitoring at Zackenberg, the climate-mediated

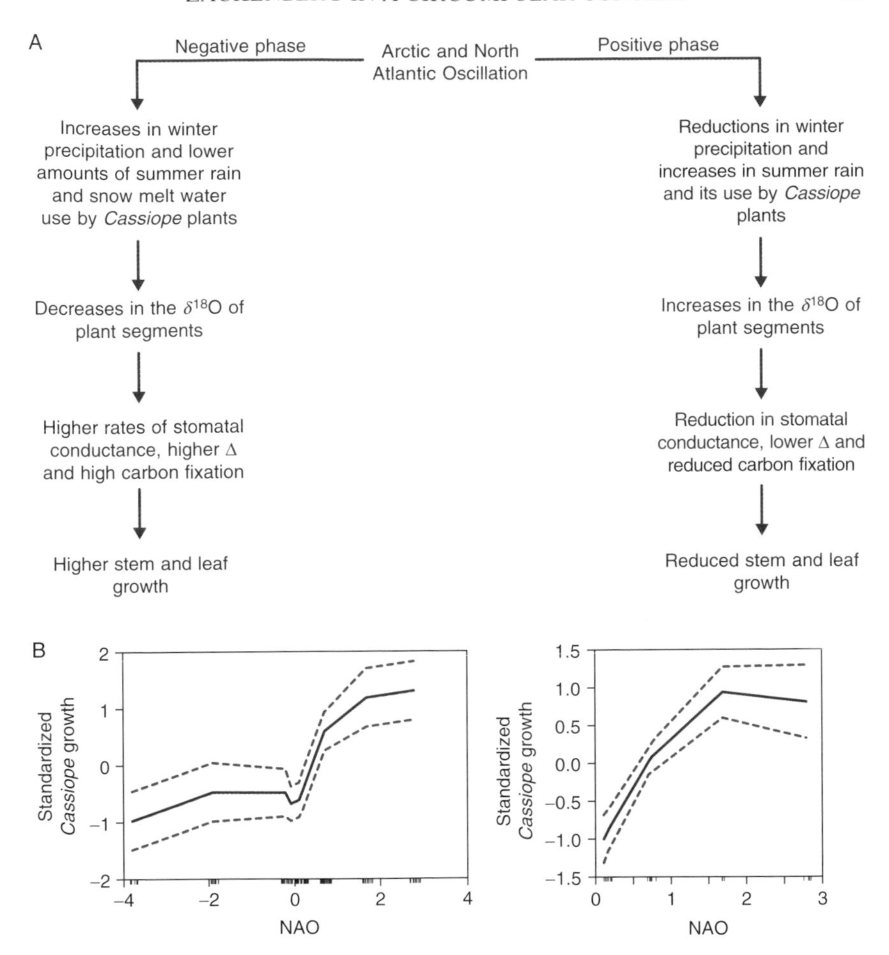

Figure 8 (A) Diagram showing the consecutive influence of negative and positive phases of the NAO/AO and growth of *C. tetragona*. Modified from Welker *et al.* (2005). (B) Annual growth of *C. tetragona* at Zackenberg in relation to the NAO. Standardised growth was estimated from the inverse relationship between the recorded changes in number of *C. tetragona* flowers and measured changes in the Normalised Vegetation Difference Index (NDVI) at *C. tetragona* sample plots (model R^2 values 0.59–0.71, $p < 0.01$). Current-year numbers of flowers are generally negatively correlated with current-year shoot elongation and leaf number (e.g., Johnstone and Henry, 1997). Right figure: annual standardised growth of *C. tetragona* for NAO > 0.

growth of *C. tetragona* observed here suggests that its growth is positively related to the presence of snow and, hence, large-scale climate change mediated by the NAO as described by Welker *et al.* (2005). Following winters characterised by increasingly higher NAO values, the growth of *C. tetragona* at Zackenberg increased non-linearly (Figure 8B) similar to the relationship between inter-annual variations in winter NAO and the amount of snow recorded at Zackenberg (Figure 5). In fact, as would be predicted from the insignificant relationship between the negative NAO indices and the amount of snow at Zackenberg (Figure 5A), negative NAO indices had no significant influence on the growth of *C. tetragona* at Zackenberg (Figure 8B). The positive effect of increasing snow mediated by the NAO has also been found in *C. tetragona* populations at Svalbard. Up to 51% of the inter-annual variations in growth were explained by the NAO (Figure 9) and, as for the Zackenberg population, the effect of the NAO was most pronounced for high positive NAO values.

Three important issues emerge from the monitoring at Zackenberg and previous studies of *C. tetragona* discussed above. First, the growth of *C. tetragona* portrays a significant and consistent signal of large-scale climate mediated by the NAO across high-arctic populations of which the response seen at Zackenberg is comparable with the responses seen elsewhere. Second, as a circumpolar key species in arctic heath communities, the consistent response of *C. tetragona* to changes in the NAO may serve as an indicator for not only climate-induced changes in biomass (i.e., growth) in high-arctic heaths but also how heaths may interact with atmosphere through the concomitant variations in carbon fixation described by Welker *et al.* (2005). Third, assuming future NAO-related climate changes across the Arctic will be characterised by increasingly higher NAO winters (Christensen *et al.*, 2007),

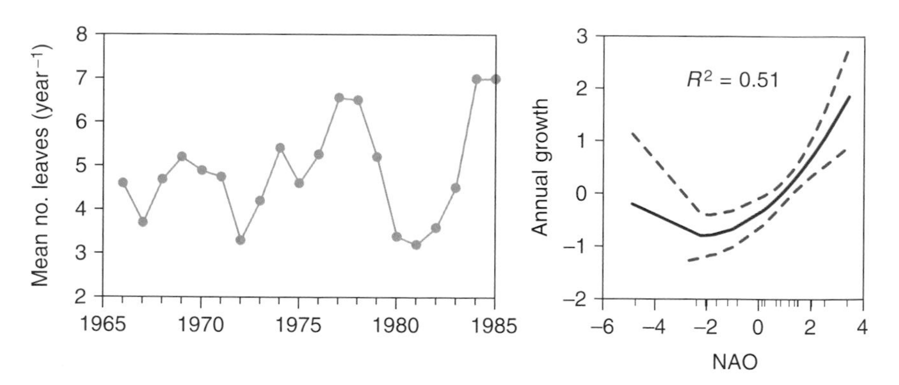

Figure 9 Annual growth of *C. tetragona* in Adventdalen, Svalbard, estimated as the mean number of leaves per year. Data are from Callaghan *et al.* (1989). Right figure: annual growth of *C. tetragona* adjusted (last year's growth; first-order autoregression) in relation to the NAO. The solid line is a non-linear GAM with 95% CIs as dotted lines; the rugplot along the *x*-axis marks the "observed" *x* values (Venables and Ripley, 1994). The GAM explained the relationship significantly better ($p < 0.04$) than a general linear model.

the future performance, growth and feedback of *C. tetragona* populations may be divided into two with respect to the dichotomous arctic changes in local winter weather conditions observed in the latest increasing phase of the NAO (Figure 6). A similar division in long-term patterns of annual growth increments observed across the Arctic has been observed in another dwarf shrub species, the arctic willow *Salix arctica*.

2. *Large-Scale Spatial Synchrony in the Growth of Salix Populations*

S. arctica is the northernmost woody plant species. It is well adapted to the arctic, sub-arctic and alpine environment with a circumpolar distribution (Böcher *et al.*, 1968). In Greenland, it is found all the way to the northern limit of land, but is replaced by the blue-grey willow *S. glauca* south of Disko on the west coast and Jameson Land on the east coast. In contrast to the taller *S. glauca*, which is a characteristic shrub of the coastal heath (Böcher *et al.*, 1968), *S. arctica* rarely grows more than 20–25 cm in height. However, despite its small size and slow annual growth (Figure 10), *S. arctica* is a long-lived plant and is known to live for more than 100 years. For example, at Zackenberg, the oldest individual recorded was 94 years old (Schmidt *et al.*, 2006), whereas a 116-year-old individual has been recorded at Qaanaaq/Thule in North Greenland (Figure 11).

Despite its huge potential for coupling climate changes with annual radial growth in woody dwarf shrub species, there have been few dendroclimatological studies of arctic *Salix* species (Schmidt *et al.*, 2006). Instead, retrospective analyses of long-lived arctic plant species have focused on describing inter-annual variation in external features such as those for *C. tetragona* (Callaghan *et al.*, 1989; Welker *et al.*, 2005). A novel approach using microscopic examination (Figure 10) has enabled within as well as across population comparisons in annual growth of arctic dwarf shrubs. This technique was successfully applied to a large sample of *S. arctica* at Zackenberg and, in contrast to *C. tetragona*, annual radial growth of *Salix* was negatively influenced by increasing early spring snow-cover. Since temperature from May through August, neither monthly nor combined seasonally, had significant effects on growth of *Salix* at Zackenberg (Schmidt *et al.*, 2006), NAO-related changes in the amount of snow during winter are probably quite important. Indeed, the variations in annual growth radii have been found to correlate directly with the winter NAO for several species throughout the Northern Hemisphere, where the reported effects are usually negative to NAO-mediated increases in snow and, hence, shortening of growth season (Mysterud *et al.*, 2003; but see Post *et al.*, 1999).

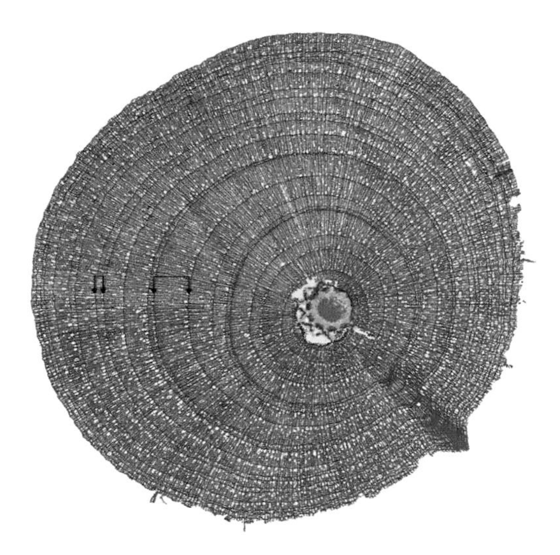

Figure 10 Enlarged microscopic crosscut section of a *Salix arctica* sampled at Qaanaaq/Thule, West Greenland, summer 2002 (M.C. Forchhammer, unpublished data). Two consecutive dark rings (summer bark) represent 1 year of radial growth, which vary tremendously from year to year as can be seen from the two sets of connected arrows. This individual was 22 years old at the time of sampling. The narrowest annual radial growth occurred in 1984 and the individual displays low autocorrelation in radial growth across years. Photo: Claudia Baittinger. For methodological details, see Schmidt *et al.* (2006).

Climate has the potential to not only synchronise population dynamics in time (Box 1) but also induce synchrony in species-specific life history traits, such as growth and reproduction. A comparison of the annual growth of *Salix* at Zackenberg with other arctic/alpine *Salix* populations (and *Betula* on Svalbard) across the Arctic revealed no distinct growth dynamics, common for all populations, although the Zackenberg and Qaanaaq/Thule populations tended to have multi-annual fluctuations in their temporal growth dynamics (Figure 11) resembling those found in the NAO winter index (Figure 3). Over the last three to four decades, the synchrony of growth patterns in these populations has changed dramatically (Figure 12A). Specifically, during the period 1965–1995, when the NAO became increasingly higher, the synchrony decreased. However, from 1995 onto 2002, as the NAO became lower, the annual growth again became more synchronous across populations (Figure 12A), illustrating a clear inverse relationship between the NAO and the degree of growth synchrony across *Salix*/*Betula* populations (Figure 12B). These dynamics are probably, as for *C. tetragona*,

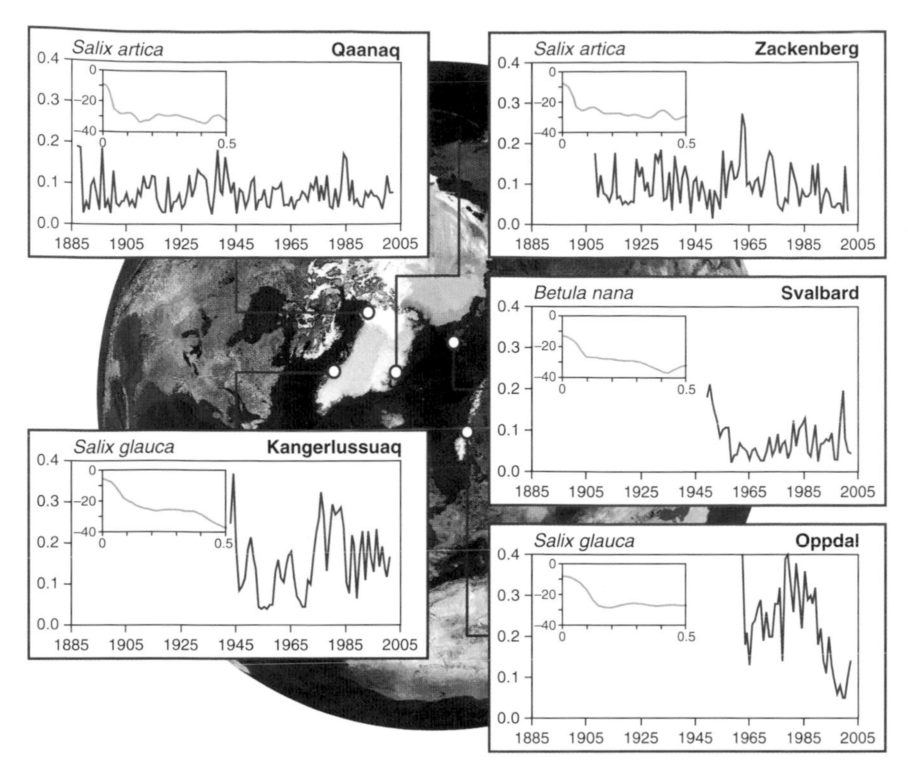

Figure 11 Selected time series of annual radial growth (mm) of *Salix arctica* (Qaanaq/Thule, Zackenberg), *Salix glauca* (Kangerlussuaq/Søndre Strømfjord, Oppdal (Norway) and *Betula nana* (Svalbard). Inset panels: smoothed periodograms of time series (see Figure 3 for details).

related to spatio-temporal changes in the NAO across the North Atlantic (Figure 4). High growth synchrony across the North Atlantic is found during periods with generally low NAO winters, in which the effects of the NAO on winter conditions in East Greenland become similar to those in northern Scandinavia (Figure 4A). In other words, the *Salix/Betula* populations experience similar NAO winters and hence similar growth conditions. In contrast, during high NAO periods they do not (Figure 4B), and, consequently, low growth synchrony is observed across populations (Figure 11B). It may be noted that the relationship between the NAO and growth synchrony is non-linear with negligible effects during low, negative NAO years. This is because as NAO becomes increasingly influential on local winter conditions during high phases (i.e., increased correlation, Figure 4A, B) so does the influence of the NAO on growth synchrony (Figure 12C).

Box 1

Coupling Ecological Responses in Space: The Climate Effect Ratio

As early as in 1953, the Australian statistician P.A.P. Moran recognised that climate may play a major role in coupling the dynamics of spatially distant populations. The theory he developed, which has become known as the Moran theorem (Moran, 1953; Royama, 1992), states that "if two regional populations have the same intrinsic (density-dependent) structure, they will be correlated under the influence of density-independent factors (such as climatic factors), if the factors are correlated between the regions." Hence, the more similar the effect of climate on any pair of population, the more correlated their dynamics should be (Post and Forchhammer, 2002).

The climatic scaling between any pair of spatially separated populations can be addressed specifically through the climate effect ratio. For example, consider two populations X and Y influenced linearly by density dependence and climate:

$$\begin{aligned} X_t &= a_0^{\{x\}} + a_1^{\{x\}} X_{t-1} + C_t^{\{x\}} \\ Y_t &= a_0^{\{y\}} + a_1^{\{y\}} Y_{t-1} + C_t^{\{y\}} \end{aligned} \quad (1.1)$$

in which X_t and Y_t are ln-transformed densities in year t, $a_0{}^{\{\bullet\}}$ and $a_1{}^{\{\bullet\}}$ are constants and $C_t{}^{\{x\}}$ and $C_t{}^{\{y\}}$ are the climatic influences on X and Y, respectively. According to the Moran theorem, if X and Y display similar and approximate log-linear density dependence in their dynamics, then the correlation between X and Y is (Royama, 1992)

$$C_t^{\{x\}} = kC_t^{\{y\}} \quad (1.2)$$

That is, the climatic influences on the two populations are related through the coefficient k, which equals the standardised correlation coefficients between the climate conditions influencing population X and Y, respectively (Sokal and Rohlf, 1995). When climate influence is mediated through the annual variation in the NAO, as reported extensively elsewhere (e.g., Forchhammer, 2001; Forchhammer and Post, 2004), the autoregressive population models in Eq. (1.1) can be written with $C_t{}^{\{x\}} = \omega_1{}^{\{x\}}\mathrm{NAO}_t$ and $C_t{}^{\{y\}} = \omega_1{}^{\{y\}}\mathrm{NAO}_t$, in which $\omega_1{}^{\{x\}}$ and $\omega_1{}^{\{y\}}$ are the regression coefficients quantifying the climatic influences of the NAO on X and Y, respectively. Isolating NAO_t and substituting into Eq. (1.2) gives

$$C_t^{\{x\}} = \frac{\omega_1^{\{x\}}}{\omega_1^{\{y\}}} C_t^{\{y\}}$$

From Eq. (1.2) it then follows that the climate effect ratio, that is, $\omega_1^{\{x\}}/\omega_1^{\{y\}}$, equals the degree of climatic correlation between the two populations k. See Post and Forchhammer (2006) for further details.

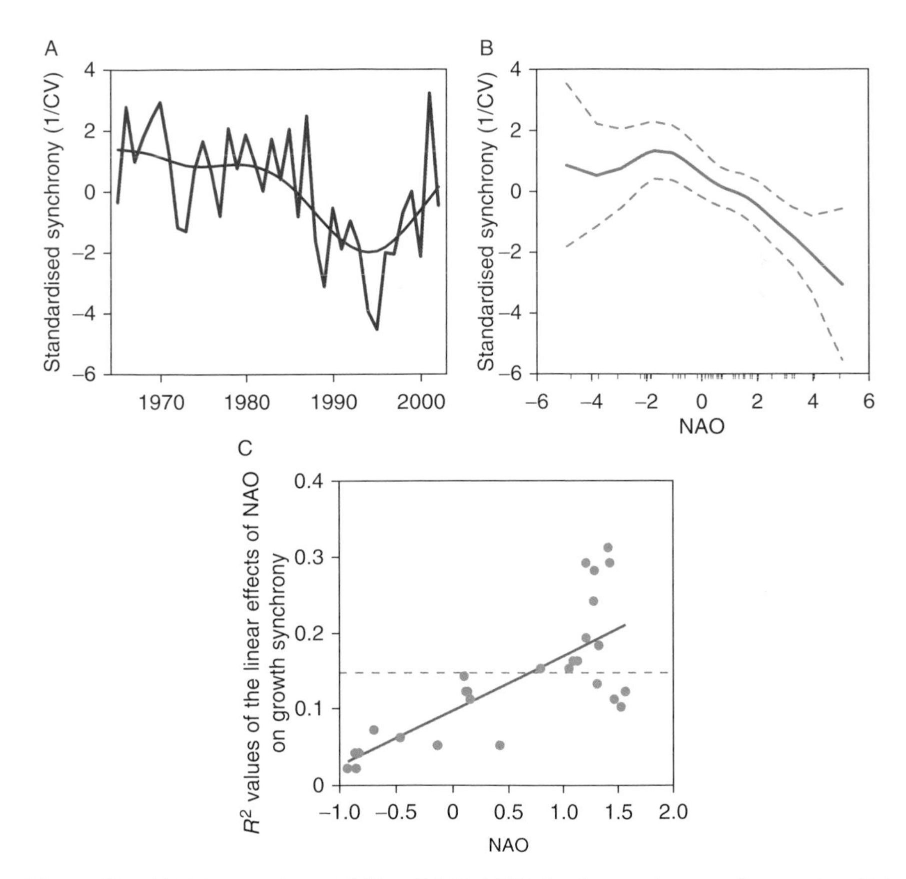

Figure 12 (A) Temporal variability (1965–2002) in the synchrony of annual radial growth of *Salix* and *Betula* from the populations in Figure 11. Red line is the most parsimonious generalised additive model (GAM) describing the temporal variation in the synchrony across all populations. (B) Annual variation of synchrony in relation to the NAO, 1965–2002 ($R^2 = 0.42$). Full line is a non-linear GAM with 95% CIs as dotted lines; the rugplot along the x-axis marks the "observed" x values. Synchrony is defined as the inverse coefficient of variation in annual radial growth, 1/CV = average/standard deviation (Post and Forchhammer, 2002). (C) Following Hinkler (2005), average values of winter NAO indices were calculated for 19-year windows for

3. Normalized Difference Vegetation Index

So far we have focused on how single key species of heath communities respond across arctic/alpine populations. In a broader context, vegetation types or even on the landscape level, the Normalised Difference Vegetation Index (NDVI) has been used successfully to study vegetational impacts of large-scale climate indices, such as the NAO/AO, the North Pacific Oscillation and the El Niño Southern Oscillation (e.g., Buermann *et al.*, 2003; Gong and Ho, 2003; Gong and Shi, 2003). The NDVI expresses the greenness of vegetation and correlates with photosynthetic activity (e.g., Tucker *et al.*, 1986; Myneni *et al.*, 1995) and may, hence, be related to seasonal as well as inter-annual variations in plant biomass production (Todd *et al.*, 1998). A detailed description of the NDVI and associated derivations is given by Ellebjerg *et al.* (2008, this volume).

Vegetation covers most of the Earth's land surface and is one of the key players in influencing the global cycles of energy, hydrology and biogeochemistry. Consequently, it has for long been a major goal to describe large-scale changes in vegetation as well as the climate mechanisms behind such changes (e.g., Zhou *et al.*, 2001; Buermann *et al.*, 2003). For this, the NDVI has proven to be a most useful index to study vegetational changes. Analyses of NDVI may be performed at any vegetational level all the way from individual vegetation plots through landscape and regional vegetation dynamics to hemispheric scales. Local-scale, species-specific plot changes in NDVI are addressed elsewhere (Ellebjerg *et al.*, 2008, this volume). Here we focus on how inter-annual changes in NDVI across the landscape at Zackenberg are influenced by the NAO and how these are related to similar interactions observed throughout the Northern Hemisphere.

At Zackenberg, there is an apparent altitudinal variation in annual mean NDVI, with highest values observed between 0 m a.s.l. and 150 m a.s.l. and decreasing up to 600 m a.s.l. (Figure 13A). This, of course, is closely related to the distribution of vegetation types where those with highest growth and biomass are found on the valley floor (Bay, 1998; Ellebjerg *et al.*, 2008, this volume). A large proportion of the inter-annual variation on the landscape level could be related to changes in the NAO (Figure 13B). In particular, for positive NAO indices there was a non-linear U-shaped relationship ($R^2_{\text{partial}} = 0.41$) where increasingly higher NAO values lead to an increase in mean

---

the "centre years" 1968–1994. For each of these periods, R^2 values of the relationship between NAO and degree of growth synchrony across the five *Salix/Betula* populations in Figure 11 are given. These are plotted as a function of the corresponding average NAO indices. Solid line is the linear regression and dots above the dashed line ($R^2 = 0.14$) are significant ($p < 0.05$).

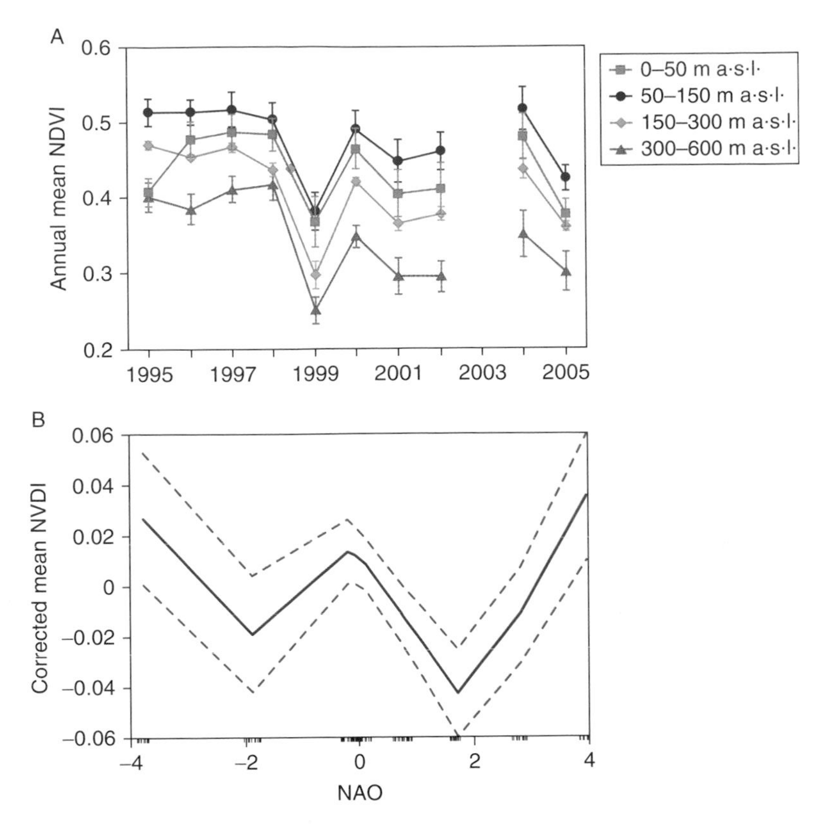

Figure 13 (A) Yearly mean NDVI for the different altitudinal regions at Zackenberg (data from M.P. Tamstorf, unpublished). There are no data available for 2003. The lowest lying altitudinal ranges did not display a significant trend ($p>0.10$), but the NDVI of vegetation types between 150–300 m a.s.l. and 300–600 m a.s.l. decreased significantly ($r = -0.48$ and $r = -0.58$, respectively) from 1995 to 2005. (B) Annual variations in mean NDVI in relation to the NAO. Variations in the annual mean NDVI were corrected for local variations in snow-cover and altitudinal range (see Table 2). Solid line is a non-linear GAM with 95% CIs as dotted lines; the rugplot along the x-axis marks the "observed" x values.

NDVI (Figure 13B). Integrating the contrasting NAO–snow interface, we see that the positive large-scale effect may be perpetuated through decreasing early snow-cover (Figure 5A).

However, the spatial analyses of the inter-annual variation in mean NDVI suggest that the effect of the NAO is highly dependent on altitude and hence the vegetation type in focus (Table 2). Whereas variations in the NDVI in the lowland vegetation types at Zackenberg are primarily influenced by snow-cover, the NDVI of vegetation types above 50 m a.s.l. is more affected by the

Table 2 Multivariate Generalised Linear Model (GLM) and Generalised Additive Model (GAM) analyses of inter-annual variations in mean Normalised Difference Vegetation Index (NDVI) at Zackenberg as the response variable

	Predictor variables			
Model type	Region	Snow-cover	NAO	Model R^2
GLM (all ranges)	**42.68**	**3.66**	0.77	0.29
GAM (all ranges)	**23.89**	1.82	**18.01**	0.66
GAM (0–50)	–	**3.38**	2.52	0.56
GAM (50–150)	–	0.31	**3.81**	0.44
GAM (150–300)	–	2.67	**10.64**	0.72
GAM (300–600)	–	1.52	**5.04**	0.52

Predictor variables are altitudinal ranges (m a.s.l.) of the valley Zackenbergdalen, snow-cover on June 10 and the NAO winter index. Altitudinal range was included in the first two full models. Since the GAM performed significantly better than GLM, GAM analyses were done for each range separately. Data from Meltofte (2006). F-values are given for each predictor influence as measure for each predictor's relative influence on NDVI. Bold values indicate significance ($p < 0.05$).

NAO (Table 2). As reported for alpine as well as temperate systems in northern Scandinavia (Pettorelli *et al.*, 2005a,b), this suggests differential large-scale climate effects across altitude (and vegetation types) at Zackenberg. In the lowland vegetation, NDVI variations are highly dependent on the topographical heterogeneity of the landscape (Hinkler, 2005) in contrast to vegetation at higher altitudes, where it is the large-scale setting of the winter by the NAO winter index which is the most influential (Table 2).

The major increase in plant growth observed in the northern high latitudes during the last 20 years (Myneni *et al.*, 1997) is tightly coupled not only to the observed temporal dynamics of the NAO indices (Figure 3A) but also to the spatial variability in the NAO–temperature/precipitation relations (Figure 6) (Buermann *et al.*, 2003). Similar relations and mechanisms are observed at Zackenberg (Figure 13B; Ellebjerg *et al.*, 2008, this volume). However, at Zackenberg, the high altitude (150–600 m a.s.l.) displayed a negative trend in NDVI over the last decade. This trend is not only found in the overall NDVI data obtained from the satellite data but also in plot measurements (Tamstorf *et al.*, in press). While winter precipitation has been variable but without significant trends, the summer temperatures and the length of the melting season have increased over the last 10 years. This has led to a decrease in the number and extent of perennial snow patches, with a resulting decline in water availability during the mid and late growing season. The effect is most pronounced at the higher levels (above 100 m a.s.l.) corresponding with the results shown in Figure 13A.

B. Life Histories: Timing of Reproduction

The temporal allocation of growth of individuals discussed above for two arctic plant species is in fact one component of many involved in the trade-offs of species life histories. For example, in the NAO-related growth in *C. tetragona* on Svalbard, current-year impact of the NAO was dependent on previous year's investment in growth (Figure 9; Forchhammer, 2002). The temporal allocation of time and energy to reproduction throughout an individual's life is probably one of the pivotal components determining lifetime reproductive success (Clutton-Brock, 1988), and for many species reproduction as early as possible when conditions are favourable may be favoured over strategies of delaying reproduction (Stearns, 1992). Since the timing of life history events such as reproduction is determined not only by evolutionary history and ecosystem embedded biological interactions but also by constraints imposed by the abiotic environment (Stearns, 1992), any consistent changes in climate are expected to affect reproduction trends (Forchhammer, 2001). Indeed, following the last two decades of winter warming, an increasing number of studies throughout the Northern Hemisphere report increasingly earlier dates of reproduction for a range of evolutionarily distinct species of plants and animals (Forchhammer *et al.*, 1998; Post and Stenseth, 1999; Root *et al.*, 2003; Menzel *et al.*, 2006; Cleland *et al.*, 2007; Høye *et al.*, 2007b). Many of these observed phenological changes have been related to the trend as well as inter-annual variability in the NAO winter index (Forchhammer *et al.*, 1998; Post *et al.*, 2001; Menzel, 2003).

Although variability exits among the most recent large-scale climatic models, one common predictive feature is the latitudinal increase in future warming where the high-arctic regions, including the area at Zackenberg, are expected to experience the fastest and greatest warming (Christensen *et al.*, 2007; Stendel *et al.*, 2008, this volume). Until recently, it was not known whether a similar latitudinal gradient in species responses to warming existed. Indeed, so far most long-term phenological studies had been confined primarily to temperate Europe and North America (e.g., Root *et al.*, 2003; Menzel *et al.*, 2006). However, using the extensive amount of data from the monitoring programme at Zackenberg, Høye *et al.* (2007b) showed that a wide range of high-arctic plant, insect, and bird species at Zackenberg displayed a considerable plasticity in their annual timing of reproduction concurrent to the pronounced directional change in spring snowmelt and summer air temperature at Zackenberg (Figure 14; Hansen *et al.*, 2008, this volume). On average, the initiation of annual breeding, that is, the date of flowering in plants, emergence of insects and egg-laying in birds, advanced by 14.5 days/decade since 1995. This average value embraced an impressive range of temporal responses within as well as across species. For example, in one plot at Zackenberg, the flowering of *S. arctica* had advanced by more

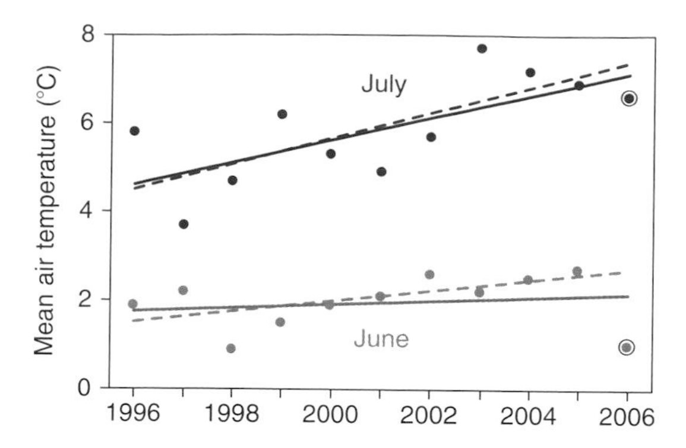

Figure 14 Temporal trends in the mean June and July air temperature (2 m a.s.l.) measured at Zackenberg 1996–2006. Solid regressions lines are based on all data (June: slope = 0.04 ± 0.05 °C/year, $R^2 = 0.05$, $p = 0.51$; July: slope = 0.25 ± 0.08 °C/year, $R^2 = 0.49$, $p = 0.01$) and the dashed regression lines exclude the circled 2006 data (June: slope = 0.11 ± 0.04 °C/year, $R^2 = 0.43$, $p = 0.04$; July: slope = 0.28 ± 0.10 °C/year, $R^2 = 0.50$, $p = 0.02$). The period analysed by the dashed regressions is the period covered by Høye *et al.* (2007b). Data from Sigsgaard *et al.* (2006) and M. Tamstorf (unpublished data).

than 3 weeks since 1995, whereas no or little advancements were observed in other plots (Høye *et al.*, 2007b). Similar ranges of timing of reproductive phenology were observed in other species. This spatial variability in phenological responses was related to the influence of the variable landscape topography at Zackenberg, which creates considerable variation in local timing of spring snowmelt (Høye *et al.*, 2007b).

Whether phenological changes in reproductive responses display trends or not, they may nevertheless still be influenced by large-scale climate. Indeed, first egg-laying dates of sanderling *Calidris alba*, which showed no significant decadal trend, clearly responded to annual changes in the NAO, where increased NAO associated with increased snow and later snowmelt (Figure 5) was associated with later initiation of egg-laying by sanderlings (Figure 15A). Similarly, reproduction in the long-tailed skua *Stercorarius longicaudus* and musk ox *Ovibos moschatus* populations at Zackenberg was influenced by the NAO (Figure 15B, C). The important point here is that species do not necessarily have to display trends in their phenology in order to display a response to variations in large-scale climate. Indeed, whereas trend analyses of species responses focus on average temporal changes, analyses employing the inter-annual dynamics of large-scale systems like the NAO also integrate variability. Along with the expected increased trends, increased variability is also expected in the future climate scenarios with more extreme events (Christensen *et al.*, 2007). Therefore, it becomes important to embrace responses to average as well as variance in climatic changes; in fact, it may be

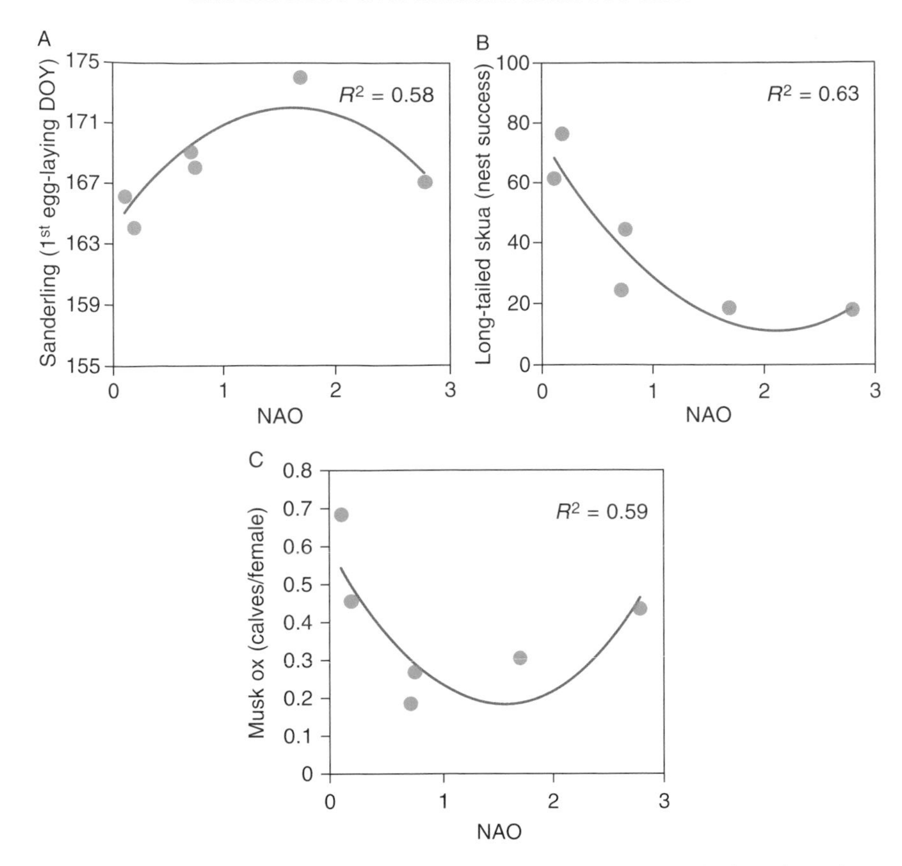

Figure 15 Inter-annual variations in reproductive performance of selected species at Zackenberg. (A) Sanderling *Calidris alba*, median first egg-laying date (day of the year June 4–24), (B) long-tailed skua *Stercorarius longicaudus*, nest success and (C) musk ox *Ovibos moschatus*, number of calves per female. All R^2 values are significant ($p > 0.05$). Data from Meltofte (2006).

argued that increased frequency of extreme weather may impose an even greater constrain on organisms than merely average changes.

C. Population Dynamics

Both short-term phenotypic responses as documented above and long-term inter-generational adaptive changes in life history strategies will lead to changes in population dynamics (Figure 7A). Since the NAO has been coupled to such changes in many species (e.g., Forchhammer, 2001; Mysterud *et al.*, 2003), we may expect to find the influence of the NAO portrayed in their population dynamics as well. In a spatial as well as temporal context, the dynamics of ungulate populations in relation to large-scale systems like the NAO are probably the best described and analysed so far.

At Zackenberg, the only ungulate species present is the musk ox, which occurs in large numbers throughout the Zackenberg area (Forchhammer *et al.*, 2008, this volume). Here, its population dynamics are highly dependent on changes in snow-cover and, hence, length of the plant growth season (Forchhammer *et al.*, 2008, this volume). The significant relations between the NAO and calves produced per female (Figure 15C) suggest a delayed influence of climate on the population level as the production and recruitment of offspring into a population depends on their survival to maturity (Forchhammer *et al.*, 2002). For musk oxen, the average age of first reproduction is their third year (e.g., Olesen *et al.*, 1994), and, indeed, long-term dynamics of the musk ox population at Zackenberg display a 3-year delayed influence of the NAO, where an increased NAO-mediated amount of snow had a delayed negative influence on population dynamics (Forchhammer *et al.*, 2002). Although this was a shared feature in all musk ox populations in Greenland, the numerical influence of the NAO decreased from south to north (Figure 16A; Forchhammer *et al.*, 2002).

Similar delayed effects of the NAO were found on the caribou *Rangifer tarandus* in West Greenland and could, as in musk oxen, be related to the fecundity and survival/recruitment of offspring (Forchhammer *et al.*, 2002).

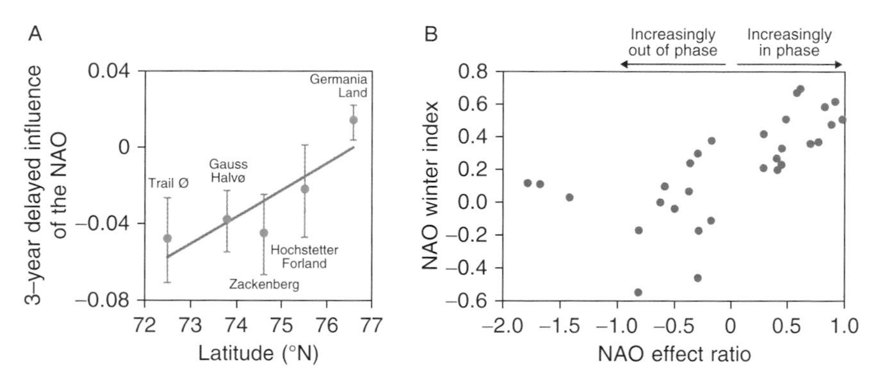

Figure 16 (A) Latitudinal variation in the 3-year delayed influence (co-variate regression coefficients) of the NAO on the dynamics of the five recognised indigenous musk ox populations in Greenland; $R^2 = 0.76$, $p < 0.05$ (adapted from Forchhammer *et al.*, 2002). (B) Relation between the degree of pairwise cross-correlation between caribou and musk ox populations in Greenland and the ratio of the effect of the NAO on the dynamics of each population ($R^2 = 0.42$, $p < 0.001$). Modified from Post and Forchhammer (2002). See also Box 1 for the description and calculation of climate effect ratios.

In fact, this delayed NAO effect was also found throughout most of the Russian caribou populations, portraying the spatial variation in the NAO–local weather relations (Post and Forchhammer, 2006). Interestingly, Post and Forchhammer (2002) observed that the higher cross-correlation between the dynamics of two arctic ungulate populations, the more similar were the effects of the NAO on these populations, that is, the NAO climate effect ratio (Box 1) approached 1 (Figure 16B). In contrast, the more the NAO effects deviated among populations, the lower cross-correlation was observed (Figure 16B). This also applied for comparisons between caribou populations in West Greenland, Finland and Russia (Post and Forchhammer, 2006). The considerable geographical area in which the NAO affects arctic ungulate populations emphasises the ability of large-scale climate systems to synchronise the dynamics of physically separated populations and, thereby, their roles as potentially important drivers affecting the probability of multi-population extinctions (Post and Forchhammer, 2006).

V. COMMUNITY-LEVEL RESPONSES

A. Consumer–Resource Interactions

Changes in climatic conditions affect simultaneously all the organisms embedded in a community or an ecosystem. Therefore, any correlation between climate and the performance of a single species may also embrace changes in its interaction with other species such as its competitors and consumers (see also Berg *et al.*, 2008, this volume; Forchhammer *et al.*, 2008, this volume; Schmidt *et al.*, 2008, this volume). For instance, in addition to the direct effects of the NAO on the reproduction and population dynamics reported above, a previous study from Zackenberg showed that an indirect influence of the NAO on the musk ox population occurred through its interaction with its main plant forage, *S. arctica* (Forchhammer *et al.*, 2005). At Zackenberg, the inter-annual variation in snow-cover is mediated through the NAO (Figure 5). Such changes in snow-cover may, through effects on both musk ox density and *S. arctica* growth dynamics, affect the distribution of foraging musk oxen in the valley Zackenbergdalen. Using general additive models (Venables and Ripley, 1994), Forchhammer *et al.* (2005) found complex cascading inter-trophic effects of the NAO with stepwise indirect and non-linear consequences for the spatial dynamics of musk ox foraging behaviour. Specifically, high NAO winters were associated with decreased snow-cover, which increased both the biomass and degree of spatial synchrony in growth of *Salix* the following summer (Figure 17A, B). Increased spatial synchrony of *Salix* growth, in turn, increased the spatial dispersion of musk oxen, but decreased the herd size in the same summer (Figure 18A, B), whereas no NAO-mediated effects on musk ox density affected their distribution in the landscape (Forchhammer *et al.*, 2005).

The results from Zackenberg are not unique in that large-scale climate influences have been documented in a range of communities throughout the Northern Hemisphere (e.g., Post and Stenseth, 1999). Like the musk oxen at

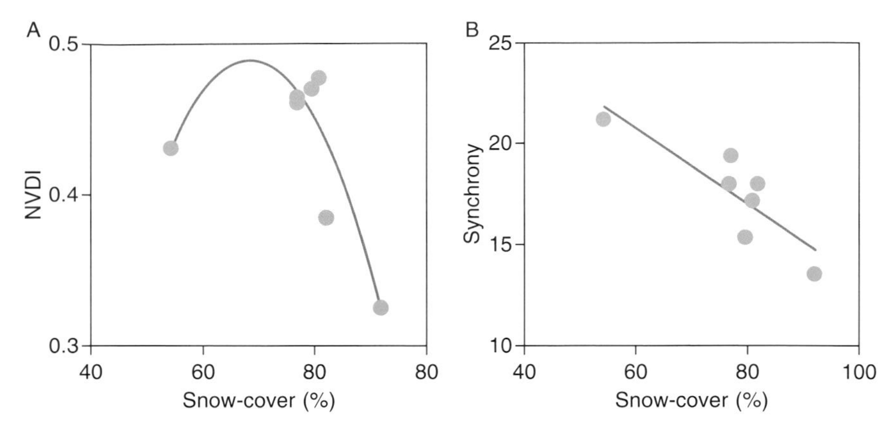

Figure 17 Annual biomass production (A) and synchrony (B) of *Salix* growth as a function of percentage of early spring snow-cover (June 10). Biomass is expressed through the Normalised Difference Vegetation Index (NDVI) and synchrony as the reverse coefficient of variation. Modified from Forchhammer *et al.* (2005).

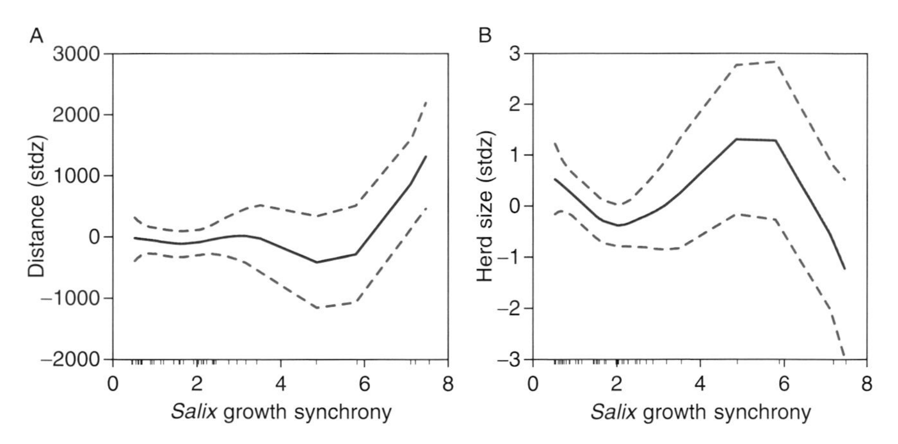

Figure 18 (A) Seasonal changes in the distance between herds as a function of the degree of synchrony (1/CV) in the growth *Salix arctica* in the study area. (B) Variations in herd size as a function of the degree of synchrony in the growth *S. arctica*. Full line is the second-order polynomial generalised additive models with 95% CIs as dotted lines; the rugplot along the *x*-axis marks the "observed" *x* values. Modified from Forchhammer *et al.* (2005).

Zackenberg, changes in the NAO affected the seasonal distribution of red deer *Cervus elaphus* populations in Norway through influences on plant biomass (expressed by the NDVI) (Pettorelli *et al.*, 2005a). Red deer populations migrate seasonally between summer areas in the highlands and winter areas in valleys. As it turns out, increasing NAO corresponded to less snow at lower altitudes but more snow at high altitudes, which, in turn, resulted in spatially more variable plant phenology, offering migrating deer a longer period of access to high-quality forage. This led to increased body mass (Pettorelli *et al.*, 2005a), which eventually affected the structure as well as dynamics of populations (Post and Stenseth, 1999).

Another example of such NAO-mediated impacts on consumer–resource interactions comes from the well-studied tri-trophic predator–prey–plant system on Isle Royale in North America (e.g., Peterson *et al.*, 1984). Here, inter-annual fluctuations in the NAO affected not only each trophic level directly (Post and Forchhammer, 2001) but also the social behaviour of wolves *Canis lupus* and, hence, their hunting efficiency, which had indirect consequences for both the moose *Alces alces* population (delayed negative) and growth (delayed positive) of the forage most important to moose, balsam fir *Abies abies* (Post *et al.*, 1999; see also Forchhammer *et al.*, 2008, this volume, for further details). The population dynamics of moose were primarily controlled by the top-down effect from wolves (33%), followed by the direct climate effect, that is, NAO (14%), and bottom-up from fir to moose (10%) (Wilmers *et al.*, 2006). In 1981, there was an outbreak of canine parvovirus in the wolf population (Peterson *et al.*, 1998), that is, a fourth trophic level was introduced to the system on Isle Royale. This had striking effects on climatic and trophic drivers of the moose population. First, in the transient dynamics, the direct effects of the NAO became the most important factor (37%). And secondly, the bottom-up effect became more important (22%) than top-down influence (1%) (Wilmers *et al.*, 2006).

B. Stability of Species Populations and Interactions

It follows, from the above, that together with the influence of density dependence and direct climatic effects on each species' population dynamics, any shift in consumer–resource interactions may greatly influence the effects of climate and, hence, the stability of both populations and communities. Conceptually, the dynamics of populations in a stochastic environment is a combined result of the stabilising influences of density dependence and destabilising influence of the extrinsic stochastic factors such as climate (e.g., May, 1973, 2001). The Zackenberg musk ox population is characterised by medium direct density dependence and a strong influence of the NAO (Figure 19A), which suggests that whereas the population may be destabilised

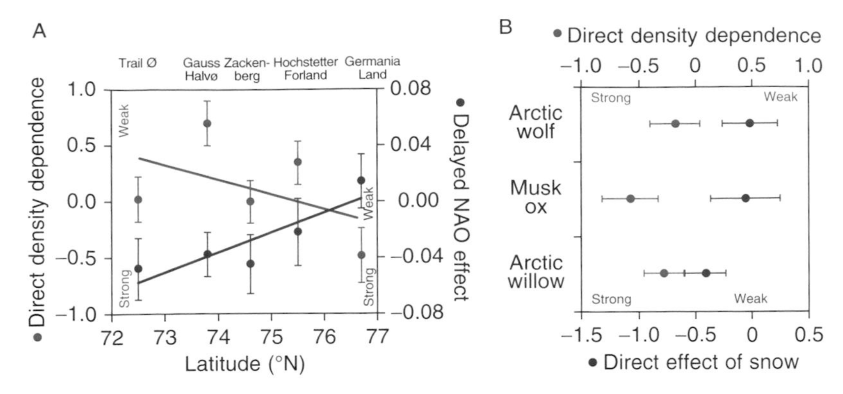

Figure 19 (A) Comparing strength of direct density dependence (first-order autoregressive coefficient) and the effect of NAO (additive co-variate coefficient) across five musk ox populations in East Greenland. Data from Forchhammer *et al.* (2002). (B) The tri-trophic system at Zackenberg embracing arctic wolf, musk ox and arctic willow after the wolf re-invasion to East Greenland in the early 1970s (Dawes *et al.*, 1986). For each trophic level is given the direct density dependence and the direct effect of snow-cover. Data from Schmidt (2006).

by the NAO, this is counteracted by a high level of density dependence in that population. This has been documented as well for caribou throughout West Greenland, Finland and Russia, where a clear inverse relationship between density dependence and the contribution of the NAO to dynamics was found across all populations studied (Post, 2005). Comparing across the musk ox populations in Northeast Greenland, the decreasing trend in the effect of the NAO and the increasing trend in direct density dependence from south to north (Figure 19A) indicate that the most stable populations are the northernmost.

Focusing on the tri-trophic system of arctic wolf *Canis lupus arctos*, musk ox and *S. arctica* at Zackenberg, we see that the strongest self-regulation, that is, direct density dependence, is found on the intermediate trophic level (musk ox), whereas the strongest influence of climate is found at the bottom trophic level of plants (Figure 19B; Schmidt, 2006). Therefore, the influence of future changes in climate on the predator–prey–plant system at Zackenberg is expected to be primarily a bottom-up process, where the influence of climate on the growth and performance of *S. arctica* is expected to spread through the system mediated by the trophic interactions of musk oxen and wolves (Schmidt, 2006). Similar system dynamics emerges from another study on the tri-trophic predator–prey–plant system on Isle Royale, where analyses similar to those at Zackenberg showed that although self-regulation was present at all three trophic levels, the greatest climatic impact was found at the bottom and top levels (Post and Forchhammer, 2001).

The stability of multi-level trophic systems and, hence, their vulnerability to climate and other stochastic environmental fluctuations have, theoretically as well as empirically, been shown to depend on the diversity of species in the systems, where increased diversity increases system stability (e.g., Pimm, 1984; Tilman and Downing, 1994; Tilman, 1996; Wilmers *et al.*, 2002). At first hand, analyses from Zackenberg support this notion. Indeed, as the wolves reappeared at Zackenberg in 1979 after 40 years of absence, the shift from a bi-trophic to a tri-trophic system was followed by an increase in stability (Figure 20; Schmidt, 2006). However, concurrent with but independent of the invasion of wolves, there was a shift in population dynamics of musk oxen and the performance of *S. arctica*, which correlated with an increased snow-cover (Schmidt, 2006). Hence, the observed change in stability may also relate to changes in the dynamic complexity (i.e., self-regulation and stochastic influence) of the individual trophic levels of musk ox and plants following a shift in the extent of snow-cover.

As conceptually demonstrated by May (1973) and later empirically by Tilman (1996), system stability increases with species diversity, whereas stability of populations does not necessarily do so. For a large temperate plant community, this dichotomy was related to extrinsic perturbations, such as shifts in climate, which affected competing species differently and, hence, increased the variability in species abundance, but at the same time stabilised plant community biomass due to decreased inter-specific competition (Tilman, 1996). Although the fundamental mechanisms shaping system stability may be similar across systems, it is reasonable to ask to what extent the

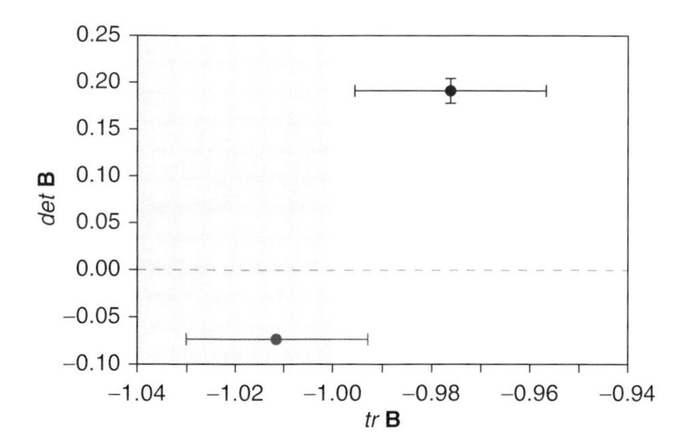

Figure 20 Community stability properties of the bi-trophic (musk ox–willow; red circle) and the tri-trophic (wolf–musk ox–willow; blue circle) system at Zackenberg. The shaded area denotes the area of stability (May, 2001). The x and y values are calculated as the trace ($tr\mathbf{B}$) and determinant ($det\mathbf{B}$) of the community matrix $\mathbf{B}$, respectively. For details, see Schmidt (2006).

spatial variations in population stabilities in the same species, as demonstrated for musk oxen in Northeast Greenland (Figure 19A), may influence system stability across different locations. To answer questions like this, the species-specific and multi-trophic monitoring approach at Zackenberg will become important.

VI. ECOSYSTEM FEEDBACK

A. Seasonal Impact on Annual Carbon Balance

In recent years, arctic terrestrial ecosystems have attracted major attention in the context of global carbon cycling (ACIA, 2005; Millenium Ecosystem Assessment, 2005). A reason for this is that arctic terrestrial ecosystems store a significant proportion of the global stock of soil organic carbon (C). In the arctic tundra proper, some 121–191 GT of C are stored, or ~12–16% of the estimated world total (McKane *et al.*, 1997; Tarnocai *et al.*, 2003). If boreal ecosystems are included, this estimate rises to almost 30%. The predicted significant climate changes and the feedbacks they engender could change the climatic conditions that have allowed the development of such large soil C stocks in the Arctic (Gorham, 1991; Shaver *et al.*, 1992; McKane *et al.*, 1997; Hobbie *et al.*, 2000). Extensive regions of the High Arctic such as much of the Zackenbergdalen that lack substantial C stocks and currently have very limited rates of atmospheric exchange could develop dynamic C cycles. Climate-driven changes in plant community structure, specifically shifts from herbaceous and cryptogamous dominance to systems dominated by ericaceous and woody species, are also likely to change ecosystem C dynamics and balance.

Arctic soils are often wet and, when waterlogged, become anoxic. Anaerobic soils often accumulate C in the form of peat (Gorham, 1991; Clymo *et al.*, 1998) and release methane (CH_4), a radiatively important trace gas (Matthews and Fung, 1987; Joabsson and Christensen, 2001; Öquist and Svensson, 2002). CH_4 flux is rarely a quantitatively important component in the ecosystem C balance, but it can play a disproportionately important role in terms of greenhouse gas forcing (gram for gram CH_4 in the atmosphere has 23 times the radiative forcing potential of CO_2).

Greening of the High Arctic and changes in the wetness of soils are two examples of climate-driven changes that in turn cause ecosystem feedback mechanisms that involve changes in greenhouse gas exchanges (CO_2 through changed carbon storage and CH_4 through changed extent of wet soil conditions), which potentially can have important inherent feedback effects in the climate system. In addition to such changes there are also feedback mechanisms associated with changes in the energy exchange as a consequence of changed vegetation composition and structure affecting reflective properties both in the summer and through impacts on snow-cover in the winter.

Within each of the four seasons, there are important, and at times very different, processes acting, which are resulting in the net effect of the individual seasons on the annual ecosystem C budget. Critical facets of early-season conditions—such as a substantial C loss during spring melt and early summer due to release of trapped CO_2, and possibly a hindered onset of photosynthesis due to dry early summer conditions—can seriously affect the annual budget. In midsummer again water deficit can be important as a limiting factor for photosynthesis, while a very warm summer has the potential to stimulate respiration (including root respiration) more than photosynthesis (in particular in dry years), so that these effects together can be very important for the annual budget (Crawford *et al.*, 1993; Marchand *et al.*, 2005; Kwon *et al.*, 2007). In the third season, a mild autumn followed by the delayed appearance of a consistent snow-cover could be critical for processes involved in C fluxes. Usually, photosynthesis will decline regardless of warm "Indian" summer conditions, which will on the contrary stimulate respiration for as long as the soils remain unfrozen (or contain free water). So a mild autumn may also be a very important triggering factor for C losses on an annual basis. Most of these seasonal aspects that may determine the annual balance are affected by NAO/AO oscillations and there may therefore also be links between these cyclic dynamics and the variations in ecosystem C balance and CH_4 emissions.

B. Carbon Flux at Zackenberg and Relations to the NAO

As discussed in detail by Grøndahl *et al.* (2008, this volume), the accumulated CO_2 during the growing season in the heath ecosystem at Zackenberg is strongly correlated with the date of snowmelt, which is related to the NAO (Figure 5). This is similar to patterns observed in peat lands of northern Scandinavia (Aurela *et al.*, 2004). However, this correlation does not necessarily translate into the annual balance, as the respiration may be strongly affected by other factors during the autumn and winter, and this may in some cases be more important for the annual budget (Vourlitis and Oechel, 1999; Johansson *et al.*, 2006). Although the sample size for investigating the interannual effects of the NAO on the net ecosystem exchange (NEE) together with other environmental factors (Grøndahl *et al.*, 2008, this volume) may be considered small, a simple comparison suggests NAO-mediated NEE dynamics, where, as would be expected from the NAO–snow relationship (Figure 5), increasing NAO winters are followed by growing seasons with higher NEE (Figure 21); that is, increased, NAO-mediated snow-cover

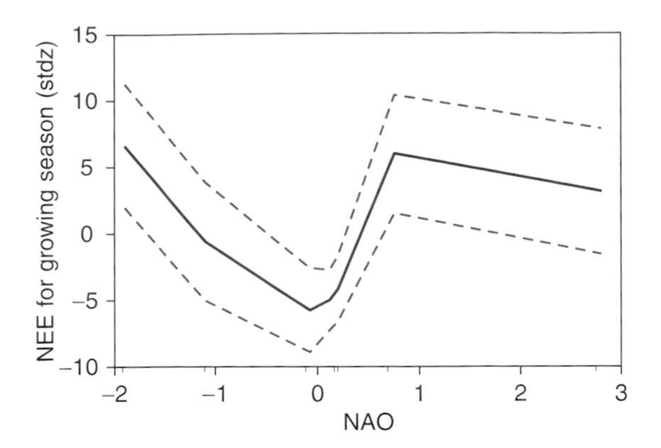

Figure 21 Inter-annual variations in the net ecosystem exchange (NEE; g C/m^2) as function of the NAO. Data from Sigsgaard *et al.* (2006) and M.P. Tamstorf (unpublished). The solid line is a non-linear GAM with 95% CIs as dashed lines. Model $R^2 = 0.92, p < 0.05$. The rugplot along the *x*-axis marks the "observed" *x* values. Note that the positive values on the secondary axis (NEE) represent growing season net carbon uptake, and also that only the increasing trend with positive NAO values is significant.

reduces the cumulative tundra uptake of carbon dioxide the following summer.

Unlike the relatively simple relationships found in relation to parameters explaining inter-annual variability in growing season carbon uptake, the controls on accumulated CH_4 emissions are seemingly more complex. Zackenberg was the first site in the circumpolar north where it was clearly demonstrated that there are important interactions between plant species composition and potential CH_4 emissions (Joabsson and Christensen, 2001; Ström *et al.*, 2003). Any effect on plant species composition is direct through a changed climate or indirect through changes in grazing pressure, and preference of grazers may therefore affect potential emissions. Such interactions between activities at different trophic levels with ecosystem functioning in the form of trace gas exchanges are so far only very poorly understood, but Zackenberg represents an ideal setting for investigating these issues further.

Preliminary data from a newly established (from 2006) automatic CH_4 flux monitoring system combined with earlier manual data are indicating that no single parameter can explain the inter-annual variations. In other wetland ecosystems, seasonal and spatial variations in CH_4 flux are often found to correlate well with the water table position (e.g., Whalen and Reeburgh, 1992; Daulat and Clymo, 1998), and wet tundra and peat land ecosystems with substantial emissions including the fen area in Zackenberg show a nice, both large-scale spatial and temporal, correlation with mean temperatures during the growing season (Christensen *et al.*, 2003). At Zackenberg, the local spatial variability in CH_4 fluxes is high (Christensen *et al.*, 2000) and as mentioned above, dependent on interactions between temperature, water table position and vascular plant species-specific functioning. It is nevertheless most likely that drier conditions will lower the CH_4 emissions both directly through restricting the anaerobic soil layers where the CH_4 is produced (and increase the oxidised zone) and indirectly through shifts towards grassland plant species that are not stimulating CH_4 emissions to the same extent as many wetland species do (Ström *et al.*, 2003).

VII. CONCLUSIONS

The monitoring in Zackenberg Basic not only complies with most of the recommendations by ACIA but also moves beyond by providing, as recommended by ICARP II (Callaghan, 2005; Prowse *et al.*, 2005), new pivotal knowledge of (1) how an entire arctic ecosystem responds to climate variability and changes and (2) how these cascade through the system as direct and indirect impacts (Figures 1 and 2). The knowledge gained from system monitoring at Zackenberg may therefore constitute a major and unique contribution to forthcoming revisions of the ACIA recommendations.

In a climatic perspective, Zackenberg is geographically located at the boarder between the two contrasting climate regions associated with the atmospheric dynamics of the NAO. As such, the relationship between local weather conditions at Zackenberg vary non-linearly with the NAO as its inter-annual as well as long-term dynamics move between the different phases (Figures 3–5; Hansen *et al.*, 2008, this volume; Hinkler *et al.*, 2008, this volume; Stendel *et al.*, 2008, this volume). Because of their integrative nature, large-scale climate systems, like the NAO, provide a suitable skeleton for studies of climate effects on different organisms at different trophic levels across the Arctic. Furthermore, this combined with an integrated system monitoring, such as Zackenberg Basic, presents a unique opportunity for describing and evaluating cumulative system responses. However, monitoring programmes with long-term perspectives, like the Zackenberg Basic, would benefit from a combined monitoring of phenotypic and genotypic responses to climate changes. Long-term observations of climate change effects will embrace both (Figure 7A), but, currently, only the phenotypic aspect is included in Zackenberg Basic.

The influence of the NAO on species dynamics and performance at Zackenberg is evident and does provide a mechanistic description of how arctic organisms respond to large-scale climate representative in a circumpolar perspective. At Zackenberg, the signature of the NAO is found in species ranging from plants to large-bodied mammals as well as in the interactions across trophic levels and system feedback dynamics to the atmosphere.

Whether system stability in a changing climate pertains to the complexity of dynamics among species at the individual trophic level and/or the complexity of food webs provides an interesting pursuit in future research (Post and Forchhammer, 2001) based on using integrative system data like those provide by Zackenberg Basic.

ACKNOWLEDGMENTS

The monitoring data used in this chapter were provided by the BioBasis programme, run by the National Environmental Research Institute, University of Aarhus, and financed by the Danish Environmental Protection Agency, Danish Ministry of the Environment. The Danish Polar Center provided access and accommodation at the Zackenberg Research Station during all the years. We extend our sincere thanks to the referee, Johannes Kollmann, who contributed significant improvements for an earlier version of the manuscript.

REFERENCES

ACIA (2005) *Arctic Climate Impact Assessment*. Cambridge University Press, New York.

AMAP (2000) *AMAP Trends and Effects Programme 1998–2003*. AMAP Report 99:7. Section C. pp 23–33.

AMAP (2005) *AMAP Workshop on Follow-up of ACIA June 15–17, 2005, Oslo, Norway*. 106 pp.

Aurela, M., Laurila, T. and Tuovinen, J.-P. (2004) Geophys. *Res. Lett* **31**, L16119.

Bamzai, A.S. (2003) *Int. J. Climatol.* **23**, 131–142.

Barnston, A.G. and Livezy, R.E. (1987) *Mon. Weather Rev.* **374**, 2901–2904.

Bay, C. (1998) *Vegetation Mapping of Zackenberg Valley, Northeast Greenland*. Danish Polar Center & Botanical Museum, University of Copenhagen. Report.

Bengtsson, L. (2005) *ICARP II: Working Group 9 Modelling and Predicting Arctic Climate and Ecosystems. Draft Science Plan.*

Böcher, T.W., Holmen, K. and Jakobsen, K. (1968) *The Flora of Greenland* P Haase & Son, Copenhagen.

Buermann, W., Anderson, B., Tucker, C.J., Dickinson, R.E., Lucht, W., Potter, C.S. and Myneni, R.B. (2003) *J. Geophys. Res.* **108**, D13.

Callaghan, T.V. (2005) ICARP II: Working Group 8 Terrestrial and Freshwater Biosphere and Biodiversity. Draft Science plan .

Callaghan, T.V., Carlsson, B.Å and Tyler, N.J.C. (1989) *J. Ecol.* **77**, 823–837.

Callaghan, T.V., Rudels, B. and Tweedie, C.E. (2007) International Review of the Zackenberg Research Station. The Danish Environmental Protection Agency Report. 32 pp.

Chapin, F.S., III, Jefferies, R.L., Reynolds, J.F., Shaver, G.R. and Svoboda, J. (1992) *Arctic Ecosystems in a Changing Climate*. Academic Press, New York.

Christensen, T.R., Friborg, T., Sommerkorn, M., Kaplan, J., Illeris, L., Soegaard, H., Nordstroem, C. and Jonasson, S. (2000) *Global Biogeochem. Cycles* **14**, 701–713.
Christensen, T.R., Joabsson, A., Ström, L., Panikov, N., Mastepanov, M., Öquist, M., Svensson, B.H., Nykänen, H., Martikainen, P. and Oskarsson, H. (2003) *Geophys. Res. Lett.* **30**, 1414.
Christensen, J.H., Hewitson, B., Busuioc, A., Chen, A., Gao, X., Held, I., Jones, R., Kolli, R.K., Kwon, W.-T., Laprise, R., Magaña Rueda, V., Mearns, L., *et al.* (2007) *Climate Change 2007: The Physical Science Basis.* Cambridge University Press.
Christoffersen, K. and Jeppesen, E. (2002) In: *Sne, is og 35 graders kulde. Hvad er effekterne af klimaændringer i Nordøstgrønland?* (Ed. by H Meltofte), pp. 79–84. Tema rapport 41, Danmarks Miljøundersøgelser.
Cleland, E.E., Chuine, I., Menzel, A., Mooney, H. and Schwartz, M.D. (2007) *TREE* **22**, 357–365.
Clutton-Brock, T.H. (ed.) (1988) Reproductive Success. Studies of Individual Variation in Contrasting Breeding Systems. University Chicago Press, Chicago.
Clymo, R.S., Turunen, J. and Tolonen, K. (1998) *Oikos* **81**, 368–388.
Crawford, R.M.M., Chapman, H.M., Abbott, R.J. and Balfour, J. (1993) *Flora* **188**, 367–381.
Dansgaard, W. (1964) *Tellus* **16**, 436–468.
Darwin, C. (1859) *On the Origin of Species*. Murray, London.
Daulat, W.E. and Clymo, R.S. (1998) *Atmos. Environ.* **32**, 3207–3218.
Dawes, P.R., Elander, M. and Ericson, M. (1986) *Arctic* **39**, 119–132.
Elton, C.S. (1958) *The Ecology of Invasions by Animals and Plants.* Methuen and Co., London.
Farquhar, G.D., Ehleringer, J.R. and Hubick, K.T. (1989) *Ann. Rev. Plant Physiol.* **40**, 503–537.
Forchhammer, M.C. (2001) In: *Climate Change Research – Danish Contributions* (Ed. by A.M.K Jørgensen, J. Fenger and K. Halsnæs), pp. 219–236. Gad, Copenhagen.
Forchhammer, M.C. (2002) In: *Sne, is og 35 graders kulde. Hvad er effekterne af klimaændringer i Grønland.* Tema rapport 41 (Ed. by H. Meltofte), pp. 15–24. Danmarks Miljøundersøgelser, Roskilde.
Forchhammer, M.C. and Post, E. (2004) *Popul. Ecol.* **46**, 1–12.
Forchhammer, M.C., Post, E. and Stenseth, N.C. (1998) *Nature* **391**, 29–30.
Forchhammer, M.C., Post, E., Stenseth, N.C. and Boertmann, D.M. (2002) *Popul. Ecol.* **44**, 113–120.
Forchhammer, M.C., Post, E., Berg, T.B., Høye, T.T. and Schmidt, N.M. (2005) *Ecology* **86**, 2644–2651.
Forchhammer, M.C., Rasch, M. and Rysgaard, S. (2007) *A Conceptual Framework for Monitoring Climate Effects and Feedback in Arctic Ecosystems.* Report to the Danish Ministry of Environment, Copenhagen.
Gong, D.Y. and Ho, C.H. (2003) *J. Geophys. Res.* **108**, D16.
Gong, D.Y. and Shi, P.J. (2003) *Int. J. Remote Sens.* **24**, 2559–2566.
Gorham, E. (1991) *Ecol. Appl.* **1**, 182–195.
Grøndahl, L., Friborg, T. and Soegaard, H. (2007) *Theor. Appl. Climatol.* **88**, 111–125.
Hallett, T.B., Coulson, T., Pilkington, J.G., Clutton-Brock, T.H., Pemberton, J.M. and Grenfell, B.T. (2004) *Nature* **430**, 71–75.
Hastie, T.J. and Tibshirani, R.J. (1990) *Generalized Additive Models.* Chapman & Hall, Boca Raton.
Havström, M., Callaghan, T.V., Jonasson, S. and Svoboda, J. (1995) *Funct. Ecol.* **9**, 650–654.

Hinkler, J. (2005) *From Digital Cameras to Large Scale Sea-Ice Dynamics – A Snow-Ecosystem Perspective*. Ph.D. thesis. National Environmental Research Institute and the University of Copenhagen, Department of Geography.
Hobbie, S.E., Schimel, J.P., Trumbore, S.E. and Randerson, J.P. (2000) *Global Change Biol.* **6**, 196–210.
Hurrell, J.W. (1995) *Science* **269**, 676–679.
Hurrell, J.W. (1996) *Geophys. Res. Lett.* **23**, 665–668.
Hurrell, J.W., Kushnir, Y., Ottersen, G. and Visbeck, M. (2003) *Geophys. Monogr.* **135**, 1–35.
Høye, T.T., Ellebjerg, S.M. and Philipp, M. (2007a) Arct. Antarct. *Alp. Res.* in press.
Høye, T.T., Post, E., Meltofte, H., Schmidt, N.M. and Forchhammer, M.C. (2007b) *Curr. Biol.* **17**, 449–451.
Joabsson, A. and Christensen, T.R. (2001) *Global Change Biol.* **7**, 919–932.
Johansson, T., Malmer, N., Crill, P.M., Mastepanov, M. and Christensen, T.R. (2006) *Global Change Biol.* **12**, 2352–2369.
Johnstone, J.F. and Henry, G.H.R. (1997) *Arct. Antarct. Alp. Res.* **29**, 459–469.
Kushnir, Y. (1999) *Nature* **398**, 289–291.
Kwon, H.-J., Oechel, W.C., Zulueta, R.C. and Hastings, S.J. (2007) *J. Geophys. Res.* accepted.
Marchand, F.L., Mertens, S., Kockelbergh, F., Beyens, L. and Nijs, I. (2005) *Global Change Biol.* **11**, 2078–2089.
Matthews, E. and Fung, I. (1987) *Global Biogeochem. Cycles* **1**, 61–86.
May, R.M. (1973) *Am. Nat.* **107**, 621–650.
May, R.M. (2001) *Stability and Complexity in Model Ecosystems*. Princeton University Press, Princeton.
McKane, R.B., Rastetter, E.B., Shaver, G.R., Nadelhoffer, K.J., Giblin, A.E., Laundre, J.A. and Chapin, F.S. III (1997) *Ecology* **78**, 1188–1198.
Meltofte, H. (2006) In: *Zackenberg Ecological Research Operations, 11th Annual Report, 2005* (Ed. by K. Caning and M. Rasch), pp. 36–76. Danish Polar Center, Ministry of Science, Technology and Innovation, Copenhagen.
Menzel, A. (2003) *Global Change* **57**, 243–263.
Menzel, A., Sparks, T.H., Estrella, N., Kockz, E., Aasa, A., Ahas, R., Alm-Kübler, K., Bissolli, P., Braslavska, O., Vriede, A., Chemielewski, F.M., Crepinsek, Z., *et al.* (2006) *Global Change Biol.* **12**, 1969–1976.
Millenium Ecosystem Assessment (2005) *Ecosystems and Human Well-Being: Current State and Trends.*. Island Press, Washington.
Moran, P.A.P. (1953) *Aust. J. Zool.* **1**, 291–298.
Myneni, R.B., Hall, F.G., Sellers, P.J. and Marshak, A.L. (1995) *Trans. Geosci. Remote Sens.* **33**, 481–486.
Myneni, R.B., Keeling, C.D., Tucker, C.J., Asrar, G. and Nemani, R.R. (1997) *Nature* **386**, 698–702.
Mysterud, A., Stenseth, N.C., Yoccoz, N.G., Ottersen, G. and Langvatn, R. (2003) *Geophys. Monogr.* **134**, 235–262.
Mølgaard, P., Forchhammer, M.C., Grøndahl, L. and Meltofte, H. (2002) In: *Sne, is og 35 graders kulde. Hvad er effekterne af klimaændringer i Nordøstgrønland?.* (Ed. by H. Meltofte), pp. 43–46. Tema rapport 41. Danmarks Miljøundersøgelser, Roskilde.
Nordström, C., Soegaard, H., Christensen, T.R., Friborg, T. and Hansen, B.U. (2001) *Theor. Appl. Climatol.* **70**, 149–166.

Oechel, W.C., Callaghan, T., Gilmanov, T., Holten, J.I., Maxwell, B., Molau, U. and Sveinbjörnsson, B. (Eds.) (1997) Global Change and the Arctic Terrestrial Ecosystems, Springer, New York.

Olesen, C.R., Thing, H. and Aastrup, P. (1994) *Rangifer* **14**, 3–10.

Öquist, M.G. and Svensson, B.H. (2002) *J. Geophys. Res.* **107**, D21.

Peterson, R.O., Page, R.E. and Dodge, K.M. (1984) *Science* **224**, 1350–1352.

Peterson, R.O., Thomas, N.J., Thurber, J.M., Vucetich, J.A. and Waite, T.A. (1998) *J. Mammal.* **79**, 828–841.

Pettorelli, N., Mysterud, A., Yoccoz, N.G., Langvatn, R. and Stenseth, N.C. (2005a) *Proc. R. Soc. B* **272**, 2357–2364.

Pettorelli, N., Weladji, R.B., Holand, Ø., Mysterud, A., Breie, H. and Stenseth, N.C. (2005b) *Biol. Lett.* **1**, 24–26.

Picton, H.D. (1984) *J. Appl. Ecol.* **21**, 869–879.

Pimm, S.L. (1984) *Nature* **307**, 321–326.

Post, E. (2005) *Ecology* **86**, 2320–2328.

Post, E. and Forchhammer, M.C. (2001) *BMC Ecol.* **1**, 5.

Post, E. and Forchhammer, M.C. (2002) *Nature* **420**, 168–171.

Post, E. and Forchhammer, M.C. (2006) *Quat. Int.* **151**, 99–105.

Post, E. and Stenseth, N.C. (1999) *Ecology* **80**, 1322–1339.

Post, E., Peterson, R.O., Stenseth, N.C. and McLaren, B.E. (1999) *Nature* **401**, 905–907.

Post, E., Forchhammer, M.C., Stenseth, N.C. and Callaghan, T.V. (2001) *Proc. R. Soc. B* **268**, 15–23.

Post, E., Pedersen, C., Wilmers, C.C. and Forchhammer, M.C. (in press) *Ecology*.

Prowse, T.D., Bøggild, C.E., Glazovsky, A.F., Hagen, J.O.M., Killingtveit, Å., Lettenmaier, D.P., Nelson, F.E., Rouse, W.R., Steffen, K., Shiklomanov, I.A., Young, K.L. and Kotlyakov, V.M. (2005) *ICARP II: Working Group 7 Terrestrial Cryospheric and Hydrological Processes and Systems.* Draft Science Plan, Copenhagen.

Rasch, M., Rysgaard, S., Meltofte, H. and Hansen, J.B. (2003) In: *Zackenberg Ecological Research Operations, 8th Annual Report, 2002.* (Ed. by K. Caning and M. Rasch), pp. 70–80. Danish Polar Center, Ministry of Research and Information Technology, Copenhagen.

Root, T.L., Price, J.T., Hall, K.R., Schneider, S.H., Rosenzweigk, C. and Pounds, J.A. (2003) *Nature* **421**, 57–60.

Royama, T. (1992) *Analytical Population Dynamics.* Chapman & Hall, London.

Sandquist, D.R. and Ehleringer, J.R. (2003) *Oecologia* **134**, 463–470.

Schmidt, N.M. (2006) *Climate, Agriculture and Density-Dependent Dynamics Within and Across Trophic Levels in Contrasting Ecosystems.* Ph.D. thesis, Royal Veterinary and Agricultural University, Copenhagen.

Schmidt, N.M., Baittinger, C. and Forchhammer, M.C. (2006) *Arct. Antarct. Alp. Res.* **38**, 257–262.

Shaver, G.R., Billings, W.D., Chapin, F.S., Giblin, A.E., Nadelhoffer, K.J., Oechel, W.C. and Rastetter, E.B. (1992) *BioScience* **42**, 433–441.

Sigsgaard, C., Petersen, D., Grøndahl, L., Thorsøe, K., Meltofte, H., Tamstorf, M. and Hansen, B.U. (2006) In: *Zackenberg Ecological Research Operations, 11th Annual Report, 2005* (Ed. by K. Caning and M. Rasch), pp. 11–35. Danish Polar Center, Ministry of Science, Technology and Innovation, Copenhagen.

Sokal, R.R. and Rohlf, F.J. (1995) *Biometry.* W.H. Freeman and Company, New York.

Stearns, S.C. (1992) *The Evolution of Life Histories.* Oxford University Press, Oxford.

Stenseth, N.C. and Mysterud, A. (2005) *J. Anim. Ecol.* **74**, 1195–1198.
Stenseth, N.C., Chan, K.-S., Tong, H., Boonstra, R., Boutin, S., Krebs, C.J., Post, E., O'Donoghue, M., Yoccoz, N.G., Forchhammer, M.C. and Hurrell, J.W. (1999) *Science* **285**, 1071–1073.
Stenseth, N.C., Mysterud, A., Ottersen, G., Hurrell, J.W., Chan, K.-S. and Lima, M. (2002) *Science* **297**, 1292–1296.
Stenseth, N.C., Ottersen, G., Hurrell, J.W., Mysterud, A., Lima, M., Chan, K.-S., Yoccoz, N.G. and Ådlandsvik, B. (2003) *Proc. R. Soc. Lond. B* **270**, 2087–2096.
Ström, L., Ekberg, A., Mastepanov, M. and Christensen, T.R. (2003) *Global Change Biol.* **9**, 1185–1192.
Sullivan, P.F. and Welker, J.M. (2007) *Oecologia* **151**, 372–386.
Tamstorf, M.P. and Bay, C. (2002) In: *Sne, is og 35 graders kulde. Hvad er effekterne af klimaændringer i Nordøstgrønland?.* (Ed. by H. Meltofte), pp. 35–41. Tema rapport 41. Danmarks Miljøundersøgelser, Roskilde.
Tamstorf, M.P., Illeris, L., Hansen, B.U. and Wisz, M. (2007) *BMC Ecol.* in press.
Tarnocai, C., Kimble, J. and Broll, G. (2003) In: *Proceedings of the Eighth International Conference on Permafrost, 21–25, Zurich Switzerland.* (Ed. by M. Phillips, S. Springman and L.U. Arenson), Vol. 2, pp. 1129–1134. Sets & Zeitlinger B.V., Lisse, The Netherlands.
Tilman, D. (1996) *Ecology* **77**, 350–363.
Tilman, D. and Downing, J.A. (1994) *Nature* **367**, 363–365.
Todd, S.W., Hoffer, R.M. and Milchunas, D.G. (1998) *Int. J. Remote Sens.* **19**, 427–438.
Tucker, C.J., Fung, I.Y., Keeling, C.D. and Gammon, R.H. (1986) *Nature* **319**, 195–199.
Turchin, P., Lorio, P.L., Taylor, A.D. and Billings, R.F. (1991) *Environ. Entomol.* **20**, 401–409.
Venables, W.N. and Ripley, B.D. (1994) *Modern Applied Statistics with S-plus.* Springer, New York.
Vibe, C. (1967) Meddr. *Grønland* **170**(5), 1–227.
Vourlitis, G.L. and Oechel, W.C. (1999) *Ecology* **80**, 686–701.
Walther, G.-R., Post, E., Convey, P., Menzel, A., Parmesan, C., Beebee, T.J.C., Fromentin, J.-M, Hoegh-Guldberg, O. and Bairlein, F. (2002) *Nature* **416**, 389–395.
Welker, J.M., Rayback, S. and Henry, G.H.R. (2005) *Global Change Biol.* **1**, 997–1002.
Whalen, S.C. and Reeburgh, W.S. (1992) *Global Biogeochem. Cycles.* **6**, 139–159.
Wilmers, C.C., Sinha, S. and Brede, M. (2002) *Oikos* **99**, 363–367.
Wilmers, C.C., Post, E., Peterson, R.O. and Vucetich, J.A. (2006) *Ecol. Lett.* **9**, 383–389.
Zhou, L., Tucker, C.J., Kaufmann, R.K., Slayback, D., Shabanov, N.V. and Myneni, R.B. (2001) *J. Geophys. Res.* **106**, 20069–20083.

Index

Advances in Ecological Research Volume 1–40

Cumulative List of Titles